ANNALS OF
THE NEW YORK ACADEMY
OF SCIENCES

Volume 939

EDITORIAL STAFF

Executive Editor
BARBARA M. GOLDMAN

Managing Editor
JUSTINE CULLINAN

Associate Editors
JOYCE HITCHCOCK
JOHN W. KENNEDY

The New York Academy of Sciences
2 East 63rd Street
New York, New York 10021

THE NEW YORK ACADEMY OF SCIENCES
(Founded in 1817)
BOARD OF GOVERNORS, September 2000–September 2001
BILL GREEN, *Chairman of the Board*
TORSTEN WIESEL, *Vice Chairman of the Board*
RODNEY W. NICHOLS, *President and CEO* [ex officio]

Honorary Life Governors
WILLIAM T. GOLDEN   JOSHUA LEDERBERG

JOHN T. MORGAN, *Treasurer*

Governors

ELEANOR BAUM	D. ALLAN BROMLEY	KAREN BURKE
LAWRENCE B. BUTTENWIESER		PRAVEEN CHAUDHARI
JOHN H. GIBBONS	MICHAEL GOLDEN	RONALD L. GRAHAM
ROBERT G. LAHITA	JACQUELINE LEO	WILLIAM J. McDONOUGH
JOHN F. NIBLACK	SANDRA PANEM	RICHARD RAVITCH
RICHARD A. RIFKIND	SARA LEE SCHUPF	JAMES H. SIMONS

HELENE L. KAPLAN, *Counsel* [ex officio]   NANCY B. EISENBERG, *Interim Secretary* [ex officio]

NEUROPROTECTIVE AGENTS

FIFTH INTERNATIONAL CONFERENCE

ANNALS OF THE NEW YORK ACADEMY OF SCIENCES
Volume 939

NEUROPROTECTIVE AGENTS

FIFTH INTERNATIONAL CONFERENCE

Edited by
William Slikker, Jr. and Bruce Trembly

The New York Academy of Sciences
New York, New York
2001

Copyright © 2001 by the New York Academy of Sciences. All rights reserved. Under the provisions of the United States Copyright Act of 1976, individual readers of the Annals *are permitted to make fair use of the material in them for teaching and research. Permission is granted to quote from the* Annals *provided that the customary acknowledgment is made of the source. Material in the* Annals *may be republished only by permission of the Academy. Address inquiries to the Permissions Department (editorial@nyas.org) at the New York Academy of Sciences.*

Copying fees: *For each copy of an article made beyond the free copying permitted under Section 107 or 108 of the 1976 Copyright Act, a fee should be paid through the Copyright Clearance Center, Inc., 222 Rosewood Drive, Danvers, MA 01923 (www.copyright.com).*

∞ *The paper used in this publication meets the minimum requirements of American National Standard for Information Sciences—Permanence of Paper for Printed Library Materials. ANSI Z39.48-1984.*

Library of Congress Cataloging-in-Publication Data

Neuroprotective agents : fifth international conference / edited by William Slikker, Jr. and Bruce Trembly.
    p. ; cm. — (Annals of the New York Academy of Sciences, ISSN 0077-8923 ; v. 939)
    Includes bibliographical references and index.
    ISBN 1-57331-352-1 (cloth : alk. paper) — ISBN 1-57331-353-X (paper : alk. paper)
    1. Nervous system—Degeneration—Prevention—Congresses. 2. Nervous Neuropharmacology—Congresses. I. Slikker, Jr., William. II. Trembly, Bruce. III. International Conference on Neuroprotective Agents: Clinical and Experimental Aspects (5th ; 2000 : Lake Tahoe, Nev.) IV. Series.
    [DNLM: 1. Nerve Degeneration—prevention & control—Congresses. 2. Neuroprotective Agents—therapeutic use—Congresses. WL 102.5 N49446 2001]
Q11 .N5 vol. 939
[RC365]
500 s—dc21
[616.8]

                                              2001032632
                                                      CIP

K-M Research/PCP
Printed in the United States of America
ISBN 1-57331-352-1 (cloth)
ISBN 1-57331-353-X (paper)
ISSN 0077-8923

ANNALS OF THE NEW YORK ACADEMY OF SCIENCES

Volume 939
June 2001

NEUROPROTECTIVE AGENTS

FIFTH INTERNATIONAL CONFERENCE[a]

Editors
WILLIAM SLIKKER, JR. AND BRUCE TREMBLY

CONTENTS

Preface. *By* WILLIAM SLIKKER, JR. AND BRUCE TREMBLY xi

Part I. Mechanisms of Excitotoxicity

Quinolinic Acid Accumulation During Neuroinflammation:
  Does it Imply Excitotoxicity? *By* T.P. OBRENOVITCH 1

Neuronal Protein Kinase Signaling Cascades and Excitotoxic Cell Death.
  By STEPHEN D. SKAPER, LAURA FACCI, AND PAUL J.L.M. STRIJBOS 11

Amino Acid Uptake and Release in Primary Astrocyte Cultures Exposed
  to Ethanol. *By* MICHAEL ASCHNER, LYSETTE A. MUTKUS, AND
  JEFFREY W. ALLEN... 23

Questions and Answers... 28

Part II. Mechanisms of Neurodegeneration and Neurogenesis

Lack of Protective Effect by Intermittent Hypoxia on MPTP-Induced
  Neurotoxicity in Mice. *By* GIN-HEUY LAI, CHAU-FONG CHEN, YU SU,
  LOW-TONE HO, AND ANYA MAAN-YUH LIN 33

Neurogenesis and Neuroprotection in the Adult Brain: A Putative Role
  for 5-Lipoxygenase? *By* HARI MANEV, TOLGA UZ, RADMILA MANEV,
  AND ZHIJING ZHANG... 45

Questions and Answers... 52

Part III. Endogenous Neuroprotectants—Adenosine Receptors

Modulation of Cyclooxygenase-2 and Brain Reactive Astrogliosis by Purinergic
  P2 Receptors. *By* ROBERTA BRAMBILLA AND MARIA P. ABBRACCHIO...... 54

---

[a]This volume is the result of a conference entitled **Fifth International Conference on Neuroprotective Agents**, held 17–21 September, 2000, in Lake Tahoe, Nevada, U.S.A.

The A_3 Adenosine Receptor Induces Cytoskeleton Rearrangement in Human Astrocytoma Cells via a Specific Action on Rho Proteins. *By* MARIA P. ABBRACCHIO, ALESSANDRA CAMURRI, STEFANIA CERUTI, FLAMINIO CATTABENI, LOREDANA FALZANO, ANNA MARIA GIAMMARIOLI, KENNETH A. JACOBSON, LETIZIA TRINCAVELLI, CLAUDIA MARTINI, WALTER MALORNI, AND CARLA FIORENTINI 63

Adenosine Extracellular Brain Concentrations and Role of A_{2A} Receptors in Ischemia. *By* FELICITA PEDATA, CLAUDIA CORSI, ALESSIA MELANI, FRANCESCA BORDONI, AND SERENA LATINI 74

Right Thing at a Wrong Time? Adenosine A_3 Receptors and Cerebroprotection in Stroke. *By* DAG K.J.E. VON LUBITZ, KIMBERLY L. SIMPSON, AND RICK C.S. LIN 85

Questions and Answers 97

Part IV. Neuroprotection in Neurosurgery

Monitoring for Neuroprotection: New Technologies for the New Millennium. *By* RUSSELL J. ANDREWS 101

Neuroprotection for the New Millennium: Matchmaking Pharmacology and Technology. *By* RUSSELL J. ANDREWS 114

Neuroprotective Role of Neurophysiological Monitoring During Endovascular Procedures in the Spinal Cord. *By* FRANCESCO SALA, YASUNARI NIIMI, ALEX BERENSTEIN, AND VEDRAN DELETIS 126

The Role of Intraoperative Neurophysiology in the Protection or Documentation of Surgically Induced Injury to the Spinal Cord. *By* VEDRAN DELETIS AND FRANCESCO SALA 137

Questions and Answers 145

Part V. Neuroprotection in Alzheimer's and Related Diseases

Neuroprotective Effects of Novel Cholinesterase Inhibitors Derived from Rasagiline as Potential Anti-Alzheimer Drugs. *By* MARTA WEINSTOCK, NATANJA KIRSCHBAUM-SLAGER, PHILIP LAZAROVICI, CORINA BEJAR, MOUSSA B.H. YOUDIM, AND SHAI SHOHAM 148

The Action of Acetyl-L-Carnitine on the Neurotoxicity Evoked by Amyloid Fragments and Peroxide on Primary Rat Cortical Neurones. *By* M.A. VIRMANI, V. CASO, A. SPADONI, S. ROSSI, F. RUSSO, AND F. GAETANI 162

Post-Stroke Dementia: Nootropic Drug Modulation of Neuronal Nicotinic Acetylcholine Receptors. *By* XILONG ZHAO, JAY Z. YEH, AND TOSHIO NARAHASHI 179

Questions and Answers 187

Part VI. Endogenous Neuroprotectants—Melatonin

Antioxidant and Antiaging Activity of *N*-Acetylserotonin and Melatonin in the *in Vivo* Models. *By* G. OXENKRUG, P. REQUINTINA, AND S. BACHURIN 190

Free Radical–Mediated Molecular Damage: Mechanisms for the Protective
   Actions of Melatonin in the Central Nervous System. *By* Russel J. Reiter,
   Dario Acuña-Castroviejo, Dun-Xian Tan, and Susanne Burkhardt . . 200

Questions and Answers. 216

Part VII. NMDA Receptor Modification and Neuroprotection

Neuroprotective and Cognition-Enhancing Properties of MK-801 Flexible
   Analogs: Structure–Activity Relationships. *By* Sergey Bachurin,
   Sergey Tkachenko, Igor Baskin, Nadegda Lermontova,
   Tatyana Mukhina, Lyudmila Petrova, Anatoliy Ustinov,
   Alexey Proshin, Vladimir Grigoriev, Nikolay Lukoyanov,
   Vladimir Palyulin, and Nikolay Zefirov 219

Effects of Chronic NMDA Receptor and Fast Sodium Channel Blockade
   during Development on the Acquisition of Visual Discriminations and
   Learning Abilities in Rhesus Monkeys. *By* Merle G. Paule, E. Jon Popke,
   Richard R. Allen, Edwin Pearson, and Tim Hammond 237

Synaptic Deprivation and Age-Related Vulnerability to Hypoxic-Ischemic
   Neuronal Injury: A Hypothesis. *By* Ann M. Marini, John Choi,
   and Robert Labutta 238

Questions and Answers. 254

Part VIII. Ischemic Stroke: Current Therapy

The Failure of Neuronal Protective Agents versus the Success of
   Thrombolysis in the Treatment of Ischemic Stroke: The Predictive
   Value of Animal Models. *By* Saran Jonas, Venkatesh Aiyagari,
   Dorice Vieira, and Miguel Figueroa 257

Low Molecular Weight Heparin and the Treatment of Ischemic Stroke: Animal
   Results, the Reasons for Failure in Human Stroke Trials, Mechanisms
   of Action, and the Possibilities for Future Use in Stroke.
   By Saran Jonas and David Quartermain. 268

Non-Pharmacologic (Physiologic) Neuroprotection in the Treatment of Brain
   Ischemia. *By* Roland N. Auer 271

T_2-Weighted MRI Correlates with Long-Term Histopathology,
   Neurology Scores, and Skilled Motor Behavior in a Rat Stroke Model.
   By Gene C. Palmer, James Peeling, Dale Corbett,
   Marc R. Del Bigio, and Thomas J. Hudzik 283

On the Relationship Between Plasma Concentrations of Drugs and Outcome
   of Stroke Studies in Laboratory Animals. *By* Stephen H. Curry 297

Combination Therapy Stroke Trial: rt-PA ± Lubeluzole.
   By James Grotta for the Trial Investigators 309

Questions and Answers. 311

Part IX. Mechanistically Based Neuroprotection

Rat Model of Autism Spectrum Disorders: Genetic Background Effects on Borna Disease Virus-Induced Developmental Brain Damage. *By* M.V. PLETNIKOV, M.L. JONES, S.A. RUBIN, T.H. MORAN, AND K.M. CARBONE 318

Antiapoptotic Properties of Rasagiline, *N*-Propargylamine-1(*R*)-aminoindan, and Its Optical (*S*)-Isomer, TV1022. *By* WAKAKO MARUYAMA, MOUSSA B.H. YOUDIM, AND MAKOTO NAOI 320

Rescue of Dying Neurons by (*R*)-Deprenyl in the MPTP-Mouse Model of Parkinson's Disease Does Not Include Restoration of Neostriatal Dopamine. *By* STEFANUS J. STEYN, KAY CASTAGNOLI, AND NEAL CASTAGNOLI, JR. ... 330

Gene Therapy for Treatment of Cerebral Ischemia Using Defective Herpes Simplex Viral Vectors. *By* MIDORI A. YENARI, THEODORE C. DUMAS, ROBERT M. SAPOLSKY, AND GARY K. STEINBERG 340

Hypothermia-Induced Ischemic Tolerance: Characteristics and Candidate Mechanisms. *By* KEVIN S. LEE, MATTHEW J. ANZIVINO, MASATOSHI YUNOKI, AND DANIEL DECKER 358

Effect of L-Carnitine Pretreatment on 3-Nitropropionic Acid–Induced Inhibition of Rat Brain Succinate Dehydrogenase Activity. *By* ZBIGNIEW K. BINIENDA, NATALYA V. SADOVOVA, ROBERT L. ROUNTREE, ANDREW C. SCALLET, AND SYED F. ALI 359

Methamphetamine-Induced Dopaminergic Neurotoxicity: Role of Peroxynitrite and Neuroprotective Role of Antioxidants and Peroxynitrite Decomposition Catalysts. *By* SYED Z. IMAM, JAMAL EL-YAZAL, GLENN D. NEWPORT, YOSSEF ITZHAK, JEAN L. CADET, WILLIAM SLIKKER, JR., AND SYED F. ALI   366

Biomarkers of 3-Nitropropionic Acid (3-NPA)-Induced Mitochondrial Dysfunction as Indicators of Neuroprotection. *By* A.C. SCALLET, P.L. NONY, R.L. ROUNTREE, AND Z.K. BINIENDA 381

Role of Peroxynitrite in Neurodegeneration and Neuroprotective Role of Antioxidants and Peroxynitrite Decomposition Catalysts. *By* SYED F. ALI, SYED Z. IMAM, GLENN D. NEWPORT, YOSSEF ITZHAK, AND WILLIAM SLIKKER, JR. 393

Questions and Answers.. 395

Part X. Novel Neuroprotective Agents

Emerging Role of Lithium as a Neuroprotective Drug: Therapeutic Implications. *By* DE-MAW CHUANG.. 404

Antiviral Medications Improve Cerebrovascular Perfusion in HIV+ Non-Drug Users and HIV+ Cocaine Abusers. *By* RONALD I. HERNING, WARREN E. BETTER, KIMBERLY TATE, AND JEAN L. CADET............. 405

Marijuana Abusers Are at Increased Risk for Stroke: Preliminary Evidence from Cerebrovascular Perfusion Data. *By* RONALD I. HERNING, WARREN E. BETTER, KIMBERLY TATE, AND JEAN L. CADET............. 413

Delayed Multidose Treatment with Nicotinamide Extends the Degree and
Duration of Neuroprotection by Reducing Infarction and Improving
Behavioral Scores up to Two Weeks Following Transient Focal Cerebral
Ischemia in Wistar Rats. *By* KENNETH I. MAYNARD, ISSAM A. AYOUB,
AND CHIUNG-CHYI SHEN 416

Antihistamine Agent Dimebon As a Novel Neuroprotector and a Cognition
Enhancer. *By* S. BACHURIN, E. BUKATINA, N. LERMONTOVA,
S. TKACHENKO, A. AFANASIEV, V. GRIGORIEV, I. GRIGORIEVA,
YU. IVANOV, S. SABLIN, AND N. ZEFIROV 425

Cell Death in Spinal Cord Injury (SCI) Requires *de Novo* Protein Synthesis:
Calpain Inhibitor E-64-d Provides Neuroprotection in SCI Lesion
and Penumbra. *By* SWAPAN K. RAY, DENISE D. MATZELLE,
GLORIA G. WILFORD, EDWARD L. HOGAN, AND NAREN L. BANIK 436

The Anti-Parkinson Drug Rasagiline and Its Cholinesterase Inhibitor
Derivatives Exert Neuroprotection Unrelated to MAO Inhibition in
Cell Culture and *in Vivo*. *By* MOUSSA B.H. YOUDIM, A. WADIA,
W. TATTON, AND MARTA WEINSTOCK 450

Questions and Answers... 459

Part XI. Summary and Conclusions

Neuroprotection: Past Successes and Future Challenges.
By WILLIAM SLIKKER, JR., SARAN JONAS, ROLAND N. AUER,
GENE C. PALMER, TOSHIO NARAHASHI, MOUSSA B.H. YOUDIM,
KENNETH I. MAYNARD, KATHRYN M. CARBONE, AND BRUCE TREMBLY ... 465

Questions and Answers... 478

Index of Contributors 481

Supported in part by:
- THE NATIONAL CENTER FOR TOXICOLOGICAL RESEARCH/FDA
- CENTRAL ARKANSAS CHAPTER OF SIGMA XI.

> The New York Academy of Sciences believes it has a responsibility to provide an open forum for discussion of scientific questions. The positions taken by the participants in the reported conferences are their own and not necessarily those of the Academy. The Academy has no intent to influence legislation by providing such forums.

Preface

This volume contains papers, abstracts, and questions/answers from the Fifth International Conference on Neuroprotective Agents held in Lake Tahoe, California, September 17–21, 2000. Previous conferences were held in Rockland, Maine in 1991; Lake George, New York in 1994 (*Annals* Volume 765); Lake Como, Italy in 1996 (*Annals* Volume 825); and Annapolis, Maryland in 1998 (*Annals* Volume 890).

It has been the aim of these conferences to bring together clinicians, regulators, and basic science researchers from many disciplines and many parts of the world in a congenial and informal setting. The clinical focus of this fifth conference tended to be more on the complexity of neuroprotection, the need for understanding the time-course effects of hypoxia/ischemia, and the need for combinational and time-course application of multiple therapies. New technologies were introduced for describing both the insult and the treatment necessary for neuroprotection, including magnetic resonance imaging (MRI), gene expression assays (genomics), and gene therapy.

Several papers were devoted to the examination of the predictive value of models for outcome measures in clinical ischemic stroke neuroprotective trials. The need to move from *in vitro* models to more predictive *in vivo* approaches was addressed. Others papers emphasized the role of endogenous agents, such as adenosine, nitric oxide, melatonin, L-carnitine, estrogens, and glycemia in acute and chronic neural injury and therapy. The wide range of presentations, from a detailed paper on the very precise use of neurophysiological monitoring to aid in spinal cord surgery to a comprehensive poster describing the role of hypothermia in enhancing bcl-2 and antioxidant enzyme concentrations, serve to illustrate the overall scientific scope of these conferences.

We are grateful to the New York Academy of Sciences for the opportunity to share these proceedings with other clinicians and scientists.

The Sixth International Conference on Neuroprotective Agents will be held on the East Coast during the Fall of 2002.

WILLIAM SLIKKER, JR.
BRUCE TREMBLY

Quinolinic Acid Accumulation During Neuroinflammation

Does it Imply Excitotoxicity?

T.P. OBRENOVITCH

*Pharmacology, School of Pharmacy,
University of Bradford, Bradford, United Kingdom*

ABSTRACT: It is often proposed that quinolinic acid (QUIN) contributes to the pathophysiology of neuroinflammation because this kynurenine pathway metabolite is a selective agonist of N-methyl-D-aspartate (NMDA) receptors, and both its brain tissue and cerebrospinal fluid concentrations increase markedly with inflammation. However, whether or not the extracellular levels of QUIN reached during neuroinflammation are high enough to promote excitotoxicity, remains unclear because QUIN is a weak NMDA receptor agonist. We have addressed this issue by evaluating the extracellular concentrations of QUIN that must be reached to initiate potentially excitotoxic changes in the cerebral cortex of rats, under normal conditions, and when superimposed on another insult. We have also examined the increase in extracellular lactate associated with the perfusion of increasing concentrations of QUIN through a microdialysis probe. The extracellular EC_{50} for induction of local depolarisation was 228 μM with QUIN alone; that is, about 30 times the levels of QUIN measured previously in immune activated brain. Furthermore, at least 20 μM extracellular QUIN needed to be reached to reduce K^+ induced spreading depression, an unexpected effect since spreading depression is inhibited by NMDA receptor antagonists. Our data suggest that, although synthesis of QUIN from activated microglia and invading macrophages can increase its extracellular concentration 10–100-fold, the levels that are reached in these conditions remain far below the concentrations of QUIN that are necessary for excessive NMDA receptor activation. However, the possibility that QUIN accumulation may be a deleterious feature of neuroinflammation cannot be ruled out at this stage.

KEYWORDS: Kynurenine pathway; Quinolinic acid; Excitotoxicity; NMDA receptor; Spreading depression; Neuroinflammation; Microdialysis; Lactate.

INTRODUCTION

Quinolinic acid (QUIN) is a metabolite of the kynurenine pathway that is often proposed to contribute to neuronal death associated with inflammatory neurological disease (e.g., HIV encephalitis) or secondary to brain injury.[1,2] Two separate lines of evidence support this hypothesis: (1) QUIN is a selective agonist of N-methyl-D-aspartate (NMDA) receptors[3] and, therefore, intracerebral administration of this compound can produce excitotoxic lesions;[4,5] (2) brain tissue and cerebrospinal fluid

---

Address for correspondence: T.P. Obrenovitch, Pharmacology, School of Pharmacy, University of Bradford, Bradford BD7 1DP, U.K. Voice: 44 1274 233359; fax: 44 1274 233363.
t.obrenovitch@bradford.ac.uk

concentrations of QUIN increase markedly with intracerebral inflammation, reflecting activation of the kynurenine pathway in invading macrophages and activated microglia.[1,6] This hypothesis is potentially relevant to a number of neurological disorders, but a key issue needs to be resolved: *are the extracellular levels of QUIN reached during brain inflammation high enough to promote excitotoxicity?* On one hand, QUIN levels in the extracellular fluid increased from 0.11 to 7.3 µM during endotoxin induced inflammation.[7] On the other hand, millimolar concentrations of QUIN are necessary to cause acute degeneration of rat striatal neurones *in vivo*, because QUIN is a relatively weak agonist of NMDA receptors.[4,8,9] The purpose of the study reported in this paper was to determine, *in vivo*, the extracellular concentrations of QUIN that must be reached in order to initiate electrophysiological changes indicative of excitotoxic stress in the cerebral cortex of rats, under normal conditions, and when superimposed on another insult (repeated spreading depression, SD). Our experimental strategy relied on microdialysis probes with an incorporated electrode, implanted in the brain of halothane anesthetized rats.[10] These devices were used to apply QUIN locally to the cortical area under study (with or without coperfusion of high K^+ for repetitive induction of SD) and to record the associated changes in the extracellular DC potential[11] and lactate.[12]

MATERIAL AND METHODS

Animal Preparation and Microdialysis

All animal procedures used were in strict accordance with the British Home Office guidelines, and specifically licensed under the Animals (Scientific Procedures) Act of 1986. Adult male Lister Hooded rats (weight, 300–350 g, Harlan U.K., Blackthorn, U.K.) were anesthetized throughout with halothane (5% for induction, 2.0–2.5% during surgery, and 1.5–1.75% for maintenance during the rest of the experiment) in $O_2:N_2O$ (1:2), with the animal breathing spontaneously. Concentric microdialysis probes incorporating a recording electrode[10] (microdialysis electrode 2-mm fiber length; ME-H2, Applied Neuroscience Ltd., London, U.K.) were implanted in the frontoparietal cortex (coordinates: 1.2 mm anterior to bregma, 2.5 mm lateral, and 2 mm deep from the dural surface)[11] and held in place by means of acrylic dental cement and two small stainless steel screws used as anchors. Unless otherwise stated, microdialysis probes were perfused with artificial cerebrospinal fluid (aCSF) (composition in mM: 125 NaCl, 2.5 KCl, 1.18 $MgCl_2$, 1.26 $CaCl_2$; pH 7.3 adjusted with 1 M NaOH) at 1 µl/min with a syringe pump (CMA/100, CMA/Microdialysis, Stockholm, Sweden). A femoral artery was catheterized for continuous monitoring of arterial blood pressure. To minimize any possible interference of halothane anesthesia with the processes under study,[14] once the surgical procedure had been completed, the depth of anesthesia was carefully controlled by monitoring the electroencephalogram (EEG) and arterial blood pressure, and the concentration of halothane in the breathing mixture held to a minimum (1.5–1.75%). Body temperature was maintained at 37.0°C throughout the experiment. At least two hours of stabilization were allowed after probe implantation.

Recording the Extracellular DC Potential and EEG

Both signals were derived from the potential between the electrode built into the microdialysis probe and an Ag/AgCl reference electrode placed under the scalp.[10] This potential was first amplified (×10) with a high impedance input preamplifier (NL834, Neurolog System, Digitimer Ltd., Welwyn Garden City, U.K.). The alternating current component in the 1–30 Hz window (about 10,000 times overall amplification) provided the EEG, and the DC component (250 times overall amplification) provided the DC potential.

Exogenous QUIN Induced Depolarization

The potency of QUIN to induce local negative shifts of the DC potential in the cortex of anesthetized rats ($n = 7$) was determined by perfusion of increasing concentrations of QUIN (1, 2, and 3 mM) through the microdialysis probe for two minutes.[11] Each perfusion of QUIN was followed by a 30–40-min recovery period (i.e., perfusion of aCSF for 30 min after 1 mM, and 40 min after 2 and 3 mM QUIN). For comparison, similar experiments ($n = 6$) were performed with NMDA (0.05, 0.1, 0.2, and 0.4 mM). The microdialysis recovery/delivery fraction, determined *in vitro* at 37.0°C, was approximately 20% for both QUIN and NMDA.

Effect of Exogenous QUIN on the Induction of SD by High K^+

K^+ induced cortical SD was selected as additional insult because its elicitation involves NMDA receptor activation, and SD markedly increases energy demand.[15,16] Our rationale was that, since SD is blocked by NMDA receptor antagonists, and SD can be triggered by the sole application of NMDA or QUIN (see FIGURE 1), superimposition of increasing concentrations of QUIN on K^+ induced SD should allow us to

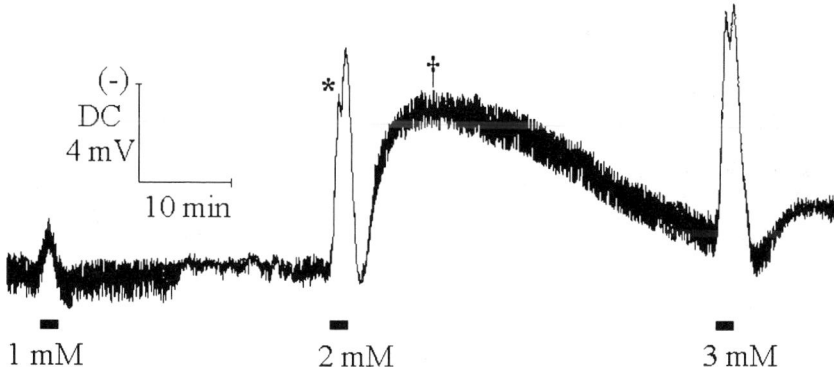

FIGURE 1. Representative changes in the extracellular direct current (DC) potential produced by QUIN in the rat frontoparietal cortex, *in vivo*. Increasing concentrations of QUIN (1–3 mM) were applied for two minutes via a microdialysis probe, and the resulting depolarization recorded precisely at the site of drug application. With 2 and 3 mM QUIN, the biphasic pattern of depolarisation (*) and secondary negative DC shift (†) indicated that the stimulus was strong enough to elicit spreading depression. In agreement with previous reports, the polarity of the DC potential was defined so that QUIN induced depolarizations produce an upward voltage deflection.

evaluate the threshold concentration of QUIN necessary to promote abnormal processes apparently involving NMDA receptor activation. In these experiments, two microdialysis probes were implanted in the cerebral cortex, symmetrically from the longitudinal suture. Repetitive SD were elicited on both sides of the cortex by switching from normal aCSF to a medium containing 130 mM K^+ (with lowered Na^+) for 20 min. Such a high concentration of K^+ is required to evoke consistent, periodic SD under these experimental conditions[17] and we had established that 130 mM K^+ was a submaximal stimulus for SD initiation.[18] Each K^+ stimulus was followed by 40 min of recovery; that is, perfusion of the microdialysis fiber with normal aCSF. Five episodes of repetitive SD were produced by perfusion of K^+ medium alone on the control side. Contralaterally, the K^+ medium was supplemented with 30, 100, and 300 µM QUIN for the second, third, and fourth SD episodes. No QUIN was applied during the first SD episode because the magnitude of the resulting SD is consistently much larger than subsequent episodes. The fifth SD episode was used to assess the reversibility of the QUIN effects.

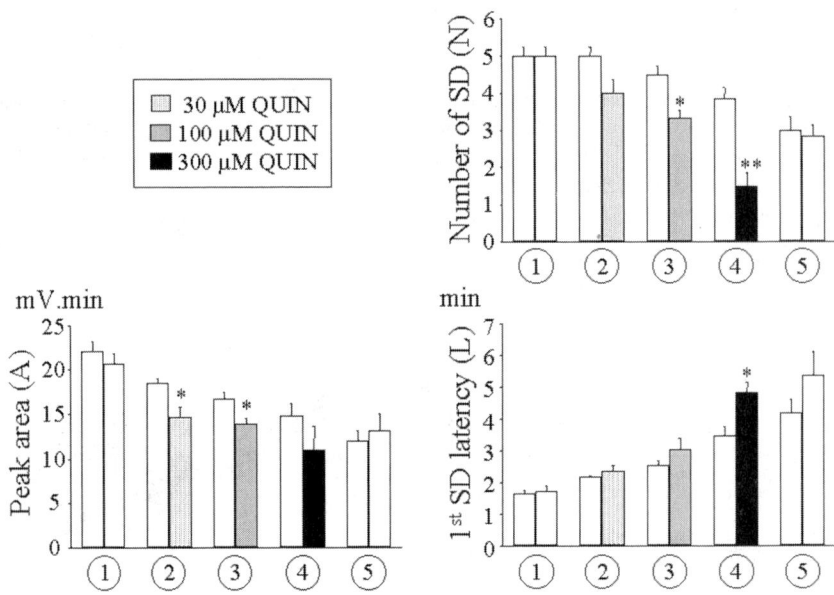

FIGURE 2. Effects of coapplication of quinolinate (QUIN) on K^+ induced cortical spreading depression (SD). Three different variables were examined: (1) number of SD elicited during each 20-min K^+ stimulus (N); (2) cumulative peak area (A, mV·min); (3) latency for first SD (L, min). On the control side (*open bars* to the left), 130 mM K^+ was perfused five times for 20 min through the probe. On the QUIN side (*right bars*), increasing concentrations of QUIN (30, 100, and 300 µM) were coperfused with high K^+ medium during the second, third, and fourth SD elicitation period. *Circled numbers* below the ordinal axis indicate the SD episode number in the experiment. *Shaded or filled bars,* coapplication of QUIN. Values are mean ± SEM; *$p < 0.05$ and **$p < 0.01$, comparison to corresponding control data by Mann–Whitney test ($n = 6$).

Effects of Increasing Extracellular Concentrations of QUIN on Cortical Extracellular Lactate

In contrast to what was expected, coperfusion of 30–300 µM QUIN did not promote K^+ induced SD, but actually inhibited it (see FIGURE 2). This finding prompted us to determine whether the effect of QUIN on SD reflected a significant challenge to the local brain energy metabolism, by studying the effects of microdialysis perfusion of QUIN (30–300 µM for 20 min, and 1–3 mM for 2 min) on extracellular lactate. Lactic acid accumulation offers a sensitive and reliable index for increased neuronal activity, especially glutamatergic activity,[19,20] and ionotropic glutamate receptor activation implies increased transmembrane movements of ions with associated increase in energy metabolism and acidosis.[21] For this purpose, our previous experimental strategy was supplemented with flow analysis of dialysate lactate by enzyme amperometry (biosensor).[12,22]

RESULTS AND DISCUSSION

Extracellular Concentration Threshold for Induction of Local Depolarization in the Normal Cerebral Cortex

With rats anesthetized with 1.25% halothane in $O_2:N_2O$ 1:2, and 2-mm dialysis probes perfused with QUIN-aCSF for two minutes at 1 µl/min, the concentration threshold for detectable responses was around 0.5 to 1 mM QUIN in the perfusion medium (FIG. 1). In comparison to NMDA,[11] the concentration response relationship for QUIN was very steep, so that 1 mM QUIN in the perfusion medium induced a small response in some experiments, and near maximal depolarization with SD elicitation in others (see FIGURES 1 and 3). These experiments confirm that QUIN is a weak NMDA receptor agonist: the apparent EC_{50} (concentration of QUIN in the dialysate producing half-maximal response) was 1.2 ± 0.3 mM (mean ± SEM; $n = 7$); that is, about one order of magnitude less potent than NMDA. The extracellular EC_{50} for QUIN induction of local depolarization (calculated from the apparent EC_{50} and the microdialysis recovery/delivery ratio) was about 250 µM—more than about 30 times greater than the extracellular QUIN levels measured in the immune activated brain of gerbils.[7] It is likely that extracellular concentrations even greater than 250 µM QUIN are required to induce neurodegeneration. First, because repeated responses of similar magnitudes were obtained in each recording session. Second, because QUIN (2 or 3 mM) consistently initiated SD (FIGS. 1 and 3) and both initiation and propagation of SD are known to depend on the efficiency of the brain parenchyma homeostatic mechanisms.[23] Furthermore, estimates of the extracellular QUIN concentrations after an injection of 200 nmol (i.e., amount of QUIN generally used to induced striatal neurodegeneration) indicated a peak level of 13.7 mM at 10–20 min postinjection and declined to 1.2 mM by two hours.[24]

Effects of Coapplication of QUIN on Cortical SD Elicited by High Extracellular K^+

Within the concentration range 30–300 µM in the perfusion medium, QUIN concentration dependently inhibited the elicitation of SD by K^+ (FIG. 2). This unexpected SD inhibition by QUIN was probably due to a direct interaction with NMDA

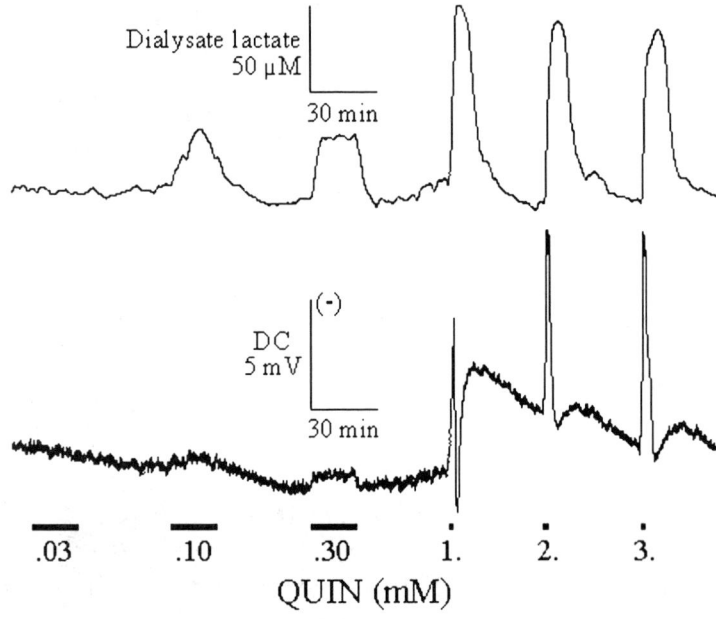

FIGURE 3. Representative changes in the extracellular DC potential (**top trace**) and dialysate lactate (**bottom trace**) produced by QUIN in the rat cortex. Both traces were recorded from the same experiment. Increasing concentrations of QUIN (0.03 to 3 mM) were perfused for 20 or 2 min via a microdialysis probe (2-mm fiber length, 1 µl/min flow rate) implanted in the frontoparietal cortex of rats anesthetized with halothane (1.5–2% in $O_2:N_2O$ 1:2). Note the complete lack of effect with 20-min perfusion of 0.03 mM QUIN. Extracellular DC potential changes were recorded precisely at the site of QUIN application. In agreement with previous reports, the polarity of the signal was reversed (−) so that negative shifts of the DC potential produced an upward deflection. Changes in lactate concentration were determined through enzyme amperometry analysis of the dialysate as it emerged from the probe outlet. The basal lactate level was 127 µM in this experiment. *Long* and *short horizontal bars* indicate perfusion of QUIN for 20 and 2 min, respectively.

receptors since the same effect was obtained with NMDA itself.[25] One possible explanation for this finding is that, although low concentrations of QUIN (and NMDA) did not produce sufficient channel gating to evoke detectable DC potential negative shifts, they were capable of altering the interactions between NMDA receptors and the glutamate efflux that is associated with SD elicitation;[26] that is, an action similar to competitive antagonism. Another possibility is that 100 and 300 µM QUIN did produce a moderate influx of Ca^{2+} through the NMDA operated cation channels, leading to glycine insensitive NMDA receptor desensitization.[27] Alternatively, moderate influx of Ca^{2+} through NMDA operated channels might have led to changes in cytoplasmic Ca^{2+} loading and/or buffering, and ultimately to alteration in the pattern of intra- and intercellular Ca^{2+} waves involved in SD initiation.[28] Intra- and intercellular Ca^{2+} waves among neural cells are now recognized as important manifestations of Ca^{2+} signalling[29,30] and Ca^{2+} waves may play a pivotal role in SD initiation.[31]

Finally, the finding that perfusion of at least 100 µM QUIN was associated with increased lactic acid formation (see next section) suggests that inhibition of K^+ induced SD could have been linked to acidosis.

Effects of Increasing Extracellular Concentrations of QUIN on Cortical Extracellular Lactate

On line monitoring of changes in extracellular lactate during microdialysis perfusion of QUIN revealed several interesting observations (FIG. 3): (1) Perfusion of 30 µM QUIN (i.e., about 6 µM calculated extracellular concentration) for 20 min had no effect on extracellular lactate in any of the 10 experiments carried out. (2) At 100 and 300 µM, QUIN produced an increase in extracellular lactate during the 20-min application, with rapid subsequent normalization. (3) Perfusion of 1, 2, and 3 mM QUIN for two minutes markedly increased dialysate lactate, to levels comparable to those associated with terminal ischemia (i.e., cardiac arrest, FIG. 3), and this change persisted well after the end of QUIN application. Note, however, that normalization of extracellular lactate was generally achieved during the subsequent 40-min recovery period. These experiments suggest that extracellular concentrations of QUIN in the low micromolar range are well tolerated by the brain parenchyma (at least in cortical regions). One could argue that some structures may be more sensitive than others, depending on the density and nature of the NMDA receptor subtypes present. This cannot be ruled out, but the striatum (where QUIN is generally microinjected to produce excitotoxic lesions) and cortex show comparable sensitivity to NMDA in our preparation (EC_{50} approximately 180 and 126 µM, respectively.[11] The fact that rats were anesthetized with halothane in the experiments reported herein could also be argued against.[14] However, equivalent responses were obtained with perfusion of 200 µM NMDA in the striatum of rats under halothane (1.5% in O_2: N_2O 1:2) anesthesia, and 150 µM NMDA in the same region when urethane (1.5 g/kg i.p.) was used instead.[32]

Exposure of Neurons to Sustained, Low Concentrations of NMDA Receptor Agonists—Excitotoxicity or Neuroprotection?

Although excessive NMDA receptors activation leads to cell death in rat cerebellar neurons *in vitro*,[33,34] pretreatment with low concentrations of NMDA promotes the survival of the same cells and can be protective against various insults, including exposure to glutamate.[35–37] Pharmacological studies and measurements of changes in intracellular Ca^{2+} showed that this type of neuroprotection is mediated by NMDA receptors, but is neither due to their desensitization nor to reduced increased intracellular Ca^{2+} during the subsequent excitotoxic stimulus.[35–37] Instead, moderate, prolonged NMDA receptor activation appears to stimulate the formation of brain derived neurotrophic factor (BDNF).[38] A similar, if not identical neuroprotective effect was recently reported for QUIN—pretreatment of primary cultures of rat cerebellar granule neurons with 10 µM QUIN for six hours suppressed the apoptosis induced by a subsequent exposure to glutamate by 68%.[39] Sustained exposure of cerebellar granule neurons to 1 or 2.5 mM QUIN did not alter their viability.[39,40] All the data outlined in this subsection are sufficient to counterbalance the earlier study of Whetsell and Schwarcz,[41] in which exposure of rat organotypic cultures for seven weeks to 100 nM QUIN produced focal degeneration of rat corticostriatal preparations, but not those of caudate nucleus.

CONCLUSIONS

Our data suggest that, although increased synthesis of QUIN from activated microglia and invading macrophages can lead to a 10–100-fold increase in its extracellular levels,[7] the latter appear to remain far below the concentration of QUIN necessary to produce excessive NMDA receptor activation and subsequent neuronal death. However, our results do not rule out the possibility that QUIN accumulation may be a deleterious feature of neuroinflammation. First, because we did not test whether the threshold concentration for QUIN induced excitotoxicity may be lowered by inflammation, a pertinent point since Lawrence *et al.* found that the cytokine interleukin-1 (IL-1) promoted excitotoxic actions (albeit those mediated by AMPA receptors).[42] Second, because QUIN accumulation could be detrimental independently from any action on glutamate receptors.[43]

ACKNOWLEDGMENTS

The contribution of my colleagues, Dr. Jutta Urenjak and Elias Zilkha (Pharmacology, Bradford School of Pharmacy) is gratefully acknowledged. I also wish to thank Richard Exley (Pharmacology, Bradford School of Pharmacy) for his help with the development of the lactate biosensor, and Dr. De-Maw Chuang (Biological Psychiatry Branch, NIMH, Bethesda) for his stimulating suggestions related to our QUIN data.

REFERENCES

1. HEYES, M.P., K. SAITO, A. LACKNER, *et al.* 1998. Sources of the neurotoxin quinolinic acid in the brain of HIV-1–infected patients and retrovirus-infected macaques. FASEB J. **12**: 881–896.
2. SINZ, E.H., P.M. KOCHANEK, M.P. HEYES, *et al.* 1999. Quinolinic acid is increased in CSF and associated with mortality after traumatic brain injury in humans. J. Cereb. Blood Flow Metab. **18**: 610–615.
3. STONE, T.W. 1993. Neuropharmacology of quinolinic and kynurenic acids. Pharmacol. Rev. **45**: 309–379.
4. SCHWARCZ, R., W.O. WHETSELL, JR. & R.M. MANGANO. 1983. Quinolinic acid: an endogenous metabolite that produces axon-sparing lesions in rat brain. Science **219**: 316–318.
5. LEVIVIER, M. & S. PRZEDBORSKI. 1998. Quinolinic acid-induced lesions of the rat striatum: quantitative autoradiographic binding assessment. Neurol. Res. **20**: 46–56.
6. ALBERATI-GIANI, D., P. RICCIARDI-CASTAGNOLI, C. KÖHLER, *et al.* 1996. Regulation of the kynurenine metabolic pathway by interferon-γ in murine cloned macrophages and microglial cells. J. Neurochem. **66**: 996–1004.

7. BEAGLES, K.E., P.F. MORRISON & M.P. HEYES. 1998. Quinolinic acid *in vivo* synthesis rates, extracellular concentrations, and intercompartmental distributions in normal and immune-activated brain as determined by multiple-isotope microdialysis. J. Neurochem. **70:** 281–291.
8. CURRAS, M.C. & R. DINGLEDINE. 1992. Selectivity of amino acid transmitters acting at N-methyl-D-aspartate and amino-3-hydroxy-5-methyl-isoxazoleproprionate receptors. Mol. Pharmacol. **41:** 520–526.
9. LEVIVIER, M. & S. PRZEDBORSKI. 1998. Quinolinic acid-induced lesions of the rat striatum: quantitative autoradiographic binding assessment. Neurol. Res. **20:** 46–56.
10. OBRENOVITCH, T.P., D.A. RICHARDS, G.S. SARNA, *et al.* 1993. Combined intracerebral microdialysis and electrophysiological recording: methodology and applications. J. Neurosci. Meth. **47:** 139–145.
11. OBRENOVITCH, T.P., J. URENJAK & E. ZILKHA. 1994. Intracerebral microdialysis combined with recording of extracellular field potential: a novel method for investigation of depolarizing drugs *in vivo*. Br. J. Pharmacol. **113:** 1295–1302.
12. EXLEY, R., E. ZILKHA, J. URENJAK, *et al.* 2001. Continuous monitoring of changes in brain extracellular lactate using microdialysis coupled to enzyme-amperometric analysis. Br. J. Pharmacol. **131**(Suppl.): 219p.
13. PAXINOS, G. & C. WATSON. 1986. The Rat Brain in Stereotaxic Coordinates. Academic Press, London.
14. MARTIN, D.C., M. PLAGENHOEF, J. ABRAHAM, *et al.* 1995. Volatile anaesthetics and glutamate activation of N-methyl-D-aspartate receptors. Biochem. Pharmacol. **49:** 809–817.
15. LAURITZEN, M. 1994. Pathophysiology of the migraine aura: The spreading depression theory. Brain **117:** 199–210.
16. OBRENOVITCH, T.P. & E. ZILKHA. 1996. Inhibition of cortical spreading depression by L-701,324, a novel antagonist at the glycine site of the N-methyl-D-aspartate receptor complex. Br. J. Pharmacol. **117:** 931–937.
17. HERRERAS, O. & G.G. SOMJEN. 1993. Analysis of potential shifts associated with recurrent spreading depression and prolonged unstable spreading depression induced by microdialysis of elevated K^+ in hippocampus of anesthetized rats. Brain Res. **610:** 283–294.
18. TAYLOR, D.L., J. URENJAK, E. ZILKHA, *et al.* 1997. Effects of probenecid on the elicitation of spreading depression in the rat striatum. Brain Res. **764:** 117–125.
19. FRAY, A.E., R.J. FORSYTH, M.G. BOUTELLE. *et al.* 1996. The mechanisms controlling physiologically stimulated changes in rat brain glucose and lactate: a microdialysis study. J. Physiol. **496:** 49–57.
20. DEMESTRE, M., M. BOUTELLE & M. FILLENZ. 1997. Stimulated release of lactate in freely moving rats is dependent on the uptake of glutamate. J. Physiol. **499:** 825–832.
21. BORDELON, Y.M., M.-F. CHESSELET, M. ERECINSKA, *et al.* 1998. Effects of intrastriatal injection of quinolinic acid on electrical activity and extracellular ion concentrations in rat striatum *in vivo*. Neuroscience **83:** 459–469.
22. ZILKHA, E., T.P. OBRENOVITCH, A. KOSHY, *et al.* 1995. Extracellular glutamate: on-line monitoring using microdialysis coupled to enzyme-amperometric analysis. J. Neurosci. Meth. **60:** 1–9.
23. KOROLEVA, V.I. & J. BUREŠ. 1980. Blockade of cortical spreading depression in electrically and chemically stimulated areas of cerebral cortex in rats. Electroencephalogr. Clin. Neurophysiol. **48:** 1–15.
24. BAKKER, M.H. & A.C. FOSTER. 1991. An investigation of the mechanisms of delayed neurodegeneration caused by direct injection of quinolinate into the rat striatum *in vivo*. Neuroscience **42:** 387–395.
25. OBRENOVITCH, T.P. 2001. New insights into the role of NMDA-receptors and nitric oxide formation in the genesis of spreading depression. *In* Pharmacology of Cerebral Ischemia 2000. J. Krieglstein & S. Klumpp, Eds.: 67–74. Medpharm, Stuttgart.

26. OBRENOVITCH, T.P. & E. ZILKHA. 1995. Changes in extracellular glutamate concentration associated with propagating cortical spreading depression. In Experimental headache models. J. Olesen & M.A. Moskowitz, Eds.: 113–117. Raven Press, New York.
27. ROSENMUND, C., A. FELTZ & G.L. WESTBROOK. 1995. Calcium-dependent inactivation of synaptic NMDA receptors in hippocampal neurons. J. Neurophysiol. **73:** 427–430.
28. WANG, Z., M. TYMIANSKI, O.T. JONES, et al. 1997. Impact of cytoplasmic calcium buffering on the spatial and temporal characteristics of intercellular calcium signals in astrocytes. J. Neurosci. **17:** 7359–7371.
29. CHARLES, A.C., S.K. KODALI & R.F. TYNDALE. 1996. Intercellular calcium waves in neurons. Mol. Cell Neurosci. **7:** 337–353.
30. DEITMER, J.W., A.J. VERKHRATSKY & C. LOHR. 1998. Calcium signalling in glial cells. Cell Calcium **24:** 405–416.
31. KUNKLER, P.E. & R.P. KRAIG. 1998. Calcium waves precede electrophysiological changes of spreading depression in hippocampal organ cultures. J. Neurosci. **18:** 3416–3425.
32. URENJAK, J. & T.P. OBRENOVITCH. 2000. Kynurenine 3-hydroxylase inhibition in rats: effects on extracellular kynurenic acid concentration and N-methyl-D-aspartate-induced depolarisation in the striatum. J. Neurochem. **75:** 2427–2433.
33. GARTHWAITE, G. & J. GARTHWAITE. 1987. Receptor-linked ionic channels mediate N-methyl-D-aspartate neurotoxicity in rat cerebellar slices. Neurosci. Lett. **83:** 241–246.
34. FAVARON, M., H. MANEV, H. ALHO, et al. 1988. Gangliosides prevent glutamate and kainate neurotoxicity in primary neuronal cultures of neonatal rat cerebellum and cortex. Proc. Nat. Acad. Sci. U.S.A. **85:** 7351–7355.
35. BALAZS, R., N. HACK, O.S. JORGENSEN, et al. 1989. N-methyl-D-aspartate promotes the survival of cerebellar granule cells: pharmacological considerations. Neurosci. Lett. **101:** 242–246.
36. CHUANG, D.M., X.M. GAO & S.M. PAUL. 1992. N-methyl-D-aspartate exposure blocks glutamate toxicity in cultured cerebellar granule cells. Mol. Pharmacol. **42:** 210–216.
37. MARINI, A.M., Y. UEDA & C.H. JUNE. 1999. Intracellular survival pathways against glutamate receptor agonist excitotoxicity in cultured neurons. Intracellular calcium responses. Ann. N.Y. Acad. Sci. **890:** 421–437.
38. MARINI, A.M., S.J. RABIN, R.H. LIPSKY, et al. 1998. Activity-dependent release of brain-derived neurotrophic factor underlies the neuroprotective effect of N-methyl-D-aspartate. J. Biol. Chem. **273:** 29394–29399.
39. SEI, Y., L. FOSSOM, G. GOPING, et al. 1998. Quinolinic acid protects rat cerebellar granule cells from glutamate-induced apoptosis. Neurosci. Lett. **241:** 180–184.
40. QIN, Z.H., R.W. CHEN, Y. WANG, et al. 1999. Nuclear factor kappaB nuclear translocation upregulates c-Myc and p53 expression during NMDA receptor-mediated apoptosis in rat striatum. J. Neurosci. **19:** 4023–4033.
41. WHETSELL, W.O. & R. SCHWARCZ. 1989. Prolonged exposure to submicromolar concentrations of quinolinic acid causes excitotoxic damage in organotypic cultures of rat corticostriatal system. Neurosci. Lett. **97:** 271–275.
42. LAWRENCE, C.B., S.M. ALLAN & N.J. ROTHWELL. 1998. Interleukin-1beta and the interleukin-1 receptor antagonist act in the striatum to modify excitotoxic brain damage in the rat. Eur. J. Neurosci. **10:** 1188–1195.
43. BEHAN, W.M., M. MCDONALD, L.G. DARLINGTON, et al. 1999. Oxidative stress as a mechanism for quinolinic acid-induced hippocampal damage: protection by melatonin and deprenyl. Br. J. Pharmacol. **128:** 1754–1760.

Neuronal Protein Kinase Signaling Cascades and Excitotoxic Cell Death

STEPHEN D. SKAPER, LAURA FACCI, AND PAUL J.L.M. STRIJBOS

Neurology Centre of Excellence for Drug Discovery,
GlaxoSmithKline Beecham Pharmaceuticals,
Harlow, Essex, United Kingdom

ABSTRACT: Perturbation of normal survival mechanisms may play a role in a large number of disease processes. Glutamate neurotoxicity, particularly when mediated by the *N*-methyl-D-aspartate (NMDA) subtype of glutamate receptors, has been hypothesized to underlie several types of acute brain injury, including stroke. Several neurological insults linked to excessive release of glutamate and neuronal death result in tyrosine kinase activation, including p44/42 mitogen activated protein (MAP) kinase. To further explore a role for MAP kinase activation in excitotoxicity, we used a novel tissue culture model to induce neurotoxicity. Removal of the endogenous blockade by Mg^{2+} of the NMDA receptor in cultured hippocampal neurons triggers a self perpetuating cycle of excitotoxicity, which has relatively slow onset, and is critically dependent on NMDA receptors and activation of voltage gated Na^+ channels. These injury conditions led to a rapid phosphorylation of p44/42 that was blocked by MAP kinase kinase (MEK) inhibitors. MEK inhibition was associated with protection against synaptically mediated excitotoxicity. Interestingly, hippocampal neurons preconditioned by a sublethal exposure to Mg^{2+}-free conditions were rendered resistant to injury induced by a subsequently longer exposure to this insult; the preconditioning effect was MAP kinase dependent. The MAP kinase signaling pathway can also promote polypeptide growth factor mediated neuronal survival. MAP kinase regulated pathways may act to promote survival or death, depending upon the cellular context in which they are activated.

KEYWORDS: Synaptic transmission; Signaling; Mitogen activated protein kinases; Excitotoxicity; Hippocampus; Neuroprotection; Neurotrophins.

INTRODUCTION

Regulation of cell survival is crucial to the normal physiology of multicellular organisms. The appropriate suppression of cell death is important in a number of contexts, ranging from sculpting of the nervous system during development to adaptive responses during adulthood. In addition, perturbation of neuronal survival mechanisms—leading to excessive cell death or survival—may play a role in a large number of disease processes. Although neuronal cell death can be triggered by a multitude of stimuli, neurodegeneration induced by the excitatory transmitter amino

---

Address for correspondence: Stephen D. Skaper, Neurology Centre of Excellence for Drug Discovery, GlaxoSmithKline Beecham Pharmaceuticals, New Frontiers Science Park North, Third Avenue, Harlow CM19 5AW, Essex, U.K. Voice: +44 1279 622350; fax: +44 1279 622555.
Stephen_Skaper-1@gsk.com

acid glutamate ("excitotoxicity"), is considered to be of particular relevance to several types of acute brain injury, ranging from cerebral ischemia to epilepsy and mechanical brain trauma.[1-3] A great deal of information is already known concerning the events that transpire at the neuronal surface during the course of excitotoxicity; however, the putative second messenger systems that mediate cell death are complex and remain ill defined. The MEK1 (MAP kinase/ERK kinase)/ERK (extracellular signal regulated kinase) pathway has been implicated in cell growth and differentiation.[4] It has been demonstrated in some systems, including neurotrophic factor dependent neurons, that the MEK/ERK pathway may have antiapoptotic effects that oppose the proapoptotic effects associated with activation of the JNK and p38 MAP kinases.[5] Recent evidence suggests that the ERK/MAP kinases can also be activated following glutamate receptor activation. Indeed, neurological insults linked to excitotoxicity, including global and focal cerebral ischemia, hypoglycemia, and kainate induced seizures, result in tyrosine kinase phosphorylation, including the ERK/MAP kinases.[6-11] p44/42 MAP kinases may have a role in excitotoxic injury mediated by synaptic mechanisms.[12-15] ERK/MAP kinase inhibitors provide neuroprotection against excitotoxicity, suggesting that this signaling cascade can mediate glutamate dependent neuronal damage. This apparent dual role of the ERK/MAP kinases in neuronal survival proposes that activation of these kinases can take place via several distinct and compartmentalized pathways. This article briefly reviews the experimental findings that support the above ideas.

NEUROTROPHIN MEDIATED NEURONAL SURVIVAL: RECEPTOR TYROSINE KINASE SIGNALING PATHWAYS

Target dependent survival of neurons is regulated by neurotrophins through activation of their cognate receptor tyrosine kinases, Trks.[16-18] Ligand occupancy induces receptor dimerization and tyrosine phosphorylation.[19] Phosphotyrosyl motifs on the Trk signaling receptor bind Shc proteins, allowing further recruitment and activation of additional cytoplasmic adapter proteins. The resultant multiprotein complex promotes binding of p21 ras GTPase which, in turn, is critical for the activation of the serine/threonine kinases Raf and MEK. Activation of ERK/MAP kinases upon Trk stimulation appears to involve two spatially distinct events: an initial ERK/MAP kinase activation by Ras in the cytoplasm, followed by ERK/MAP kinase translocation to nucleus where it phosphorylates transcription factors.[18] Dominant negative forms of Ras, Raf, and MEK block neurotrophin induced neuritogenesis, whereas constitutively active forms of these molecules promote neurite outgrowth in the absence of neurotrophin.[18,20] The latter observations attest to the functional importance of this protein kinase pathway to neurotrophic activity.

An additional signaling molecule that is important for neurotrophin dependent cell survival is phosphatidylinositol 3-kinase (PI 3-kinase).[21,22] A mechanism by which PI 3-kinase promotes survival of neurons is via activation of the serine/threonine kinase Akt/protein kinase B (PKB).[22,23] Akt/PKB inhibits apoptosis through phosphorylation of cell death regulatory molecules, such as BAD,[24] caspase 9,[25] and Forkhead transcription factors.[26] Neurotrophin deprived cells transfected with constitutively active PI 3-kinase, with receptor mutants in which PI 3-kinase is exclusively activated in response to receptor ligation, or with constitutively active Ras mutants that bind to and activate only PI 3-kinase are resistant to apoptosis.[22,27,28] Ras dependent cell survival may be mediated both by the Raf–MEK–MAP kinase pathway and Ras mediated activation of PI 3-kinase.[22]

KINASE SIGNALING CASCADES AND EXCITOTOXICITY

Although much work has focused on the role of ERK/MAP kinases on the neurotrophic actions of growth factors,[18] it has become apparent that ERK/MAP signaling pathways can play multiple roles in the activity dependent regulation of neuronal function, including synaptic plasticity, learning, and memory.[29] For example, classical conditioning in *Hermissenda* has been shown to be associated with increased ERK phosphorylation.[30] More compelling evidence for an ERK/MAP kinase dependent role in learning and memory has come from rodent behavioral studies. In three different models (fear conditioning, aversive taste learning, and spatial learning), behavioral performance was associated with increased ERK activity, and inhibition of ERK/MAP kinase signaling specifically impaired learning.[31–33]

Glutamate is the principal excitatory neurotransmitter in the brain, where it contributes to several forms of neuronal plasticity, including that involved in memory formation, sensory processing, and neurodegeneration. Although glutamate is essential to normal nervous system function, excessive glutamatergic activity can precipitate in excitotoxicity.[2,29] The mechanisms by which stroke-induced neuronal death occur are complex and are likely to depend upon the severity and duration of ischemic insult and an elaborate interplay between ischemic death initiators, such as excitotoxicity, oxidative stress, DNA damage, and inflammatory responses.[2,34] Excitotoxic damage has been associated with alterations in a number of signaling systems, including the protein kinase C cascades,[35] the Ca^{2+}/calmodulin dependent protein kinase cascade,[35,36] and the nitric oxide signaling system.[37]

The ERK/MAP kinases are phosphorylated in the hippocampus in response to global brain ischemia.[6] Tyrosine kinase inhibitors, such as genistein, decrease ERK phosphorylation and protect against neuronal damage in a model of global cerebral ischemia,[6] although their non-selectivity does not provide unequivocal evidence linking ERK/MAP kinase activation and ischemic cell injury. A more recent study has reported that the MEK/ERK pathway is also activated during focal cerebral ischemia.[11] Treatment of mice 30 min before ischemia with the MEK1-specific inhibitor PD98059 reduced focal infarct volume at 22 h after ischemia by 55% following transient occlusion of the middle cerebral artery. This was accompanied by a reduction in phospho-ERK1/2 immunohistochemical staining (see FIGURE 1).[11] MEK1 inhibition also resulted in reduced brain damage 72 h after ischemia, with focal infarct volume reduced by 36%.[11] Changes in tyrosine phosphorylation of ERK/MAP kinase in the rat hippocampus during and following severe hypoglycemia have also been described.[10]

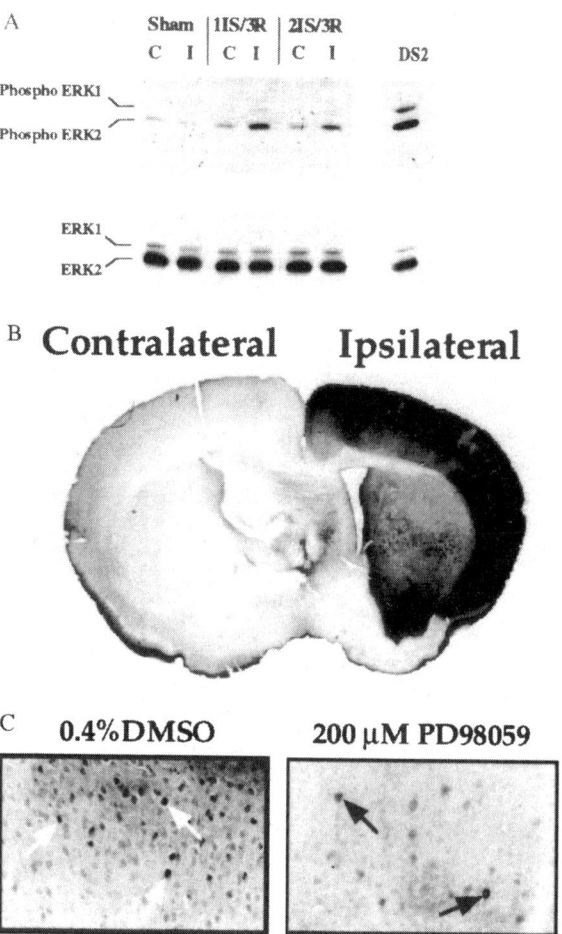

FIGURE 1. (A) Increase in ERK1/2 phosphorylation after ischemia and reperfusion. Western blot analysis (phosphospecific ERK1/2 antibodies) was performed on lysates from the contralateral (*C*) and ipsilateral/ischemic (*I*) side of the brain after one hour of ischemia (*1IS*) following three minutes of reperfusion (*3R*) or two hours of ischemia (2IS) followed by three minutes of reperfusion (3R). DS2 total cell lysate was used as control for ERK1/2 phosphorylation. **(B)** Increased ERK1/2 phosphorylation in the cortical brain region after one hour focal cerebral ischemia and three minutes reperfusion. Immunohistochemistry was performed with phosphospecific ERK1/2 antibodies. Contralateral (nonischemic) and ipsilateral (ischemic) sides of the brain are indicated. (×100.) **(C)** Pretreatment of mice with PD98059 leads to decreased ERK1/2 phosphorylation in the cortical region of the brain after two hours of focal cerebral ischemia and three minutes of reperfusion. Immunohistochemistry of brain sections was performed with the phosphospecific ERK1/2 antibodies. Nuclear staining is indicated by *arrows*. (×400.) (Reproduced from Alessandrini *et al.* Ref. 11, Fig.1, with permission from the National Academy of Sciences, U.S.A.)

A MODEL FOR SYNAPTICALLY MEDIATED EXCITOTOXICITY IN HIPPOCAMPAL NEURONS

The precise cellular mechanisms that lead to glutamate neurotoxicity remain unclear, however evidence suggests that excessive synaptic activity may be involved. Prolonged epileptiform activity is associated with neurotoxicity. Disruption of glutamatergic afferents to the affected region or application of NMDA receptor antagonists reduces seizure induced neuronal loss.[3] *In vitro* removal of glutamate receptor blockade or elimination of Mg^{2+} from the extracellular medium, in order to alleviate Mg^{2+}-block of the NMDA receptor, induces excessive synaptic transmission[38] and NMDA receptor mediated neuronal degeneration in hippocampal and cortical cell cultures.[39–42]

To further explore these ideas, we used a novel tissue culture model to induce neurotoxicity.[43] This model involves removing the endogenous blockade of the NMDA receptor by magnesium, which triggers a self perpetuating cycle of excitotoxicity that has relatively slow onset and is critically dependent on NMDA receptors and activation of TTX-sensitive sodium channels (see FIGURE 2). This type of cell death is similar to that described in our previous publications, where we applied NMDA for a brief period to activate this neurotoxic pathway.[37,44] In analogy with these earlier studies, neuronal death was mediated by NMDA receptors and sensitive to sodium channel blockade (see FIGURE 3).

Under Mg^{2+}-free, glycine supplemented conditions (mast cell derived) histamine directly enhanced the sensitivity of hippocampal neurons to synaptically mediated excitotoxicity involving NMDA receptors.[43] This receptor activation presumably occurs at synaptic sites and may be akin to events occurring in status epilepticus, in which neurons are apparently killed as a result of massive waves of excitatory synaptic activity.[45] Histamine potently modulates the gating of NMDA channels in the hippocampus.[46,47] These findings suggest that, in addition to histamine modulation of physiological processes, histamine may also participate in the pathophysiology of

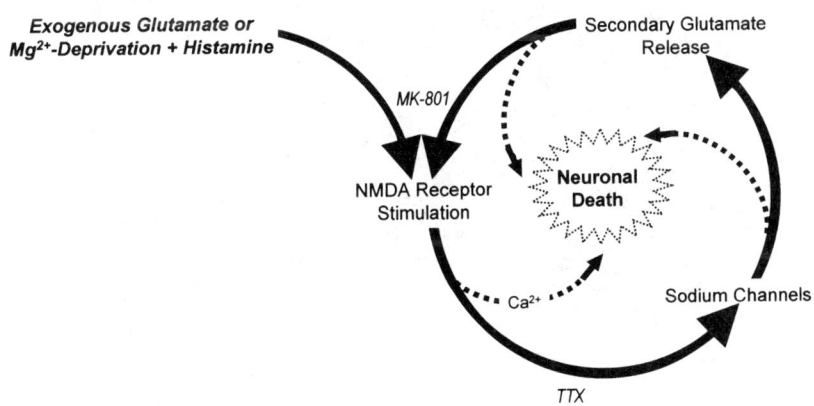

FIGURE 2. Excitotoxicity model. Schematic representation of synaptically mediated excitotoxic neuronal injury. TTX, tetrodotoxin.

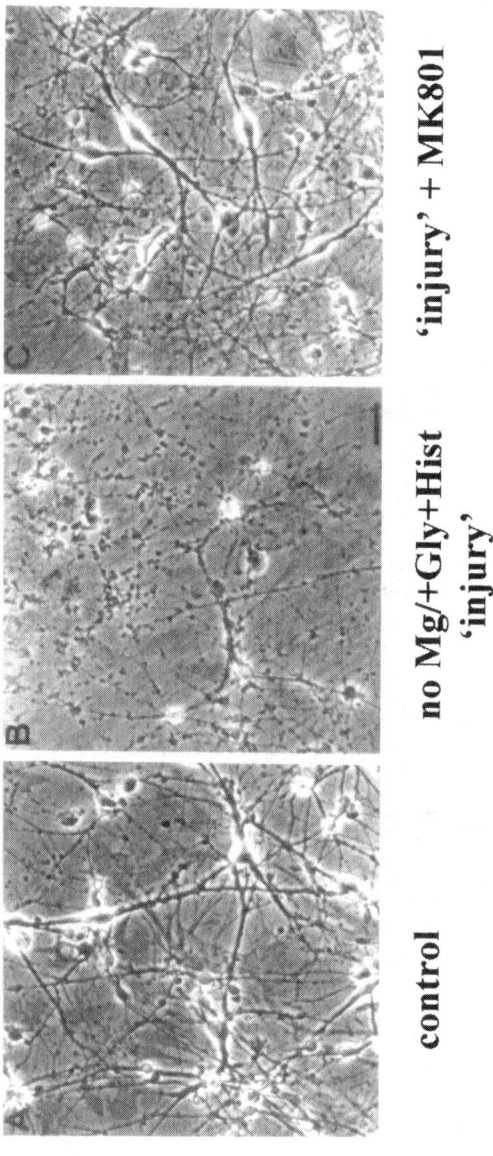

FIGURE 3. Exposure of rat hippocampal neurons (14 days in culture) to 15 min (22°C) of Mg^{2+}-free Locke's solution (pH 7.0) containing glycine (0.1 μM) and histamine (30 μM) results in cell death 24 h later. Inclusion of the NMDA receptor antagonist MK-801 (1 μM) during the time of Mg^{2+} deprivation prevented the ensuing neurodegeneration. See Reference 42 for further details.

disorders that involve synaptically mediated excitotoxicity; for example, in stroke, where extracellular K^+ increases and neurons depolarize, resulting in increased glutamate release and relief of the Mg^{2+} block of NMDA receptors.

The histaminergic system originates from the posterior hypothalamus and innervates the entire CNS with widely arborizing projections.[48] A novel potentiating effect of histamine on NMDA mediated synaptic transmission in hippocampal cultures has been reported elsewhere.[46,47] The present results demonstrate that histamine is also capable of enhancing NMDA mediated synaptic activity leading to excitotoxic cell death.[43] Similar to the potentiating effect of histamine on NMDA currents,[46,47] the effects of histamine on synaptic transmission induced death occurred in the absence of an interaction with archetypal histamine receptors, since its actions were not mimicked by histamine H_1, H_2, or H_3 receptor agonists or blocked by their respective receptor antagonists.[43] This may suggest the involvement of an atypical histamine receptor or a direct effect of histamine on the NMDA receptor.

Mast cells occur within the CNS of many species.[49,50] Up to 50% of whole brain histamine levels and 90% of the thalamic histamine is attributable to the presence of these cells.[51] Isolated brain mast cells secrete histamine,[52] which could modulate neuronal events. When peritoneal mast cells were coapplied with the hippocampal cultures and stimulated with antigen under Mg^{2+}-free conditions, a greater degree of neuronal death was observed, which relied critically on mast cell derived histamine.[43] It has been proposed that histamine release is involved in the pathogenesis of thiamine deficiency-induced neuronal necrosis within the thalamus.[53] The distribution of mast cells and histamine terminals within the thalamus bears a striking resemblance to the pattern of thiamine deficiency induced lesions in the same brain area.[54] The long lasting increase in neuronal histamine release induced by cerebral ischemia[55] could also play a role in excitotoxic brain injury.

Histamine enhancement of hippocampal neuron injury under conditions of synaptic activity mediated excitotoxicity appeared to be potentiated by extracellular acidity. A pH dependent facilitation of synaptic transmission by histamine in hippocampus *in vitro* has been reported elsewhere.[56] Increased synaptic activity causes local acidification,[57] such that extracellular pH may be transiently lowered during normal brain function[58] and, more pertinently, during hypoxia/ischemia or epileptiform activity. Under these conditions histamine could act to maintain transmission or to aggravate excitotoxicity.

ACTIVATION OF THE ERK/MAP KINASE CASCADE IN NEURONAL ISCHEMIC PRECONDITIONING

Cerebral ischemic preconditioning is a phenomenon in which brief episodes of sublethal ischemia induce a robust resistance against the deleterious effects of subsequent, prolonged lethal ischemia.[59–61] Ischemic tolerance has been described in a variety of other organ systems, including the heart.[62] In the brain, ischemic tolerance is mediated largely by activation of the NMDA subtype of glutamate receptors.[61] The importance of ischemic preconditioning in providing protection against ischemic damage is well established; however, the intracellular mechanisms responsible for

TABLE 1. Preconditioning protects against synaptically mediated excitotoxicity

Treatment	Percent Survival
Mg^{2+}-free (5')	98
Mg^{2+}-free (15')	45
Mg^{2+}-free (5') → Mg^{2+}-free (15') : 2 h	49
Mg^{2+}-free (5') → Mg^{2+}-free (15') : 24 h	81
Mg^{2+}-free (5') + U0126 → Mg^{2+}-free (15') : 24 h	65

NOTE: Hippocampal neurons (14–16 days in culture) were first incubated five minutes (22°C) in Mg^{2+}-free Locke's solution (with 0.1 µM glycine and 30 M histamine, pH 7.0) (Mg^{2+}-free). Cultures were then returned to their original medium for 2 h or 24 h, after which they were subjected to a 15-min Mg^{2+}-free incubation. Additional cultures were subjected to a single 5-min or 15-min Mg^{2+}-free incubation and then returned to their original medium for 24 h. Neuronal survival was quantified 24 h after the final Mg^{2+}-free incubation in all treatment groups. Values are mean ± SD ($n = 3$). Survival is expressed relative to Mg^{2+}-containing buffer (100%).

this phenomenon remain ill defined. Stimulation of glutamate receptors and Ca^{2+} influx result in phosphorylation of ERK1/2.[63] In addition, ischemic preconditioning has been linked to changes in protein phosphorylation, including increased phosphorylation of ERK/MAP kinases.[8,63,64] Several studies have now demonstrated the functional consequences of these changes, with the ERK/MAP kinase being required for neuronal excitotoxic preconditioning (see TABLE 1).[43,65] The events linking the ERK/MAP kinase pathway to resistance to neuronal excitotoxic damage are not known but may involve ERK induced phosphorylation and inactivation of proapoptotic proteins.[66] The effect of ERK signaling on neuronal survival in ischemic preconditioning could also reside in the induction of nuclear gene expression. One of the characteristic features of ERKs is their ability to translocate to the nucleus, where they phosphorylate and activate transcription factors, thereby regulating gene transcription.[67]

CONCLUDING REMARKS

Research during the past few years has demonstrated the potential of ERK signaling cascades for regulating diverse neuronal processes, such as differentiation, synaptic plasticity, and cell death. Recent advances have increased our understanding of ERK/MAP kinase signaling pathways at the cellular level. Multiple second messengers, such as Ca^{2+}, cyclic AMP, and nitric oxide may control ERK signaling through small G proteins like p21ras and Rap1, and may explain how ERKs regulate activity dependent neuronal events. In mammalian systems an antiapoptotic role for ERKs has been proposed, and ERKs can protect certain populations of neurons against specific insults. Intriguingly, accumulating evidence indicates that sustained overstimulation of ERK signaling cascades may actually promote neuronal cell death. Signal transduction mechanisms in neurons are arguably more complex than in other cell types due to the striking morphological specializations of neurons. For example, spatially and distinct roles have been described for a single Trk effector, PI 3-kinase, during neurotrophin signaling in sympathetic neurons.[68] It is not unlikely that both the stimulation of ERKs and their downstream actions varies depending on their subcellular localization, particularly with respect to the postsynaptic signaling and nuclear actions of ERKs.

REFERENCES

1. ROTHMAN, S.M. & J.W. OLNEY. 1987. Excitotoxicity and the NMDA receptor. Trends Neurosci. **10:** 299–302.
2. CHOI, D.W. 1988. Glutamate neurotoxicity and diseases of the nervous system. Neuron **1:** 623–634.
3. CLIFFORD, D.B., C.F. ZORUMSKI & J.W. OLNEY. 1989. Ketamine and MK-801 prevent degeneration of thalamic neurons induced by focal cortical seizures. Exp. Neurol. **105:** 272–279.
4. SEGER, R. & E.G. KREBS. 1995. The MAPK signaling cascade. FASEB J. **9:** 726–735.
5. XIA, Z., M. DICKENS, J. RAINGEAUD, et al. 1995. Opposing effects of ERK and JNK-p38 MAP kinases on apoptosis. Science **270:** 1326–1331.
6. KINDY, M.S. 1993. Inhibition of tyrosine phosphorylation prevents delayed neuronal death following cerebral ischemia. J. Cereb. Blood Flow Metab. **13:** 372–377.
7. GASS, P., M. KIESSLING & H. BADING. 1993. Regionally selective stimulation of mitogen activated protein (MAP) kinase tyrosine phosphorylation after generalized seizures in the rat. Neurosci. Lett. **162:** 39–42.
8. HU, B.-R. & T. WIELOCH. 1994. Tyrosine phosphorylation and activation of mitogen-activated protein kinase in the rat brain following transient cerebral ischemia. J. Neurochem. **62:** 1357–1367.
9. KIM, Y.S., K.S. HONG, Y.S. SEONG, et al. 1994. Phosphorylation and activation of mitogen-activated protein kinase by kainic acid-induced seizure in rat hippocampus. Biochem. Biophys. Res. Comm. **202:** 1163–1168.
10. KURIHARA, J., B.-R. HU & T. WIELOCH. 1994. Changes in tyrosine phosphorylation of mitogen-activated protein kinase in the rat hippocampus during and following severe hypoglycemia. J. Neurochem. **63:** 2346–2348.
11. ALESSANDRINI, A., S. NAMURA, M.A. MOSKOWITZ, et al. 1999. MEK1 protein kinase inhibition protects against damage resulting from focal cerebral ischemia. Proc. Natl. Acad. Sci. U.S.A. **96:** 12866–12869.
12. MURRAY, B., A. ALESSANDRINI, A.J. COLE, et al. 1998. Inhibition of the p44/42 MAP kinase pathway protects hippocampal neurons in a cell-culture model of seizure activity. Proc. Natl. Acad. Sci. U.S.A. **95:** 11975–11980.
13. PROBERT, A.W., R. NATH, S.S. BASU, et al. 1998. The MEK inhibitor, PD-98059, reduces phosphorylation of site 4/5 Synapsin 1 and protects against oxygen/glucose deprivation-mediated neuronal injury. (Abstr.) Soc. Neurosci. **24:** 1569.
14. SKAPER, S.D., K.L. MURKITT, L. FACCI, et al. 1999. Potentiation of network activity-induced hippocampal neuron injury by histamine: a role for the p44/42 MAP kinase pathway. (Abstr.) Soc. Neurosci. **25:** 289.
15. STATON, P.C., J.P. HUGHES, M.G. WILKINSON, et al. 2000. Regulators and effectors of ERK/MAPK activation in synaptically mediated excitotoxicity in hippocampal neurones. (Abstr.) Soc. Neurosci. **26:** 1883.
16. LEVI-MONTALCINI, R. 1987. The nerve growth factor 35 years later. Science **237:** 1154–1162.
17. BOTHWELL, M. 1995. Functional interactions of neurotrophins and neurotrophin receptors. Annu. Rev. Neurosci. **18:** 223–253.
18. SEGAL, R.A. & M.E. GREENBERG. 1996. Intracellular signaling pathways activated by neurotrophic factors. Annu. Rev. Neurosci. **19:** 463–489.
19. WEISS, A. & J. SCHLESSINGER. 1998. Switching signals on or off by receptor dimerization. Cell **94:** 277–280.

20. SKAPER, S.D. & F.S. WALSH. 1998. Neurotrophic molecules: strategies for designing effective therapeutic molecules in neurodegeneration. Mol. Cell. Neurosci. **12:** 179–193.
21. DUDEK, H., S.A. DATTA, T.F. FRANKE, et al. 1997. Regulation of neuronal survival by the serine-threonine protein kinase Akt. Science **275:** 661–665.
22. DATTA, S.R., A. BRUNET & M.E. GREENBERG. 1999. Cellular survival: a play in three Akts. Genes Dev. **13:** 2905–2927.
23. FRANKE, T.F., D.R. KAPLAN & L.C. CANTLEY. 1997. PI3K: downstream AKTion blocks apoptosis. Cell **88:** 435–437.
24. DATTA, S.R., H. DUDEK, X. TAO, et al. 1997. Akt phosphorylation of BAD couples survival signals to the cell-intrinsic death machinery. Cell **91:** 231–241.
25. CARDONE, M.H., N. ROY, H.R. STENNICKE, et al. 1998. Regulation of cell death protease caspase-9 by phosphorylation. Science **282:** 1318–1321.
26. BRUNET, A., A BONNI, M.J. ZIGMOND, et al. 1999. Akt promotes cell survival by phosphorylating and inhibiting a forkhead transcription factor. Cell **96:** 857–868.
27. RODRIGUEZ-VICIANA, P., P.H. WARNE, B. VANHAESEBROECK, et al. 1996. Activation of phosphoinositide 3-kinase by interaction with Ras and by point mutation. EMBO J. **15:** 2442–2451.
28. PHILPOTT, K.L, M.J. MCCARTHY, A. KLIPPEL, et al. 1997. Activated phosphatidylinositol 3-kinase and Akt kinase promote survival of superior cervical neurons. J. Cell Biol. **139:** 809–815.
29. GREWAL, S.S., R.D. YORK & P.J.S. STORK. 1999. Extracellular-signal-regulated kinase signalling in neurons. Curr. Opin. Neurobiol. **9:** 544–553.
30. CROW, T., J.J. XUE-BIAN, V. SIDDIQI, et al. 1998. Phosphorylation of mitogen-activated protein kinase by one-trial and multi-trial classical conditioning. J. Neurosci. **18:** 3480–3487.
31. BLUM, S., A.N. MOORE, F. ADAMS, et al. 1999. A mitogen-activated protein kinase cascade in the CA1/CA2 subfield of the dorsal hippocampus is essential for long-term spatial memory. J. Neurosci. **19:** 3535–3544.
32. BERMAN, D.E., S. HAZVI, K. ROSENBLUM, et al. 1998. Specific and differential activation of mitogen-activated protein kinase cascades by unfamiliar taste in the insular cortex of the behaving rat. J. Neurosci. **18:** 10037–10044.
33. ATKINS, C.M., J.C. SELCHER, J.J. PETRAITIS, et al. 1998. The MAPK cascade is required for mammalian associative learning. Nat. Neurosci. **1:** 602–609.
34. DIRNAGL, U., C. IADECOLA & M.A. MOSKOWITZ. 1999. Pathobiology of ischaemic stroke: an integrated view. Trends Neurosci. **22:** 391–397.
35. CARDELL, M. & T. WIELOCH. 1993. Time course of the translocation and inhibition of protein kinase C during complete cerebral ischemia in the rat. J. Neurochem. **61:** 1308–1314.
36. WIELOCH, T., B.-R. HU, A. BORIS-MOLLER, et al. 1996. Intracellular signal transduction in the postischemic brain. Adv. Neurol. **71:** 371–387.
37. STRIJBOS, P.J.L.M. 1998. Nitric oxide in cerebral ischemic neurodegeneration and excitotoxicity. Crit. Rev. Neurobiol. **12:** 223–243.
38. MODY, I., J.D.C. LAMBERT & U. HEINEMANN. 1987. Low extracellular magnesium induces epileptiform activity and spreading depression in rat hippocampal slices. J. Neurophysiol. **57:** 869–888.
39. ABELE, A.E., K.P. SCHOLZ, W.K. SCHOLZ, et al. 1990. Excitotoxicity induced by enhanced excitatory neurotransmission in cultured hippocampal pyramidal neurons. Neuron **4:** 413–419.
40. FURSHPAN, E.J. & D.D. POTTER. 1989. Seizure-like activity and cellular damage in rat hippocampal neurons in cell culture. Neuron **3:** 199–207.
41. ROSE, K., C.W. CHRISTINE & D.W. CHOI. 1990. Magnesium removal induces paroxysmal neuronal firing and NMDA receptor-mediated neuronal degeneration in cortical cultures. Neurosci. Lett. **115:** 313–317.
42. SKAPER, S.D., B. ANCONA, L. FACCI, et al. 1998. Melatonin prevents the delayed death of hippocampal neurons by enhanced excitatory neurotransmission and the nitridergic pathway. FASEB J. **12:** 725–731.

43. SKAPER, S.D., L. FACCI, W.J. KEE, et al. 2001. Potentiation by histamine of synaptically-mediated excitotoxicity in hippocampal neurons: a possible role for mast cells. J. Neurochem. **76:** 47–55.
44. STRIJBOS, P.J.L.M., N. LEACH & J. GARTHWAITE. 1996. Vicious cycle involving Na$^+$ channels, glutamate release, and NMDA receptors mediates delayed neurodegeneration through nitric oxide formation. J. Neurosci. **16:** 5004–5013.
45. SLOVITER, R.S. 1987. Decreased hippocampal inhibition and a selective loss of interneurons in experimental epilepsy. Science **235:** 73–76.
46. BEKKERS, J.M. 1993. Enhancement by histamine of NMDA-mediated synaptic transmission in the hippocampus. Science **261:** 104–106.
47. VOROBJEV, V.S., I.N. SHARONOVA, I.B. WALSH, et al. 1993. Histamine potentiates N-methyl-D-aspartate responses in acutely isolated hippocampal neurons. Neuron **11:** 837–844.
48. WADA, H., N. INAGAKI, A. YAMATODANI, et al. 1991. Is the histaminergic neuron system a regulatory center for whole-brain activity? Trends Neurosci. **14:** 415–418.
49. THEOHARIDES, T.C. 1990. Mast cells: the immune gate to the brain. Life Sci. **46:** 607–617.
50. SILVER, R., A.J. SILVERMAN, L. VITKOVIC, et al. 1996. Mast cells in the brain: evidence and functional significance. Trends Neurosci. **19:** 25–31.
51. GOLDSCHMIDT, R.C., L.B. HOUGH, S.D. GLICK, et al. 1985. Rat brain mast cells: contribution to brain histamine levels. J. Neurochem. **44:** 1943–1947.
52. COCCHIARA, R., G. ALBEGGIANI, N. LAMPIASI, et al. 1999. Histamine and tumor necrosis factor alpha production from purified rat brain mast cells mediated by substance P. NeuroReport **10:** 575–583.
53. LANGLAIS, P.J., S.X. ZHANG, G. WEILERSBACHER, et al. 1994. Histamine-mediated neuronal death in a rat model of Wernicke's encephalopathy. J. Neurosci. Res. **38:** 565–574.
54. PANULA, P., U. PIRVOLA, S. AUVINEN, et al. 1989. Histamine-immunoreactive nerve fibers in the rat brain. Neuroscience **28:** 585–610.
55. ADACHI, N., Y. ITOH, R. OISHI, et al. 1992. Direct evidence for continuous histamine release in the striatum of conscious freely moving rats produced by middle cerebral artery occlusion. J. Cereb. Blood Flow Metab. **12:** 477–483.
56. YANOVSKY, Y., K. REYMANN & H.L. HAAS. 1995. pH-dependent facilitation of synaptic transmission by histamine in the CA1 region of mouse hippocampus. Eur. J. Neurosci. **7:** 2017–2020.
57. SOMJEN, G.G. 1984. Acidification of interstitial fluid in hippocampal formation caused by seizures and by spreading depression. Brain Res. **311:** 186–188.
58. BUZASKI, G., H.L. HAAS & E.G. HENDERSON. 1987. LTP induced by physiologically relevant stimulus patterns. Brain Res. **435:** 331–333.
59. GLAZIER, S.S., D.M. O'ROURKE, D.I. GRAHAM, et al. 1994. Induction of ischemic tolerance following brief focal ischemia in rat brain. J. Cereb. Blood Flow Metab. **14:** 545–553.
60. CHEN, J., S.H. GRAHAM, R.L. ZHU, et al. 1996. Stress proteins and tolerance to focal cerebral ischemia. J. Cereb. Blood Flow Metab. **16:** 566–577.
61. GRABB, M.C. & D.W. CHOI. 1999. Ischemic tolerance in murine cortical cell culture: critical role for NMDA receptors. J. Neurosci. **19:** 1657–1662.
62. FERRARI, R., C. CECONI, S. CURELLO, et al. 1999. Ischemic preconditioning, myocardial stunning, and hibernation: basic aspects. Am. Heart J. **138:** S61–S68.
63. XIA, Z., H. DUDEK, C.K. MIRANTI, et al. 1996. Calcium influx via the NMDA receptor induces immediate early gene transcription by a MAP kinase/ERK-dependent mechanism. J. Neurosci. **16:** 5425–5436.
64. SHAMLOO, M., A. RYTTER & T. WIELOCH. 1999. Activation of the extracellular signal-regulated protein kinase cascade in the hippocampal CA1 region in a rat model of global cerebral ischemic preconditioning. Neuroscience **93:** 81–88.
65. GONZALEZ-ZULUETA, M., A.B. FELDMAN, L.J. KLESSE, et al. 2000. Requirement for nitric oxide activation of p21^{ras}/extracellular regulated kinase in neuronal ischemic preconditioning. Proc. Natl. Acad. Sci. U.S.A. **97:** 436–441.

66. BERGMANN, A., J. AGAPITE, K. MCCALL, *et al.* 1998. The Drosophila gene hid is a direct molecular target of Ras-dependent survival signaling. Cell **95:** 331–341.
67. GOTOH, Y. & E. NISHIDA. 1995. Activation mechanism and function of the MAP kinase cascade. Mol. Reprod. Dev. **42:** 486–492.
68. KURUVILLA, R., YE, HAIHONG & D.D. GINTY. 2000. Spatially and functionally distinct roles of the PI3-K effector pathway during NGF signaling in sympathetic neurons. Neuron **27:** 499–512.

Amino Acid Uptake and Release in Primary Astrocyte Cultures Exposed to Ethanol

MICHAEL ASCHNER, LYSETTE A. MUTKUS, AND JEFFREY W. ALLEN

Department of Physiology and Pharmacology, and Interdisciplinary Program in Neuroscience, Wake Forest University School of Medicine, Winston-Salem, North Carolina, U.S.A.

KEYWORDS: Taurine; Aspartate; Astrocyte; Ethanol.

Neurophysiological and pathological effects of EtOH are mediated in part through the glutamatergic system.[1–4] The acute intoxicating effects of EtOH have been attributed to its ability to block voltage gated Ca^{2+} and Na^+ channels and N-methyl-D-aspartate (NMDA) glutamate receptor cation channels, and to facilitate $GABA_A$ receptor Cl^- channels.[2,4] Acute EtOH exposure inhibits NMDA stimulated neuronal activity in a dose dependent manner.[5–7] At maximal NMDA concentrations, EtOH completely blocks NMDA induced neurotoxicity and acts as a noncompetitive antagonist by reducing Ca^{2+} influx via NMDA receptors.[4,8] Thus, acute exposure to EtOH appears to protect neurons from NMDA induced excitotoxicity.

In contrast to acute EtOH exposure, chronic EtOH exposure with prolonged inhibition of NMDA receptors by EtOH results in development of a phenomenon.[9,10] The mechanisms implicated in this supersensitivity include increased density of glutamate and MK-801 binding,[11,12] increased NMDA stimulated Ca^{2+} influx,[13,14] and upregulation of the NMDAR1 receptor subunit.[15] Decreased GABA stimulated Cl^- flux, diminished Na^+/Ca^{2+} exchange, and increased voltage dependent Ca^{2+} influx are also likely to contribute to NMDA stimulated neuronal activity following chronic EtOH exposure.[16]

CNS damage in a number of pathological states (e.g., hypoxia, seizures, hypoglycemia, and hepatic encephalopathy), neurodegenerative disorders (Parkinson's disease and Huntington's disease), and aging is thought to be partly due to excessive stimulation of neuronal glutamate gated ion channels.[17] The origin of glutamate (and its analog, aspartate) has been tacitly assumed to be presynaptic nerve endings. However, it is known that astrocytes remove extracellular glutamate by a Na^+ dependent mechanism.[18] This transport has a likely stoichiometry of 1 glutamate and 3 Na^+ transported inward, and 1 K^+ transported outward to offset the negative charge of glutamate. In the presence of ammonia, glutamate is metabolized to glutamine by the astrocyte specific enzyme glutamine synthetase (GS),[19,20] maintaining $[glutamate]_o$ at 0.3 µM.[21,22] This represents a 10,000-fold gradient versus $[glutamate]_i$ (3 mM).

Address for correspondence: Michael Aschner, Ph.D., Department of Physiology and Pharmacology, Wake Forest University School of Medicine, Medical Center Boulevard, Winston-Salem, NC 27157-1083, U.S.A. Voice: 336-716-8530; fax: 336-716-8501.
maschner@wfubmc.edu

This glutamate–glutamine pathway constitutes the pool of brain glutamate originally described by Berl et al.[23] Astrocytes also efficiently remove extracellular taurine by a Na^+ dependent mechanism.[24] This transport system generates and maintains a $[taurine]_i/[taurine]_o$ ratio of about 10,000, and has a likely stoichiometry of 1 taurine and 2 Na^+ ions transported inward, to generate and maintain the observed taurine gradient.[24] Release from both the glutamate and taurine pools occurs as a result of astrocytic swelling.[25,26] In its exaggerated form, astrocytic swelling is deleterious, and can be viewed as a pathological extension of more limited and controlled volume changes that are otherwise part of the normal homeostatic function of astrocytes.

In the study reported here we have tested astrocytic responses to EtOH vis-a-vis amino acid uptake and release as a sequelæ of EtOH induced cell swelling. Since EtOH rapidly equilibrates across cellular membranes, making both the intracellular and extracellular compartments hyperosmotic but isotonic respective to each other, we hypothesized that exposure to hyperosmotic EtOH solution would be sufficient to stimulate uptake of osmoregulatory amino acids such as taurine and aspartate[27] (see FIGURE 1), and that their release would be increased during EtOH withdrawal due to astrocytic swelling.

Primary cultures of neonatal rat cortical astrocytes were isolated and grown as described elsewhere.[28] Cells were treated with ± 100 mM EtOH for 24 or 96 hours. Following treatment, cells were washed three times with HEPES buffer ± EtOH. Uptake of ^{3}H-taurine or ^{3}H-D-aspartate (1 µCi/ml) in HEPES buffer ± EtOH was allowed to proceed for one minute. Cells were washed in 4°C buffer then lysed in 1 N NaOH. Radioactivity was determined and corrected for protein content.

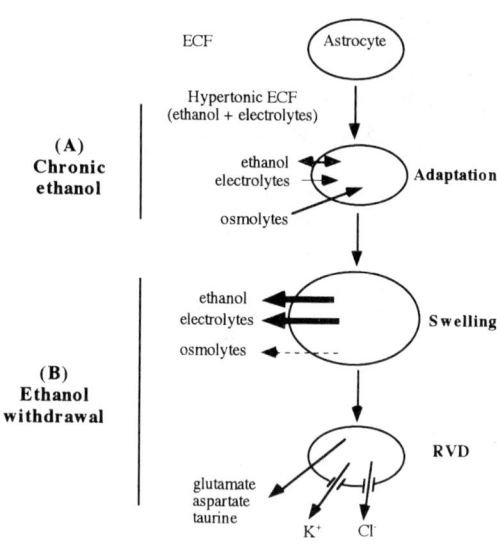

FIGURE 1. Sequence of events involved in the adaptation of astrocytes to chronic hypertonic stress associated with ethanol exposure. For details refer to the text.

EtOH exposure for 96 hours produced significant increases in both ^{3}H-taurine ($p < 0.01$) and ^{3}H-D-aspartate (an analog of L-glutamate) uptake in astrocytes ($p < 0.05$). EtOH exposure for 24 hours produced no significant increase in uptake (data not shown). These studies suggest that chronic (96 h) but not acute (no more than 24 h) EtOH exposure leads to increased uptake of osmoregulatory amino acids. EtOH withdrawal following 24 h of exposure produced significant increases in ^{3}H-taurine efflux at all measurement times ($p < 0.01$, ANOVA with Tukey-Kramer multiple comparisons test). After 96 h of exposure, EtOH withdrawal produced ^{3}H-taurine efflux greater than control cells at all measurement times except for the first minute of uptake ($p < 0.01$, ANOVA with Tukey-Kramer multiple comparisons test). Taurine efflux was highly correlated with changes in astrocyte cellular volume, suggesting that EtOH withdrawal produces astrocytic swelling and subsequent regulatory volume decrease (RVD). ^{3}H-D-aspartate efflux was not altered following EtOH withdrawal after 24 or 96 h of EtOH exposure.

To test if the lack of increased aspartate efflux was a result of increased transporter function, aspartate uptake was blocked with 1 mM threo-β-hydroxyaspartate (THA). Following THA treatment, ^{3}H-D-aspartate efflux was unaltered in EtOH exposed cells as compared to control cells, suggesting that EtOH withdrawal does not promote aspartate efflux. These data suggest that taurine, but not aspartate, is released following EtOH withdrawal. In addition, these findings suggest that, following EtOH withdrawal, taurine and aspartate are released via separate channels, not the non-specific, volume expansion sensitive, outwardly rectifying (VSOR) channel (previously termed volume sensitive organic channel or VSOAC) that is activated during severe hypotonic stress.

Taken together, the study suggests that chronic exposure to EtOH, even in the absence of hypernatremia, is an osmotic stressor in astrocytes and stimulates accumulation of osmoregulatory amino acids. EtOH withdrawal produces taurine, but not aspartate release. This amino acid release profile is different from that seen during hypotonic induced swelling (data not shown), suggesting a taurine release mechanism other than via the non-specific, volume expansion sensitive, outwardly rectifying (VSOR) channel. The significance of the increased taurine efflux from astrocytes during EtOH withdrawal remains to be determined. It is consistent with previous studies in which it has been demonstrated that osmoregulation and postischemic glutamate surge suppression (PIGSS) are important mechanisms in the neuroprotective properties of taurine.[29,30] In conclusion, during chronic EtOH withdrawal astrocytes are likely to serve a neuroprotective role by releasing intracellular taurine, which, as previously suggested, may function to increase uptake of excitatory amino acids, or attenuate NMDA neurotoxicity, thus inhibiting neuronal excitability.

ACKNOWLEDGMENTS

This study was supported in part by PHS Grant NIAA11617 awarded to M.A. J.W.A. was supported by a NIAAA training Grant T32 AA07565.

REFERENCES

1. GONZALES, R.A. & L.C. ROPER. 1993. Ethanol effects of NMDA-stimulated levels of extracellular neurotransmitters by *in vivo* microdialysis. Alcohol Alcohol **2:** 371–376.
2. FITZGERALD, L.W. & E.J. NESTLER. 1995. Molecular and cellular adaptations in signal transduction pathways following ethanol exposure. Clin. Neurosci. **3:** 165–173.
3. TSAI, G., D.R. GASTFRIEND & J.T. COYLE. 1995. The glutamatergic basis of human alcoholism. Am. J. Psychiat. **152:** 332–340.
4. GRANT, K.A. & D.M. LOVINGER. 1995. Cellular and behavioral neurobiology of alcohol: receptor-mediated neuronal processes. Clin. Neurosci. **3:** 155–164.
5. LOVINGER, D.L., G. WHITE & F.F. WEIGHT. 1990. NMDA receptor mediated synaptic excitation selectively inhibited by ethanol in hippocampal slice from adult brain. J. Neurosci. **10:** 1372–1379.
6. WEIGHT, F.F., D.M. LOVINGER & G. WHITE. 1991. Alcohol inhibition of NMDA channel function. Alcohol Alcohol Supp. **1:** 163–169.
7. HOFFMAN, P.L., C.S. RABE, F. MOSES & B. TABAKOFF. 1989. NMDA receptor and ethanol; inhibition of calcium flux and cyclic GMP production. J. Neurochem. **52:** 1937–1940.
8. CHANDLER, L.J., N. GUZMAN, C. SUMNERS & F.T. CREWS. 1994. Induction of nitric oxide synthase in brain glial cells: possible interaction with ethanol in reactive neuronal injury. *In* National Institute on Alcohol Abuse and Alcoholism Research Monograph, Alcohol and Glial Cells. NIH Publication No. 94-3742, Bethesda, MD, Monograph 27, 195–214.
9. BUCK, K.J. & R.A. HARRIS. 1991. Neuroadaptive responses to chronic ethanol. Alcohol Clin. Exp. Res. **15:** 460–470.
10. LOVINGER, D.M. 1993. Excitotoxicity and alcohol-related brain damage. Alcohol. Clin. Exp. Res. **17:** 19–27.
11. GRANT, K.A., P. VALVERIUS, M. HUDSPITH & B. TABAKOFF. 1990. Ethanol withdrawal seizures and the NMDA receptor complex. Eur. J. Pharmacol. **176:** 289–296.
12. GULYA, K., K.A. GRANT, P. VALVENIUS, *et al.* 1991. Brain regional specificity and time-course of changes in the NMDA receptor-ionophore complex during ethanol withdrawal. Brain Res. **547:** 129–134.
13. IORIO, K.R., L. REINLIB, B. TABAKOFF & P.L. HOFFMAN. 1992. Chronic exposure of cerebellar granule cells to ethanol results in increased *N*-methyl-D-aspartate receptor function. Mol. Pharmacol. **41:** 1142–1148.
14. IORIO, K.R., B. TABAKOFF & P.L. HOFFMAN. 1993. Glutamate-induced neurotoxicity is increased in cerebral granule cells exposed chronically to ethanol. Eur. J. Pharmacol. **248:** 209–212.
15. TREVISAN, L., L.W. FITZGERALD, N. BROSE, *et al.* 1994. Chronic ingestion of ethanol upregulates NMDAR1 receptor subunit immunoreactivity in rat hippocampus. J. Neurochem. **62:** 1635–1638.
16. CREWS, F.T. & L.J. CHANDLER. 1993. Excitotoxicity and the neuropathology of ethanol. *In* Alcohol Induced Brain Damage. W.A. Hunt & S.J. Nixon, Eds.: 355–371. National Institutes of Alcohol and Alcoholism Research Monograph 22. NIH Publication No. 93-3549. Bethesda, MD, National Institutes of Health.
17. COYLE, J.T. & P. PUTTFARCKEN. 1994. Oxidative stress, glutamate, and neurodegenerative disorders. Science **262:** 689–695.
18. HERTZ, L. 1979. Functional interactions between neurons and astrocytes 1. Turnover and metabolism of putative amino acid transmitters. *In* Progress in Neurobiol. **13:** 277–323. Pergamon, Oxford.
19. MARTINEZ-HEMANDEZ, A., K.P. BELL & M.D. NORENBERG. 1977. Glutamine synthetase: glial localization in brain. Science **195:** 1356–1358.
20. NORENBERG, M.D. & A. MARTINEZ-HERNANDEZ. 1979. Fine structural localization of glutamine synthetase in astrocytes of rat brain. Brain Res. **161:** 303–310.
21. SCHOUSBOE, A. & I. DIVAC. 1979. Differences in glutamate uptake in astrocytes cultured from different brain regions. Brain Res. **177:** 407–409.

22. WANIEWSKI, R.A. & D.L. MARTIN. 1986. Exogenous glutamate is metabolized to glutamine and exported by primary astrocyte cultures. J. Neurochem. **47:** 304–313.
23. BERL, S., A. LAJTHA & A. WAELSCH. 1961. Amino acid and protein metabolism. VI. Cerebral compartments of glutamic acid metabolism. J. Neurochem. **7:** 186–197.
24. MARTIN, D.L. 1992. Synthesis and release of neuroactive substances by glial cells. Glia **5:** 81–94.
25. KIMELBERG, H.K. & M.D. NORENBERG. 1989. Astrocytes. Sci. Amer. **260:** 66–76.
26. VITARELLA, D., D.J. DIRISIO, H.K. KIMELBERG & M. ASCHNER. 1994. Potassium and taurine release are highly correlated with regulatory volume decrease in neonatal primary rat astrocyte cultures. J. Neurochem. **63:** 1143–1149.
27. STRANGE, K. 1992. Regulation of solute and water balance and cell volume in the central nervous system. J. Am. Soc. Nephrol. **3:** 12–27.
28. ASCHNER, M., M. GANNON & H.K. KIMELBERG. 1992. Manganese (Mn) uptake and efflux in cultured rat astrocytes. J. Neurochem. **58:** 730–735.
29. LO, E.H., P. BOSQUE-HAMILTON & W. MENG. 1998. Inhibition of poly(ADP-ribose) polymerase: reduction of ischemic injury and attenuation of N-methyl-D-aspartate–induced neurotransmitter dysregulation. Stroke **29:** 830–836.
30. KHAN, S.H., A. BANIGESH, A. BAZIANI, *et al.* 2000. The role of taurine in neuronal protection following transient global forebrain ischemia. Neurochem. Res. **25:** 217–223.

Part I: Mechanisms of Excitotoxicity

GENERAL COMMENT

From Dr. DeMaw Chuang

This is a comment on the excellent talk of Dr. Obrenovitch. Quinolinic acid (QA) is a partial agonist for NMDA receptors and its receptor sensitivity is brain region dependent. For example, QA does not induce excitotoxicity in cerebellar granule cells up to mM concentrations and is neuroprotective against glutamate excitotoxicity in these cells. In contrast, NMDA receptors in the striatum are highly sensitive to QA. Degeneration of medium sized neurons in the striatum readily occurs following infusion of low nonamolar doses of QA. Therefore, it may be of interest to extend your studies on the cerebral cortex to other brain areas particularly the striatum.

ANSWER: You are correct. Some regions of the brain may well be more vulnerable than others to NMDA-receptor agonists, depending on receptor density and subtype composition. However, striatum and cortex showed comparable sensitivity when NMDA was perfused through a microdialysis electrode implanted in these regions (EC_{50} for NMDA-induced depolarization was about 180 and 126 µM for striatum and cortex, respectively). Thus, I am fairly confident that what we have found in the cortex can be extrapolated to the striatum.

QUESTIONS FOR DR. SKAPER

From Dr. Narahashi

Is the large increase in spontaneous activity by Mg^{2+}-free solutions due to an increase in transmitter release as a result of imbalance between CA^{2+} and Mg^{2+}?

ANSWER: The large increase in spontaneous activity of hippocampal neurons is likely due to relief of NMDA receptor gated cation channels upon removal of extracellular Mg^{2+}, leading to depolarization, activation of voltage sensitive Na^+ channels and presynaptic release of glutamate. The neurons undergo intense asynchronous firing as assessed electrophysiologically and rapid, prolonged Ca^{2+} transients. These effects, as well as the subsequent neurodegeneration, are blocked by antagonists of both voltage operated Na^+ channels and NMDA, but not kainate/AMPA receptors.

From Dr. Scallet

Would ligands or modulators of TRK-A receptors (such as growth factors or estrogenic compounds, respectively) be expected to switch the pathways of phosphorylations such that MAP–kinase inhibitors may shift between neuroprotection or neurotoxicity? That is, how do growth factor or estrogens influence the balance between the different pathways of phosphorylation (after TRK-A activation) that you described in the first part of your talk?

ANSWER: There are circumstances where growth factors, such as neurotrophins, promote neuronal survival by activation of the ERK/MAP kinase pathway. In such instances MAP kinase kinase inhibitors would be expected to reduce cell survival. Although an antiapoptotic role for ERKs in neurons has been proposed, recent reports suggest paradoxically that overstimulation of ERK signaling cascades actually promote necrotic cell death. Indeed, this appears to be the case in the *in vitro* model of excitotoxicity described here and in experimental models of cerebral ischemia, where inhibitors of MAP kinase kinase are clearly neuroprotective. Neurons represent a highly polarized cell type. It is not unlikely that both the stimulation of ERKs and their downstream actions will vary based on their subcellular localization, particularly with respect to the postsynaptic signaling and nuclear actions of ERKs. The challenge will now be to address the mechanisms by which these pathways coordinate the generation of signal specificity.

From Dr. Bowyer

Relative to the fact that E18 cells are used in your studies are there changes in the MAP, MEK kinase pathways over the developing animal from prenatal to adult?

ANSWER: Our studies have been carried out exclusively with embryonic tissue. I am not aware of reports in which the MEK kinase pathway has been examined in neurons as a function of developmental state of the animal. MEK kinase activation has been described in adult mice following cerebral ischemia, with protection afforded by the MEK inhibitor PD98059 (Alessandrini *et al.*, PNAS, 1999). Similar observations were also reported at the 2000 Society for Neuroscience meeting. It would, thus, appear that MEK kinase activation by excitotoxic mechanisms can occur in both embryonic and adult neurons.

From Dr. Abbracchio

Concerning the role of MAPK in injury, we have data supporting the fact that the MAPK pathway in involved in induction of the COX^2 gene in astrocytes, an event that may contribute to neurodegeneration in chronic main inflammatory diseases, which may also be characterized by excessive astrogliosis. Do you have any ideas of the events down stream of MAPK activation that may contribute to cell deaths in your experimental mode? Moreover, what is the time window of the protection elicited *in vivo* by inhibitors of the MAPK in the stroke model, and are you aware of any other *in vivo* data with these compounds in other models of neurodegeneration?

ANSWER: The nature of upstream regulators and downstream effectors of ERK/MAP kinase activation in synaptically mediated excitotoxicity in hippocampal neurons is ongoing, and will be the subject of a future publication. Initial studies indicate that there is activation of both the MAP kinase substrate MSK1, and the MSK1 substrate CREB. Furthermore, injury is reduced by tool inhibitors of B-Raf, suggesting a role for this component. These findings have been reported in a poster from our group (Staton *et al.*, Society for Neuroscience abstracts, 2000). The MAP kinase kinase inhibitor PD98059 is reportedly neuroprotective in a mouse model of cerebral ischemia in pretreatment (Alessandrini *et al.*, PNAS, 1999). I believe others have described similar protective effects of PD98059 (and U0126, another MAP kinase kinase inhibitor) in both mouse and rat brain ischemia, as well as improvement in outcome when given postocclusion (2000 Society for Neuroscience meeting). I am not aware of other models of neurodegeneration where these compounds have been used.

QUESTIONS FOR DR. ASCHNER

From Dr. Narahashi

Does alcohol cause swelling to occur in neurons as well?

ANSWER: Yes, there is no reason to believe that neurons are not to be subjected to the same osmotic changes that I have described. Nevertheless, given differential distribution of various osmolytes, the release of amino acids and other osmolytes will likely differ between the neurons and astrocytes.

From Dr. Marini

Are there any taurine inhibitors or antagonists?

ANSWER: No, to the best of my knowledge there are no taurine inhibitors or antagonists.

From Dr. Banik

At what level of alcohol do you think astrocytes or even neurons and oligodendrocytes become susceptible to die?

ANSWER: At the concentrations that we use in our studies, up to 100 mM of ethanol (109 mM Ethanol equals 0.5%) there appears to be no cell death, and as I have shown in the presentation, the effects on cell swelling are fully reversible with time.

From Dr. Jonas

In your very interesting work you call a seven-day exposure of astrocytes in culture a chronic situation. In humans seven days of drinking would be a "binge". "Chronic" usually means years of drinking. Would you comment?

ANSWER: Obviously, *in vitro* studies offer both advantages and disadvantages. They offer a reductionist approach where cellular mechanisms can be teased apart much easier compared to *in vivo* studies. I used the term "chronic" exposure to describe a seven-day exposure, because we are limited with the culture to a relatively short exposure paradigm. Once we isolate the astrocytes they are grown to confluence within three to four weeks. Experiments are done one week later. It would obviously be very interesting to culture the astrocytes for longer duration, but unfortunately after five to six weeks in culture the cells, even when undisturbed, commence to die.

Dr. John Brust of Columbia University has shown that most alcohol related seizures seen in the Harlem Hospital emergency room occur in association with intoxication not withdrawals. Also, seizures are rare in the alcoholic withdrawal state of delirium tremors. Do you have astrocyte related thoughts?

ANSWER: Neurological diseases, such as seizures and stroke, have been reported to occur in temporal relation to alcohol intoxication as well as withdrawal. Clinical studies also report an increase in the prevalence of alcohol withdrawal related seizures in patients with a history of multiple detoxifications, and withdrawal symptom severity, inclusive of delirium and seizures increase with advancing age. Therefore, both alcohol intoxication and withdrawal should be evaluated in relationship to the onset seizures.

Longitudinal MRI studies show the appearance of brain atrophy in chronic alcoholics, with reversion toward a more normal appearance after suitable periods of abstinence. Can you comment?

ANSWER: I am aware of a number of cases or subgroups where chronic alcohol abusers had several weeks of abstinence and general supportive treatment that led to regression of neurological signs, as evidenced by MRI (Iwinska-Buksowicz *et al.*, 1999; Bergman *et al.*, 1998). The reduction in drinking habits was associated with cognitive improvement in both of these studies (Iwinska-Buksowicz *et al.*, 1999), but no significant difference in MRI variables including T1-relaxation time were noted by Bergman *et al.*, 1998. Given that white matter lesions are also associated with smoking, lower education, hypertension, systolic blood pressure, and pulse pressure, just to name a few, I am not sure that the issue of "reversion toward a more normal appearance" upon alcohol abstinence in chronic alcoholics has been sufficiently and systematically documented when all these variables are taken into account.

QUESTIONS FOR DR. OBRENOVITCH

From Dr. Banik

In autoimmune demyelinative diseases, do activated T-cells produce increased levels of quinolinic acid? If so, what role does quinolinic acid have on oligodendrocyte or neuronal death?

ANSWER: I am not aware of any data related to the contribution of activated T-cells to the accumulation of quinolinic acid that occurs during neuroinflammation. All our data, so far, do not support the notion that quinolinic acid accumulation during neuroinflammation may be excitotoxic. Whether or not quinolinic acid accumulation is detrimental to brain cells for other reason(s) remains to be clarified.

From Dr. Auer

Does quinolinic acid play a role in any of the numerous neuroinflammatory diseases characterized by white blood cell infiltration of brain and induction of indoleamine 2,3-dioxygenase? Diseases such as meningitis or septic encephalopathy (characterized by microabscesses 50 to 200 μm diameter)

ANSWER: Our study was an attempt to resolve this issue, and our data do not support the notion that quinolinic acid accumulation associated with neuroinflammation adds an excitotoxic component to the pathology. So far, all the investigators working in this area have focused their attention on the accumulation of quinolinic acid, that is, the product formed by indolamine 2,3-dioxygenase. In contrast, the potential consequences of decreased brain levels of L-tryptophan (i.e. the enzyme substrate, but also the precursor of 5-hydroxytryptamine, 5-HT) have been completely overlooked. The possibility that, in some neuroinflammatory diseases, induction of indolamine 2,3-dioxygenase may promote depression through reduction of L-tryptophan availability for 5-HT synthesis, is an interesting hypothesis.

From Dr. Aschner

I have noted in one of your slides a bullet dealing with the neuroprotective potential of quinolic acid. My questions is whether levels of this amino acid have been measured in any model of neurodegeneration, and whether it is possible, given your studies that quinolinic acid levels, at least in the area that you looked at are insufficient for generating excitotoxicity and that it might play a neuroprotective role. (For example, by reducing sensitivity of NMDA receptors of activation).

ANSWER: Yes, quinolinic acid levels have been measured and found to increase markedly with neuroinflammation. From our data, we perceive quinolinic acid accumulation as potentially more neuroprotective than excitotoxic. Neuroprotection associated with sustained, moderate levels of NMDA-receptor agonists may be linked to NMDA-receptor desensitization and/or the release and activation of neurotrophic factors (see Ann Marini's communication).

Lack of Protective Effect by Intermittent Hypoxia on MPTP-Induced Neurotoxicity in Mice

GIN-HEUY LAI,[a] CHAU-FONG CHEN,[b] YU SU,[c]
LOW-TONE HO,[a] AND ANYA MAAN-YUH LIN[a,c]

[a]*Department of Medical Research and Education,*
Veterans General Hospital-Taipei, Taiwan

[b]*Department of Physiology, National Taiwan University, Taipei, Taiwan*

[c]*Department of Biopharmaceutics, National Yang-Ming University, Taiwan*

ABSTRACT: In contrast to acute ischemia and subsequent reperfusion that produce excess free radicals, intermittent hypoxia (IH) is reported to play an important role in upregulation of antioxidative defensive mechanisms. In the study we report here, the neuroprotective effect of IH was evaluated using intraperitoneal injection of 1-methyl-4-phenyl-1,2,3,6-tetrahydropyridine (MPTP) in ICR mice. Adult male ICR mice were subjected to 380 mmHg in an altitude chamber for 15 hours/day for 14 or 28 days. MPTP decreased striatal dopamine content in normoxic mice. However, IH did not significantly alter the MPTP-induced depletion of striatal dopamine content. Furthermore, MPTP had no effect on GSH content but reduced GSH/GSSG ratio in mouse striatum. IH altered neither GSH content nor MPTP-induced reduction in GSH/GSSG ratio. Although MPTP had no effect on striatal SOD activity in normoxic mouse, IH increased SOD activity in the saline and MPTP groups. Neither MPTP nor IH affected GPx in mouse striatum. Furthermore, in our *ex vivo* study, both the autooxidation and iron induced lipid peroxidation of cortical homogenates were lower in the IH-treated group than those of the normoxic group, indicating a reduced oxidative status after IH treatment. In conclusion, exposure to IH has been suggested to be beneficial in preventing iron-induced oxidative injuries in biological organisms, and our data support this notion in that IH not only decreased iron-induced lipid peroxidation but also increased antioxidative defense enzyme activity in mouse brain. Furthermore, the lack of neuroprotective effect by IH of MPTP-induced depletion of striatal dopamine content suggests that oxidative stress may not be the only mechanism for the MPTP-induced neurotoxicity.

KEYWORDS: Intermittent hypoxia; MPTP; Neurotoxicity.

INTRODUCTION

One of the popular hypotheses for CNS impairment is that it is due to oxidative stress, which can occur in the normal aging process and in the development of CNS

Address for correspondence: Low-Tone Ho and Anya Maan-Yuh Lin, Department of Medical Research and Education, Veterans General Hospital-Taipei, Taipei, Taiwan. Voice: 886-2-28712121 ext. 2688; fax: 886-2-28751562.

diseases.[1-19] Oxidative stress may result from an imbalance between free radical generation and antioxidative defensive systems.[1-19] Clinical studies have reported accumulation of iron and zinc in patients with Parkinson's disease[18] and increases in transition metal concentrations may generate excess free radicals.[2-6,9,17] Furthermore, reduced antioxidative defense systems are thought to be crucial to the etiology of Parkinson's disease, including diminished content of glutathione (GSH) and reduced activities of superoxide dismutase and catalase.[14,19] Therefore, excess free radical generation and/or reduced capacity of antioxidative defensive systems may cause biological damages.[1-19] Oxidative injuries may occur on several levels, including lipid peroxidation, protein peroxidation, and DNA crosslinkage that may lead to cell death.[1-19]

A significant body of studies has focused on neuroprotective strategies, including enhanced antioxidative enzyme systems and supplementation of antioxidants.[2-4,11,15,20-22] Repeated moderate oxidative stress by intermittent hypoxia has been suggested to be protective, not only in renal[23] and cardiovascular systems,[20,24-26] but also in CNS.[20-22] For example, intermittent hypoxia reportedly prevented apoptotic formation during renal ischemia/reperfusion[23] and the development of atherosclerosis in rabbits,[25] and it is reported to reduce the duration of severe ventricular arrhythmia in acute ischemia and reperfusion.[26] In contrast to acute ischemia and subsequent reoxygenation, both of which can result in secondary reperfusion insults,[27-28] prior intermittent hypoxia has been found to reduce the oxidative damage by local infusion of 1-methyl-4-phenyl-1,2,3,6-tetrahydropyridine (MPTP) in rat striatum, indicating that intermittent hypoxia may protect CNS against oxidative injury.[20] Mechanisms underlying the protective effect of intermittent hypoxia have been proposed.[20,21] Both *in vivo* and *in vitro* studies have shown that antioxidative defense systems in brain,[20,22] in heart, and in many other organs[20,24-26,29-30] are altered in response to intermittent hypoxia. Furthermore, intermittent hypoxia modulated expression of transcription factors, including NF-κB and AP-1.[21]

Because accumulating oxidative stress is thought to result in progressive deterioration in Parkinson's disease, the antioxidative action of intermittent hypoxia may decrease the progression of this disease. In the study reported in this paper, intraperitoneal administration of MPTP, a dopamine selective neurotoxin[13,31-34] was employed to produce an experimental mouse model of Parkinson's disease. Much research has shown that systemic administration of MPTP can mimic Parkinson like symptoms in primates[13] as well as in mice.[7,13,31-34] In the brain, MPTP is metabolized to 1-methyl-4-phenylpyridinium (MPP^+) by monoamine oxidase$_B$ in astrocytes.[35] MPP^+ accumulates in astrocytes or in dopaminergic neurons via the dopamine reuptake system.[36] Intracellularly, MPP^+ is concentrated in mitochondria where it may block NADH-CoQ10 reductase (complex I) of the respiratory chain.[37] Several proposed mechanisms of MPP^+ action have been suggested, including an oxidative stress by free radical generation. For example, it has proposed that MPP^+ elicits striatal dopamine overflow, and when oxidized *in situ,* dopamine is found to produce free radicals.[38] In support of this notion, nitric oxide and other antioxidants have been found to prevent MPP^+-induced neurotoxicity by scavenging free radicals.[39] In addition to the MPTP-induced depletion of striatal dopamine content, two oxidative indicators, auto-oxidation and iron-induced lipid peroxidation, were evaluated. Antioxidative defensive systems, including antioxidants and antioxidative enzyme activities, were also evaluated.

METHODS

In Vivo *Study*

Male ICR mice, weighing 20–30 g, were used. These animals were maintained according to the guidelines established in *Guide for the Care and use of Laboratory Animals* prepared by the Committee on Care and Use of Laboratory Animals of the Institute of Laboratory Animal Resources Commission on Life Sciences, National Research Council, U.S.A. (1985). Mice were randomly separated into two groups. One group was subjected to 380 mmHg in an altitude chamber for 15 hours/day for 28 days and the other was maintained in the normoxic condition. In each group, mice were treated daily with MPTP or saline intraperitoneally for five days. At the end of the injection period, animals were sacrificed 24 hours after the final injection. Brains were rapidly removed and chilled in saline (0°C). Regional dissections were performed, immediately frozen in liquid nitrogen, and stored at −70°C until analysis.

HPLC-EC Analysis of Striatal Dopamine Content

Dissected striata were thawed and homogenized in chilled 0.1N perchloric acid. A procedure using a high pressure liquid column coupled with electrochemical detection was used to quantify dopamine content in striatum.[40] Statistical comparisons were made using one way ANOVA followed by a Bonferroni multiple comparison procedure as a post hoc analysis.

Fluorescence Assay of Lipid Peroxidation of Cortical Homogenates

Dissected cortical tissues were thawed and homogenized in ice cold Ringer's solution using a homogenizer and treated for either autooxidation or iron-induced lipid peroxidation, as follows. Autooxidation: cortical homogenates (50 mg/ml, 400 µl) were incubated at 37°C for four hours. Iron-induced lipid peroxidation: cortical homogenates were incubated at 37°C for four hours following an addition of iron (1 µM). A 400-µl sample was transferred to a tube containing 300 µl chloroform and 100 µl methanol. The slurry was mixed and then kept on ice for 15 minutes. After centrifugation ($8000g$) for five minutes, an aliquot (400 µl) of chloroform extract was transferred to another tube containing 100 µl methanol and was scanned using a spectrofluorometer (Aminco Bowman-2, U.S.A.). The lipid peroxidation was determined by measuring fluorescent Schiff base products of malonaldehydes and its dihydropyridine polymers, which emit fluorescence at 426 nm when activated by UVa at 356 nm.[41,42] Statistical comparisons were made using an unpaired Student's *t*-test.

Measurement of the GSH/GSSG Ratio in Brain Tissues

The ratio of reduced glutathione/oxidized glutathione (GSH/GSSG) was measured according to the method of Griffith.[43] Frozen substantia nigras were thawed in five volumes of ice cold 0.2 N PCA and homogenized by sonication. Tissue homogenates were centrifuged at $12,000\,g$ for 15 minutes. Aliquots of the supernatants were added to 0.1 M phosphate buffer (pH 7.5) containing 0.6 mM DTNB and 0.2 mg/ml NADPH, to comprise a total volume of 980 µl. After mixing, GSH reductase (20 U/ml, 20 µl) was added to initiate the assay. The formation of 5-thio-2-nitrobenzoic acid was measured spectrophotometrically at 412 nm for two minutes. The amount of total GSH was determined from a standard curve with known quantities of GSH standard. 2-Vinylpyridine was added to derive the reduced form of glutathione 60 minutes prior to the measurement of the GSSG content. Aliquots of supernatants were assayed for GSH.

Measurement of Enzyme Activity in Brain Tissues

Superoxide Dismutase (SOD)

Total SOD activity was assayed according to the method of Misra and Fridovich.[44] Tissue homogenates were centrifuged at $1,000\,g$ for 10 minutes. Aliquots of the supernatants were added to 0.1M phosphate buffer (pH 7.8) containing 0.2 mM xanthine and 0.3 mM epinephrine. Xanthine oxidase was diluted appropriately, so that an assay mixture without the SOD source yielded 0.03 OD/min at 320 nm. The reaction was started by the addition of xanthine oxidase (total reaction volume, 1 ml) and absorbance was measured at 320 nm continuously for three minutes.

Glutathione Peroxidase (GPx)

Selenium-dependent GPx activity was determined using glutathione reductase and NADPH.[45] Tissue homogenates were centrifuged at $12,000\,g$ for 15 min. Aliquots of the supernatants were added to 0.1 M phosphate buffer (pH 7.8) containing 1 mM NaN_3, 1 mM GSH, 0.2 mM NADPH, and 0.24 U GSH reductase. The reaction was started by adding 2.5 mM H_2O_2 100 µl (total reaction volume was 1 ml) and was measured at 340 nm for five minutes.

RESULTS AND DISCUSSION

Effect of Intermittent Hypoxia on Dopamine Content in Mouse Striatum

Intermittent hypoxia appeared to increase the basal dopamine content in mouse striatum. The dopamine content of the saline group in the normoxic mice was 1.18 ± 0.12 µmole/mg protein (mean ± S.E.M., $n = 20$). The striatal dopamine content of two-week and 4-week intermittent hypoxic mice was 1.28 ± 0.10 µmole/mg protein ($n = 20$) and 1.43 ± 0.16 µmole/mg protein ($n = 20$, $p < 0.05$ compared to that of normoxic mice), respectively. Our data suggest that intermittent hypoxia may enhance the function of nigrostriatal dopaminergic system.

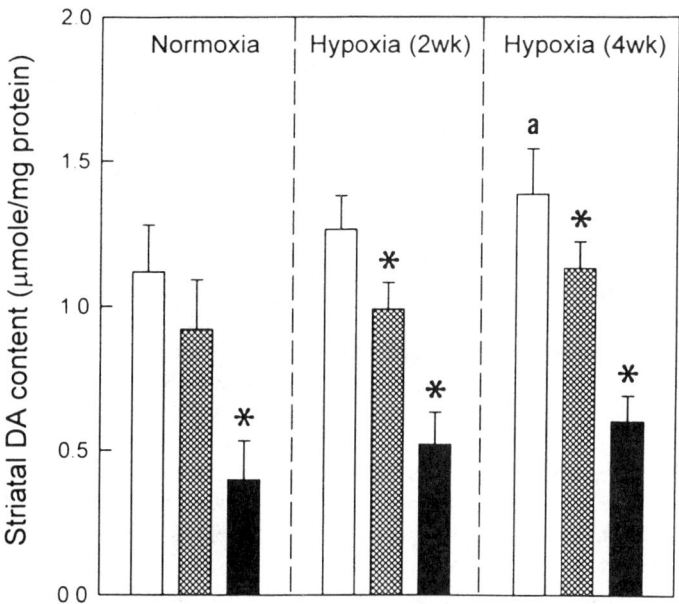

FIGURE 1. Effect of intermittent hypoxia on MPTP-induced reduction in striatal dopamine in mice. MPTP or saline was injected intraperitoneally, daily for five days. Striatal dopamine content was determined using HPLC-EC detection. Values shown are mean ± S.E.M. ($n = 6$–12). $^a p < 0.05$ in the saline group after intermittent hypoxia compared to that in the normoxic group; *$p < 0.05$ in MPTP group compared to the saline controls in normoxic group and in intermittent hypoxia by one way ANOVA followed by *post hoc* analyses. ☐, saline; ▧, MPTP (20 mg/kg); ■, MPTP (30 mg/kg).

Effect of Intermittent Hypoxia on MPTP-Induced Striatal Dopamine Depletion

Both normoxic and intermittent hypoxic mice daily received an intraperitoneal injection of MPTP for five consecutive days. MPTP (20 or 30 mg/kg) dose dependently decreased striatal dopamine content in normoxic mice (see FIGURE 1). However, the depletion by MPTP of striatal dopamine content was not significantly altered after intermittent hypoxia (FIG. 1). The striatal dopamine depletion by MPTP (30 mg/kg) was 36% of the saline control in the normoxic group, whereas that of MPTP group was 41% of the saline control after two weeks of intermittent hypoxia. Furthermore, MPTP induced striatal dopamine depletion was 43% of the saline control group after four weeks of intermittent hypoxia (FIG. 1). These data suggest that intermittent hypoxia did not significantly inhibit MPTP induced neurotoxicity in mice. In contrast to the MPTP study reported here, our preliminary study found that four-week intermittent hypoxia markedly abolished oxidative injuries induced by intranigral infusion of iron in rat brain, including attenuation of an elevated lipid peroxidation in the infused substantia nigra and a decreased dopamine content in the ipsilateral striatum (paper submitted). One possibility for the differential effects of intermittent hypoxia may be that intermittent hypoxia primarily protects against damage induced by free radicals. Iron is known to generate hydroxyl radicals via Fenton's reaction, and thus results in oxidative injuries by this mechanism.[15] In contrast, oxidative stress may not be the only mechanism for neurotoxicity induced by MPTP in work reported here. In support of this notion, several antioxidant drugs, including PBN, ethoxyquin, DL-6,8-thioctic acid failed to overcome MPTP induced neurotoxicity in C57BL/6 mice.[34] Furthermore, MPTP treatment has been reported to deplete striatal dopamine content without interfering mitochondrial complex activity and indices of oxidative damage in the maromset.[46] Using a microarray assay, MPTP treatment was found to alter the expression of 57 genes. Many of these genes are known to be altered by oxidative stress, inflammation, neurotrophic factors, iron-related mechanisms, glutamate receptors, transporters, apoptotic cell cycles, and hormone receptors (personal communication, Dr. M. Youdim). These alternate mechanisms for MPTP action may explain why intermittent hypoxia abolished iron-induced oxidative injuries but not MPTP-induced depletion of striatal dopamine content in our study. Another differential effect of intermittent hypoxia on neurotoxicity induced by local application or systemic administration of MPTP was demonstrated in rats and mice, respectively. Intermittent hypoxia reportedly prevented neurotoxicity by intrastriatal infusion of MPTP in rat brain,[20] but not the neurotoxicity by systemic MPTP in mice in our study. The mechanism underlying this discrepancy is still under investigation.

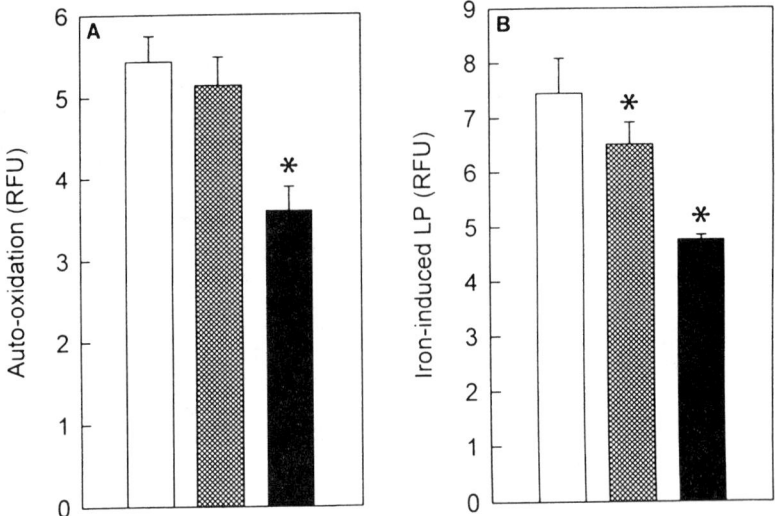

FIGURE 2. Effects of intermittent hypoxia on auto-oxidation and iron-induced lipid oxidation in cortical homogenates. *$p < 0.05$ in the intermittent hypoxic group compared to the normoxic group by an unpaired Student's *t*-test. ▢, normoxia; ▨, IH (2 weeks); ■, IH (4 weeks).

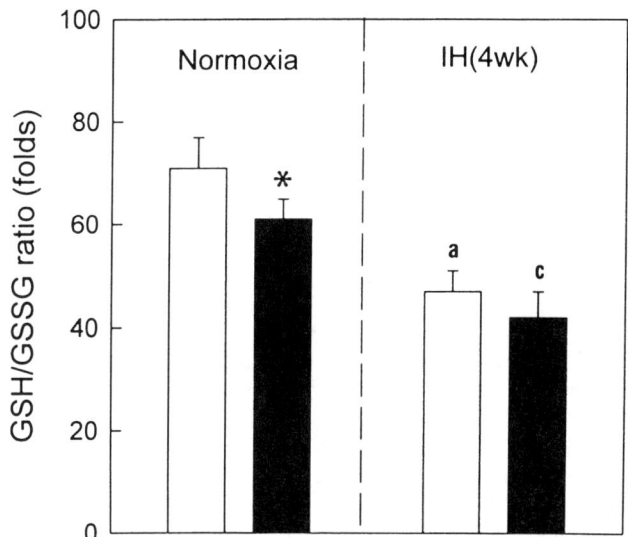

FIGURE 3. Effects of intermittent hypoxia on GSH/GSSG ratio in striatum in mouse brain. $*p < 0.05$ in MPTP group compared to the saline control in the normoxic group. $^{a}p < 0.05$ in the saline control, and $^{c}p < 0.05$ in MPTP group in the intermittent hypoxic group compared to values in the normoxic group by one way ANOVA followed by *post hoc* analyses. ☐, saline; ■, MPTP.

Effect of Intermittent Hypoxia on Autooxidation and Iron-Induced Lipid Peroxidation

Ex vivo studies, including autooxidation and iron-induced lipid peroxidation, were employed to evaluate the antioxidative effect of intermittent hypoxia in the mouse cortex. Incubation of cortical homogenates at 37°C for four hours significantly

TABLE 1. Effects of intermittent hypoxia (IH) on GSH/GSSG ratio, superoxide dismutase activity and glutathione peroxidase activity in mouse brains[a]

	GSH/GSSG (folds)		Superoxide dismutase (unit/mg tissue)		Glutathione peroxide (unit/mg tissue)	
	Normoxia	IH	Normoxia	IH	Normoxia	IH
Cortex	71±3	60±4[b]	2.2±0.1	2.1±0.1	5.9±0.1	6.5±0.2[b]
Hippocampus	53±3	45±1[b]	1.2±0.2	1.1±0.1	6.9±0.1	6.7±0.3
Cerebellum	68±3	62±5	2.1±0.1	2.0±0.1	5.4±0.1	5.6±0.1

[a]MPTP was infused intraperitoneally for five days. Mice were sacrificed one day after withdrawal of the drug. Values are mean ± S.E.M. ($n = 8$–10).
[b]$p < 0.05$ in the intermittent hypoxia group compared to that of the normoxic controls by unpaired Student's *t*-test analyses.

increased the formation of peroxidized lipids compared to those incubated at 0°C (basal level). The autooxidation of cortical homogenates was lower in the intermittent hypoxia-treated group compared to the normoxic group (see FIGURE 2). Furthermore, addition of iron (1 µM) profoundly increased lipid peroxidation in cortical homogenates following a four-hour incubation at 37°C in the normoxic group. However, iron induced lipid peroxidation was attenuated in the intermittent hypoxia-treated group (FIG. 2). These data support our previous study in suggesting that intermittent hypoxia prevents oxidative stress by iron.[20]

Effect of Intermittent Hypoxia on the GSH/GSSG Ratio

Systemic application of MPTP has been reported to decrease GSH content and increase GSSG/GSH ratio in mouse striatum,[31,47] representing an attenuation of antioxidative defense systems. Our preliminary study showed that iron not only induced oxidative injuries, but also decreased the GSH/GSSG ratio in the infused SN (manuscript submitted). In the study reported here, although intraperitoneal injection of MPTP did not alter the GSH content, it did decrease the GSH/GSSG ratio in mouse striatum in the MPTP group, indicating an increased oxidative status for this group (see FIGURE 3). After four-week intermittent hypoxia, the GSH/GSSG ratio was not altered in the saline groups. Furthermore, the MPTP induced reduction in GSH/GSSG ratio was not reversed in the intermittent hypoxia-treated group (FIG. 3).

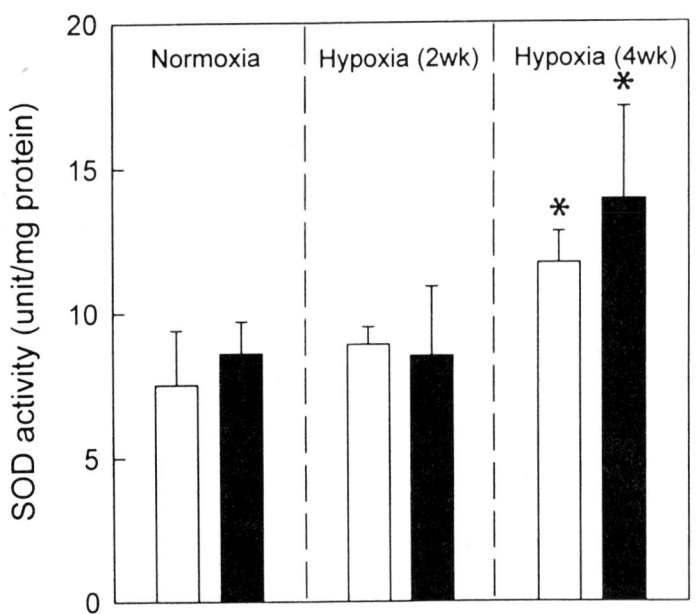

FIGURE 4. Effects of intermittent hypoxia on SOD in striatum in mouse brain. *$p < 0.05$ in the four-week intermittent hypoxic group compared to the normoxic group by one way ANOVA followed by *post hoc* analyses. ▢, saline; ■, MPTP.

The effect of intermittent hypoxia on GSH/GSSG ratio of other brain regions was investigated. Intermittent hypoxia did not increase the GSH/GSSG ratio in cortex, hippocampus, or cerebellum of mouse brain. The results are shown in TABLE 1.

Effect of Intermittent Hypoxia on SOD Activity

Superoxide dismutase, a scavenger enzyme for superoxide anion radicals, reduces superoxide to hydrogen peroxide. Our preliminary data showed a decreased SOD activity in response to oxidative stress by intranigral infusion of iron (paper submitted). However, systemic MPTP administration has been reported to either increase[20,32,48] or not affect SOD activity.[49] In our MPTP study, SOD activity was not significantly increased in the striatum (see FIGURE 4). Four-week intermittent hypoxia enhanced SOD activity in mouse striatum. Intermittent hypoxia did not alter SOD activity in the cortex, hippocampus, or cerebellum.

Effect of Intermittent Hypoxia on GPx Activity

No significant change in striatal GPx, a scavenging enzyme for hydroperoxides, has been reported following systemic MPTP administration.[32,48] Similarly, in this study, GPx was not altered by intraperitoneal injection of MPTP. The GPx activity of saline control was 11.2 ± 0.4 unit/mg protein versus 10.6 ± 0.3 unit/mg protein of MPTP group ($n = 10$). Furthermore, intermittent hypoxia did not change GPx activity. After four weeks of intermittent hypoxia, the GPx activities were 10.8 ± 0.8 unit/mg protein of saline control and 11.1 ± 1.1 unit/mg protein of MPTP group, respectively ($n = 10$). However, intermittent hypoxia appeared to enhance GPx in mouse cortex, but not in hippocampus or cerebellum (TABLE 1).

SUMMARY AND CONCLUSION

In the study we report here, several lines of evidence support the preventive action of intermittent hypoxia against oxidative injuries in CNS. Our *in vitro* data showed that autooxidation and iron-induced lipid peroxidation were attenuated in cortical homogenates of intermittent hypoxia-treated animals. Furthermore, our preliminary study found that iron induced oxidative injuries were abolished in rat brain after intermittent hypoxic treatment (paper submitted). Several antioxidative defensive systems improve in response to intermittent hypoxia. Since attenuation of autooxidation and iron-induced lipid peroxidation were observed in cortical homogenates of intermittent hypoxia-treated mice, the lack of prevention by intermittent hypoxia of MPTP-induced neurotoxicity may be due to the MPTP action that is not oxidative related. Together with our previous studies, in which several antioxidants were shown to successfully prevent oxidative injuries, our data here suggest that intermittent hypoxia may offer a potential treatment for preventing CNS degenerative diseases.

ACKNOWLEDGMENTS

The authors express their gratitude to Dr. Chai C.Y., at the Institute of Biomedical Sciences Academia Sinica, for his encouragement and support. Special thanks are due to Dr. R.K. Freund at the University of Colorado, Health Sciences Center, Denver, CO., U.S.A., for editing this paper. This study was supported by Grants NSC88-2413-B-075-009-M35 and VGH89-438, Taipei, Taiwan.

REFERENCES

1. AMES, B.N., M.K. SHIGENAGA & T.M. HAGEN. 1993. Oxidants, antioxidants and the degenerative disease of aging. Proc. Natl. Acad. Sci. U.S.A. **90:** 7915–7922.
2. LIN, A.M.Y. & L.T. HO. 2000. Melatonin suppresses iron-induced neurodegeneration in rat brain. Free Rad. Biol. Med. **28:** 904–911.
3. LIN A.M.Y., B.Y. CHYI, S.D. WANG, et al. 1999 Carboxyfullerene prevents iron-induced oxidative stress in rat brain. J. Neurochem. **72:** 1634–1640.
4. LIN, A.M.Y., L.T. HO & T.-Y. LUH. 1999. The antioxidative effect of carboxyfullerenes (C3 / D3) on iron-induced oxidative injury in CNS. Ann. N.Y. Acad. Sci. **890:** 340–351.
5. LIN, A.M.Y., C.H. YANG & C.Y. CHAI. 1998. Striatal dopamine dynamics are altered following an intranigral infusion of iron in adult rats. Free Rad. Biol. Med. **24(6):** 988–993.
6. BEN-SHACHAR, D. & M.B.H. YOUDIM. 1991. Intranigral iron injection induces behavioral and biochemical "Parkinsonism" in rats. J. Neurochem. **57:** 2133–2135.
7. CHIUEH C.C., R.M. WU, K.P. MOHANAKUMAR, et al. 1994. In vivo generation of hydroxyl radicals and MPTP-induced dopaminergic toxicity in the basal ganglia. Ann. N.Y. Acad. Sci. **738:** 25–36.
8. COHEN, G. & P. WERNER. 1994. Free radicals, oxidative stress and neurodegeneration. In Neurodegenerative Diseases. D.B. Calne, Ed.: 139–162. WB Saunders, Philadelphia.
9. HALLIWELL, B. & J.M. GUTTERIDGE. 1984. Oxygen toxicity, oxygen radicals, transition metals and diseases. Biochem. J. **219:** 1–14.
10. HALLIWELL, B. 1992. Reactive oxygen species and the central nervous system. J. Neurochem. **59:** 1609–1623.
11. HEO, J.H., D.G. KIM, H.R. BAHNG & J.S. KIM. 1995. Dimethylthiourea prevents MPTP-induced decreased in [^{3}H]dopamine uptake in rat striatal slices. Brain Res. **671:** 321–324.
12. YOUDIM M.B., D. BEN-SHACHAR, G. ESHEL, et al. 1993. The neurotoxicity of iron and nitric oxide relevance to the etiology of Parkinson's disease. In Advances in Neurology. H. Narabyashi, T. Nagatsu, N. Yanagisawa & Y. Mizuno, Eds.: 259–266. Raven Press, New York.
13. ZIGMOND, M.J. & E.M. STRICKER. 1989. Animal models of Parkinsonism using selective neurotoxins: clinical and basic implication. Int. Rev. Neurobiol. **31:** 1–79.
14. PERRY, T.L., D.V. GODIN & A. HANSEN. 1982. Parkinson's disease: a disorder due to nigral glutathione deficiency. Neurosci. Lett. **33:** 305–310.
15. RAUHALA, P., A.M.Y. LIN & C.C. CHIUEH. 1998. Neuroprotection by s-nitroglutathione of brain dopamine neurons from oxidative stress. FASEB J. **12:** 165–173.
16. SCHULZ, J.B., D.R. HENSHAW, D. SIWEK, et al. 1995. Involvement of free radicals in excitotoxicity in vivo. J. Neurochem. **64:** 2239–2247.
17. SENGSTOCK, G.J., C.W. OLANOW, A.J. DUNN, et al. 1994. Progressive changes in striatal dopaminergic markers, nigral volume, and rotational behavior following iron infusion into the rat substantia nigra. J. Neurol. **130:** 82–94.
18. DEXTER, D.T., F.R. WELLS, A.J. LEES, et al. 1989. Increased nigral iron content and alteration in other metal ions occurring in brain in Parkinson's disease. J. Neurochem. **52:** 1830–1836.

19. SOFIC, E., K.W. LANG, K. JELLINGER & P. RIEDERER. 1992. Reduced and oxidized glutathione in the substantia nigra of patients with Parkinson's disease. Neurosci. Lett. **142:** 128–130.
20. GULYAEVA, N.V. & E.N. TKATCHOUK. 1998. Antioxidant effects of interval hypoxia training in rat brain. 12th European Society of Neurochemistry Meeting, S31C.
21. FLISS H., T.E. COMAS, C.E. HOWLETT, IV., et al. 1998. Antioxidant effects of interval hypoxia training on transcription factors in rat heart and brain. 12th European Society of Neurochemistry Meeting, S31D.
22. DUAN, C.L., F.S YAN, X.Y. SONG & G.W. LU. 1999. Changes of superoxide dismutase, glutathione peroxidase and lipid peroxide in the brain of mice preconditioned by hypoxia. Biol. Sig. Receptor. **8:** 256–260.
23. CHIEN, C.T., C.F. CHEN, S.M. HSU, et al. 1999. Protective mechanism of preconditioning hypoxia attenuates apoptosis formation during renal ischemia/reperfusion phase. Transplant. Proceed. **31:** 2012–2013.
24. ZHUANG, J. & Z. ZHOU. 1999. Protective effect of intermittent hypoxic adaptation on myocardium and its mechanisms. Biol. Sig. Receptors. **8:** 316–322.
25. VOVC, E. 1998. Arrhythmic effect of adaptation to intermittent hypoxia. Folia Medica **40:** 51–54.
26. MEERSON, FZ., E.E. USTINOVA & E.B. MANUKHINA. 1989. Prevention of cardiac arrhythmias by adaptation to hypoxia: regulatory mechanisms and cardiotropic effect. Biomed. Biochim. Acta. **48:** S83–88.
27. TRAYSTMAN, R.J., J.R. KIRSCH & R.C. KOEHLER. 1991. Oxygen radical mechanisms of brain injuries following ischemia and reperfusion. J. Appl. Physiol. **71:** 1185–1195.
28. SIESJO, B.K. 1993. A new perspective on ischemic brain damage? Prog. Brain Res. **96:** 1–9.
29. MEERSON, F.Z., IVB ARKHIPENKO, I.I. ROZHITSKAIA, et al. 1992. Opposite effects on antioxidant enzymes of adaptation to continuous and intermittent hypoxia. Biull. Ekspéri. Biol. Med. **114:** 14–15.
30. NAKANISHI, K, F. TAJIMA, A. NAKAMURA, et al. 1995. Effect of hypoxia on antioxidant enzymes in rats. J. Physiol. **489:** 869–876.
31. OISHI, T., E. HASEGAWA & Y. MURAI. 1993. Sulfhydryl drugs reduce neurotoxicity of 1-methyl-4-phenyl-1,2,3,6-tetrahydropyridine in the mouse. J. Neural. Transm. **6:** 45–52.
32. HUNG, H.C. & E.H. LEE. 1998. MPTP produces differential oxidative stress and antioxidative responses in the nigrostriatal and mesolimbic dopaminergic pathways. Free Radic. Biol. Med. **24:** 76–84.
33. THIFFAULT, C., N. AUMONT, R. QUIRION & J. POIRIER. 1995. Effect of MPTP and L-deprenyl on antioxidant enzymes and lipid peroxidation levels in mouse brain. J. Neurochem. **65:** 2725–2733.
34. BONHOMME, N., P. GREVE, C. CUISINIER & P. LESTAGE. 1999. Lack of protective effect of antioxidant drugs against MPTP neurotoxicity in C57BL/6 mice. (Abstr.) Twenty-ninth Annual Meeting, Society for Neuroscience, Miami Beach, Florida.
35. CHIBA, K., A.J. TREVOR & N. CASTAGNOLI, JR. 1984. Metabolism of the neurotoxic tertiary amine, MPTP by brain monoamine oxidase. Biochem. Biophys. Res Commun. **120:** 574–587.
36. JAVITCH, J.A., R.J. D'MATO, S.M. STRITTMATTER & S.H. SNYDER. 1985. Parkinsonism-inducing neurotoxin, MPTP: uptake of the metabolite MPP$^+$ by dopamine neurons explains selective toxicity. Proc. Natl. Acad. Sci. U.S.A. **82:** 2173–2177.
37. BATES, T.E., S. HEALES, JR., S.E.C. DAVIES, et al. 1994. Effects of 1-methyl-4-phenylpyidium on isolated rat brain mitochondria: evidence for a primary involvement of energy depletion. J. Neurochem. **63:** 640–648.
38. CHIUEH, C.C., G. KRISHNA, P. TULSI, et al. 1992. Intracellular microdialysis of salicylic acid to detect hydroxyl radical generation through dopamine autoxidation in the caudate nucleus effect of MPP$^+$. Free Rad. Biol. Med. **13:** 581–583.
39. TSAI, M.J. & E.H.Y. LEE. 1998. NO donors protect cultured astrocytes from MPP$^+$-induced toxicity. Free Rad. Biol. Med. **24:** 705–713.

40. CHIUEH, C.C., Z. ZUKOWAKA-GROJEC, K.L. KIRK & I.J. KOPIN. 1983. 6-Fluorocatecholamines as false adrenergic neurotransmitters. J. Pharmacol. Exp. Ther. **225:** 529–533.
41. KIKUGAWA, K., T. KATO, M. BEPPU & A. HAYSAKA. 1989. Fluorescent and crosslinked protein formed by free radical and aldehyde species generated during lipid peroxidation. Adv. Exper. Med. Biol. **266:** 345–356.
42. MOKANAKUMAR, K.P., A. DEBARTOLOMEIS, R.M. WU, *et al.* 1994. Ferrous citrate complex and nigral degeneration: evidence for free radical formation and lipid peroxidation. Ann. N.Y. Acad. Sci. **738:** 392–399.
43. GRIFFITH, O.W. 1980. Determination of glutathione and glutathione disulfide using glutathione reductase and 2-vinylpyridine. Anal. Biochem. **106:** 207–212.
44. MISRA, H.P.& I. FRIDPVICH. 1989. The role of superoixde anion in the autoxidation of epinephrine and a simple assay for superoxide dismutase. J. Biol. Chem. **264:** 7761–7764.
45. FLOHE L. & W.A. GUNZLER. 1984. Assay of glutathione peroxidase. Methods Enzymol **105:** 114–121.
46. GERLACH M., M. GOTS, A. DIRR, *et al.* 1996. Acute MPTP treatment produced no changes in mitochondrial complex activities and indices of oxidative damage in the common marmoset *ex vivo* one week after exposure to the toxin. Neurochem. Internat. **28:** 41–49.
47. FERRARO, T.N., G.T. GOLDEN, M. DE MATTEI, *et al.* 1986. Effect of MPTP on levels of glutathione in the extrapyramidal system of the mouse. Neuropharmacol. **25:** 1071–1074.
48. THIFFAULT, C., N. AUMONT, R. QUIRION & J. POIRIER. 1995. Effect of MPTP and L-deprenyl on antioxidant enzymes and lipid peroxidation levels in mouse brain. J. Neurochem. **65:** 2725–2733.
49. PIKARSKY, E., E. MELAMED, J. ROSENTHAL, *et al.* 1987. The neurotoxin MPTP does not affect striatal superoxide dismutase activity in mice. Neurosci. Lett. **82:** 327–331.

Neurogenesis and Neuroprotection in the Adult Brain

A Putative Role for 5-Lipoxygenase?

HARI MANEV, TOLGA UZ, RADMILA MANEV, AND ZHIJING ZHANG

The Psychiatric Institute, Department of Psychiatry,
University of Illinois at Chicago, Chicago, Illinois, U.S.A.

ABSTRACT: 5-Lipoxygenase (5-LOX) and cyclooxygenase-2 (COX-2) are two enzymes that are critical for the synthesis of eicosanoids, the inflammatory metabolites of arachidonic acid. Both 5-LOX and COX-2 are expressed in the brain, including in CNS neurons. The physiologic role of these proteins in neuronal functioning is not clear. In non-neuronal tissues these two enzymes often assume similar roles: in addition to their function in inflammation, 5-LOX and COX-2 appear to be associated with cell proliferation, that is, with tumor growth. High 5-LOX expression has been noticed in the proliferating brain or pancreatic tumor cells; reduction in tumor cell proliferation and/or destruction of tumor cells was achieved with 5-LOX inhibitors. Proliferation of immature neurons/neuroblasts is an important component of mitotic neurogenesis. We investigated the role of 5-LOX in proliferation using cultures of human neuronal precursor cells, NT2. We found that these cells express 5-LOX mRNA and we used ^{3}H-thymidine incorporation as a measure of cell proliferation; this was reduced by treating the cultures with 5-LOX inhibitor AA-861. We propose that the 5-LOX pathway plays a crucial role in mitotic neurogenesis. Additional studies should explore whether 5-LOX may participate in neurogenesis related pathologies and whether it should be considered a target for procedures aimed at altering neurogenesis for therapeutic purposes.

KEYWORDS: 5-Lipoxygenase (5-LOX); Leukotrienes; Cell proliferation; Neurogenesis; Stroke.

INTRODUCTION

5-Lipoxygenase (5-LOX) and cyclooxygenase-2 (COX-2) are two enzymes that are involved in the metabolism of arachidonic acid into inflammatory metabolites, leukotrienes and prostaglandins, respectively. The presence of 5-LOX and COX-2 mRNAs and proteins has been documented in the central nervous system (CNS), including neurons.[1,2] An overexpression and/or overactivation of these two proteins in CNS has been related to neurodegenerative pathologies.[3–6] Particularly interesting is the observation that 5-LOX gene expression in CNS is upregulated during aging[5,7] and by glucocorticoids,[8] pointing to possible important hormonal influences

---

Address for correspondence: Hari Manev, M.D., Ph.D., The Psychiatric Institute, Department of Psychiatry, University of Illinois at Chicago, 1601 West Taylor Street, MC912, Chicago, IL 60612, U.S.A. Voice: 312-413-4558; fax: 312-413-4569.
HManev@psych.uic.edu

on 5-LOX and in its relation to CNS pathologies. Numerous recent studies have explored a possibility that drugs acting on 5-LOX and/or COX-2 may provide the pharmacological means for neuroprotection (for a review, see Ref. 9).

The discovery of the capacity of the adult human brain to generate new neurons[10] has raised hope that neurogenesis in the adult CNS might provide a novel target for therapeutic treatment.[11] Although most studies on the role of 5-LOX/COX-2 in CNS pathologies refer to the "proinflammatory" role of these proteins, another important action of 5-LOX and COX-2 has been identified in cell proliferation.[12–14] Since cell proliferation is a crucial step during neurogenesis in the adult brain[15] it is important to elucidate whether 5-LOX and/or COX-2 participate in neurogenesis. Our studies with neuronal precursors and 5-LOX are the first steps in this direction.

NEUROGENESIS IN THE ADULT BRAIN

Neurogenesis was first observed in the brain of adult rodents a long time ago.[16] It was argued, however, that it could not occur in the adult human brain.[17] After the work by Eriksson et al.,[10] this dogma has lost some strength. It is believed that two brain regions, the hippocampal dentate gyrus and the subventricular zone, are sites where the residing neuronal precursors could produce new neurons. A steady stream of reports over the past several years has provided evidence of adult neurogenesis in various animal species and under various experimental conditions. Interestingly, the rate of adult neurogenesis can be affected by a variety of exogenous stimuli, some of them very prosaic, such as physical exercise.[18]

Neurogenesis is the outcome of a dynamic process that includes a fine tuning between cell proliferation and cell death or survival. That is to say, that a continuous proliferation of neuronal precursors is kept in balance by an accompanying cell death. It was suggested that programmed cell death (apoptosis) might have an important regulatory function by eliminating supernumerous cells from neurogenic regions, and that the balance between cell proliferation and apoptosis provides the adult brain with a powerful self renewal mechanism.[19]

ADULT NEUROGENESIS IN RESPONSE TO STROKE

A recent report by Magavi et al.[11] points to an important new element in the puzzle of adult neurogenesis. These authors found that endogenous neural precursors can be induced in situ to differentiate into mature neurons, in regions of the adult mammalian neocortex that do not normally undergo neurogenesis. Moreover, these new neurons were shown to be capable of forming appropriate connections. Thus, it was proposed that neuronal replacement therapies for neurodegenerative disease and brain injury might be possible through manipulation of endogenous neuronal precursors.[11]

In two models of brain ischemia, the transient global ischemia in gerbils and the photothrombotic stroke in rats, authors have investigated whether neurogenesis is stimulated in response to injury.[20,21] In the study with the transient global ischemia, neurogenesis was increased in the dentate gyrus; newborn cells with a neuronal phenotype were first seen 26 days after ischemia and survived for at least seven months.[20] In the model of photothrombotic stroke, the lesion was produced in the cortex; about 3–5% of cells labeled with bromodeoxyuridine (BrdU, a marker of proliferating cells) were also positive for neuronal markers and were found seven and 100 days after stroke in the area surrounding the core of the injury.[21] Both studies suggest that new neurons can be generated in the adult CNS as a potential pathway for brain repair and functional neuroprotection.

5-LIPOXYGENASE: ANY ROLE IN STROKE AND/OR NEUROGENESIS?

Several investigators have suggested a possible involvement of the 5-LOX pathway in the pathobiology of brain ischemia.[3,22] Among different leukotrienes, cysteinyl leukotrienes (cysLT) appear to be the major product of 5-LOX enzymatic activity in the brain. CysLTs are produced when the brain 5-LOX is activated either by seizures,[4] ischemia,[3] or during the blood–brain cell contact.[23] However, in addition to its well described role as an enzyme, 5-LOX might affect cell functioning via 5-LOX specific non-enzymatic actions (see FIGURE 1). Thus, as reviewed recently,[7] the interaction of 5-LOX and 5-LOX activating protein (FLAP) enables the enzymatic

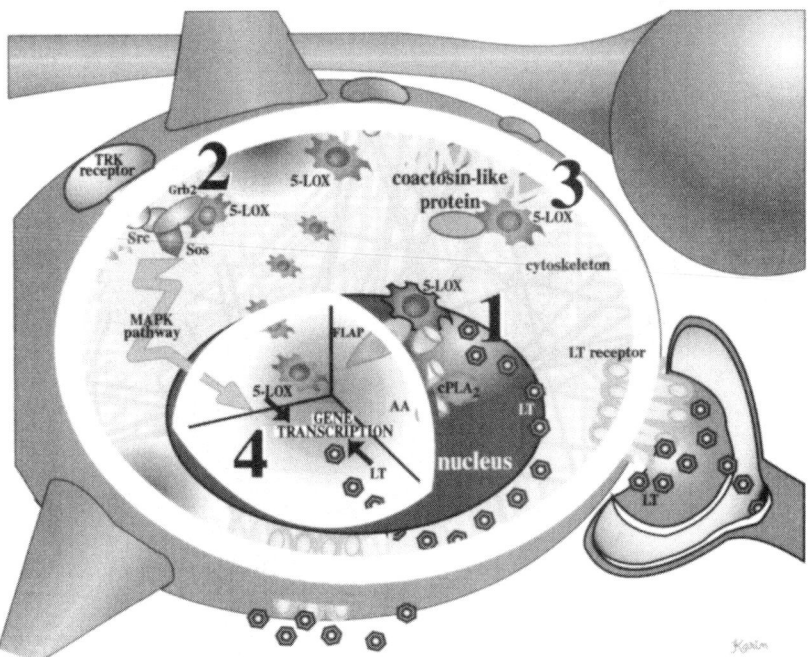

FIGURE 1. Putative enzymatic and non-enzymatic actions of 5-LOX protein (modified, with permission, from Manev *et al.*, Ref. 7). In resting cells, 5-LOX appears to be predominantly a cytosolic protein that, depending on the state of its phosphorylation, is capable of binding 5-LOX activating protein (FLAP) (①), receptor bound protein 2 (Grb2) (②), coactosin like protein that appears to be linked to the cytoskeleton (③), or it may translocate into the nucleus (④). *AA*, arachidonic acid; *LT*, leukotriene.

activity of 5-LOX that catalyzes the insertion of oxygen into free arachidonic acid (AA) and initiates the formation of leukotrienes. The arachidonic acid that participates in LT formation presumably originates in the nuclear membrane, from which it is released by an action of cytosolic phospholipase A_2 (cPLA$_2$) (FIG. 1, ①). Leukotrienes could affect the cell in which they are produced (e.g., by affecting the nucleus and/or by interacting with membrane associated leukotriene receptors), or they could leave the cell and affect neighboring cells (neurons? or glia?) via metabotropic G-protein coupled leukotriene receptors. The putative non-enzymatic actions of 5-LOX include its interaction with the intracellular signaling system of tyrosine kinase receptors for growth factors, with the coactosin like proteins of the cytoskeleton, and, possibly, the nuclear 5-LOX import (FIG. 1). The contribution of enzymatic versus the non-enzymatic action of 5-LOX to the pathology of stroke has not yet been investigated.

In addition to its role in inflammation, 5-LOX has also been studied for its involvement in cell proliferation. Hence, it has been established that inhibitors of 5-LOX reduce the *in vitro* proliferation of bone marrow cells and of cell lines from various other sources. For example, MK-886, a putative FLAP inhibitor, reduced proliferation and induced apoptotic cell death in cultures of prostate PC-3 cells;[24] MK-886 or AA-861 (another selective 5-LOX inhibitor) inhibited the proliferation

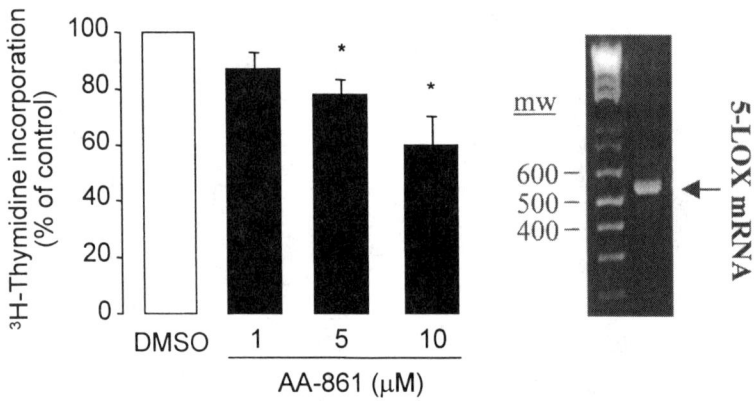

FIGURE 2. 5-Lipoxygenase inhibitor AA-861 reduces proliferation of NT2 human neuronal precursor cells. The NT2 cells were plated into 24-well plates at a density 50,000 cells/well/0.5 ml medium (D-MEM/F-12 containing 10% fetal bovine serum; 2 mM glutamine; 100 U/ml penicillin; 100 µg/ml streptomycin).[27] They were grown overnight. AA-861 (Biomol) was dissolved in dimethylsulfoxide (DMSO) and added directly into the medium (1 µl/well); controls were treated with DMSO ($n = 6$ wells per group; repeated in three different culture preparations). Three hours later, ^{3}H-thymidine (1 µCi/ml; Amersham; TRK120) was added to all wells and the reaction was stopped one hour later; cells were detached and processed.[28] Results are expressed as percent of control (mean ± S.E.M); *$p < 0.05$ compared with control (ANOVA and Dunnett's test). RT-PCR assay[5] with primers designed specifically for human 5-LOX (5'-GCA ACA CCG ACG TAA AGA ACT GGA; 5'-TTA TAC AGC AAG CAG ATG GGA GCG)[34] was used to identify 5-LOX mRNA expression in NT2 cells (indicated by the *arrow* in the **right panel**).

of prostate cancer cells.[25] 5-LOX inhibitors also reduced the proliferation of pancreatic cancer cells; interestingly, only cancerous but not normal pancreatic cells exhibited high 5-LOX expression.[13]

In our pilot studies, we noticed that the expression of 5-LOX mRNA and protein is very high in proliferating cultures of rat neuronal precursors and that 5-LOX expression decreases when these precursors differentiate into neurons. We also noticed that 5-LOX inhibitors exert an inhibitory action on the proliferation of the rat neuronal precursors.[26]

In the study we report here, we used the NT2 culture of human neural precursors (a gift from D.R. Grayson).[27] First, we confirmed that the proliferating NT2 cells express 5-LOX mRNA (see FIGURE 2). Second, we used the ^{3}H-thymidine assay[28] to assess the proliferation of NT2 cells and to test the action of the 5-LOX inhibitor AA-861 on ^{3}H-thymidine incorporation. We found that this 5-LOX inhibitor induced a concentration dependent inhibition of proliferation of human neural precursors (FIG. 2). These results and our preliminary findings[26] indicate that 5-LOX might be necessary to support the proliferation of neuronal precursors, and that it could be involved in mitotic neurogenesis. It remains to be elucidated whether the adverse effect of 5-LOX inhibition on cell proliferation as opposed to its direct neuroprotective action[5] has a beneficial or an adverse action with respect to brain physiology and pathology.

POSSIBILITIES FOR THERAPEUTIC ALTERATIONS OF NEUROGENESIS IN THE ADULT BRAIN

At least two different approaches appear to be available for altering neurogenesis in the adult CNS: (1) surgical, and (2) pharmacological. The surgical approach involves transplantation of neuronal precursors, for example, "stem cells". Current research in this area is directed toward identifying the best sources of stem cells; it indicates that these cells could be harvested either from fetal derived cells or from the bone marrow.[29] Hence, Mezey and Chandross[29] demonstrated that some of the neurons and glia in the adult CNS can be derived from bone marrow. With respect to stroke, Li et al.[30] transplanted adult bone marrow non-hematopoetic cells into the striatum of mice after embolic middle cerebral artery occlusion. Using the method noted above of BrdU labeling of proliferating cells, they found evidence that the transplanted cells survived and improved functional recovery after stroke.

The search for pharmacological tools for altering the adult neurogenesis is still in its infancy. Recently, Aberg et al.[31] showed that peripheral infusion of IGF-I (insulin-like growth factor-I, a growth-promoting peptide hormone) increases progenitor cell proliferation and selectively induces neurogenesis in the progeny of adult neural progenitor cells. Previously, it was suggested that the mitogenic activity of IGF might involve the 5-LOX pathway.[32] Interestingly, IGF-I also has potential for the treatment of stroke.[32] We propose that a better understanding of molecular mechanisms responsible for neurogenesis, for example, 5-LOX, will help in discovering novel neuroprotective therapies.

CONCLUSIONS

Neurogenesis appears to be the intrinsic property of the adult brain; a fine balance between cell proliferation and cell death provides the adult CNS with a powerful self renewing mechanism. Neurogenesis is triggered in response to brain injury/stroke, possibly as an endogenous protective mechanism. It is proposed that increasing adult neurogenesis, for example, surgically by cell transplantation or pharmacologically, may be helpful in treating neurodegeneration/stroke. We propose that the 5-LOX pathway may be necessary to maintain cell proliferation and that further characterization of its role in neurogenesis might help to find novel neuroprotective therapies.

ACKNOWLEDGMENT

This work was in part supported by NIH-NIA Grant RO1-AG15347 (H.M.).

REFERENCES

1. LAMMERS, C.H., P. SCHWEITZER, P. FACCHINETTI, et al. 1996. Arachidonate 5-lipoxygenase and its activating protein: prominent hippocampal expression and role in somatostatin signaling. J. Neurochem. **66:** 147–152.
2. O'BANION, M.K. & J.A. OLSCHOWKA. 1999. Localization and distribution of cyclooxygenase-2 in brain tissue by immunohistochemistry. Method. Mol. Biol. **120:** 55–66.
3. OHTSUKI, T., M. MATSUMOTO, Y. HAYASHI, et al. 1995. Reperfusion induces 5-lipoxygenase translocation and leukotriene C4 production in ischemic brain. Am. J. Physiol. **268:** H1249–H1257.
4. SIMMET, T. & B. TIPPLER. 1991. On the relation between cerebral cysteinyl-leukotriene formation and epileptic seizures. Brain Res. **540:** 283–286.
5. UZ, T., C. PESOLD, P. LONGONE & H. MANEV. 1998. Aging-associated up-regulation of neuronal 5-lipoxygenase expression: putative role in neuronal vulnerability. FASEB J. **12:** 439–449.
6. O'BANION, M.K. 1999. Cyclooxygenase-2: molecular biology, pharmacology, and neurobiology. Crit. Rev. Neurobiol. **13:** 45–82.
7. MANEV, H., T. UZ, M. SUGAYA & T. QU. 2000. Putative role of neuronal 5-lipoxygenase in an aging brain. FASEB J. **14:** 1464–1469.
8. UZ, T., Y. DWIVEDI, P.D. SAVANI, et al. 1999. Glucocorticoids stimulate inflammatory 5-lipoxygenase gene expression and protein translocation in the brain. J. Neurochem. **73:** 693–699.
9. SUGAYA, K., T. UZ, V. KUMAR & H. MANEV. 2000. New anti-inflammatory treatment strategy in Alzheimer's disease. Jpn. J. Pharmacol. **82:** 85–94.
10. ERIKSSON, P.S., E. PERFILIEVA, T. BJORK-ERIKSSON, et al. 1998. Neurogenesis in the adult human hippocampus. Nat. Med. **4:** 1313–1317.
11. MAGAVI, S.S., B.R. LEAVITT & J.D. MACKLIS. 2000. Induction of neurogenesis in the neocortex of adult mice. Nature **405:** 951–955.
12. PASQUALE, D. & G. CHIKKAPPA. 1993. Lipoxygenase products regulate proliferation of granulocyte-macrophage progenitors. Exp. Hematol. **21:** 1361–1365.
13. DING, X.Z., P. IVERSEN, M.W. CLUCK, et al. 1999. Lipoxygenase inhibitors abolish proliferation of human pancreatic cancer cells. Biochem. Biophys. Res. Commun. **261:** 218–223.
14. JOKI, T., O. HEESE, D.C. NIKAS, et al. 2000. Expression of cyclooxygenase 2 (COX-2) in human glioma and *in vitro* inhibition by a specific COX-2 inhibitor, NS-398. Cancer Res. **60:** 4926–4931.

15. FUCHS, E. & E. GOULD. 2000. Mini-review: *in vivo* neurogenesis in the adult brain: regulation and functional implications. Eur. J. Neurosci. **12:** 2211–2214.
16. ALTMAN, J. & G.D. DAS. 1965. Autoradiographic and histological evidence of postnatal hippocampal neurogenesis in rats. J. Comp. Neurol. **124:** 319–335.
17. RAKIC, P. 1985. DNA synthesis and cell division in the adult primate brain. Ann. N.Y. Acad. Sci. **457:** 193–211.
18. VAN PRAAG, H., G. KEMPERMANN & F.H. GAGE. 1999. Running increases cell proliferation and neurogenesis in the adult mouse dentate gyrus. Nat. Neurosci. **2:** 266–270.
19. BIEBL, M., C.M. COOPER, J. WINKLER & H.G. KUHN. 2000. Analysis of neurogenesis and programmed cell death reveals a self-renewing capacity in the adult rat brain. Neurosci. Lett. **291:** 17–20.
20. LIU, J., K. SOLWAY, R.O. MESSING & F.R. SHARP. 1998. Increased neurogenesis in the dentate gyrus after transient global ischemia in gerbils. J. Neurosci. **18:** 7768–7778.
21. GU, W., T. BRANNSTROM & P. WESTER. 2000. Cortical neurogenesis in adult rats after reversible photothrombotic stroke. J. Cereb. Blood Flow Metab. **20:** 1166–1173.
22. MINAMISAWA, H., A. TERASHI, Y. KATAYAMA, *et al.* 1988. Brain eicosanoid levels in spontaneously hypertensive rats after ischemia with reperfusion: leukotriene C4 as a possible cause of cerebral edema. Stroke **19:** 372–377.
23. WINKING, M., R.M. HELDT & T. SIMMET. 1996. Thrombin stimulates activation of the cerebral 5-lipoxygenase pathway during blood-brain cell contact. J. Cereb. Blood Flow Metab. **16:** 737–745.
24. ANDERSON, K.M., T. SEED, M. VOS, *et al.* 1998. 5-Lipoxygenase inhibitors reduce PC-3 cell proliferation and initiate nonnecrotic cell death. Prostate. **37:** 161–173.
25. GHOSH, J. & C.E. MYERS. 1997. Arachidonic acid stimulates prostate cancer cell growth: critical role of 5-lipoxygenase. Biochem. Biophys. Res. Commun. **235:** 418–423.
26. MANEV, H., T. UZ & R. MANEV. 2000. Putative role of neuronal 5-lipoxygenase (5-LOX) in proliferation of immature cerebellar granule cells. J. Neurochem. **74:** S64.
27. LEE, V.M. & P.W. ANDREWS. 1986. Differentiation of NTERA-2 clonal human embryonal carcinoma cells into neurons involves the induction of all three neurofilament proteins. J. Neurosci. **6:** 514–521.
28. FISZMAN, M.L., L.N. BORODINSKY & J.H. NEALE. 1999. GABA induces proliferation of immature cerebellar granule cells grown in vitro. Brain Res. Dev. Brain Res. **115:** 1–8.
29. MEZEY, E. & K.J. CHANDROSS. 2000. Bone marrow: a possible alternative source of cells in the adult nervous system. Eur. J. Pharmacol. **405:** 297–302.
30. LI, Y., M. CHOPP, J. CHEN, *et al.* 2000. Intrastriatal transplantation of bone marrow nonhematopoietic cells improves functional recovery after stroke in adult mice. J. Cereb. Blood Flow Metab. **20:** 1311–1319.
31. ABERG, M.A., N.D. ABERG, H. HEDBACKER, *et al.* 2000. Peripheral infusion of IGF-I selectively induces neurogenesis in the adult rat hippocampus. J. Neurosci. **20:** 2896–2903.
32. SANDY, J., M. DAVIES, S. PRIME & R. FARNDALE. 1998. Signal pathways that transduce growth factor-stimulated mitogenesis in bone cells. Bone **23:** 17–26.
33. DORE, S., S. KAR, W.H. ZHENG & R. QUIRION. 2000. Rediscovering good old friend IGF-I in the new millenium: possible usefulness in Alzheimer's disease and stroke. Pharm. Acta Helv. **74:** 273–280.
34. DIXON, R.A., R.E. JONES, R.E. DIEHL, *et al.* 1988. Cloning of the cDNA for human 5-lipoxygenase. Proc. Natl. Acad. Sci. U.S.A. **85:** 416–420.

Part II: Mechanisms of Neurodegeneration and Neurogenesis

QUESTIONS FOR DR. MANEV

From Dr. Youdim

I found your paper very interesting in light of the fact we have replicated the increased inflammatory responses as a step toward neurodegeneration in neurodegenerative diseases and have avoided the use of COX_2 inhibition. However, recent laboratory and clinical studies have been disappointing. Would you advocate LOX inhibitors as neuroprotectants or even combination of COX_2 and LOX inhibitors.

ANSWER: Our data indicate that 5-LOX and COX_2 pathways could be considered two parallel mechanisms of arachidonic acid metabolism that, possibly, are involved in neurodegenerative pathology. Thus, a combined pharmacological inhibition of both 5-LOX and COX_2 might be advantageous. However, there are only few clinically available specific 5-LOX inhibitors that could be tested, and it is not known whether they cross the blood–brain barrier. An additional issue relevant for neurodegenerative disorders, such as Alzheimer's disease, is the putative contribution of 5-LOX-related pharmacogenetic variability due to the polymorphism in 5-LOX promoter (H. Manev, 2000. 5-Lipoxygenase gene polymorphism and onset of Alzheimer's disease. Med. Hypotheses **54:** 75–76; T. Qu, et al. 2001. 5-Lipoxygenase (5-LOX) promoter polymorphism in-patients with early and late onset Alzheimer's disease. J. Neuropsych. Clin. Neurosci., in press.) Alternatively, if diminished adult neurogenesis is a significant component of neurodegenerative pathology, one might propose that drugs capable of inhibiting neurogenesis, for example, 5-LOX inhibitors, should not be used as neuroprotective therapies.

From Dr. Abbracchio

You mentioned the fact that glucocorticoids increase the expression of 5-LOX. Do they have a similar effect on COX_2 as well? Could you comment on this effect?

ANSWER: Initially, we observed the stimulatory effect of glucocorticoids on 5-LOX expression in the rat brain *in vivo* (T. Uz, et al., 1999. Glucocorticoids stimulate inflammatory 5-lipoxygenase gene expression and protein translocation in the brain. J. Neurochem. **73:** 696–699). Recently, we confirmed these findings in the *in vitro* studies with neuronal cultures and found that the stimulatory effect of dexamethasone on 5-LOX requires the involvement of glucocorticoid receptors (Uz and Manev, in preparation). We did not measure the expression of COX_2, but others have found that glucocorticoids inhibit rather than stimulate the expression of this enzyme in CNS (J. Koistinaho, *et al.*, 1999. Expression of cyclooxygenase-2 mRNA after global ischemia is regulated by AMPA receptors and glucocorticoids. Stroke **30:** 1900–1905).

You also raised the fascinating hypothesis that 5-LOX may have biological effects which are fully independent of its enzymatic activity (i.e., fully independent of eicosamoids synthesis). Could 5-LOX act as an early gene? Do you have any ideas as to which genes that may be replicated by 5-LOX?

ANSWER: There are several possibilities for 5-LOX to act as a regulatory protein including its putative interactions with the nucleus, the cytoskeleton, and the signaling system of tyrosine kinases (for a review see H. Manev, *et al.*, 2000. Putative role of neuronal 5-lipoxygenase in an aging brain. FASEB J. **14**:1464–1469).

I was intrigued by the possible role of 5-LOX in neurogenesis. Has anybody looked at 5-LOX expression in neuroblasts of either the subventricular area or hippocampus of adult animals that have undergone an ischemic or traumatic insult?

ANSWER: To my knowledge, there are no published reports on 5-LOX expression in neuroblasts or in the proliferating brain cells after trauma or ischemia.

The finding that antidepressants increase neurogenesis is the brain is amazing. Are there any epidemiological studies demonstrating that chronic treatment with anti-depressants somehow protects against stroke induced neurodegeneration? Is the incidence of associated neurodegenerative disease somehow decreased in these patients?

ANSWER: In our recent report (H. Manev, *et al.*, 2001. Antidepressants alter cell proliferation in the adult brain *in vivo* and in neural cultures *in vitro*. Eur. J. Pharmacol. **411:** 67–70) we point out that antidepressants affect cell proliferation both in the adult brain *in vivo* and in neuronal cultures *in vitro*. Whether the effect of antidepressants on adult neurogenesis is therapeutically relevant, or, maybe, responsible for side effects should be investigated.

QUESTION FOR DR. LIN

From Dr. Youdim

I am really surprised that you did not see damage in striatal GSH in your MPTP treated mice. Could you explain this in light of previous publications?

ANSWER: MPTP has been widely used to induce Parkinsonian-like syndromes in several animal modes, including mice and monkeys. Several studies have reported that systemic MPTP deceased GSH content in SN in mice (Ferraro *et al.*, 1986; Hung and Lee, 1998). Due to the limited mass of SN, our study did not show the effect of MPTP on GSH content in SN. In terms of MPTP effect on striatal GSH content, our data have found that MPTP did not alter striatal GSH content in ICR mice. Several studies support our data in that striatal GSH content was not significantly affected by systemic MPTP in mice (Ferraro *et al.*, 1986; Hung and Lee, 1998) and monkeys (Gerlach *et al.*, 1996).

Modulation of Cyclooxygenase-2 and Brain Reactive Astrogliosis by Purinergic P2 Receptors

ROBERTA BRAMBILLA AND MARIA P. ABBRACCHIO

Department of Pharmacological Sciences, University of Milan, Italy

ABSTRACT: Astroglial cells respond to trauma and ischemia with reactive gliosis, a reaction characterized by increased astrocytic proliferation and hypertrophy. Although beneficial to a certain extent, excessive gliosis may be detrimental, contributing to neuronal death in neurodegenerative diseases. We have tested the hypothesis that ATP may act as a trigger of reactive gliosis in an *in vitro* model (rat brain primary astrocytes) where reactive astrogliosis can be quantified as elongation of astrocytic processes. Challenge of cells with the ATP analog α,βmethyleneATP (α,βmeATP) resulted in concentration dependent elongation of astrocytic processes, an effect that was fully counteracted by the non-selective ATP/P2 receptor antagonists suramin and pyridoxalphosphate-6-azophenyl-2',4'-disulphonic acid (PPADS). Signalling studies revealed that α,βmeATP-induced gliosis is mediated by a novel G-protein-coupled receptor (a P2Y receptor) coupled to an early release of arachidonic acid. Challenge of cells with α,βmeATP also resulted in an increase of inducible cyclooxygenase-2 (COX-2), the activity of which has been reported to be pathologically increased in a variety of neurodegenerative diseases characterized by inflammation and astrocytic activation. Induction of COX-2 by α,βmeATP was causally related to reactive astrogliosis, since the selective COX-2 inhibitor NS-398 prevented both the purine-induced elongation of astrocytic processes and the associated COX-2 increase. Preliminary data on the putative receptor-to-nucleus pathways responsible for purine-induced gliosis suggest that induction of the COX-2 gene may occur through the protein kinase C/mitogen activated protein kinase system, and may involve the formation of activated AP-1 transcription complexes. We speculate that antagonists selective at this novel P2Y receptor subtype may represent a novel class of neuroprotective agents able to slow down neurodegeneration by counteracting the inflammatory events contributing to neuronal cell death.

KEYWORDS: ATP; Inflammatory gliosis; Neurodegenerative diseases.

INTRODUCTION

As well as its well known role in cellular metabolism, ATP has been recognized to play a crucial role in neurotransmission[1,2] and to act as a powerful signal for secretory exocrine and endocrine cells (for a review see Ref. 3). In both central and peripheral nervous system, ATP is co-stored with "classical" neurotransmitters in presynaptic vesicles, from which it is released on depolarization, and mediates a variety of different functions by activating specific postsynaptic membrane receptors.[3] It has been also recognized that, in a similar way to "classical" neurotransmitters, such as acetylcholine (Ach), GABA, glutamate, and serotonin (5-HT), the effects of extracellular ATP are mediated by two distinct receptor classes, ligand gated ion channels (P2X receptors) and G-protein coupled receptors (P2Y receptors) (see TABLE 1).[3–6] It is believed that P2X receptors mediate the fast ionotropic responses to ATP, whereas P2Y receptors are responsible for the slower "metabotropic" effects induced by this nucleotide. To date, at least seven distinct P2X receptors and 12 P2Y receptors have been cloned from a variety of different tissues (TABLE 1).

Both P2X and P2Y receptors are present in mammalian brain.[7] Although P2X receptors have been implicated in fast neuro–neuronal communication,[8] P2Y receptors have been mainly found on astroglial cells.[7] Following traumatic and hypoxic insults to the brain, these cells undergo profound changes, including increased proliferation, phenotypic changes, and cellular hypertrophy.[9,10] The latter is characterized by increased cellular size, increased expression of the specific astroglial protein glial fibrillary acidic protein (GFAP) and emission of long GFAP positive cellular processes.[9–12] There is still debate about the functional significance of this reaction. Activated (reactive) astrocytes acquire the capability of synthesizing and releasing neurotrophins and pleiotrophins,[13] which may be crucial for regeneration and functional recovery. Moreover, the morphological changes typical of reactive astrocytes

TABLE 1. Ionotropic and metabotropic receptor families and subtypes for several neurotransmitters, including ATP

Neurotransmitter	Receptors	
	Fast Ionotropic (Ligand-gated Channels)	Slow Metabotropic (G Protein coupled)
ACh	nicotinic muscle type neuronal type	muscarinic $M_1 \rightarrow M_2$
Glutamate	AMPA kainate NMDA	$MGlu_1 \rightarrow MGlu_7$
GABA	$GABA_A$	$GABA_B$
5-HT	$5\text{-}HT_3$	$5\text{-}HT_{1A\text{-}F}$ $5\text{-}HT_{2A\text{-}C}$ $5\text{-}HT_4$ $5\text{-}HT_{5A\text{-}B}$ $5\text{-}HT_6$ $5\text{-}HT_7$
ATP	P2X $P2X_1 \rightarrow P2X_7$	P2Y $P2Y_1 \rightarrow P2Y_{12}$

(see above) have been hypothesized to contribute to isolating the injured areas from recovering ones, hence preventing the propagation of the necrotic damage from the ischemic core to the penumbra (see also Ref. 14). However, activated astrocytes also release a variety of potentially toxic metabolites, including nitric oxide and arachidonic acid metabolites. Moreover, the formation of the gliotic scar, although initially beneficial, may eventually prevent the regeneration of damaged neurites. Hence, excessive gliosis, a feature of several acute and chronic neurodegenerative diseases that are characterized by a marked inflammatory component,[15-18] may be detrimental, and may even contribute to brain damage. This raises the need for identifying the endogenous factors that regulate reactive astrogliosis, in order to develop novel pharmacological strategies to control this phenomenon.

In our laboratory we have recently developed an *in vitro* model of reactive astrogliosis (primary astrocytes obtained from either the corpus striatum or the cortex of seven-day-old rats), where the morphological signs of astroglial activation resembling *in vivo* gliosis (i.e., elongation of astrocytic processes) can be correlated with changes in the expression and/or activity of proteins putatively involved in this reaction.[12] Astrocytic cultures are initially established in complete medium, and at day two in culture, shifted to serum free medium for additional 24 h, before a brief (1 or 2 h) challenge with potential "gliotic" factors. After challenge, cultures are washed, placed in drug free medium and maintained in culture up to day five, when they are fixed, stained with an anti-GFAP antibody and analyzed by fluorescence microscopy. Analysis is performed by quantifying the mean length of GFAP positive astrocytic processes by means of a computerized image analysis system. The morphological analysis is paralleled by biochemical evaluations performed at different times in culture aimed at assessing the role of specific proteins/enzymes and/or transductional pathways in the induction of reactive gliosis (for more details see also below and Ref. 12). In initial experiments, we have demonstrated that classic gliotic triggers, such as basic fibroblast growth factor (bFGF), induce a marked and concentration dependent elongation of astrocytic processes,[11] hence validating the model for the identification of novel, as yet uncharacterized, gliotic agents.

INVOLVEMENT OF ATP AND P2 RECEPTORS IN REACTIVE ASTROGLIOSIS

ATP Analogs Act as Gliotic Agents

Among the various endogenous factors that have been implicated in the onset and maintenance of reactive astrogliosis, ATP and related nucleotides play a major role. Massive amounts of ATP are released in hypoxic or injured brain areas, as a consequence of both release from the cytosolic ATP pool, and as a result of nucleic acid degradation from dying cells. This leads to a notable elevation of local ATP concentrations that can last for days (or even weeks) after the toxic stimulus.[13,19,20] Moreover, as emphasized above, astroglial cells do express P2 receptors, mainly belonging to the P2Y class. On this basis, we challenged astrocytic cultures with various hydrolysis resistant ATP analogs and found that, in a similar way to bFGF, they were able to induce a marked concentration dependent elongation of GFAP positive astrocytic processes.[11] Among the various analogs tested, α,βmethyleneATP

(α,βmeATP) was equipotent with the natural nucleotide ATP in inducing the formation of reactive astrocytes, and was hence used for all the subsequent experiments. A specific role for P2 receptors was suggested by the fact that two non selective P2 receptor antagonists such as suramin and pyridoxalphosphate-6-azophenyl-2',4'-disulphonic acid (PPADS) could completely prevent the astrocytic elongation induced by α,βmeATP.

ATP Induced Gliosis is Mediated by a Novel P2 Receptor

To assess the nature of the P2 receptor involved in the effect described above, we performed experiments with pertussis toxin, an inhibitor of G-proteins. Pretreatment with pertussis toxin fully prevented α,βmeATP-induced effects,[11] suggesting that a G-protein coupled P2Y receptor may be responsible for α,βmeATP induced gliosis. This was further confirmed by the demonstration that oxidized ATP, a P2X receptor antagonist, did not modify α,βmeATP induced effects,[21] hence ruling out a role for ligand gated receptors.

Hence, the coupling of the novel gliotic P2 receptor was characterized. This receptor did not seem to share any of the transduction mechanisms of the already known P2Y receptors that have been demonstrated to either stimulate phospholipase C or inhibit adenylyl cyclase (for a review see Ref. 22). Activation of this receptor by α,βmeATP did not result in either changes in intracellular Ca^{2+} concentrations,[11] which ruled out the involvement of phospholipase C, or modulation of adenylyl cyclase activity,[22] hence ruling out a role for cAMP. Activation of this receptor resulted in an early release of arachidonic acid, suggesting coupling to phospholipase A_2.[22] The importance of this transduction pathway in α,βmeATP induced effects was further confirmed by the fact that the phospholipase A_2 inhibitor mepacrine could both prevent the purine induced arachidonic acid release and the associated elongation of astrocytic processes. Moreover, challenge of cells with arachidonic acid mimicked the gliotic effects induced by α,βmeATP.[11]

The possible relevance of the effects described above to human diseases characterized by chronic reactive astrogliosis is supported by the demonstration that exposure of human cells of the astroglial lineage (ADF astrocytoma cells) to α,βmeATP also results in marked concentration dependent elongation of astrocytic processes (see FIGURE 1).[21] Furthermore, on the human cells, α,βmeATP-induced gliosis could be completely prevented by the concomitant exposure to the P2 receptor antagonist suramin.[21]

The peculiar transduction pathway associated with the gliotic receptor, together with its rather atypical pharmacological profile (high sensitivity to α,βmeATP, which is quite unusual for a P2Y receptor) suggests that we may be dealing with a novel P2Y receptor subtype specifically associated with the induction of reactive gliosis.

Induction of Cyclooxygenase-2 is a Prerequisite for ATP-Induced Gliosis

After demonstrating that activation of the novel gliotic P2 receptor is followed by rapid release of arachidonic acid, we focused our attention on cyclooxygenases (COXs). This was justified by several experimental data. First, arachidonic acid is a substrate for both constitutive COX-1 and inducible, mitogen-activated COX-2.[23]

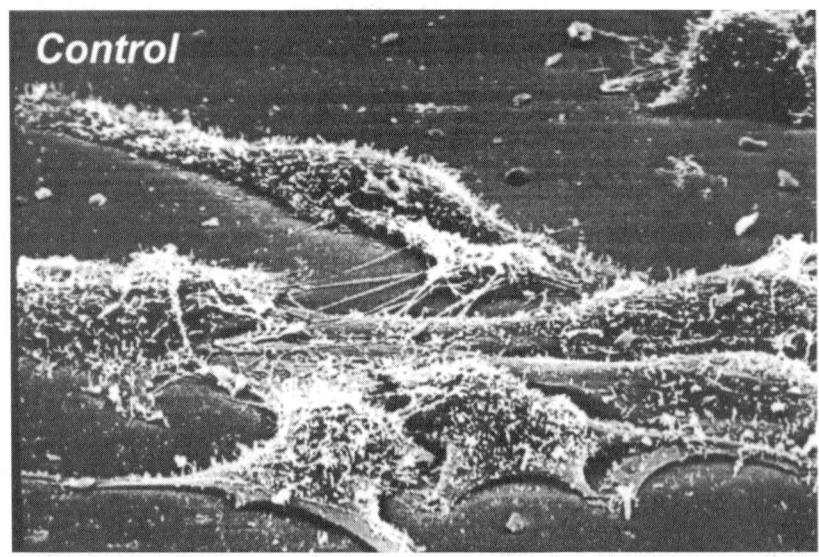
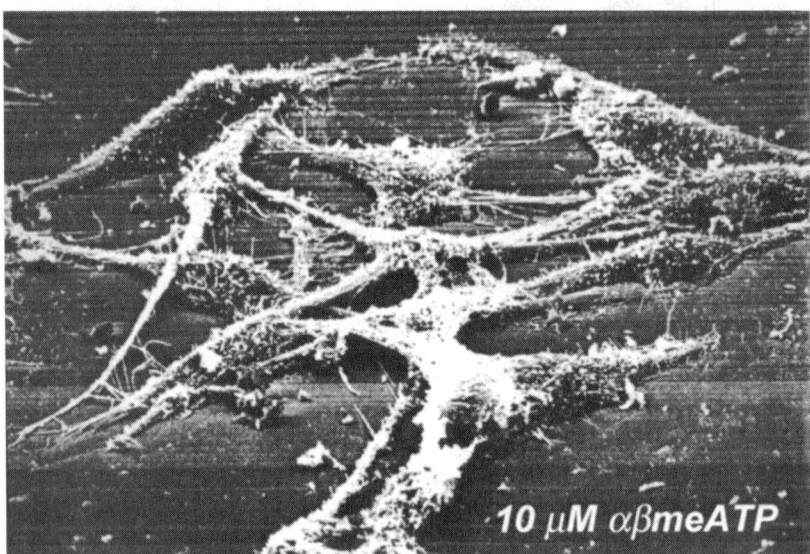

FIGURE 1. Scanning electron microscopy of ADF astrocytoma cells. Note the marked elongation of astrocytic processes on challenge of cells with α,βmeATP. Reproduced, with permission, from Reference 21.

Second, besides acting as a substrate for COX-2 in the synthesis of prostanoids (mainly PGE_2 in astrocytes[18]) arachidonic acid (or its metabolites) have been reported to induce the transcription of the COX-2 gene,[24–26] likely via the protein kinase C/mitogen activated kinase (MAPK) pathway.[13,24] Third, excessive and chronic COX-2 induction has been reported for various neurological disorders, including chronic pain and inflammation,[16] acute (e.g., ischemia[15,27]) as well as chronic (e.g.,

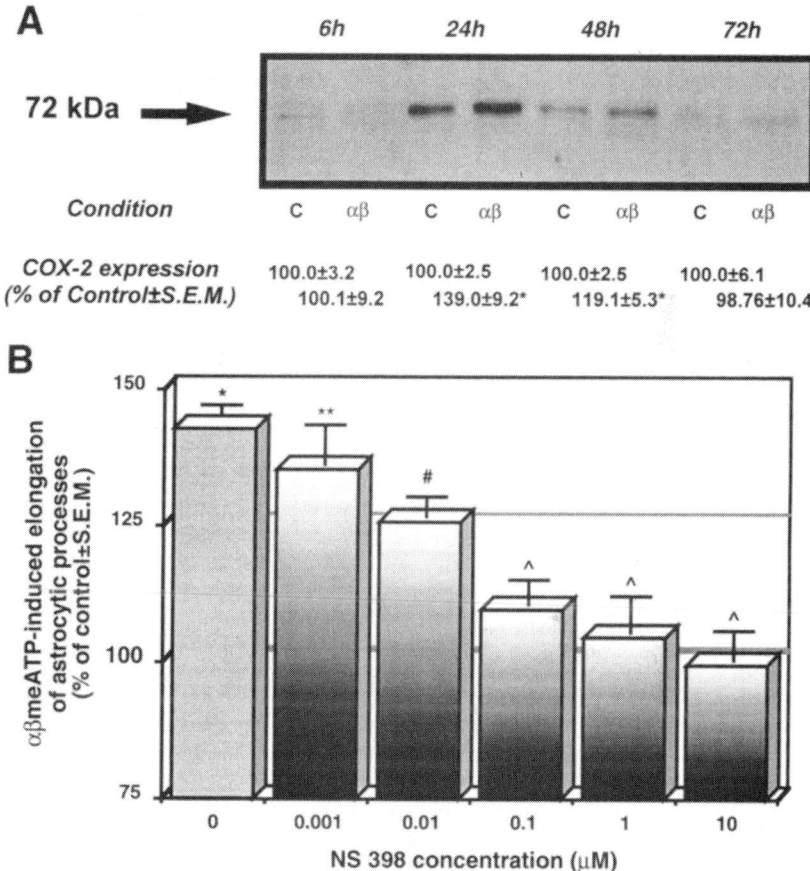

FIGURE 2. Involvement of COX-2 in P2Y receptor mediated reactive astrogliosis. (**A**) Immunoblot analysis of protein lysates obtained at various times from rat primary astrocytes challenged with α,βmeATP for two hours. COX-2 protein expression was significantly increased at 24 and 48 h (*$p < 0.05$, one-way ANOVA, Scheffe's F-test). (**B**) Immunofluorescence analysis of GFAP positive astrocytic processes after a two-hour exposure to α,βmeATP. α,βmeATP induced elongation of processes, measured three days after the challenge with the ATP analog, was concentration-dependently prevented by NS-398 (*$p < 0.05$ vs. control; **$p < 0.05$ vs control and not different from α,βmeATP alone; #$p < 0.05$ vs. control and α,βmeATP alone; ^$p < 0.05$ vs. α,βmeATP alone not different from control, one-way ANOVA, Scheffe's F-test). Reproduced from Reference 12.

Alzheimer's disease[17]) neurodegenerative diseases characterized by a marked inflammatory component and activation of astroglial cells.[15–18] All this evidence prompted us to verify whether astrocytic elongation by α,βmeATP may occur through induction of COX-2. Challenge of astrocytes with α,βmeATP indeed resulted in an increase of the COX-2 protein, as demonstrated by the Western blot analysis shown in FIGURE 2, with a maximal increase 24 h after challenge. This effect could also be prevented by exposure of cultures to non selective P2 receptor antagonists.[21] Moreover, the selective COX-2 inhibitor NS-398 completely abolished α,βmeATP induced astrocytic activation, suggesting that induction of COX-2 in our experimental model is not an epiphenomenon, but is causally related to the development of reactive astrogliosis (FIG. 2).

CONCLUSIONS

The data presented above support the involvement of ATP in the induction and maintenance of reactive astrogliosis, and suggest that this may occur via activation of a novel P2Y receptor subtype. Our data also demonstrate that ATP induced gliosis is mediated by induction of COX-2, a finding that may have intriguing implications for neurodegenerative diseases, where excessive COX-2 activation and reactive gliosis play a major role. We can speculate that antagonists selective for the gliotic P2 receptor may represent a novel class of antineurodegenerative agents of potential interest in both acute and chronic neurological diseases. By reducing the extent of gliosis evoked by ATP, such antagonists may indeed slow down the progression of these diseases by counteracting the inflammatory events contributing to neurodegeneration. The hope of developing selective antagonists reasonably devoid of side effects is supported by the quite peculiar and atypical pharmacological profile of this receptor, which makes it different from all the other known P2 receptors. The cloning of this novel receptor is expected to represent an important step in our understanding of the gliotic effects induced by ATP, and will indeed represent the *sine qua non* condition to synthesize and develop selective ligands for this receptor subtype.

We are currently working on the elucidation of the transductional mechanisms responsible for purine induced gliosis. An intracellular pathway that may possibly be involved is represented by the MAPK system, and the AP-1 transcription complex, both of which can promote transcription of the COX-2 gene. The involvement of this transductional cascade is supported by the following evidence: (1) As mentioned above, stimulation of the novel gliotic receptor results in an early release of arachidonic acid, which may then activate an arachidonate sensitive protein kinase C. We previously demonstrated that activators of protein kinase C (i.e., phorbol esters) can mimic the gliotic response induced by purine analogs in our model, and, conversely, inhibitors of this enzyme (i.e., H7) completely abolish α,β-meATP induced effects.[11] (2) Protein kinase C participates to the activation of MAP kinases,[13] which in turn can directly phosphorylate and activate the AP-1 complex or promote the transcription of c-fos and c-jun—the main components of the AP-1 complex.[28] The involvement of the latter in purine-induced gliosis is suggested by previous data from our laboratory demonstrating that, one–two hours after challenge with α,βmeATP, astrocytes show a marked nuclear accumulation of Fos and Jun proteins that can hence dimerize to form the AP-1 complex.[11] (3) There is an AP-1 consensus sequence on the COX-2 promoter.[29]

To further substantiate this hypothesis, we are currently evaluating the activity of MAP kinases, in particular the ERK1/ERK2 cascade, that participate in c-fos and c-jun induction, and hence AP-1 formation.[28] The involvement of COX-2 will be confirmed by reverse transcription polymerase chain reaction (RT-PCR) using specific primers for the COX-2 gene. Moreover, we plan to test the ability of various MAP kinase inhibitors to prevent both α,βmeATP induced gliosis and the associated induction of the COX-2 gene. The increased DNA binding of AP-1 complexes, as well as of other transcription factors, will be determined by electrophoretic mobility shift assay (EMSA) in nuclear extracts obtained from α,βmeATP treated cells. We envisage that the characterization of the receptor–to–nucleus cascade responsible for ATP induced gliosis will disclose additional molecular targets for the pharmacological modulation of reactive astrogliosis, and hence contribute to the identification of novel therapeutic strategies to neurodegenerative diseases.

AKNOWLEDGMENTS

The authors are grateful to Prof. Geoffrey Burnstock (Autonomic Neuroscience Institute, London, U.K.), Prof. Flaminio Cattabeni (University of Milan, Milan, Italy), Dr. Walter Malorni (Istituto Superiore di Sanità, Rome, Italy), and to Dr. Joseph T. Neary, Dr. Paul Schiller, and Dr. Gianluca D'Ippolito (University of Miami, FL) for useful discussion. This work was partially supported by the European Union BIOMED 2 Programme BMH4 CT96-0676 to MPA. RB is the recipient of a fellowhip issued by Searle Farmaceutici in collaboration with the Italian Pharmacological Society on: "Caratterizzazione di un nuovo recettore P2Y per ATP associato ad induzione di cicloossigenasi-2 e a gliosi reattiva."

REFERENCES

1. BURNSTOCK, G. 1972. Purinergic nerves. Pharmacol. Rev. **24**: 509–581.
2. BURNSTOCK, G. 1978. A basis for distinguishing two types of purinergic receptors. In Cell Membrane Receptors for Drugs and Hormones: A Multidisciplinary Approach. R.W. Straub & L. Bolis, Eds.: 107–118. Raven Press, New York.
3. ABBRACCHIO, M.P. & G. BURNSTOCK. 1998. Purinergic signalling: pathophysiological roles. Jpn. J. Pharmacol. **78**: 113–145.
4. DUBYAK, G.R. 1991. Signal transduction by P2-purinergic receptors for extracellular ATP. Am. J. Respir. Cell Biol. **4**: 295–300.
5. ABBRACCHIO, M.P. & G. BURNSTOCK. 1994. Purinoceptors: are there families of P2X and P2Y purinoceptors? Pharmacol. Ther. **64**: 445–475.
6. RALEVIC, V. & G. BURNSTOCK. 1998. Receptors for purines and pyrimidines. Pharm. Rev. **50**: 413–492.
7. ABBRACCHIO, M.P. 1997. ATP in brain function. In Purinergic Approaches in Experimental Therapeutics. K.A. Jacobson & M.F. Jarvis, Eds.: 383–404. Wiley & Sons, New York.
8. NORTH, R.A. 1996. P2X purinoceptors plethora. Semin. Neurosci. **8**: 187–194.
9. HATTEN, M.E., R.K.H. LIEM, M.L. SHELANSKI & C.A. MASON. 1991. Astroglia in CNS injury. Glia **4**: 233–243.

10. RIDET, J.L., S.K. MALHOTRA, A. PRIVAT & F.H. GAGE. 1997. Reactive astrocytes: cellular and molecular cues to biological function. Trends Neurosci. **20:** 250–257.
11. BOLEGO, C., S. CERUTI, R. BRAMBILLA, et al. 1997. Characterization of signaling pathways involved in ATP and basic fibroblast growth factor-induced astrogliosis. Br. J. Pharmacol. **121:** 1692–1699.
12. BRAMBILLA, R., G. BURNSTOCK, A. BONAZZI, et al. 1999. Cyclo-oxygenase-2 mediates P2Y-receptor-induced reactive astrogliosis. Br. J. Pharmacol. **126:** 563–567.
13. NEARY J.T., M.P. RATHBONE, F. CATTABENI, et al. 1996. Trophic actions of extracellular nucleotides and nucleosides on glial and neuronal cells. Trends Neurosci. **19:** 13–18.
14. VON LUBITZ, D.K.J.E., W. YE, J. MCCLELLAN & R.C.-S. LIN. 1999. Stimulation of adenosine A_3 receptors in cerebral ischemia. Neuronal death, recovery, or both? Ann. N.Y. Acad. Sci. **890:** 93–106.
15. OHTSUKI, T., K. KITAGAWA, K. MANDAL, et al. 1996. Induction of cyclooxygenase-2 mRNA in gerbil hippocampal neurons after transient forebrain ischemia. Brain Res. **736:** 353–356.
16. DOLAN, S., P.J. O'SHAUGHNESSY & A.M. NOLAN. 1998. Up-regulation of COX-2 and prostaglandin EP3 receptor mRNA expression in spinal cord in a clinical model of chronic inflammation. Eur. J. Neurosci. **10**(Suppl. 10): 77.
17. HO, L., C. PIERONI, D. WINGER, et al. 1999. Regional distribution of cyclooxygenase-2 in the hippocampal formation in Alzheimer's disease. J. Neurosci. Res. **57:** 295–303.
18. BLOM, M.A.A., M.G. VAN TWILLERT, S.C. DE VREIS, et al. 1997. NSAIDs inhibit the IL-β1-induced IL-6 release from human post-mortem astrocytes: the involvement of prostaglandin E_2. Brain Res. **777:** 210–218.
19. NEARY, J.T. & M.P. ABBRACCHIO. 2001. Trophic roles of purines and pyrimidines. In Handbook of Experimental Pharmacology, Purinergic and Pryrimidinergic Signalling. M.P. Abbracchio & M. Williams, Eds.: 305–308. Springer-Verlag, Heidelberg.
20. ABBRACCHIO, M.P & F. CATTABENI. 1999. Brain adenosine receptors as targets for therapeutic intervention in neurodegenerative deseases. Ann. N.Y. Acad. Sci. **890:** 79–92.
21. BRAMBILLA, R., S. CERUTI, W. MALORNI, et al. 2000. A novel gliotic P2 receptor mediating cyclooxygenase-2 induction in rat and human astrocytes. J. Autonom. Nerv. Syst. **81:** 3–9.
22. KING, B.F., G. BURNSTOCK, J.L. BOYER, et al. 2001. The P2Y receptors. In The IUPHAR Compendium of Receptor Characterization and Classification. IUPHAR Media, London. In press.
23. WU, K.H. 1996. Cyclooxygenase 2 induction: molecular mechanisms and pathophysiological roles. J. Lab. Clin. Med. **128:** 242–245.
24. BARRY, O.P., M.G. KAZANIETZ, D. PRATICO' & A.F. GARRET. 1999. Arachidonic acid in platelet microparticles up-regulates cyclooxygenase-2-dependent prostaglandin formation via protein kinase C/mitogen-activated protein kinase-dependent pathway. J. Biol. Chem. **274:** 7545–7556.
25. TORDJMANN, C., F. COGE, N. ANDRE, et al. 1995. Characterization of cyclooxygenase 1 and 2 expression in mouse resident peritoneal macrophages in vitro; interactions of non steroidal anti-inflammatory drugs with COX-2. Biochem. Biophys. Acta **1256:** 249–256.
26. TAKAHASHI, Y., Y. TAKETANI, T. ENDO, et al. 1994. Studies on the induction of cyclooxygenase isoenzymes by various prostaglandins in a mouse osteoblastic cell line with reference to signal transduction pathways. Biochem. Biophys. Acta **1212:** 249–256.
27. PLANAS, A.M., M.A. SORIANO, E. RODRIGUEZ-FERRE' & I. FERRER. 1995. Induction of cyclooxygenase-2 mRNA and protein following transient focal ischemia in the rat brain. Neurosci. Lett. **200:** 187–190.
28. KARIN, M. 1995. The regulation of AP-1 activity by mitogen-activated protein kinase cascade. Procl. Natl. Acad. Sci. U.S.A. **92:** 7686–7689.
29. FLETCHER, B.S., D.A. KUJUBU, D.M. PERRIN & H.R. HERSCHMAN. 1991. Structure of the mitogen-inducible TIS10 gene and demonstration that TIS10-encoded protein is a functional prostaglandin G/H synthase. J. Biol. Chem. **276:** 4338–4344.

The A_3 Adenosine Receptor Induces Cytoskeleton Rearrangement in Human Astrocytoma Cells via a Specific Action on Rho Proteins

MARIA P. ABBRACCHIO,[a] ALESSANDRA CAMURRI,[a] STEFANIA CERUTI,[a] FLAMINIO CATTABENI,[a] LOREDANA FALZANO,[b] ANNA MARIA GIAMMARIOLI,[b] KENNETH A. JACOBSON,[c] LETIZIA TRINCAVELLI,[d] CLAUDIA MARTINI,[d] WALTER MALORNI,[b] AND CARLA FIORENTINI[b]

[a]*Department of Pharmacological Sciences, University of Milan, Milan, Italy*

[b]*Istituto Superiore di Sanità, Rome, Italy*

[c]*Molecular Recognition Section, Laboratory of Bioorganic Chemistry, National Institute of Diabetes, Digestive and Kidney Diseases, National Institutes of Health, Bethesda, Maryland, U.S.A.*

[d]*Department of Psychiatry, Neurobiology, Pharmacology and Biotechnology, University of Pisa, Pisa, Italy*

ABSTRACT: In previous studies, we have demonstrated that exposure of astroglial cells to A_3 adenosine receptor agonists results in dual actions on cell survival, with "trophic" and antiapoptotic effects at nanomolar concentrations and induction of cell death at micromolar agonist concentrations. The protective actions of A_3 agonists have been associated with a reinforcement of the actin cytoskeleton, which likely results in increased resistance of cells to cytotoxic stimuli. The molecular mechanisms at the basis of this effect and the signalling pathway(s) linking the A_3 receptor to the actin cytoskeleton have never been elucidated. Based on previous literature data suggesting that the actin cytoskeleton is controlled by small GTP-binding proteins of the Rho family, in the study reported here we investigated the involvement of these proteins in the effects induced by A_3 agonists on human astrocytoma ADF cells. The presence of the A_3 adenosine receptor in these cells has been confirmed by immunoblotting analysis. As expected, exposure of human astrocytoma ADF cells to nanomolar concentrations of the selective A_3 agonist 2-chloro-N^6-(3-iodobenzyl)-adenosine-5'-N-methyluronamide (Cl-IB-MECA) resulted in formation of thick actin positive stress fibers. Preexposure of cells to the C3B toxin that inactivates Rho-proteins completely prevented the actin changes induced by Cl-IB-MECA. Exposure to the A_3 agonist also resulted in significant reduction of Rho-GDI, an inhibitory protein known to maintain Rho proteins in their inactive state, suggesting a potentiation of Rho-mediated effects. This effect was fully counteracted by the concomitant exposure to the selective A_3 receptor antagonist MRS1191. These results suggest that the reinforcement of the actin cytoskeleton induced by A_3 receptor agonists is mediated by an interference with the activation/inactivation cycle of Rho proteins, which may, therefore, represent a biological target for the identification of novel neuroprotective strategies.

KEYWORDS: Adenosine; A_3 receptor; Neuroprotection; Rho proteins

Address for correspondence: Professor Maria P. Abbracchio, Department of Pharmacological Sciences, University of Milan, Via Balzaretti 9, 20133 Milan, Italy. Voice: +39-02-20488310; fax: +39-02-29404961.
Mariapia.Abbracchio@unimi.it

INTRODUCTION

The actions of adenosine are mediated by four G-protein coupled membrane receptors, the A_1, A_{2A}, A_{2B}, and A_3 receptors.[1] Although the existence of the A_1, A_{2A}, and A_{2B} receptors was postulated before they were cloned, the A_3 receptor was discovered by cloning, initially from rat and subsequently from human tissues.[2] The development of selective agonists, such as N^6-(3-iodobenzyl)-adenosine-5'-N-methyluronamide (IB-MECA)[3] and its 2-chloro-derivative (Cl-IB-MECA),[4] and more recently, antagonists, such as MRS1191 (3-ethyl-5-benzyl-2-methyl-6-phenyl-4-phenylethynyl-1,4-(±)-dihydropyridine-3,5-dicarboxyla-te),[5] and MRS1220 (9-chloro-2-(2-furyl)-5-phenyl-acetylamino-[1,2,4]-triazolo-[1,5-c]-quinazoline)[6] has played a crucial role in defining the putative pathophysiological roles of this receptor.

The A_3 receptor is involved in inflammation,[2] hypotension and mast cell degranulation,[7] ischemic heart preconditioning.[2] Although expressed to quite low levels in the brain,[8] this receptor has also been implicated in behavioral depression[8] and modulation of cerebral ischemic damage.[9] Use of selective A_3 agonists also revealed that this receptor profoundly affects cell survival, by promoting both cell protection and cell death, depending upon the cell type and/or the agonist concentrations. At low concentrations in the nanomolar range, A_3 agonists reduce hypoxic heart damage[10] and protect HL-60, U-937 cells, and mammalian astrocytes from apoptosis.[11,12] In astroglial cells, increased resistance to apoptosis was associated with a reinforcement of the actin cytoskeleton and with the intracellular redistribution of the antiapoptotic protein Bcl-x_L.[12,13] At high concentrations in the micromolar range, these same agonists markedly impaired cell cycle progression[14] and induced death of cerebellar granule neurons,[15] HL-60 cells,[16] human lymphocytes,[17] and astroglial cells.[12] These dual actions confirm the concept that adenosine may represent a signal of both life and death for its target cells, simply depending on specific pathophysiological conditions.[18]

A hypothesis that may reconcile the opposite effects of adenosine on brain cell survival and give them a pathophysiological significance has recently been raised.[19] In traumatic and ischemic brain, large amounts of adenosine are released from nucleic acids of dying cells as a result of increased neurotransmitter release and breakdown of nucleotides and nucleosides.[20] Hence, concentrations of adenosine in the ischemic brain are believed to depend on the extent of cellular damage; that is, to attain their highest values within the ischemic "core" area and to gradually decrease progressively in the direction of the "penumbra" area.[19] Hence, within the core itself, adenosine is likely to reach levels that fully activate the A_3 receptor, resulting in the full range of its destructive effects to favour the elimination of irreversibly damaged cells and to save space and energy for those cells that retain the ability to recover.[19] In contrast, in the penumbral area, the progressively lower concentrations of adenosine would lead to a milder, "subthreshold", stimulation of the A_3 receptor, to result in more benign actions, such as increased resistance of cells to stress, antiapoptotic effects, and recovery of damaged neurons via a potentiation of the astrocytic support to these cells.[19]

The molecular mechanisms underlying the protective effects mediated by the A_3 receptor remain obscure. As mentioned above, in human astrocytoma cells, the cytoskeletal rearrangement induced by nanomolar A_3 agonist concentrations is characterized by a marked increase of F-actin stress fibers.[12,13] Since actin polymerization is crucially controlled by small GTP binding proteins of the Rho family,[21] in the study reported here we tested whether the cytoprotective effects induced by A_3 agonists may be mediated by a specific action on this system.

Under basal, unstimulated conditions, Rho proteins are found in the cytoplasm, where they are maintained in their inactive GDP bound state by guanine nucleotide dissociation inhibitors (Rho-GDIs)[21] (see FIGURE 1). Exchange of GDP with GTP results in release of Rho-GDI, binding of Rho to a family of stimulatory proteins (Rho-GDS) and migration of the activated Rho protein to the membrane (FIG. 1).

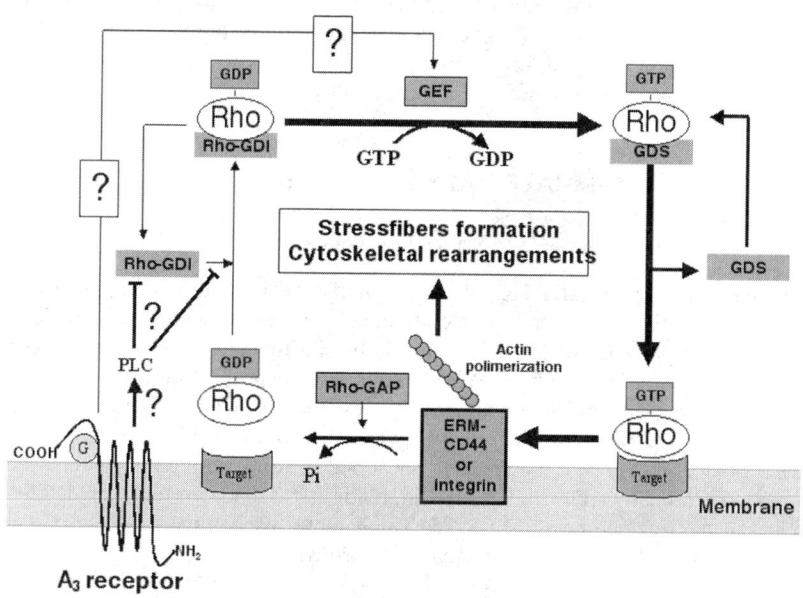

FIGURE 1. Schematic representation of the putative relationship between the adenosine A_3 receptor and Rho proteins. Under basal, unstimulated conditions, Rho proteins are maintained in their GDP bound inactive state by Rho-GDI. Stimulation of GDP exchange with GTP via guanine nucleotide exchange factors (GEF) results in release of Rho-GDI, binding of Rho to Rho-GDS and migration of the activated Rho protein to the membrane. Here, Rho can interact with its targets (mainly kinases) that in turn activate a number of proteins, including integrins, CD44, and proteins involved in regulation of the actin cytoskeleton, such as ERM (Ezrin/Radixin/Moesin). Interaction with actin promotes its polymerization and the formation of F-actin stress fibers (as an example, see FIG. 3 b). Return of Rho protein to its inactive state is favoured by proteins that promote the hydrolysis of GTP to GDP (Rho-GAPs) and binding to Rho-GDI. These results suggest that stimulation of the A_3 receptor, likely through PLC,[24] can influence the activation/inactivation cycle of Rho proteins. This could be achieved by either a direct stimulation of GEFs and/or by modulation of Rho-GDI. Data show that exposure to A_3 agonists reduces Rho-GDI availability, by either reducing its expression or by promoting conformational changes that decrease its ability to bind to Rho. This would lead to a potentiation of Rho-mediated effects (*thick arrows*). See text for further details.

Here, activated Rho can interact with a variety of different targets, mainly kinases, that in turn activate a number of proteins, including those regulating the actin cytoskeleton, such as specific membrane associated proteins (e.g., ERM, Ezrin/Radixin/Moesin).[22] As a result of such interactions, actin fibers organize to form filamentous structures, termed stress fibers. Return of Rho proteins to their inactive state is favored by proteins that promote the hydrolysis of GTP to GDP (Rho-GAPs) and binding to Rho-GDI (FIG. 1). The activation/inactivation cycle of Rho proteins is crucially regulated by guanine nucleotide exchange factors (GEFs) that can in turn be activated by several signals merging on Rho proteins. Rho links membrane receptors, activated by extracellular factors such as lysophosphatidic acid (LPA), bombesin, or thrombin, to the formation of actin stress fibers and focal adhesion contacts,[22] through various signalling pathways, including tyrosine kinases, cAMP and phospholipaseC/protein kinase.[23] It is presently unknown whether activation of the A_3 receptor can result in signalling to Rho proteins. However, this receptor has been previously shown to activate phospholipase C (PLC) in the brain.[24]

MATERIALS AND METHODS

Cell Culture and Treatment

Human astrocytoma ADF cells were grown at 37°C in humidified atmosphere as previously described.[25] After 24 h in culture, cells were exposed to Cl-IB-MECA in the absence or presence of MRS 1191 for 48 h prior to analysis. In selected experiments, before addition of Cl-IB-MECA, cells were exposed to the chimeric toxin C3B (0.5 microg/ml) for three hour, and then the toxin was maintained in the medium for the remainder of the experiment (which lasted 48 h). This toxin consists of the C3 isoenzyme linked to the binding portion of the diphtheria toxin, and has been shown to specifically ADP-ribosylate Rho, rendering it inactive (for more details on the effects of C3B, see Refs. 23 and 26) The A_3 selective agonist Cl-IB-MECA and antagonist MRS1191 were synthesized as described elsewhere.[4,5]

Analytical Cytology

For actin staining, ADF cells were fixed with 3.7% formaldehyde in PBS (pH 7.4) for 20 min at room temperature. After washing in the same buffer, cells were permeabilized with 0.5% Triton X-100 (Sigma) in PBS for five minute at room temperature. Cells were stained with fluorescein–phalloidin (Sigma) at 37°C for 30 min. Finally, after washing, all the samples were mounted with glycerol-PBS (2:1) and observed with a Nikon Microphot fluorescence microscope.

Western Blot Analysis

Detection of the A_3 Adenosine Receptor

Cells at subconfluency were washed, scraped and incubated in lysis buffer (9.1 mM $Na_2H_2PO_4$; 1.7 mM Na_2HPO_4; 150 mM NaCl, pH 7.4; 0.5% sodium deoxycholate; 1% Nonidet P-40; 0.1% SDS, containing proteinase inhibitors) for 60 min at 4°C. After centrifugation, proteins were assayed in the soluble fraction and lysates

(1 mg) incubated overnight at 4°C with an antibody raised against the human A_3 receptor (Alfa-Diagnostic, San Antonio, TX, U.S.A.; 4 µg/ml). The immunocomplex was precipitated with Protein-A sepharose (50 µg) for two hours at 4°C. The beads were washed with buffer (150 mM NaCl, 10 mM Tris, 1% NP-40) three times, and bound proteins solubilized in Laemmli buffer. For immunoblotting, equivalent amounts of protein (typically 100 µg/samples) were resolved on 12% (w/v) sodium-dodecylsulphate (SDS) polyacrilamide gels. Resolved proteins were transferred to nitrocellulose and incubated with the primary anti-A_3 receptor antibody (1 µg/ml overnight at 4°C). After extensive washing with TBS (10 mM Tris-HCL, 150 mM NaCl, pH 8) containing 0.05% Tween-20, the nitrocellulose membrane was incubated for 120 min at room temperature with horseradish peroxidase (HRP) goat anti-rabbit conjugated secondary antibody diluted to 1:2,000 in Blotto. After several washes, reactive proteins were visualized by an enhanced chemiluminescence protocol ECL (Amersham Pharmacia Biotech). Immunoblotting was quantified by densitometric scanning of films exposed in the linear range. For these experiments, CHO cells transfected with the human A_3 adenosine receptor cDNA were used as a positive control.[14]

Detection of Rho-GDI

Samples, containing about 20 µg protein each, were then loaded on 11% sodium-dodecylsulphate (SDS) polyacrilamide gels and blotted onto nitrocellulose filters. Filters were then incubated with rabbit polyclonal antibody anti-Rho-GDI (1:4,000, Santa Cruz Biotechnology), followed by a secondary antirabbit antibody (peroxidase conjugated, 1:4000), and reactive proteins visualized and quantified as described above. In selected experiments, changes of Rho-GDI were also confirmed by immunoprecipitation. In this case, samples, containing about 30 µg protein, prepared as described above, were incubated with anti-Rho-GDI polyclonal antibody overnight at 4°C (1µg/sample). Rho-GDI immunoprecipitates were then incubated with protein-A-agarose (Santa Cruz Biotechnology, one hour, room temperature) and the immune complexes were centrifuged. The supernatants were then collected and added of sample buffer (187.5 mM Tris-HCl, 6% SDS, 30% glycerol, 15% B-mercaptoethanol, 0.003% bromophenol blue), whereas the precipitates were washed with PBS and resuspended in water and sample buffer. Samples from both supernatants and precipitates were loaded on 11% sodium-dodecylsulphate polyacrylamide gels and processed as described above.

RESULTS

The presence of the A_3 adenosine receptor in human astrocytoma ADF cells has been confirmed by inmmunoblotting analysis with a specific antibody raised against the human A_3 receptor, and by using CHO cells expressing this receptor subtype to high levels as a positive control.[14] In both CHO and ADF cells, a specific immunoreactive protein band with an apparent molecular weight of 36 kDa corresponding to that of the A_3 receptor[1] could be detected (see FIGURE 2).

As expected, exposure of ADF cells to the A_3 receptor agonist Cl-IB-MECA (100 nM for 48 h) resulted in marked cytoskeletal rearrangement, as demonstrated by the appearance of thick F-actin positive stress fibers (compare FIGURE 3 b with control, FIGURE 3 a). Inhibition of Rho, as a result of exposure of cells to the C3B toxin, provoked retraction of the cell body and breakdown of the actin cytoskeleton (FIG. 3 c). This effect is not associated with a permanent disruption of the actin cytoskeleton, since C3B toxin-treated cells still respond to agents (e.g., cytotoxic necrotizing factor-1[27]) that stimulate actin polymerization. Hence, C3B toxin is widely used to assess the involvement of Rho in cellular function.[23] Addition of Cl-IB-MECA to C3B toxin treated astrocytoma cells could not reproduce the typical actin changes normally induced by the A_3 agonist (compare FIGS. 3 d and 3 b), suggesting that integrity of Rho is needed to mediate such effects. To evaluate whether Cl-IB-MECA can affect the expression of Rho-GDI, we performed Western blot analysis with a specific anti-Rho-GDI antibody. As expected,[28] Rho-GDI was detected as a specific protein band with a molecular weight of 28 kDa (see FIGURE 4 A). A 48-h exposure of cells to 100 nM Cl-IB-MECA resulted in a marked reduction of this protein band with respect to control cells (FIG. 4 A). A specific role for the A_3 receptor in this effect was demonstrated by the ability of the selective A_3 antagonist MRS1191 to fully prevent Cl-IB-MECA-induced reduction of the 28-kDa protein band (FIG. 4 A).

These data have also been confirmed by an immunoprecipation technique. Cell homogenates have been incubated with the anti-Rho-GDI antibody, and the immunocomplex quantified in precipitates. The results indicated in FIGURE 3 B show that,

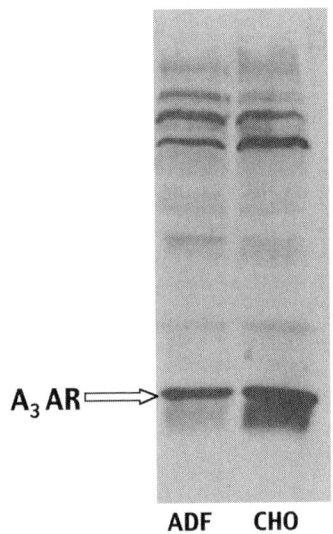

FIGURE 2. Detection of the A_3 adenosine receptor (A_3AR) by immunoblotting analysis in human astrocytoma ADF cells. After cell lysis, the A_3 receptor was detected by immunoprecipitation with a specific antibody and Protein A-sepharose, followed by protein resolution on 11% SDS polyacrilamide gels as described in MATERIALS AND METHODS. CHO cells transfected with the human A_3 receptor cDNA were used as a positive control.[14] Under such conditions, the A_3 receptor can be detected in both ADF and CHO cells as a specific immunoreactive protein band with an apparent molecular weight of 36 kDa (*arrow*).

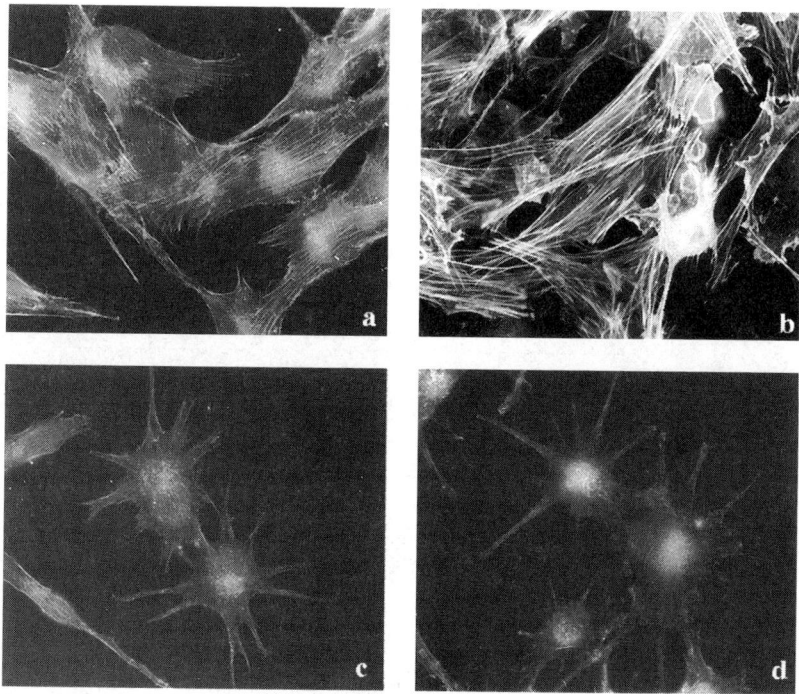

FIGURE 3. Inhibition of Rho prevented Cl-IB-MECA-induced reinforcement of the actin cytoskeleton. In comparison to control cells (**a**), exposure of ADF cells to 100 nM Cl-IB-MECA for 48 h (**b**) resulted in marked formation of thick actin stress fibers. Exposure of ADF cells to C3B toxin (**c**) provoked the retraction of the cell body and the breakdown of the actin cytoskeleton. Pretreatment of cells with C3B toxin for three hours before exposure to Cl-IB-MECA (**d**) fully prevented the actin changes induced by the A_3 agonist.

under the experimental conditions used (see MATERIALS AND METHODS), immunoprecipitation was complete, with no detectable residual immunoreactivity for Rho-GDI in the supernatants of both control and treated cells. A notable reduction of the amount of Rho-GDI was also detected with this methodology after exposure of cells to Cl-IB-MECA (FIG. 4 B), hence confirming the results described in FIGURE 4 A.

DISCUSSION

The adenosine A_3 receptor has been reported to be expressed at low levels in brain-derived tissues.[1,8] For this reason, we deemed it important to confirm its presence in the human astrocytoma cells used in this study. Immunoblotting analysis performed with a specific antibody raised against the human A_3 receptor confirmed that these cells do indeed express the receptor protein to significant levels, hence validating this experimental model for studying the functional effects mediated by this receptor in cells of the astroglial lineage. The present data also demonstrate for the first time that activation of the adenosine A_3 receptor results in signaling to the small G-protein Rho, possibly through the phospholipase C pathway.[24] Moreover, they suggest that the previously reported reinforcement of the actin cytoskeleton on exposure to selective A_3 agonists[12,13] is mediated by an interference with the activa-

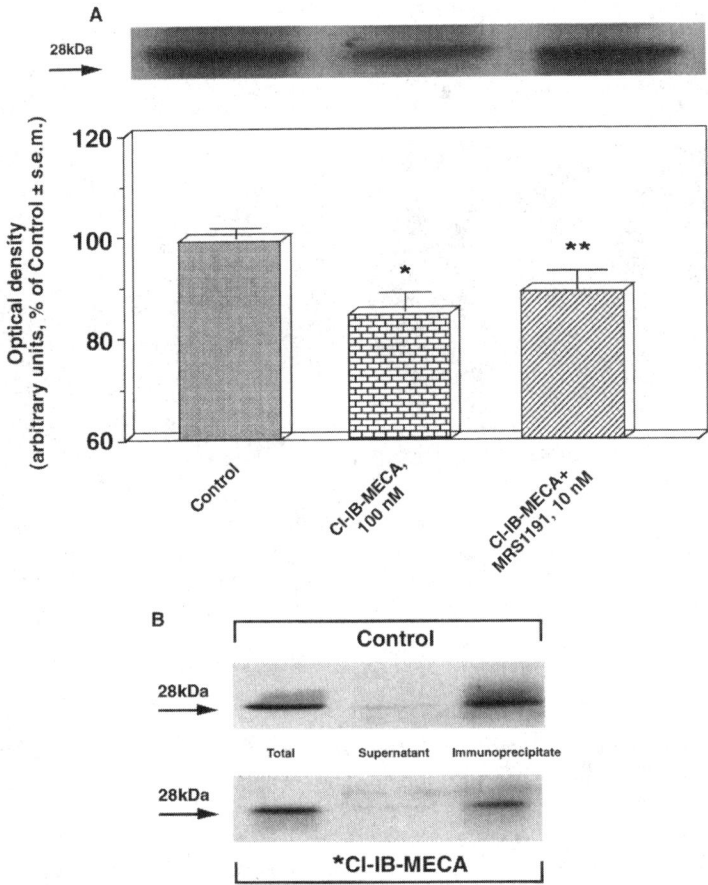

FIGURE 4. Reduction of Rho-GDI in cultures exposed to Cl-IB-MECA. Cells were exposed to 100 nM Cl-IB-MECA for 48 h in the absence or presence of MRS1191, as indicated. For the Rho-GDI immunoblotting analysis shown in **A**, cells were lysed, homogenized, proteins separated by SDS-PAGE, and Rho-GDI identified by immunoblot analysis as described in MATERIALS AND METHODS. For the Rho-GDI immunoprecipitation experiments shown in **B**, cell homogenates were incubated with the anti-Rho-GDI antibody overnight, and immune complexes separated by centrifugation as described in MATERIALS AND METHODS. Thus, SDS-PAGE and immunoblot analysis were performed on both supernatants and precipitates. $*p < 0.05$ with respect to control, $**p < 0.05$ with respect to control, and Cl-IB-MECA alone, one way ANOVA (Fisher test).

tion/inactivation cycle of Rho-proteins. This conclusion is based on the following evidence: (1) inhibition of Rho with the specific C3B toxin completely abolishes the actin changes induced by the A_3 selective agonist Cl-IB-MECA; and (2) exposure of cells to Cl-IB-MECA under experimental conditions that are associated with cell protection (i.e., 100 nM for 48 h) resulted in a reduction of Rho-GDI, as demonstrated by both standard Western blot analysis and by immunoprecipitation. We do not know at present whether the detected reduction of Rho-GDI is a result of diminished protein synthesis, of increased protein turn over, and/or induction of conformational changes that decrease its ability to bind to both Rho (FIG. 1) and to the anti-Rho-GDI antibody. Experiments specifically aimed at quantifying the expression of Rho-GDI (i.e., reverse transcriptase polymerase chain reaction) will enable us to shed light on the mechanisms underlying this effect. However, this result indirectly suggests a potentiation of Rho mediated effects. A lower availability of the inhibitory protein necessary to keep Rho at a basal, unstimulated state may in fact result in a higher percentage of GDS bound Rho, which in turn would lead to increased Rho activation and stimulation of actin polymerization (FIG. 1). To confirm this hypothesis, we plan to used the immunoprecipitation technique to quantify GDI-bound Rho proteins in immunoprecipitates after incubation of cell homogenates with the anti-Rho-GDI antibody. If stimulation of the A_3 receptor increases the percentage of Rho activation, smaller amounts of Rho proteins are expected to coimmunoprecipitate with Rho-GDI in A_3 agonist-treated cells. We also plan to identify the Rho protein(s) involved in this effect by using antibodies that specifically recognize the various members of this family (i.e., Rho-A, Rho-B, Rho-C, rac1/2, cdc42, etc).[21] Of course, we cannot rule out the possibility that activation of the A_3 receptor leads to a direct stimulation of GEFs, which would directly potentiate Rho (FIG. 1). In this case, the detected reduction of Rho-GDI represents a compensatory mechanism aimed at counteracting the excessive activation of the system.

Our data also imply that Rho activation via a subthreshold stimulation of the A_3 receptor (as that attained by nanomolar agonist concentrations) is associated with cell protection; that is, inhibition of apoptosis.[12] It is not clear at present whether induction of cell death upon a robust activation of this receptor (as that attained with micromolar agonist concentrations) also occurs through modulation of Rho proteins. By using other experimental paradigms,[29] it has been demonstrated that prolonged inhibition of Rho is indeed associated with induction of apoptosis. It has been hypothesized that the dual effects induced by the A_3 receptor may depend on either the state of receptor activation and/or induction of receptor desensitization by high agonist concentrations.[19,20] We suggest that, depending upon the agonist concentrations, the degree of receptor activation may regulate the balance between Rho activation/inactivation, which in turn regulates the susceptibility to cell death.

The elucidation of the molecular mechanisms forming the basis of the protective effects induced by A_3 agonists may have important pathophysiological implications. Besides their established role in cardioprotection[30] and their putative contribution to neuronal recovery in brain ischemic penumbra,[19,20] these effects may also play a role in the development of brain preconditioning, a phenomenon according to which a brief ischemic attack can protect the brain from a subsequent, and lethal, stronger ischemic insult.[31] The demonstration that such protection is mediated by modulation of Rho proteins activity may disclose further biological targets for the identification of novel antiischemic and neuroprotective strategies.

ACKNOWLEDGMENTS

This work was supported by the Italian National Research Council (CNR) Contributo di Ricerca No. 98.01047.CT04 to MPA and by the Ministero dell'Universita' e della Ricerca Scientifica e Tecnologica (MURST), Cofinanziamento di ricerche di interesse nazionale 1999 to FC on "Recettori purinergici e neuroprotezione". The authors are grateful to Dr. Charly Klotz (University of Wurzburg, Germany) for kindly providing CHO cells tranfected with the human A_3 adenosine receptor.

REFERENCES

1. FREDHOLM, B.B., M.P. ABBRACCHIO, G. BURNSTOCK, et al. 1994. Nomenclature and classification of purinoceptors. Pharm. Rev. **46:** 143–156.
2. LINDEN, J. 1994. Cloned adenosine A_3 receptors: pharmacological properties, species-differences and receptor functions. Trends Pharmacol. Sci. **15:** 298–306.
3. GALLO-RODRIGUEZ, C., X.-D. JI, N. MELMAN, et al. 1994. Structure–activity-relationships of N^6-benzyladenosine-5'-uronamides as A_3-selective adenosine agonists. J. Med. Chem. **37:** 636–646.
4. KIM, H.O., X.D. JI, S.M. SIDDIQI, et al. 1994. 2-Substitution of N^6-benzyladenosine-5'-uronamides enhances selectivity for A_3-adenosine receptors. J. Med. Chem. **37:** 3614–3621.
5. JACOBSON, K.A., K.S. PARK, J.I. JIANG, et al. 1997. Pharmacological characterization of novel A_3 adenosine receptor-selective antagonists. Neuropharmacol. **36:** 1157–1165.
6. KIM, Y.C., X.D. JI & K.A. JACOBSON. 1996. Derivatives of the triazoloquinazoline adenosine antagonist (CGS15943) are selective for the human A_3 receptor subtype. J. Med. Chem. **39:** 4142–4148.
7. HANNON, J.P., H.J. PFANNKUCHE & J.R. FOZARD. 1995. A role for mast cells in adenosine A_3 receptor-mediated hypotension in the rat. Br. J. Pharmacol. **115:** 945–952.
8. JACOBSON, K.A., O. NIKODIJEVIC, D. SHI, et al. 1993. A role for central A_3-adenosine receptors: mediation of behavioral depressant effects. FEBS Lett. **336:** 57–60.
9. VON LUBITZ, D.K.J.E., R.C.S. LIN, P. POPIK, et al. 1994. Adenosine A_3 receptor stimulation and cerebral ischemia. Eur. J. Pharmacol. **263:** 59–67.
10. STAMBAUGH, C., J.L. JIANG, K.A. JACOBSON & B.T. LIANG. 1997. Novel cardioprotective function of adenosine A_3 receptor during prolonged simulated ischemia. Am. J. Physiol. **273:** H501–H505.
11. YAO, Y., Y. SEI, M.P. ABBRACCHIO, et al. 1997. Adenosine A_3 receptor agonists protect HL-60 cells and U-937 cells from apoptosis induced by A_3 antagonists. Biochem. Biophys. Res. Commun. **232:** 317–322.
12. ABBRACCHIO, M.P., S. CERUTI, R. BRAMBILLA, et al. 1998. Adenosine A_3 receptors and viability of astrocytes. Drug Dev. Res. **45:** 379–386.
13. ABBRACCHIO, M.P., G. RAINALDI, A.M. GIAMMARIOLI, et al. 1997. The A_3 adenosine receptor mediates cell spreading, reorganization of actin cytoskeleton, and distribution of Bcl-x_L. Studies in human astroglioma cells. Biochem. Biophys. Res. Commun. **241:** 297–304.
14. BRAMBILLA, R., F. CATTABENI, S. CERUTI, et al. 2000. Activation of the A_3 Adenosine Receptor Affects Cell Cycle Progression and Cell Growth. N-S Arch. Pharmacol. **361:** 224–234.

15. SEI, Y., D.K.J.E. VON LUBITZ, M.P. ABBRACCHIO, et al. 1997. Adenosine A_3 receptor agonist-induced neurotoxicity in rat cerebellar granule neurons. Drug Dev. Res. **40:** 267–273.
16. KOHNO, Y., Y. SEI, M. KOSHIBA, et al. 1996. Induction of apoptosis in HL-60 human promyelocytic leukemia cells by selective adenosine A_3 receptor agonists. Biochem. Biophys. Res. Commun. **219:** 904–910.
17. BARBIERI, D., M.P. ABBRACCHIO, S. SALVIOLI, et al. 1997. Apoptosis by 2-chloro-2′-deoxy-adenosine and 2-chloro-adenosine in human peripheral blood mononuclear cells. Neurochem. Int. **32:** 493–504.
18. JACOBSON, K.A., C. HOFFMANN, F. CATTABENI & M.P. ABBRACCHIO. 1999. Adenosine-induced cell death: evidence for receptor-mediated signalling. Apoptosis, **4:** 197–211.
19. VON LUBITZ, D.K.J.E., W. YE, J. MCCLELLAN & R.C.S. LIN. 1999. Stimulation of adenosine A_3 receptors in cerebral ischemia. Neuronal death, recovery, or both? Ann. N.Y. Acad. Sci. **890:** 93–106
20. ABBRACCHIO, M.P. & F. CATTABENI. 1999. Brain adenosine receptors as targets for therapeutic intervention in neurodegenerative diseases. Ann. N.Y. Acad. Sci. **890:** 79–92.
21. HALL, A. 1998. Rho GTPases and the actin cytoskeleton. Science **279:** 509–514.
22. BISHOP, A.L. & A. HALL. 2000. Rho GTPases and their effector proteins. Biochem. J. **348:** 241–255.
23. SEASHOLTZ, T.M., M. MAJUMDAR & J. HELLER BROWN. 1999. Rho as a mediator of G-protein-coupled receptor signaling. Mol. Pharmacol. **55:** 949–956.
24. ABBRACCHIO, M.P., R. BRAMBILLA, S. CERUTI, et al. 1995. G-protein-dependent activation of phospholipase C by adenosine A_3 receptors in rat brain. Mol. Pharmacol. **48:** 1038–1045.
25. MALORNI, W., G. RAINALDI, R. RIVABENE & M.T. SANTINI. 1994. Different susceptibilities to cell death induced by t-butylhydroperoxide could depend upon cell histiotype-associated growth features. Cell Biol. Toxicol. **10:** 207–218.
26. AULLO, P., M. GIRY, M.R. POPOFF, et al. 1993. A chimeric toxin to study the role of the 21 kDa GTP binding protein rho in the control of actin microfilament assembly. EMBO J. **12:** 921–931.
27. FIORENTINI, C., A. FABBRI, G. FLATAU, et al. 1997. Escherichia coli cytotoxic necrotizing factor 1 (CNF1), a toxin that activates the rho GTPase. J. Biol. Chem. **272:** 19532–19537.
28. HIRAO, M., N. SATO, T. KONDO, et al. 1996. Regulation mechanism of ERM (Erzin/Radixin/Moesin) protein/plasma membrane association: possible involvement of phosphatidylinositol turnover and Rho-dependent signaling pathway. J. Cell Biol. **135:** 37–51.
29. FIORENTINI, C., A. FABBRI, L. FALZANO, et al. 1998. Clostridium difficile toxin B induces apoptosis in intestinal cultured cells. Infect. Immun. **66:** 2660–2665.
30. LIANG, B.T. & K.A. JACOBSON. 1998. A physiological role of the adenosine A_3 receptor: sustained cardioprotection. Proc. Natl. Acad. Sci. U.S.A. **95:** 6995–6999.
31. KITAGAWA, K., M. MATSUMOTO, M. TAGAYA, et al. 1990. "Ischemic tolerance" phenomenon found in the brain. Brain Res. **528:** 21–24.

Adenosine Extracellular Brain Concentrations and Role of A_{2A} Receptors in Ischemia

FELICITA PEDATA, CLAUDIA CORSI, ALESSIA MELANI, FRANCESCA BORDONI, AND SERENA LATINI

Department of Preclinical and Clinical Pharmacology, University of Florence, Viale Pieraccini, Florence, Italy

ABSTRACT: Various experimental approaches have been used to determine the concentration of adenosine in extracellular brain fluid. The cortical cup technique or the microdialysis technique, when adenosine concentrations are evaluated 24 hours after implantation of the microdialysis probe, are able to measure adenosine in the nM range under normoxic conditions and in the µM range under ischemia. *In vitro* estimation of adenosine show that it can reach 30 µM at the receptor level during ischemia, a concentration able to stimulate all adenosine receptor subtypes so far identified. Although the protective role of A_1 receptors in ischemia seems consistent, the protective role of A_{2A} receptors appears to be controversial. Both A_{2A} agonists and antagonists have been shown to be neuroprotective in various *in vivo* ischemia models. Although A_{2A} agonists may be protective, mainly through peripherally mediated effects, A_{2A} antagonists may be protective through local brain mediated effects. It is possible that A_{2A} receptors are tonically activated following a prolonged increase of adenosine concentration, such as occurs during ischemia. A_{2A} receptor activation desensitizes A_1 receptors and reduces A_1 mediated effects. Under these conditions A_{2A} receptor antagonists may be protective by potentiating all the neuroprotective A_1 mediated effects, including decreased neurotoxicity due to reduced ischemia induced glutamate outflow.

KEYWORDS: Adenosine; Extracellular brain concentrations; Role of A_{2A} receptors; Ischemia.

BRAIN ADENOSINE METABOLISM

Adenosine is considered to be a neurotransmitter/neuromodulator under physiological conditions and a neuroprotective metabolite. The latter role is exerted by adenosine under hypoxic/ischemic conditions, in which breakdown of ATP occurs and there is a mismatch between degradation and resynthesis of ATP. The main source of intracellular adenosine is represented by ATP, which is degraded via adenylate kinase to ADP and AMP, and then adenosine is formed via cytosolic 5′-nucleotidase. Adenosine may also derive from the degradation of S-adenosylhomocysteine (SAH) through SAH-hydrolase. Although SAH-hydrolase has been localized by immunocytochemistry in several brain regions,[1] this pathway does not represent a consistent

---

Address for correspondence: Professor Felicita Pedata, Department of Preclinical and Clinical Pharmacology, University of Florence, Viale Pieraccini 6, 50139 Florence, Italy. Voice: +39-55-4271262; fax: +39-55-4271280.

pedata@server1.pharm.unifi.it

source of adenosine in the brain.[2,3] Another source of adenosine may be from intracellular conversion of cAMP to AMP through phosphodiesterase, and then degradation of AMP to adenosine.[4] Adenosine formed intracellularly escapes from cells through a bidirectional nucleoside transporter that transfers adenosine following its concentration gradient. Furthermore, adenosine may also derive from released ATP, which is dephosphorylated to AMP that, in turn, is converted to adenosine by extracellular 5'-nucleotidases.[5,6] Adenosine may also derive from released cAMP, which is degraded by ecto-phosphodiesterases.[7,8] Under "quasi-physiological stimulation" conditions, such as *in vitro* electrical stimulation, adenosine derives as such from cells.[9–11] On the other hand, the origin of adenosine formed under hypoxic/ischemic conditions is quite controversial, since both intracellular[11,12] and extracellular formation[13] has been observed.

ADENOSINE CONCENTRATIONS UNDER NORMOXIC AND ISCHEMIC CONDITIONS

Adenosine is physiologically present in the extracellular brain fluid in nanomolar concentrations. Phillis,[14] using the *in vivo* cortical cup technique, estimated an adenosine concentration of 30–50 nM in the rat cerebral cortex. The microdialysis technique also permits the determination of *in vivo* brain adenosine concentrations. Pazzagli et al.[15] have demonstrated that, 24 h after microdialysis fiber implantation, adenosine levels are 20 times lower than on the first day of operation, at the time when a plateau is reached. Determinations made 24 h after the microdalysis fiber implantation are probably more reliable estimates of adenosine concentrations because cellular metabolism has recovered from the stress of surgery. This is indicated by the recovery of glucose metabolism that increases soon after surgery and by the normalization of cerebral blood flow.[16] Adenosine concentrations determined in the rat 24 h after surgery and corrected for *in vitro* recovery of the microdialysis probe are 40–210 nM in the striatum[15,17–19] and 109 nM in the cortex.[20]

Following focal ischemia, induced in the rat by medial cerebral artery occlusion, a peak adenosine concentration of 3 µM has been measured in the striatum 24 h after probe implantation. At that time the effect of ischemia is not mixed with the trauma induced by microdialysis probe insertion.[19] It is worth to mentioning that adenosine concentrations as high as 24–40 µM have been measured by microdialysis after induction of global ischemia in the rat and gerbil.[21,22] However these values were reached on the first day following microdialysis tube implantation. Latini et al.[23] used a pharmacodynamic elaboration of the reversal exerted by the selective antagonist of A_1 adenosine receptors, 1,3-dipropyl-8-cyclopentylxanthine (DPCPX), on depression of the e.p.s.ps in the CA1 region induced by *in vitro* ischemic stimulus in hippocampal slices. This *in vitro* methodology allows adenosine concentrations at the receptor level to be measured. Adenosine concentrations calculated under normoxic conditions are 180–240 nM, which is about the same concentration range as that calculated by Dunwiddie[24] in the same *in vitro* preparation (120–200 nM). Thus, it may be assumed that, under normoxic conditions, adenosine concentrations estimated by microdialysis are indicative of those that tonically stimulate receptors. During *in vitro* ischemic stimulus, adenosine peak concentrations of 30 µM are reached in the CA1 region. This represents a 150-fold increase over normoxic values.

This is higher than that calculated in the superfusion fluid collected from hippocampal slices (4–6-fold)[9,25] or *in vivo* from the hippocampus (19–23-fold)[21,26] and from the striatum (15-fold).[19] However, it should be recalled that the time discrimination (15 sec) allowed by this *in vitro* approach is greater than that achieved with the microdialysis technique, where the sampling periods are more likely to represent the equilibrium reached in the extracellular fluid after partial reuptake and degradation of released adenosine. It is likely that an adenosine concentration of 30 µM represents the concentration of adenosine acting on receptors before diffusion and equilibration in the extracellular fluid. Therefore, the degree of receptor stimulation is greater than that extrapolated by the microdialysis adenosine measurement after ischemia. Concentrations of adenosine of 30 µM can stimulate all the adenosine receptor subtypes that have been cloned and identified in the central nervous system: the higher affinity (nM range) A_1 and A_{2A} receptors, and the lower affinity (µM range) A_3 and A_{2B} receptors.[27] Among these receptors the A_1 and A_{2A} receptors have been the most studied for their neuroprotective role in ischemia. The neuroprotective role of A_1 receptors is mainly attributed to inhibition of a neuronal Ca^{2+} uptake in the presynaptic terminal, which *per se* is protective under ischemia and also contributes to the reduction of presynaptic release of various neurotransmitters.[28] Among transmitters reduction of excitatory amino acids that are known to be involved in the rise of ischemic damage,[29] is relevant to the neuroprotective effect of adenosine. Acting on A_1 receptors, adenosine also hyperpolarizes cells, thus reducing neuronal excitability and firing rate,[30] with a consequent reduction in cellular metabolism, energy consumption, and hypothermia. All of these effects contribute to reduced ischemic brain damage.

ROLE OF A_{2A} ADENOSINE RECEPTORS IN BRAIN ISCHEMIA

Although the role of A_1 receptors in cerebral ischemia appears consistent, the role of A_{2A} receptors is controversial and needs to be clarified. A crucial role of A_{2A} receptors in the development of ischemic damage was demonstrated by the finding that A_{2A} receptor knockout mice are protected from cortical and striatal damage, as well as from the associated neurological deficit induced by transient middle cerebral artery occlusion.[31] However both A_{2A} agonists and antagonists have been shown to exert protective effects in various ischemia or neurotoxicity animal models. Preischemic administration of A_{2A} antagonists, 8-(3-chlorostyril) caffeine (CSC), 4-amino-8-chloro-1-phenyl[1,2,4]triazolo[4,3-α]quinoxaline (CP 66713), and 9-chloro-2-(2-furanyl)-5,6dihydro-[1,2,4]-triazolo[1,5]quinazolin-5-imine monomethanesulfonate (CGS 15943), reduced the ischemia induced injury assessed by locomotor activity and histopathological assay of CA1 pyramidal cell injury in the gerbil model of transient ischemia induced by bilateral carotid occlusion.[32–34] Postischemic treatment with the A_{2A} antagonist 7-(2-phenylethyl)-5-amino-2-(2-furyl)-pyrazolo-[4,3-*e*]-1,2,4-triazolo[1,5-*c*]pyrimidine (SCH 58261) has been shown to reduce brain injury consequent to hypoxia/ischemia induced by transient occlusion of the left carotid artery in immature rats[35] and to reduce cortical infarct volume in a model of permanent focal cerebral ischemia induced by electrocoagulation of medial cerebral artery (MCA) after craniectomy.[36]

A_{2A} receptors are expressed on neurons at high levels in the striatum[37] and at lower level in the cortex and hippocampus.[38,39] Selective agonists of A_{2A} receptors have been shown to stimulate glutamate outflow from the ischemic cortex[40,41] and from the striatum under normoxic conditions.[42] Corsi et al.[43] have recently demonstrated that the selective A_{2A} receptor antagonist SCH 58261 decreases both spontaneous and K^+-evoked striatal glutamate outflow in young rats, suggesting that adenosine exerts a tonic stimulatory action on excitatory amino acid outflow through A_{2A} receptor stimulation. The protective effects of A_{2A} receptor antagonists against ischemic injury may thus be attributed to their ability to reduce excitatory amino acids outflow.

Unexpectedly, the A_{2A} selective agonist 2-[p-(2-carboxyethyl)-phenethylamino]-5'-N-ethylcarboxamidoadenosine (CGS 21680) has also been shown to protect neurons against forebrain ischemia in the gerbil;[44] and the A_{2A} agonist 2-[(2-aminoethylamino) carbonylethylphenylethylamino]-5'-N-ethylcarboxamido adenosine (APEC) has been shown to improve the recovery of postischemic blood flow and animal survival, without preserving the survival of hippocampal gerbil neurons subjected to transient global ischemia.[34] A heterogeneous distribution of A_{2A} adenosine receptors on cell types other than neurons could account for several receptor mediated protective effects of A_{2A} agonists. Adenosine A_{2A} receptors located on endothelial and smooth muscle cells of the cerebral vasculature may account for cerebral vasodilation and increased oxygen supply.[14,45] A_{2A} receptors on platelets may account for inhibition of platelet aggregation and antithrombotic activity.[46,47] A_{2A} receptors on neutrophils may account for inhibition of adhesion to endothelial cells and ensuing production of free radicals.[48,49]

Jones et al.[50,51] have provided insight into the possible protection mechanism of A_{2A} receptors agonists and antagonists. They have shown, using compounds and a dosing regimen that is selective for A_{2A} adenosine receptors, that systemic administration of both the A_{2A} agonist CGS 21680 and antagonist 4-(2-[7-amino-2-(2-furyl)-{1,2,4}-triazolo{2,3-a}{1,3,5}triazin-5-ylamino]ethyl)phenol (ZM 241385) protects against the hippocampal neuronal damage induced by intrahippocampal injection of the excitotoxin kainate.[51] However, when injected directly in the hippocampus, only the A_{2A} antagonist ZM 241385 significantly reduced kainate induced neuronal damage.[50] They hypothesised that protective effects exerted by the agonist are not locally exerted on hippocampal neuronal cells, but may be mediated by a number of mechanisms that are mainly consequent to systemic administration, including increased blood flow to compromised hippocampal neurons, reduction in the degree of infiltration by peripheral leukocytes into the central nervous system, inhibition of platelet aggregation, and even reduction of cerebral glucose metabolism and requirement of depleted nutrients during ischemia. On the other hand, the intrahippocampal injection of A_{2A} antagonists directly protects hippocampal neuronal cells by reducing excitatory amino acid release and, thus, the excitotoxic damage.[50,51]

The reduced sensitivity of the A_1 receptors brought about by A_{2A} receptor activation when adenosine concentration increases has been proposed to explain the protective effect of the intrahippocampal administered A_{2A} antagonist ZM 241385.[50] In rat striatal synaptosomes A_{2A} receptor activation by CGS 21680 reduces the affinity of A_1 receptors for their agonist.[52] Reduced sensitivity of A_1 receptors, brought about by A_{2A} receptor activation has been proposed to explain the lack of the neuroprotective effect of the A_1 receptor agonist N^6-(R-phenylisopropyl)-adenosine (R-PIA) against the hippocampal neuronal damage induced by the neurotoxin kainate.[50] Latini et al.[53] have recently confirmed that the e.p.s.ps depression induced by an *in vitro* ischemic stimulus is selectively mediated by adenosine A_1 receptors and that a prolonged application of CGS 21680 reduces e.p.s.ps ischemic depression, an effect that is antagonized by the selective A_{2A} antagonist ZM 241385. Furthermore, prolonged application of adenosine on hippocampal slices also reduces e.p.s.ps depression induced by the *in vitro* ischemic stimulus, an effect that is antagonized by ZM 241385. It was, therefore, suggested that a tonic activation of A_{2A} receptors, following prolonged stimulation, could induce a desensitization of A_1 receptors and the ensuing decrease of A_1 mediated responses in the CA1 region of the hippocampus.[53] This mechanism may explain the protection by A_{2A} receptor antagonists against kainate induced excitotoxicity and may be of relevance under *in vivo* ischemic conditions when adenosine levels rise considerably and the release is sustained over time.[19] Thus, the neuroprotective effects of A_{2A} antagonists observed with *in vivo* models of brain ischemia[32–36] may be ascribed to a potentiation of all the protective effects mediated by A_1 receptors.

The protective effects of adenosine A_{2A} antagonists can also be attributed to effects on glial cells. Adenosine A_{2A} receptors are present on microglial cells but not on astrocytes.[54] Under pathological conditions microglial cells start to proliferate and are responsible for the inflammatory responses that follow both acute and chronic brain diseases. Several cellular and molecular markers of microglial activation, such as upregulation of cyclo-oxygenase-2 (COX-2), cytokines or prostaglandins have been described,[55] and involvement of adenosine A_{2A} receptors in these processes has been investigated. It has been shown that A_{2A} receptor activation stimulates the proliferation of rat microglial cells in culture[56] and upregulates COX-2 gene expression in microglia, triggering the production of prostaglandins.[54] In this respect blockade of A_{2A} receptors may be protective by preventing the stimulatory effect of adenosine on resting microglial cells. In accord with this hypothesis, it has been recently demonstrated that the A_{2A} antagonist 1,3-dimethyl-7-propylxanthine (DMPX) protects hippocampal neurons in culture from death induced by microglial secretory products.[57] Although A_{2A} receptor antagonists may be regarded as protective for their microglial effect, it should be mentioned that prolonged elevation of the intracellular cAMP level, as can be obtained after A_{2A} receptor activation, results in the inhibition of microglial cells activation induced by pathological stimuli (see Ongini and Shubert Ref. 58 for a review). Thus, the extent to which A_{2A} receptor block or stimulation contributes to antiflammatory effects due to inhibition of microglial activation remains to be clarified.

AN *IN VIVO* MODEL OF BRAIN ISCHEMIA FOR STUDYING THE ROLE OF A_{2A} ADENOSINE RECEPTORS

Several animal models have been used to study cerebral ischemia[59] in order to understand its pathophysiology and identify therapeutic strategies for minimizing the severity of the ischemic damage. Whereas global ischemia typically results in neuronal injury within vulnerable brain regions, focal ischemia brings about a localized brain infarction. Focal ischemia induced by middle cerebral artery occlusion (MCAO) in the rat[60] has gained increasing acceptance as a model for hemispheric

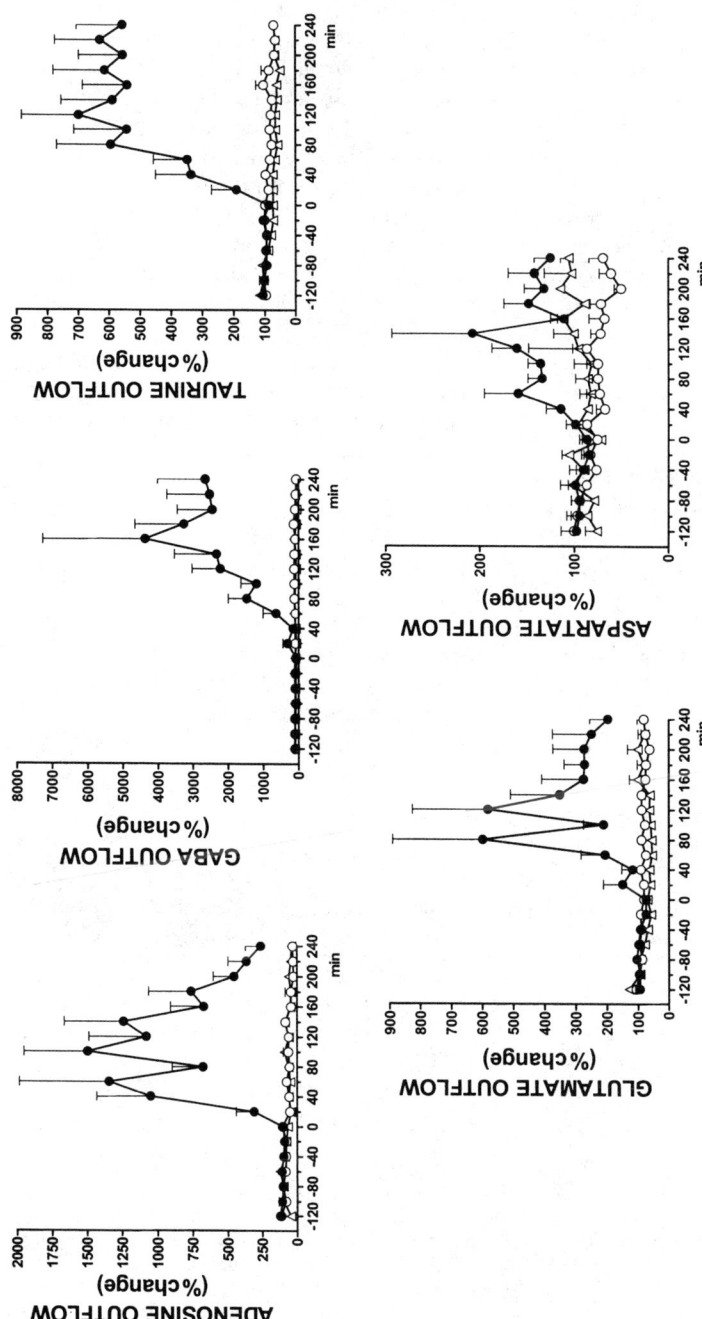

FIGURE 1. Adenosine, GABA, taurine, glutamate, and aspartate outflow from the rat striatum during permanent MCAO. Data are expressed as the percentage variation of spontaneous outflow, which in each rat group is the mean of the first seven samples collected between −120 and 0 minutes. Each point is the mean ± S.E.M. of n rats. ●, MCAO hemisphere ($n = 16$); ○, contralateral hemisphere ($n = 10$); △, sham operated rats ($n = 4$). (Reprinted from Melani et al., Ref. 19, with permission from *Stroke*).

infarction in humans.[59] After permanent occlusion, a cortical and striatal infarct with temporal and spatial evolution occurs[61] within the vascular territory supplied by the MCA. Melani et al.[19] have recently published a method for inducing focal cerebral ischemia associated with the microdialysis technique. This intraluminal approach of MCAO in combination with microdialysis offers several advantages: (1) the method is less invasive in comparison to MCAO induced after craniectomy;[62] (2) MCAO is performed 24 h after microdialysis probe implantation, thus the effects of ischemia are not mixed with those of microdialysis surgery; (3) for the first time this method is associated with the microdialysis technique and transmitter outflow is evaluated before and during ischemia in awake and freely moving animals; and (4) drug neuroprotective properties, as evaluated by histological assessment of the ischemic area and neurological deficit, may be related to neurochemical changes in the same animal. Following focal cerebral ischemia induced by the suture technique, the peak outflow of the excitatory amino acids, glutamate and aspartate in the striatal region, is increased 2.5-fold in comparison to preischemic outflow. GABA outflow increases slowly, but dramatically, 140-fold, adenosine increases 15-fold, and taurine 6-fold. Except for adenosine, the outflow of all transmitters remains high for four hours after induction of ischemia (see FIGURE 1).

Striatal damage, as histologically assessed 24 h after MCAO, was correlated with both striatal transmitter outflow and with the neurological deficit that, 24 h after MCAO, is severe (see TABLE 1). There is a significant correlation between the three parameters: neurotransmitter outflow, histological damage, and neurological deficit. The possibility of evaluating neurological deficit, histological damage, and different transmitter outflow in the same animal may be of value when studying drug neuroprotective effects against focal ischemia and trying to relate them to neurochemical changes. In particular, such a methodological approach may help to clarify the mechanism of action of the neuroprotective effect of A_{2A} agonists and antagonists in ischemia.

TABLE 1. Correlations between transmitter efflux and neurological deficit and histopathological outcome 24 hours after MCAO[a]

	Neurological Score	Ischemic Striatal Damage
Adenosine	0.49	0.47
Glutamate	0.24	0.33
Aspartate	0.36	0.44
GABA	0.38	0.34
Taurine	0.48	0.58

NOTE: Correlation (r^2, $n = 20$, $p < 0.03$) was calculated between the mean percent outflow of transmitters and neurological score determined in each rat for sham operated rats ($n = 4$) and permanent MCAO rats ($n = 16$). Correlation (r^2, $n = 12$, $p < 0.05$) was calculated between mean percent outflow of transmitter and ischemic striatal damage (calculated as percentage of the volume of the ipsilateral hemisphere) determined in each rat for sham operated rats ($n = 4$) and permanent MCAO rats ($n = 8$). (Reproduced from Melani, et al., Ref. 19, with permission from Stroke.)

CONCLUSIONS

Various methods for estimating adenosine concentrations in the extracellular brain fluid were reviewed. The importance of evaluating adenosine concentrations in microdialysis experiments 24 h after microdialysis fiber implantation is stressed, because after 24 hours the cellular metabolism has recovered from the trauma of surgery and its effect is not compounded with that of ischemia. Under normoxic conditions adenosine concentrations range from 40 to 210 nM in the cortex. Under ischemia, adenosine levels reach the µM range and, by using an *in vitro* approach, it was calculated that 30 µM concentrations are reached at the receptor level in the ischemic hippocampus. The neuroprotective effects of A_{2A} adenosinergic compounds are reviewed. Several actions may account for the neuroprotective effects of A_{2A} agonists in ischemia: increased oxygen supply by vasodilatory action, decreased platelet aggregation and thus antithrombotic activity, decreased neutrophil adhesion to endothelial cells, and reduced free radical formation. Protective effects of A_{2A} antagonists are attributed to local brain reduction of excitotoxic damage due to reduced glutamate outflow and microglial activation. A recently developed technique of medial cerebral artery occlusion associated with the microdialysis technique may contribute to understanding of the neuroprotective mechanism of action of A_{2A} adenosinergic compounds. In the same experimental animal, neuroprotective effects, evaluated on the ischemia induced neurological deficit and histological damage, can be related to modifications of excitatory amino acid outflow, and thus to excitotoxic damage.

ACKNOWLEDGMENTS

Financial support of MURST ex 40% 1999 is gratefully acknowledged. The authors are grateful to Dr. Angela Monopoli (Schering-Plough Research Institute, Milan, Italy) for useful discussion and to Mary Forrest for review of the English.

REFERENCES

1. PATEL, B.T. & N. TUDBALL. 1986. Localization of S-adenosylhomocysteine hydrolase and adenosine deaminase immunoreactivities in rat brain. Brain Res. **370:** 250–264.
2. LATINI, S., C. CORSI, F. PEDATA & G. PEPEU. 1995. The source of brain adenosine outflow during ischemia and electrical stimulation. Neurochem. Int. **27:** 239–244.
3. PAK, M.A., H.L. HAAS, U.K.M. DECKING & J. SCHRADER. 1994. Inhibition of adenosine kinase increases endogenous adenosine and depresses neuronal activity in hippocampal slices. Neuropharmacol. **33:** 1049–1053.
4. CRAIG, C.G., S.D. TEMPLE & T.D. WHITE. 1994. Is cyclic AMP involved in excitatory amino acid-evoked adenosine release from rat cortical slices? Eur. J. Pharmacol. **269:** 79–85.
5. TERRIAN, D.M., P.G. HERNANDEZ, M.A. REA & R.I. PETERS. 1989. ATP release, adenosine formation, and modulation of dynorphin and glutamic acid release by adenosine analogues in rat hippocampal mossy fiber synaptosomes. J. Neurochem. **53:** 1390–1399.
6. RICHARDSON, P.J., S.J. BROWN, E.M. BAILYES & J.P. LUZIO. 1987. Ectoenzymes control adenosine modulation of immunoisolated cholinergic synapses. Nature **327:** 232–234.

7. ROSENBERG, P.A. & M.A. DICHTER. 1989. Extracellular cAMP accumulation and degradation in rat cerebral cortex in dissociated cell culture. J. Neurosci. **9:** 2654–2663.
8. ROSENBERG, P.A. & Y. LI. 1995. Vasoactive intestinal peptide regulates extracelluar adenosine levels in rat cortical cultures. Neurosci. Lett. **200:** 93–96.
9. PEDATA, F., S. LATINI, A.M. PUGLIESE & G. PEPEU. 1993. Investigations into the adenosine outflow from hippocampal slices evoked by ischemic-like conditions. J. Neurochem. **61:** 284–289.
10. CUNHA, R.A., E.S. VIZI, J.A. RIBEIRO & A.M. SEBASTIAO. 1996. Preferential release of ATP and its extracellular catabolism as a source of adenosine upon high- but not low-frequency stimulation of rat hippocampal slices. J. Neurochem. **67:** 2180–2187.
11. LLOYD, H.G.E., K. LINDSTROM & B.B. FREDHOLM. 1993. Intracellular formation and release of adenosine from rat hippocampal slices evoked by electrical stimulation or energy depletion. Neurochem. Int. **23:** 173–185.
12. MEGHJI, P., J.B. TUTTLE & R. RUBIO. 1989. Adenosine formation and release by embryonic chick neurons and glia in cell culture. J. Neurochem. **53:** 1852–1860.
13. KOOS, B.J., L. KRUGER & T.F. MURRAY. 1997. Source of extracellular brain adenosine during hypoxia in fetal sheep. Brain Res. **778:** 439–442.
14. PHILLIS, J.W. 1989. Adenosine in the control of cerebral circulation. Cerebrovasc. Brain Metab. Rev. **1:** 26–54.
15. PAZZAGLI, M., F. PEDATA & G. PEPEU. 1993. Effect of K^+ depolarization, tetrodotoxin, and NMDA receptor inhibition on extracellular adenosine levels in rat striatum. Eur. J. Pharmacol. **234:** 61–65.
16. BENVENISTE, H., J. DREJER, A. SCHOUSBOE & N.H. DIEMER. 1987. Regional cerebral glucose phosphorylation and blood flow after insertion of a microdialysis fiber through the dorsal hippocampus in the rat. J. Neurochem. **49:** 729–734.
17. BALLARIN, M., B.B. FREDHOLM, S. AMBROSIO & N. MAHY. 1991. Extracellular levels of adenosine and its metabolites in the striatum of awake rats: inhibition of uptake and metabolism. Acta Physiol. Scand. **142:** 97–103.
18. PAZZAGLI, M., C. CORSI, S. FRATTI, et al. 1995. Regulation of extracellular adenosine levels in the striatum of aging rats. Brain Res. **684:** 103–106.
19. MELANI, A., L. PANTONI, C. CORSI, et al. 1999. Striatal outflow of adenosine, excitatory amino acids, gamma-aminobutyric acid, and taurine in awake freely moving rats after middle cerebral artery occlusion. Correlations with neurological deficit and histopathological damage. Stroke **30:** 2448–2455.
20. PAZZAGLI, M., C. CORSI, S. LATINI, et al. 1994. In vivo regulation of extracellular adenosine levels in the cerebral cortex by NMDA and muscarinic receptors. Eur. J. Pharmacol. **254:** 277–282.
21. DUX, E., J. FASTBOM, U. UNGERSTED, et al. 1990. Protective effect of adenosine and a novel xanthine derivative propentophylline on the cell damage after bilateral carotid occlusion in the gerbil hippocampus. Brain Res. **516:** 248–256.
22. HAGBERG, H., P. ANDERSSON, J. LACAREWICZ, et al. 1987. Extracellular adenosine, inosine, hypoxanthine, and xanthine in relation to tissue nucleotides and purines in rat striatum during transient ischemia. J. Neurochem. **49:** 227–231.
23. LATINI, S., F. BORDONI, F. PEDATA & R. CORRADETTI. 1999. Extracellular adenosine concentrations during in vitro ischaemia in rat hippocampal slices. Br. J. Pharmacol. **126:** 729–739.
24. DUNWIDDIE, T.V. & L. DIAO. 1994. Extracellular adenosine concentrations in hippocampal brain slices and the tonic inhibitory modulation of evoked excitatory responses. J. Pharmacol. Exp. Ther. **268:** 537–545.
25. LATINI, S., F. BORDONI, R. CORRADETTI, et al. 1998. Temporal correlation between adenosine outflow and synaptic potential inhibition in rat hippocampal slices during ischemia-like conditions. Brain Res. **794:** 325–328.
26. ANDINE', P., K.A. RUDOLPHI, B.B. FREDHOLM & H. HAGBERG. 1990. Effect of propentophylline (HWA 285) on extracellular purine and excitatory amino acids in CA1 of rat hippocampus during transient ischemia. Br. J. Pharmacol. **100:** 814–818.
27. FREDHOLM, B.B., M.P. ABBRACCHIO, G. BURNSTOCK, et al. 1994. Nomenclature and classification of purinoceptors. Pharmacol. Rev. **46:** 143–156.

28. VON LUBITZ, D.K.J.E. 1999. Adenosine and cerebral ischemia: therapeutic future or death of a brave concept? Eur. J. Pharmacol. **365:** 9–25.
29. CHOI, D.W. & S.M. ROTHMAN. 1990. The role of glutamate neurotoxicity in hypoxic-ischemic neuronal death. Ann. Rev. Neurosci. **13:** 171.
30. GREENE, R.W. & H.L. HASS. 1991. The electrophysiology in the mammalian central nervous system. Progr. Neurobiol. **36:** 329–341.
31. CHEN, J.-F., Z. HUANG, J. MA, et al. 1999. A2A adenosine receptor deficiency attenuates brain injury induced by transient focal ischemia in mice. J. Neurosci. **19:** 9192–9200.
32. GAO, Y. & J. W. PHILLIS. 1994. CGS 15943, an adenosine A_2 receptor antagonist, reduces cerebral ischemia injury in the mongolian gerbil. Life Sci. **55:** PL 61–65.
33. PHILLIS, J.W. 1995. The effects of selective A_1 and A_2 adenosine receptor antagonists on cerebral ischemic injury in the gerbil. Brain Res. **705:** 79–84.
34. VON LUBITZ, D.K.J.E., R.C.S. LIN & K.A. JACOBSON. 1995. Cerebral ischemia in gerbils: effects of acute and chronic treatment with adenosine A_{2A} receptor agonist and antagonist. Eur. J. Pharmacol. **287:** 295–302.
35. BONA, E., U. ADEN, E. GILLAND, et al. 1997. Neonatal cerebral hypoxia-ischemia: the effect of adenosine receptor antagonists. Neuropharmacology **36:** 1327–1338.
36. MONOPOLI, A., G. LOZZA, A. FORLANI, et al. 1998. Blockade of adenosine A_{2A} receptors by SCH 58261 results in neuroprotective effects in cerebral ischaemia in rats. NeuroReport **9:** 3955–3959.
37. SCHIFFMANN, S.N., O. JACOBS & J.-J. VANDERHAEGHEN. 1991. Striatal restricted adenosine A_2 receptor (RDC8) is expressed by enkephalin but not by substance P neurons: an *in situ* hybridization histochemistry study. J. Neurochem. **57:** 1062–1067.
38. JOHANSSON, B., V. GEORGIER, F.E. PARKINSON & B.B. FREDHOLM. 1993. The binding of the adenosine A_2 receptor selective agonist [^{3}H]CGS 21680 to rat cortex differs from its binding to rat striatum. Eur. J. Pharmacol. **247:** 103–110.
39. ROSIN, D.L., A. ROBEVA, R.L. WOODARD, et al. 1998. Immunohistochemical localization of adenosine A_{2A} receptors in the rat central nervous system. J. Comp. Neurol. **401:** 163–186.
40. O'REGAN, M.H., L.M. SIMPSON, L.M. PERKINS & J.W. PHILLIS. 1992. The selective A_2 adenosine receptor agonist CGS 21680 enhances excitatory transmitter amino acid release from the ischemic rat cerebral cortex. Neurosci. Lett. **138:** 169–172.
41. SIMPSON, R.E., M.H. O'REGAN, L.M. PERKINS & J.W. PHILLIS. 1992. Excitatory transmitter amino acid release from the ischemic rat cerebral cortex: effects of adenosine receptor agonists and antagonists. J. Neurochem. **58:** 1683–1690.
42. POPOLI, P., P. BETTO, R. REGGIO & G. RICCIARELLO. 1995. Adenosine A_{2A} receptor stimulation enhances striatal extracellular glutamate levels in rats. Eur. J. Pharmacol. **287:** 215–217.
43. CORSI, C., M. MELANI, L. BIANCHI & F. PEDATA. 2000. Striatal A_{2A} adenosine receptor antagonism differentially modifies striatal glutamate outflow *in vivo* in young and aged rats. NeuroReport **11:** 1–5.
44. SHEARDOWN, M.J. & L.J.S. KNUTSEN. 1996. Unexpected neuroprotection observed with the adenosine A_{2a} receptor agonist CGS21680. Drug Develop. Res. **39:** 108–114.
45. IBAYASHI, S., A.C. NGAI, J.R. MENO & H.R. WINN. 1991. Effects of topical adenosine analogs and forskolin on rat pial arterioles *in vivo*. J. Cereb. Blood Flow Metab. **11:** 72–76.
46. SANDOLI, D., P.J.S. CHIU, M. CHINTALA, et al. 1994. *In vivo* and *ex vivo* effects of adenosine A_1 and A_2 receptor agonists on platelet aggregation in the rabbit. Eur. J. Pharmacol. **259:** 43–49.
47. LEDENT, C., J.-M. VAUGEOIS, S.N. SCHIFFMANN, et al. 1997. Aggressivenes, hypoalgesia and high blood pressure in mice lacking the adenosine A_{2A} receptor. Nature **388:** 674–678.
48. CRONSTEIN, B.N. 1994. Adenosine, an endogenous anti-inflammatory agent. J. Appl. Physiol. **76:** 5–13.

49. JORDAN, J.E., Z.-Q. ZHAO, H. SATO, et al. 1997. Adenosine A_2 receptor activation attenuates reperfusion injury by inhibiting neutrophil accumulation, superoxide generation and coronary endothelial adherence. J. Pharmacol. Exp. Ther. **280:** 301–309.
50. JONES, P.A., R.A. SMITH & T.W. STONE. 1998. Protection against hippocampal kainate excitotoxicity by intracerebral administration of an adenosine A_{2A} receptor antagonist. Brain Res. **800:** 328–335.
51. JONES, P.A., R.A. SMITH & T.W. STONE. 1998. Protection against kainate-induced excitotoxicity by adenosine A_{2A} receptor agonists and antagonists. Neuroscience **85:** 229–237.
52. DIXON, A.K., L. WIDDOWSON & P.J. RICHARDSON. 1997. Desensitisation of the adenosine A_1 receptor by the A_{2A} receptor in the rat striatum. J. Neurochem. **69:** 315–321.
53. LATINI, S., F. BORDONI, R. CORRADETTI, et al. 1999. Effect of A_{2A} adenosine receptor stimulation and antagonism on synaptic depression induced by *in vitro* ischaemia in rat hippocampal slices. Br. J. Pharmacol. **128:** 1035–1044.
54. FIEBICH, B., K. BIBER, K. LIEB, et al. 1996. Cyclooxygenase-2 expression in rat microglia is induced by adenosine A_{2a}-receptors. Glia **18:** 152–160.
55. GEBICKE-HAERTER, P.J., D. VAN CALKER, W. NORENBERG & P. ILLES. 1996. Molecular mechanisms of microglial activation. Neurochem. Int. **29:** 1–12.
56. GEBICKE-HAERTER, P.J., F. CHRISTOFFEL, J. TIMMER, et al. 1996. Both adenosine A_1- and A_2-receptors are required to stimulate microglia proliferation. Neurochem. Int. **29:** 37–42.
57. FLAVIN, M.P. & L.T. HO. 1999. Propentofylline protects neurons in culture from death triggered by macrophage or microglial secretory products. J. Neurosci. Res. **56:** 54–59.
58. ONGINI, E. & P. SCHUBERT. 1998. Neuroprotection induced by stimulating A_1 or blocking A_{2A} adenosine receptors: an apparent paradox. Drug Develop. Res. **45:** 387–393.
59. GINSBERG, M.D. & R. BUSTO. 1989. Rodent models of cerebral ischemia. Stroke **20:** 1627–1642.
60. TYSON, G.W., G.M. TEASDALE, D.I. GRAHAM & J. MCCULLOCH. 1984. Focal cerebral ischemia in the rat: topography of hemodynamic and histopathological changes. Ann. Neurol. **15:** 559–567.
61. GARCIA, J.H., K.F. LIU & K.L. HO. 1995. Neuronal necrosis after middle cerebral artery occlusion in Wistar rats progresses at different time intervals in the caudoputamen and the cortex. Stroke **26:** 636–643.
62. ZEA LONGA, E., P.R. WEINSTEIN, S. CARLSON & R. CUMMINS. 1989. Reversible middle cerebral artery occlusion without craniectomy in rats. Stroke **20:** 84–91.

Right Thing at a Wrong Time? Adenosine A_3 Receptors and Cerebroprotection in Stroke

DAG K.J.E. VON LUBITZ,[a] KIMBERLY L. SIMPSON,[b] AND RICK C.S. LIN[b]

[a]*Emergency Medicine Research Laboratories, Department of Emergency Medicine, University of Michigan Health System, Ann Arbor, Michigan, U.S.A.*

[b]*Department of Anatomy, University of Mississippi Medical Center, Jackson, Mississippi, U.S.A.*

ABSTRACT: The involvement of adenosine A_3 receptors in normal and pathologic functions of the brain remains to be defined. Previous studies have shown that chronic preischemic administration of the agonist [N^6-(3-iodobenzyl)-5'-N-methylcarboxoamidoadenosine or IB-MECA) results in a significant protection of neurons in selectively vulnerable brain regions and in an equally significant reduction of the subsequent mortality. Acute administration of the drug, on the other hand, resulted in a pronounced worsening of these parameters. We now report that the effect of administration of IB-MECA depends on the timing of treatment with respect to the onset of the focal insult, and provide the first data supporting speculation that treatment with adenosine A_3 receptor agonists may decrease the infarct size following focal brain ischemia.[1,2] Treatment with IB-MECA administered 20 min prior to transient middle cerebral ischemia ($MCAO_t$ = 30 min) resulted in a significant increase of the infarct size ($p < 0.01$), whereas administration 20 min after ischemia resulted in statistically significant decrease of the infarct volume. Postischemic treatment results in improved neuronal preservation, decreased intensity of reactive gliosis, and pronounced reduction of microglial infiltration. The data indicate that the effects of adenosine A_3 receptor stimulation depend on the differential impact of these receptors on both neuronal and non-neuronal elements of the cerebral tissue, for example, astrocytes, microglia, and vasculature.

KEYWORDS: Adenosine A_3 Receptors; Stroke; Focal cerebral ischemia; Neuroprotection; Cerebral infarction; Physiology; Pharmacology; Treatment; Mice.

INTRODUCTION

Despite nearly a decade since the discovery of the latest member of the adenosine receptor family,[3,4] the biological role of adenosine A_3 receptors is still ill defined. Studies have shown that stimulation of this receptor results in degranulation of mast cells, histamine release, vasoconstriction, and hypotension.[5–7] Other authors have demonstrated the potent inhibitory effect of adenosine A_3 receptor activation on the processes accompanying inflammation, for example, neutrophil degranulation[8] or

---

Address for correspondence: Dag K.J.E. von Lubitz, Department of Emergency Medicine, TC/B1354/0303, University of Michigan Health System, 1500 E. Medical Center Drive, Ann Arbor, MI 48109-0303, U.S.A. Voice: +1 734 936 6020; fax: +1 734 9369414.
dvlubitz@umich.edu

inhibition of eosinophil migration.[9] Adenosine A_3 receptors also appear to activate certain types of Cl^- channels of the non-pigmented epithelium.[10]

The bulk of experimental work that has attempted to define the role of adenosine A_3 receptors in these pathological processes has concentrated on ischemia of the heart and brain. Several authors have shown that preischemic exposure of either cardiac myocytes or isolated hearts to adenosine A_3 receptor agonists results in protection against ischemic damage,[11–13] indicating that these receptors may be involved in the cardioprotective effects of preconditioning. These results are starkly opposed by the effects of the acute preischemic treatment on the outcome of global cerebral ischemia, where both neuronal damage and postischemic mortality are very significantly aggravated by the agonists of adenosine A_3 receptor.[14] However, protective effects of adenosine A_3 receptor agonist treatment have been reported that are consequent to chronic administration of N^6-(3-iodobenzyl)-5'-N-methylcarboxoamidoadenosine (IB-MECA).[14] Involvement of adenosine A_3 receptors in these effects has been confirmed by preliminary reports of the intensely cerebroprotective effect of preischemic exposure to the adenosine A_3 receptor antagonist.[15,16]

Although the damaging, involvement of adenosine A_3 receptors in global ischemia has been demonstrated,[14] nothing is known about the effects of either pre- or postischemic stimulation of these receptors in focal brain ischemia. Based largely on the circumstantial evidence, it has recently been theorized that postischemic stimulation of adenosine A_3 receptors may actually result in cerebroprotection.[1,16] The unquestionable yet rather confusing involvement of adenosine A_3 receptors in ischemic brain damage warrants extension of previous studies to focal ischemia. Stimulation of adenosine A_3 receptors appears to affect at least two transducing systems, and some of the affected pathways appear to be clearly involved in stroke induced pathology.[17,18] Hence, the data obtained during studies of experimental focal ischemia may also shed further light on the pathology of stroke in humans.

MATERIALS AND METHODS

Animals and Drugs

CD-1 ($n = 60$) male mice (35 g; Charles River Laboratories, Wilmington, MA) were used. The animals were randomly separated into three experimental groups ($n = 20$/group). Treated animals were injected i.p. with 1.0 mg/kg N^6-(3-iodobenzyl)-5'-N-methylcarboxoamidoadenosine (IB-MECA; RBI, Natick, MA) dissolved in Emulphor (Rhône-Poulenc, Cranbury, NJ) and saline as described elsewhere.[14] The drug was administered either 20 min prior to, or 20 min following middle cerebral artery occlusion. The control group received i.p. injections of the vehicle.

Surgery

Anesthesia and Temperature Maintenance

Presurgical anesthesia was initiated with 4% isoflurane carried in the 70:30 mixture of nitrous oxide and oxygen, and maintained at 1.5% isoflurane. The gas was administered through a face mask (Kent Scientific, Litchfield, CT). Body temperature was maintained at 37.5°C throughout the surgery and during recovery using a rectal temperature probe and a heating pad (Kent Scientific, Litchfield, CT).

Surgical Procedure

The model of filament induced transient middle cerebral artery occlusion was used. Following a ventromedial incision of the neck, the left common carotid artery was exposed under surgical microscope (Vision Engineering, New Milford, CT). The occipital branches of the external carotid and the terminal lingual and maxillary branches were then isolated and coagulated. Subsequently, the internal carotid was exposed and its pterygopalatine branch ligated close to its origin using a 6-0 silk suture. The procedure ensured that only the intracranial ramus of the common carotid artery remained patent. A 2-cm length of 5-0 blunt tipped (through heat application) nylon suture was then introduced into the external carotid and gently advanced into the internal branch of the artery. Resistance and slight bending of the suture indicated that the tip lodged in the proximal segment of the anterior cerebral artery. Typically, the distance traveled by the tip (i.e., between the tip and the bifurcation of the common carotid artery) was 10–11 mm. After ensuring that the tip was firmly lodged in the artery, the suture was anchored for 30 min. After withdrawing the suture, the wound neck was sutured, and the animal was left to recover. To prevent postsurgical infection, a mixture of neomycin, polymycin, B-sulphate, and bacitracin (Triple-Antibiotic Ointment, E. Fougera & Co., Melville, NY) was topically applied to the wound.

Verification of the Occlusion

The completeness of the artery occlusion was verified using a laser-Doppler technique (Perimed, North Royalton, OH). Prior to the occlusion, the cortical blood flow was measured transcranially by means of a 1-mm diameter probe apposed to the skull surface over the area of the expected infarct.[14] Five minutes after the occlusion the probe was advanced to the same place, and the continuous flow measurement was made over the next five minutes. Consistent depression of the postischemic flow by at least 95%, compared to the preischemic values, indicated successful occlusion of the artery.

Determination of the Infarct Size

Seven days after the occlusion, the animals were anesthetized with Nembutal (50 mg/kg) and decapitated. Following removal, the brains were sectioned into 1-mm thick coronal slices that were then immersed in a warm (37°C) 2% solution of 2,3,5-triphenyl-tetrazolium chloride (TTC) in 1X PBS. After 20 min, the slices were gently washed with several rinses of 2% buffered (pH 7.4), 2% solution of paraformaldehyde in PBS, followed by additional fixation in fresh paraformaldehyde solution. The slices were then placed under an automated scanning system (EXPRESSION 636, Epson, Japan) and scanned at 720 d.p.i. The volume of the infarct was determined using NIH Image analyzing software. In the animals that showed signs of secondary cerebral hemorrhages (seen only in the pretreatment group), the area affected was considered part of the total infarct zone.

Histology

Separate groups of animals ($n = 20$) were used for the assessment of histological damage and immunocytochemistry. Seven days after ischemia, the animals were anesthetized and perfused with buffered paraformaldehyde (4.5%, pH 7.4). After fixation, the brains were removed and cut on a freezing microtome into 25 μm slices. Slices from the penumbra zone (determined by means of laser-Doppler measurement during ischemia) were subjected to either to Nissl or GFAP and lectin immunocytochemical[19] staining to indicate the extent of neuronal preservation in the hippocampus, and the extent of astrocyte and microglia activation in the hippocampus and the cortex of the penumbra zone.

Statistics

Infarct volume data were analyzed using GraphPad/Inplot (GraphPazd Software, San Diego, CA) software. Statistical parameters were determined by means of ANOVA followed by the Student–Newman–Keuls test with $p < 0.05$ indicating significant difference.

RESULTS

At seven days postischemia, the differences in the extent of the infarcted brain tissue were easily perceived (see FIGURE 1). Apart from the clearly demarcated infarct, the gross appearance of coronally sectioned brains in controls and posttreated animals was unremarkable. In the pretreated group, several brains (4/7) showed widespread intracerebral hemorrhages and partial collapse of the tissue (FIG. 1). Comparison of infarct volumes (see FIGURE 2) showed significant reduction of infarct volume in the post treated animals.

Even in the absence of quantitative determination of neuronal preservation, the extent of the hippocampal damage in the pretreated group was slightly worse than in the controls (see FIGURE 3 A and B). We doubt, however, that neuronal counting would reveal any statistical differences. In the posttreated group, both the thickness and the appearance of the pyramidal neurons in the lateral CA1 segment of the penumbral hippocampus appeared to be normal (FIG. 3 C). The impact of pre- versus postischemic treatment with IB-MECA became fully apparent when GFAP and lectin stains were compared.

The extent of GFAP staining in the controls and pretreated animals (see FIGURE 4 A and B) was fully comparable to that observed in the gerbils exposed to IB-MECA preceding global cerebral ischemia.[19] As in the previous studies, the astrocytes in the pretreated animals were characterized by sturdy processes containing an intense load of GFAP (FIG. 4 B). In the animals treated after ischemia, the appearance of astrocytes changed dramatically. The GFAP containing processes were slender, their elongated ramifications creating a delicate lattice (FIG. 4 C) rather than the coarse network characteristic of the pretreated animals and, to a lesser degree, control mice as well (FIG. 4 A, B, and C).

Microglial infiltration of the penumbral cortex showed similar pattern to that of reactive gliosis. As expected, a significant presence of activated microglia was

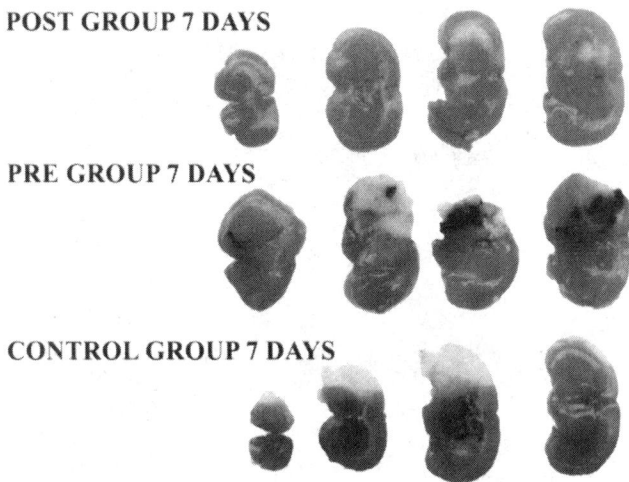

FIGURE 1. Slices of brains removed seven days after ischemia. Note signs of extensive intracerebral hemorrhaging in the striatum and the temporal cortex of the animals treated with IB-MECA prior to the occlusion.

observed in the control animals (FIG. 4 D). In the pretreated animals, the number of microglia reacting to the lectin stain was much higher than in the controls (FIG. 4 E). The surprising finding was the rarity of activated microglia in the penumbra of the posttreated group (FIG. 4 F).

Another unexpected finding was the effect of exposure to IB-MECA on the hillar region of the dentate gyrus in the contralateral hemisphere, where posttreatment with the drug (see FIG. 5 A and C) resulted in a reduction of the changes that were observed following preischemic administration of the agonist (FIG. 5 B and D).

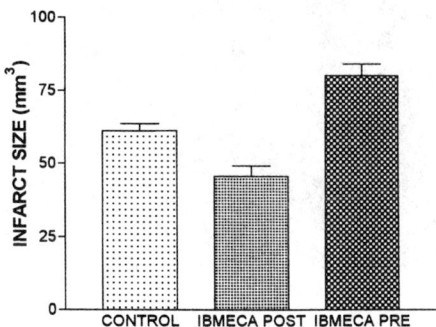

FIGURE 2. Administration of IB-MECA prior or post temporary MCAO. Infarct size in the controls (CTRL), animals pre-treated with IB-MECA (PRE) and post-treated (POST) with the drug. ▭, control; ▓, IBMECA post; ▓, IBMECA pre.

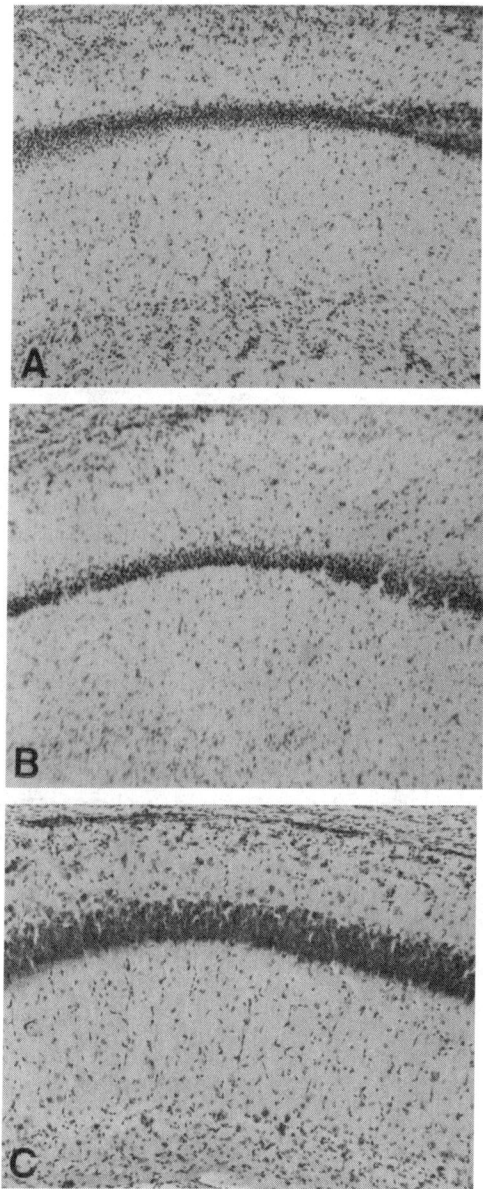

FIGURE 3. The effect of focal ischemia (**A**), IB-MECA prior to focal ischemia (**B**), and IB-MECA after focal ischemia (**C**). Treatment following MCAO occlusion results in a pronounced sparing of the hippocampal neurons in the hippocampus adjacent to the lesion. The damage in the two other groups is, essentially, identical.

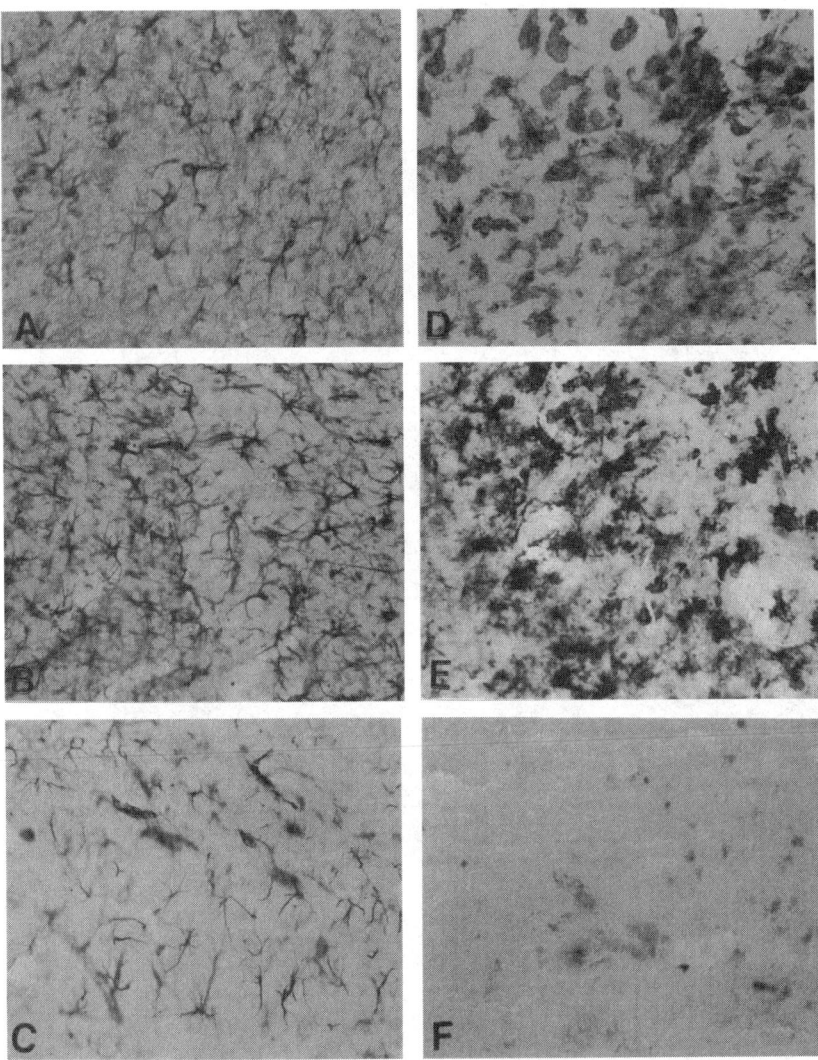

FIGURE 4. Focal ischemia (**A**) and either pretreatment with IB-MECA (**B**) or administration of the drug after removal of the occluding filament (**C**) and their effect on GFAP immunostaining (**A–C**) or microglial infiltration (**D**, ischemia alone; **E**, pretreatment; **F**, posttreatment) within the penumbral zone of the cortex. Note that posttreatment with IB-MECA markedly changes the appearance of astrocytic processes (compare with **A** and **B**) and practically eliminates microglial elements from the affected volume of the brain.

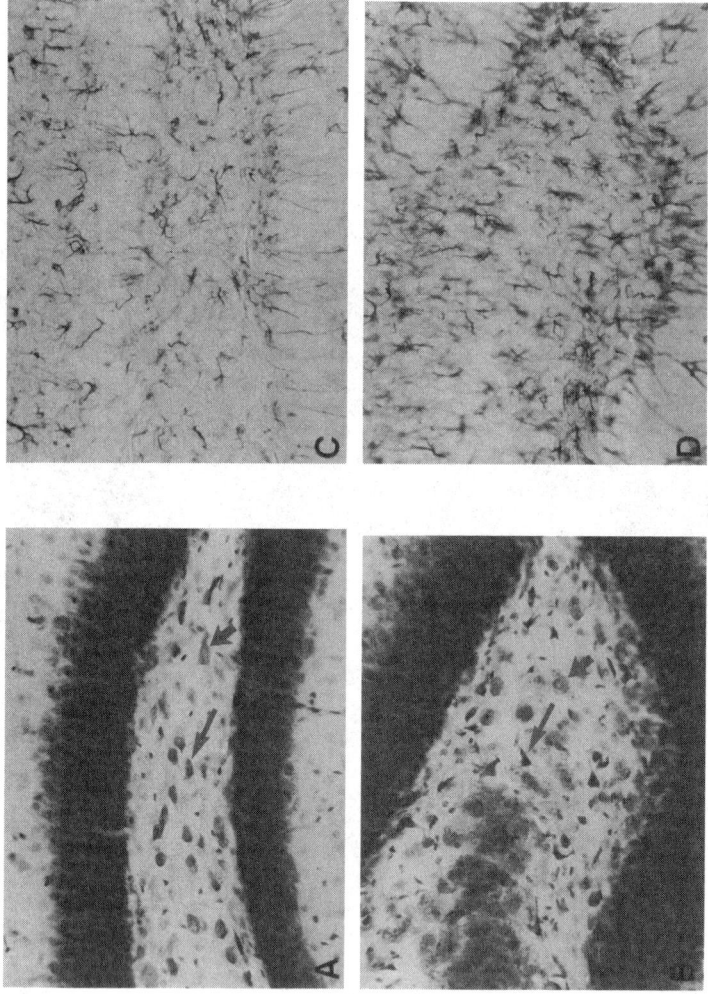

FIGURE 5. Hilus of gyrus dentatus on the contralateral side. **A**, brain of one of the animals posttreated with IB-MECA (Nissl stain). The *thin arrow* points at one of the few damaged neurons, the *thick arrow* at an astrocyte. **C**, GFAP stain from another brain slice in the same animal. Note that the appearance of astrocytes is very similar to those in FIGURE 4 C. **B** represents the same region in the animal pretreated with IB-MECA. The intensity of the collateral damage is far more intense and the astrocytes resemble (**D**) those seen in the ipsilateral cortex (compare to FIG. 4 B).

DISCUSSION

Although the relative importance of the adenosine A_3 receptor in the operation of several cell, tissue, and organ types is the subject of intense speculation,[1,20,21] none of the existing data allow definitive conclusions. To the contrary, many of the recently published studies reveal consistencies within one system, merely to be contradicted by the results obtained in another (for recent reviews see Refs. 16, 20, and 21). The complex intracellular effects evoked by the stimulation of adenosine A_3 receptors (e.g., see Refs. 21–24) are unquestionably among the underlying sources of the ongoing mystery to which the present paper adds, rather than subtracts.

As in global ischemia,[14] acute exposure to IB-MECA prior to middle cerebral artery occlusion (focal ischemia) results in a significant increase of the infarcted brain volume. Since, contrary to our studies of global ischemia,[14] we did not measure postischemic blood flow during the experiments described here, it is unknown whether drug related amplification of the regional disturbances of cerebral blood perfusion contribute to the aggravation of damage in the animals pretreated with IB-MECA. It has been shown, however, that a 200 µg/kg dose of a more selective and potent adenosine A_3 receptor agonist 2-Cl-IB-MECA (2-chloro-N^6-(3-iodobenzyl)adenosine-5'-N-methylcarboxamide) administered i.v. in conscious rats results in the complete release of vascular histamine stores.[6] Thus, since histamine plays a critical role in the formation of brain oedema following focal cerebral ischemia,[25,26] we can not exclude the possibility that the increased infarct volume in animals pretreated with IB-MECA is, at least in part, influenced by the stimulation of the adenosine A_3 receptors located on mast cells.[5,27] Histamine induced oedema involves a nitric oxide component,[28,29] and our previous studies[15,16] demonstrated increased immunocytochemical reactivity of nitric oxide synthase in the pyramidal and radiatum strata of the hippocampus following preischemic injection of IB-MECA in gerbils. However, activation of nitric oxide synthase was entirely independent of the interaction between adenosine A_3 receptors and histamine releasing mechanisms,[2,16] and the immunocytochemical reaction product appeared to concentrate at the neuronal rather than vascular components of the brain.[16,19] Hence, it is quite likely that the destructive impact of preischemically administered IB-MECA involves both circulatory and neuronal components. Finally, astrocytes and microglia are also affected by IB-MECA as shown by our latest studies,[30] indicating that the degree of astro- and microglial activation depends on the timing of treatment with respect to the insult itself.

The multiplicity of cellular and organs systems affected by stimulation of adenosine A_3 receptors, and the powerful effects elicited by such stimulation,[1,16,18,20,21] stand in contrast to the relative paucity of these receptors in all studied cell/organ systems.[31] Yet, although an ever increasing number of adenosine A_3 receptor mediated effects is being reported, their mechanistic aspects remain unclear.[1,16,21] In view of these uncertainties, any attempt at an explanation of the different outcomes between pre- and posttreated animals is also highly tentative and needs extensive experimental corroboration.

It has been suggested[14] that worsening of the outcome of cerebral ischemia induced by acute preischemic administration of adenosine A_3 agonist may represent the cumulative effect of several adverse events triggered by the drug immediately prior to the occlusion ("priming effect"). Events such as release of inflammatory mediators and concomitant degradation of the blood brain barrier integrity, and deleterious attenuation of the cerebral blood flow,[6,25–27,32] may be of primary importance. Combined with the increased influx of Ca^{2+} (both passive and through the voltage regulated calcium channels) and the liberation of internal calcium stores elicited by adenosine A_3 receptor stimulation,[22] the overall impact of these events would, indeed, predispose the brain to a significantly increased susceptibility to ischemia. On the other hand, when administered following a focal insult, IB-MECA induces cerebroportection in focal ischemia when the drug is given following transient occlusion of the middle cerebral artery.[30] Whether this cerebroprotective impact of adenosine A_3 receptor agonists is related to astrocyte activation, direct neuroprotective effect, or both[16,18,19,33,34] is unclear. The results of immunocytochemical studies presented in this paper confirm the presence of complex neuronal and glial effects induced by postischemic stimulation of adenosine A_3 receptors. Moreover, the latter observations offer strong support for the recently published hypothesis on the role of cerebral adenosine A_3 receptors as a part of the "adenosine based cerebroprotective complex".[1,16] This paper, and also our previous studies of adenosine A_3 receptor impact on the outcome of focal and global ischemia,[14,16,30] indicate significant involvement of these receptors in the pathology resulting from the arrested cerebral blood supply. However, the relationship between adenosine A_3 receptor stimulation and the outcome of an ischemic event is not straightforward. The presented data have confirmed initial hypothetical assumptions that the extent of the subsequent damage depends on the timing of adenosine A_3 receptor activation versus the onset of ischemia.[30] Most likely, the degree of receptor activation[18,21,33] may be an important factor as well. In summary, although our results provide further illumination of the complexity of A_3 receptor elicited effects, they do not provide a definitive solution to the baffling role of these receptors in the generation of stroke damage. As in many other studies of this still untreatable disease, we too must meekly conclude that "further extensive experiments are necessary" in order to understand the exact nature of the participating mechanisms and the factors determining the dual role of adenosine A_3 receptors in the brain.

REFERENCES

1. VON LUBITZ, D.K.J.E. 1999. Adenosine and cerebral ischemia: therapeutic future or death of a brave concept? Eur. J. Pharmacol. **371:** 85–102.
2. VON LUBITZ, D.K.J.E., *et al.* 1999. J. Cereb. Blood Flow.
3. MEYERHOF, W.R., R. MÜLLER-BRECHLIN & D. RICHTER. 1991. Molecular cloning of novel putative G protein coupled receptor expressed during rat spermiogenesis. FEBS Lett. **284:** 155–160.
4. ZHOU, Q.Y., C.Y. LI, M.E. OLAH, *et al.* 1992. Molecular cloning and characterization of an adenosine receptor: the A_3 receptor. Proc. Natl. Acad. Sci. U.S.A. **89:** 7432–7436.
5. SHEPHERD, R.K., J. LINDEN & B.R. DULING. 1996. Adenosine-induced vasoconstriction in vivo. Role of the mast cell and A_3 adenosine receptor. Circ. Res. **78:** 627–634.
6. VAN SCHAICK, E.A., K.A. JACOBSON, H.O. KIM, *et al.* 1996. Hemodynamic effects and histamine release elicited by the selective adenosine A_3 receptor agonist 2-Cl-IB-MECA in conscious rats. Eur. J. Pharmacol. **308:** 311–314.

7. FOZARD, J.R. & A.M. CARRUTHERS. 1993. Adenosine A_3 receptors mediate hypotension in the angiotensin II-supported circulation of the pithed rat. Br. J. Pharmacol. **109:** 3–5.
8. BOUMA, M.G., T.M. JEUNHOMME, D.L. BOYLE, et al. 1997. Adenosine inhibits neutrophil degranulation in activated human whole blood: involvement of adenosine A_2 and A_3 receptors. J. Immunol. **158:** 5400–5408.
9. KNIGHT, D., X. ZHENG, C. ROCCHINI, et al. 1997. Adenosine A_3 receptor stimulation inhibits migration of human eosinophils. J. Leukoc. Biol. **62:** 465–468.
10. MITCHELL, C.H., K. PETERSON-YANTORNO, D.A. CARRE, et al. 1999. A_3 adenosine receptors regulate Cl- channels of nonpigmented ciliary epithelial cells. Am. J. Physiol. **276**(3-1): C659–666.
11. STAMBAUGH, K., K.A. JACOBSON, J.-L. JI & B.T. LIANG. 1997. A novel cardioprotective function of adenosine A_1 and A_3 receptors during prolonged simulated ischemia. J. Physiol. (Heart Circ. Physiol.) **42:** H501–H505.
12. DOUGHERTY, C., J. BARUCHA, P.R. SCHOFIELD, et al. 1998. Cardiac myocytes rendered ischemia resistant by expressing the human adenosine A_1 and A_3 receptor. FASEB J. **12:** 1785–1792.
13. HILL, R.J., J.J. OLEYNEK, W. MAGEE, et al. 1998. Relative importance of adenosine A_1 and A_3 receptors in mediating physiological or pharmacological protection from ischemic myocardial injury in the rabbit heart. J. Mol. Cell Cardiol. **30:** 579–585.
14. VON LUBITZ, D.K.J.E., R.C.S. LIN, P. POPIK, et al. 1994. Adenosine A_3 receptor stimulation and cerebral ischemia. Eur. J. Pharmacol. **263:** 59–67.
15. VON LUBITZ, D.K.J.E., R.C.S. LIN & K.A. JACOBSON. 1997. Adenosine A_3 receptors and protection against cerebral ischemic damage in gerbils. Soc. Neurosci. Abstr. 23/2, 745.16.
16. VON LUBITZ, D.K.J.E. 1999. Stimulation of adenosine A_3 receptors in cerebral ischemia: neuronal death, recovery, or both? Ann. N.Y. Acad. Sci. **890:** 93–105.
17. ABBRACCHIO, M.P., R. BRAMBILLA, S. CERUTTI, et al. 1995. G-protein dependent activation of phospholipase C by adenosine A_3 receptors in the rat brain. Mol. Pharmacol. **48:** 1038–1045.
18. ABBRACCHIO, M.P., S. CERUTI, R. BRAMBILLA, et al. 1997. Modulation of apoptosis by adenosine in the central nervous system: a possible role for the A_3 receptor. Ann. N.Y. Acad. Sci. **825:** 11–22.
19. VON LUBITZ, D.K.J.E., R.C.S. LIN, M.BOYD, et al. 1999. Chronic administration of adenosine A_3 receptor agonist and cerebral ischemia: neuronal and glial effects. Eur. J. Pharmacol. **367:** 157–163.
20. JACOBSON, K.A. 1999. Adenosine A_3 receptors: novel ligands and paradoxical effects. Trends Pharmacol. Sci. **19:** 185–191.
21. JACOBSON, K.A., C. HOFFMANN, F. CATTABENI & M.P. ABBRACCHIO. 1999. Adenosine-induced cell death: evidence for receptor-mediated signaling. Apoptosis **4:** 197–211.
22. KOHNO, Y., X.-D. JI, S.D. MAWHORTER, et al. 1996. Activation of A_3 adenosine receptors on human eosinophils elevates intracellular calcium. Blood **88:** 5569–3574.
23. CERUTI, S., D. BARBIERI, C. FRANCESCHI, et al. 1996. Effects of adenosine A_3 receptor agonists on induction of cell protection at low and cell death at high concentrations. Drug Dev. Res. **37:** 177.
24. ABBRACCHIO, M.P., G. RAINALDI, A.M. GIAMMARIOLI, et al. 1997. The A_3 adenosine receptor mediates cell spreading, reorganization of actin cytoskeleton, and distribution of Bcl-XL: studies in human astroglioma cells. Biochem. Biophys. Res. Comm. **24:** 297–304.
25. JOO, F., J. KOVACS, P. SZRDAHELYIS, et al. 1994. The role of histamine in brain oedema formation. Acta Neurochir. Suppl. **60:** 76–78.
26. NEMETH, L., M.A. DELL, A. FALUS, et al. 1998. Cerebral ischemia reperfusion-induced vasogenic brain edema formation in rats: effect of an intracellular histamine receptor antagonist. Eur. J. Pediatr. Surg. **8:** 216–219.
27. FOZARD, J.R., H.J. PFANKUCHE & H.J. SCHUURMAN. 1996. Mast cell degranulation following adenosine A_3 receptor activation in rats. Eur. J. Pharmacol. **298:** 293–297.
28. MAYHAN, W.G. 1996. Role of nitric oxide in histamine-induced increases in permeability of the blood-brain-barrier. Brain Res. **743:** 70–76.

29. SZABO, A., J. KASZAKI, M. BOROS & S. NAGY. 1997, Possible relationship between histamine and nitric oxide release in the postischemic flow response following mesenteric ischemia of different durations. Shock **7:** 376–382.
30. YE, W., J. MCCLELLAN & D.K.J.E. VON LUBITZ. 1999. Involvement of adenosine A_3 receptor in generation of experimental brain damage following middle cerebral occlusion in mice. Acad. Emerg. Med. **6:** 261.
31. JI, X.-D., D.K.J.E.VON LUBITZ, M.E. OLAH, et al. 1994. Species differences in ligand affinity at central A_3 adenosine receptors. Drug Dev. Res. **33:** 51–59.
32. HALLENBECK, J.M. 1996. Significance of the inflammatory response in brain ischemia. Acta Neurochir. Supp. **66:** 27–31.
33. SEI, Y., D.K.J.E. VON LUBITZ, M.P. ABBRACCHIO, et al. 1997. Adenosine A_3 receptor agonist-induced neurotoxicity in rat cerebellar granule neurons. Drug Dev. Res. **40:** 267–273.
34. YAO, Y., Y. SEI, M.P. ABBRACCHIO, et al. 1997. Adenosine A_3 receptor agonists protect HL-60 and U-937 cells from apoptosis induced by A_3 antagonists. Biochem. Biophys. Res. Comm. **232:** 317–322

Part III: Endogenous Neuroprotectants—Adenosine Receptors

COMMENT FROM DR. ANDREWS

This is a comment for Dr. von Lubitz with respect to the increase in microglia seen in the dentate gyrus contralateral to the MCA-O. We found in our swine model for brain retraction ischemia that during unilateral brain retraction not only was there a decrease ipsilaterally in CBF and evoked potential amplitude, but also a *contralateral increase* in CBF and EP amplitude. This contralateral effect was lost when the corpus callosum was sectioned (Neurosci. Lett. **154**: 9–12, 1993). I suspect your findings regarding microgila contralaterally are further support for the notion that the contralateral hemisphere in unilateral ischemia models is *not* a good control, given the effects of transcallosal or transhemispheric diaschisis (Stroke **22**: 943–949, 1991).

QUESTIONS FOR DR. ABBRACCHIO

From Dr. Marini

One word of caution about knockout animals that the COX_2 inhibitor was still found to protect against ischemia in COX_2 knockout animals.

ANSWER: I think that this is an important point. The ability of COX_2 inhibitors to still protect against ischemia in knockout animals may depend on several reasons. It has been demonstrated for a number of other knockout models that there are compensation mechanisms in these animals, so that when you knock out a gene, its functions are vicariated by other genes belonging to the same family. Hence, COX_2 knockout animals may express other proteins involved in the same function. Another likely explanation would be that COX_2 inhibitors also have additional mechanisms of action unrelated to inhibition of COX_2 and still resulting in neuroprotection.

From Dr. Slikker

What is the importance of A3 receptor effects on apoptosis following ischemia and penumbral development?

ANSWER: There are no definitive experimental data at present on this issue. There is a fascinating hypothesis raised by Dr. von Lubitz (Von Lubitz, *et al.* Ann. N.Y. Acad. Sci. **890**: 93–106, 1999), according to which A3 receptor mediated formation of an apoptotic ring around the ischemic core may reduce the propagation of damage to the surrounding penumbra area. In fact, at variance from necrosis, death by apoptosis is not characterized by inflammation and release of cytotoxic mediators. This hypothesis is also consistent with the concept that in some cases apoptosis may indeed be beneficial (Neary *et al.*, Trends Neurosci. **19**: 13–18, 1996).

QUESTIONS FOR DR. PEDATA

From Dr. Slikker

Have you examined for peripheral effects following those treatments that resulted in increases in taurine microdialysis data (for example, blood urea nitrogen (BUN) or plasma ammonia)?

ANSWER: Our assessment of taurine concentrations in microdialysis studies was limited to the model of ischemia induced by medial cerebral artery occlusion and we did not check for models, such as hyperammonemic animals, which may induce change in brain extracellular taurine concentrations.

From Dr. Auer

Do you think your finding in cortex (protection) versus striatum (less protection) support the notion of two kinds of infarction, one autolytic (striatum) and the other one metabolic (cortex)? The idea is that autolytic infarction is less amenable to treatment than metabolic infarction and in the latter, adenosine antagonist administration may have benefit.

ANSWER: In the model of focal cerebral ischemia induced by middle cerebral artery occlsion (MCA-O) by suture technique, the striatum represents the "ischemic core" with minimal post-occlusion blood flow and the cortex represents an "ischemic penumbra", where some blood flow persists via a collateral supply. Neurons in the penumbral area are electrically silent, but they can be salvaged if flow is reestablished or if neuroprotective measures are undertaken. In the first hours following cerebral ischemia, Ca^{2+} overload, glutamate excitotoxicity, free-radical damage and energy depletion, secondary to ischemic neuronal depolarization are the main destructive mechanisms threatening the ischemic penumbra. Since the ischemic penumbra is a region at risk for infarction and in which glutamate excitotoxicity is most marked, we can presume that adenosine A_{2A} antagonism, by reducing glutamate outflow in the early stages of ischemia, offers protection against damage progression. Conversely, it is generally accepted that the infarct in the ischemic core is not, or is less, amenable with a pharmacological approach and we can suppose that, despite reduced excitotoxicity brought about by A_{2A} antagonists, the lack of a sufficient blood flow causes a progressive neuronal necrosis, which can be defined autolytic.

From Dr. Narahashi

SCH 58261 inhibited GABA outflow almost completely as compared with outflow of other transmitters, such as glutamate and aspartate. What is the reason for such a potent inhibition of GABA outflow?

Do you think that the neuroprotective action of A_{2A} antagonists is associated solely or partly with antagonism of glutamate? You are getting differential effects in the cortex and striatum? We know that glutamate antagonists are not neuroprotective in the MPTP model and yet A_{2A} antagonists are. What is the mechanism? In our studies with Ken Jacobson, we find that A_{2A} antagonists are neuroprotective in PC-12 cells and neuroblatomas cells in response to H_2O_2 or 6-hydroxy dopamine.

Yet A_{2A} antagonists are not antioxidant. Do preliminary results suggest involvement of Ca^{2+}?

ANSWER: Preliminary experiments indicate that A_{2A} antagonism strongly reduces GABA outflow induced by MCA-O. GABA shows the largest outflow increase of the aminoacids in our and also different models of *in vivo* ischemia. Adenosine A_{2A} receptors strongly regulate the GABA outflow. We have recently observed, in a microdialysis study (Corsi *et al.*, NeuroReport **10:** 3933, 1999) that the selective A_{2A} agonist CGS 21680 approximately doubles striatal GABA outflow. Furthermore, glutamate released under ischemia is known to exert a stimulatory tonus on all striatal circuits including GABA neurons. Thus, the clear decrease of GABA outflow induced by A_{2A} antagonism could result from different effects. Preliminary data indicate that A_{2A} antagonism is protective against the histopathological outcome in the cortex and not in the striatum. The cortex, in this model of MCA-O, represents an "ischemic penumbra" that is most sensitive to glutamate excitoxicity. On the other hand, since it has been reported that A_{2A} receptor stimulation desensitize adenosine A_1 receptors, we may envisage that A_{2A} antagonism is protective also because it potentiates all the neuroprotective A_1-mediated effects. A reduction of the development of inflammatory processes may also be taken into consideration to explain the A_{2A} antagonism protective effect.

It is generally accepted that the main protective effect of the A_{2A} antagonism, in the MPTP model of Parkinson disease, is the potentiation of the dopamine effects mediated by the D_2 receptors on the GABA-enkephalin output neurons from the striatum. We did not perform direct Ca^{2+} measurement in our experiments.

QUESTION FOR DR. ABBRACCHIO

From Dr. Youdim

I agree with Dr. Abbracchio that not enough attention has been paid to reactive microglia observed at the site of lesions in neurological degenerative diseases. We have observed both in Parkinson's disease and its animal model of 6-hydroxy dopamine significant reactive gliosis with microglia containing significant amounts of iron. This is also seen with kianic acid induced neurodegeneration. What is fascinating is that if we induce nutritional iron deficiency in rats, where brain iron falls by some 30%, neither kinate or 6-hydroxy dopamine is able to induce neurodegeneration or reactive microglyosis, suggesting that a microglia exacerbate significantly the process of neural degeneration, which may involve iron.

ANSWER: I am glad to hear that you have data obtained in your experimental models supporting the importance of glial cells in the development of cerebral damage. Evidence is accumulating that glial cells (both astroglia and microglia) may indeed contribute to neuronal death both in acute (e.g., ischemia) and in chronic neurodegenerative diseases. The emerging idea is that damaged neurons may release substances that activate glial cells. On one side, activation of these cells is known to lead to beneficial effects aimed at favoring neuronal recovery; on the other hand, intense or prolonged glial cell activation (as that hypothesized to occur after a strong acute damage or as a consequence of chronic release of activating factors from neurons) is accompanied by release of cytotoxic substances that may themselves contribute to neuronal damage. I think that in the near future we will hear more about this neuron–glia cell interaction, and that inhibition of glial cell activation may represent a novel neuroprotective strategy in several diseases.

Monitoring for Neuroprotection

New Technologies for the New Millennium

RUSSELL J. ANDREWS

NASA Ames Research Center, Moffett Field, California, U.S.A.

Department of Neurosurgery, Stanford University Medical Center, Stanford, California, U.S.A.

Division of Neurosurgery, Texas Tech University Health Sciences Center, El Paso, Texas, U.S.A.

ABSTRACT: Monitoring for neuroprotection, like surgery, has placed an emphasis on minimal or non-invasiveness. Monitoring of parameters that truly reflect the degree of injury to the nervous system is another goal. Thus, two themes for the coming decade in neuromonitoring will be: (1) less-invasive monitoring; and (2) parameters that more closely reflect the etiological factors in ischemic or other neuroinjury. In this paper, we review neuromonitoring techniques and devices that can be used readily in the operating room or intensive care unit setting. Those that require transport of the patient to a special facility (e.g., for computed tomography or magnetic resonance imaging/spectroscopy) and those that have been in standard practice for neuromonitoring (e.g., electrophysiological monitoring—EEG, evoked potentials) are not considered. The two techniques considered in detail are (1) continuous multiparameter local brain tissue monitoring with microprobes, and (2) non-invasive continuous local brain tissue oxygenation monitoring by near infrared spectroscopy. Both techniques have been cleared by the Food and Drug Administration (FDA) for clinical use. The rationale for their use, the nature of the devices, and clinical results to date are reviewed. It is expected that both techniques will gain wide acceptance during the coming decade; further advances in neuromonitoring that can be expected further into the twenty-first century are also discussed.

KEYWORDS: Brain ischemia; Cerebral oxygenation; Neuromonitoring; Neuroprotection, Near infrared spectroscopy.

INTRODUCTION

Monitoring of the nervous system, in both the operating room (OR) and the intensive care unit (ICU), progressed significantly during the past decade—the decade of the brain. However, much greater strides in monitoring for neuroprotection in these clinical settings are being made by implementation of techniques developed in the laboratory during the last couple of decades of the twentieth century. The first decade

Address for correspondence: Russell J. Andrews, Division of Neurosurgery, Texas Tech University, Health Sciences Center, 4800 Alberta, El Paso, TX 79905, U.S.A., Voice: 915-545-6676; fax: 915-545-7584.

andrews_rj@compuserve.com

of the new millennium will continue to see new technologies applied to minimally or non-invasive monitoring of the nervous system in the OR and ICU.

Clinical monitoring has traditionally consisted of "we monitor what we can measure", not necessarily what would be most relevant to avoiding brain injury. This has been for technical reasons—frequently what we would like to measure is difficult to do in the OR or ICU setting. Much of the advance in neuromonitoring during the next decade will consist of improvement in the parameters monitored, for example, local brain tissue blood flow and local brain tissue oxygenation.

Two themes for the coming decade in neuromonitoring will be: (1) less invasive monitoring and (2) use of parameters that more closely reflect the etiological factors in ischemic or other neuroinjury.

This paper considers only neuromonitoring techniques and devices that can be used readily in the OR or ICU setting, and not those requiring transport of the patient to a special facility (e.g., single photon emission computed tomography [SPECT] or magnetic resonance spectroscopy [MRS]). Additionally, we do not consider advances in electrophysiological monitoring, which is the topic of another paper elsewhere in this volume.

THE STATE OF THE ART: MAP, ICP, CPP, CBF

During the 1990s, the standard of care in major trauma centers for head injury monitoring included:

1. Continuous measurement of mean arterial pressure (MAP) by means of a catheter placed in the radial artery.

2. Continuous measurement of intracranial pressure (ICP) by means of a catheter placed into the ventricle of the brain or a probe (customarily fiberoptic) in the brain parenchyma.

3. Calculation of the cerebral perfusion pressure (CPP = MAP − ICP) while maintaining the CPP greater than 70 mm Hg as a goal.

4. In many centers, the use of transcranial doppler (TCD) or other techniques to measure cerebral blood flow (CBF).

Reviewing these techniques briefly, intracranial pressure (ICP) is measured by a small catheter (approximately 3 mm diameter) placed into the ventricular system and connected to a pressure transducer. Alternatively, a 1-mm diameter fiberoptic probe or strain gauge can be inserted into the brain parenchyma. Either technique yields a pressure within the relatively fixed volume of the skull that is interpreted as a resistance countering the systemic arterial blood pressure (MAP), the difference being the cerebral perfusion pressure (CPP)—that is, CPP = MAP − ICP.

Transcranial Doppler (TCD) employs a Doppler probe placed over the thin bone of the temporal skull to assess the blood flow in the middle cerebral artery (see FIGURE 1).[1] Although of questionable accuracy in measuring absolute flow, TCD is quite reliable in detecting the marked declines in CBF that may produce cerebral ischemia in settings such as brain swelling following severe closed head injury or intraoperative brain retraction. Other techniques for measuring CBF in the OR and ICU involve placing a probe either on the brain surface or into the brain parenchyma, but these techniques have not gained widespread clinical acceptance.

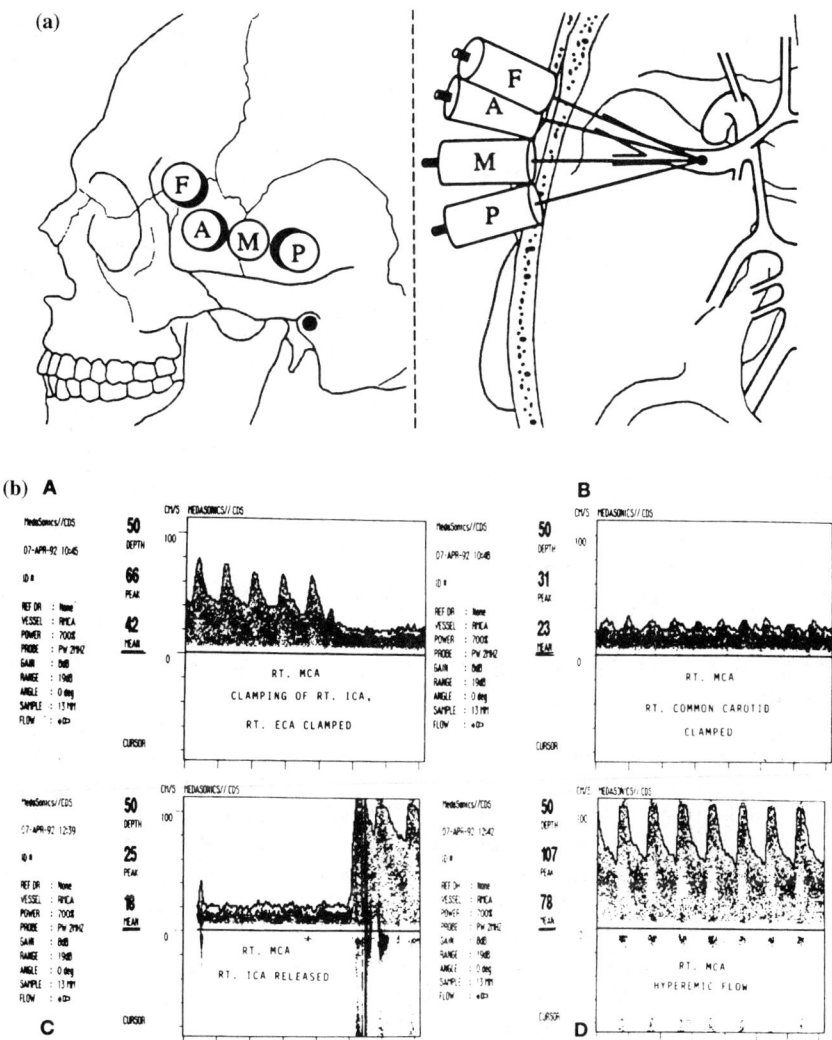

FIGURE 1. (a) The transtemporal "acoustic window" for transcranial Doppler cerebral blood flow monitoring of the middle cerebral artery has four areas—frontal, anterior, middle, and posterior—with transducer angulation varying according to which window is being used. (b) Characteristic changes in middle cerebral artery flow velocity during carotid endarterectomy. A. Crossclamp of the internal carotid artery leads to an abrupt decrease in flow velocity. B. Slight increase in pulsatility with continuous crossclamp, most likely a result of increased flow via collaterals. C. Release of clamp is followed by reactive hyperemia. The vertical streaks represent air emboli. D. Persistent hyperemia after release. (Reprinted from Ref. 1 with permission from Lippincott Williams & Wilkins.)

CLINICAL NEUROMONITORING: YEAR 2000 TO 2010

Two devices representing two different techniques of neuromonitoring are discussed. The first received Food and Drug Administration (FDA) approval in 1999. Because it is marketed by a large medical equipment and supply company (which also markets a popular ICP monitor), this device should come into relatively widespread use—particularly for patients, such as those with severe head injuries, who currently are followed in the ICU with ICP monitoring.

The second device also has FDA approval for clinical use, but although it has been available for several years, it has not achieved significant clinical use. As more experience and data are acquired—and the device becomes more robust and user friendly—its advantages of non-invasiveness and relatively inexpensive cost per patient should result in more widespread use. Perhaps a major reason for its relatively slow penetration into the neuromonitoring arena is that it is not marketed by a major medical equipment company. However, like TCD (which is also marketed by several companies that are not among the "major players" in medical equipment companies) this device should assume a greater role in neuromonitoring year-by-year. There are some similarities between pharmacologic neuroprotection and neuromonitoring in that the "politics" of (1) companies with large market share, and (2) failure to collaborate among companies have greatly altered (i.e., retarded) progress toward optimal patient benefit.[2]

Continuous Local Brain PO_2, PCO_2, pH, and Temperature Monitoring

The Neurotrend Cerebral Tissue Monitoring System (Codman/Johnson & Johnson, Raynham, MA) is an example of a medical device where the time from conceptualization to clinical implementation and FDA approval has been relatively brief. The device was initially developed for continuous arterial blood gas monitoring by several competing companies in the early 1990s. During beta site testing by one of these companies, the present author obtained one of the devices and demonstrated that it could be used to monitor pH and PCO_2 *in vivo* in the cerebrospinal fluid (CSF) of the brain of a large animal (swine).[3]

The concept and the device were rapidly adopted by several groups interested in neuromonitoring for brain ischemia in the OR and ICU settings.[4-12] The device underwent clinical trials during the late 1990s that resulted in FDA approval for neuromonitoring in humans in 1999. The Codman/Johnson & Johnson Neurotrend Cerebral Tissue Monitoring System is illustrated in FIGURE 2.

The sensor is a 0.5 mm diameter catheter (of a size that fits through a 20 gauge intravenous catheter for continuous radial artery blood gas monitoring). A schematic of the sensor is shown in FIGURE 3.[4] Fluorescent dyes (chosen to respond to oxygen, carbon dioxide, and hydrogen [pH])—or alternatively, in the case of oxygen, a Clark electrode—and a temperature sensor are included in the 0.5 mm catheter. The device is placed through a small hole in the skull, customarily into the frontal cortex, to a depth of two to three centimeters.

The results of brain tissue monitoring in severe head injury patients have been quite impressive. One study found significant differences in the mean values for brain PO_2, PCO_2, and pH ($p < 0.01$) between patients with a good outcome ($n = 8$) and those who died or remained vegetative ($n = 10$): PO_2 was 39 versus 19 mm Hg,

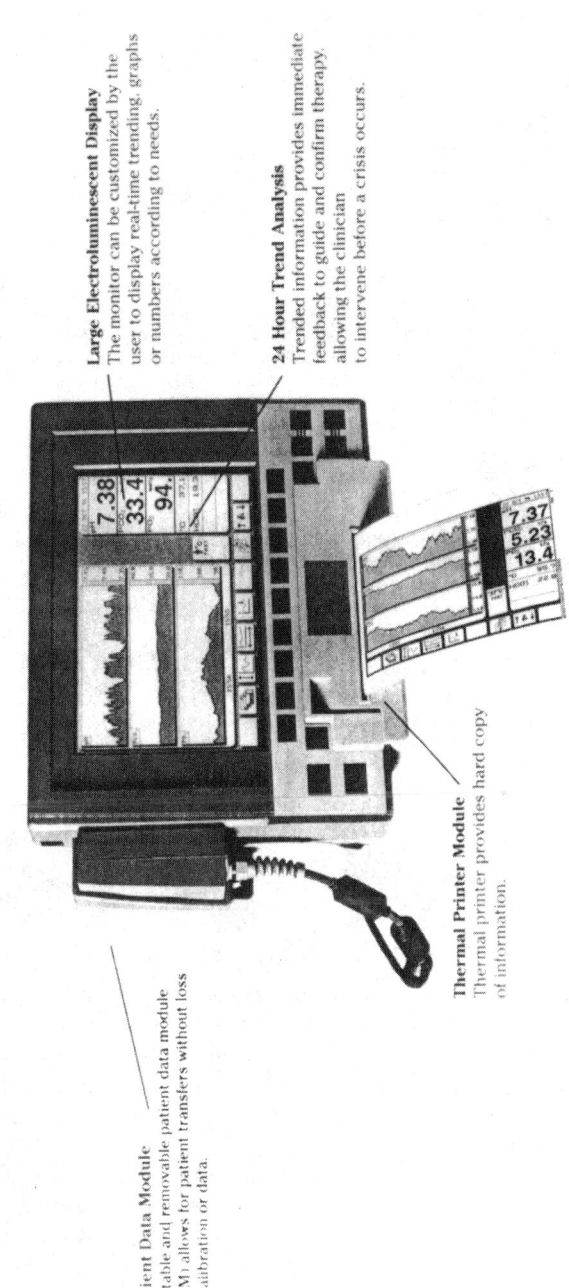

FIGURE 2. The Codman Neurotrend Cerebral Tissue Monitoring System. Local brain tissue temperature, pH, PO_2, PCO_2 are monitored continuously. (Reprinted with permission from Codman, Raynham, MA.)

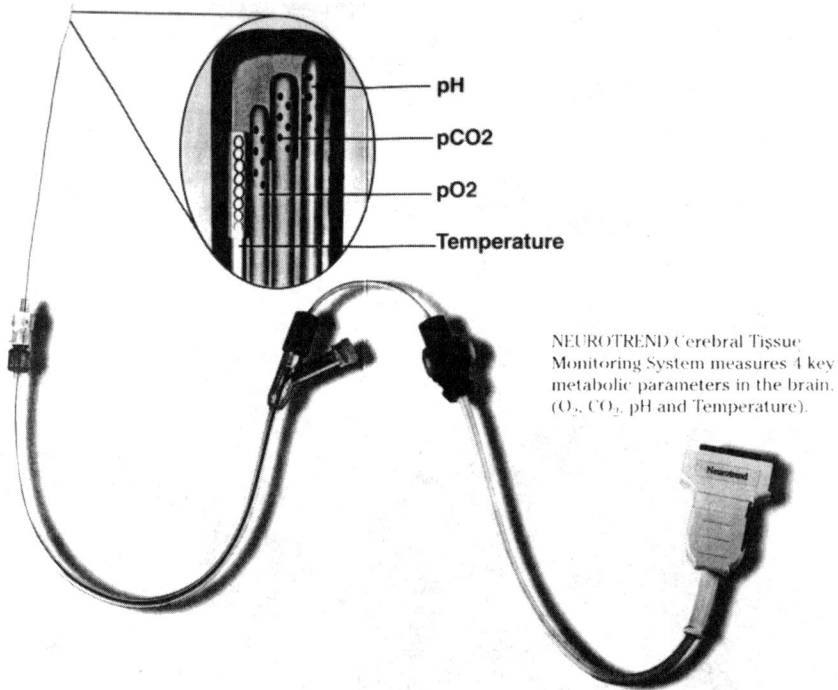

FIGURE 3. The multiparameter sensor probe of the Neurotrend System, approximately 0.5 mm in diameter. (Reprinted with permission from Codman, Raynham, MA.)

PCO_2 was 50 versus 64 mm Hg, and pH was 7.14 versus 6.85, respectively.[9] The same study showed an example of the transient changes in pH, PCO_2, and PO_2 seen in a severe head injury patient who experienced a generalized seizure (confirmed by electroencephalography) (see FIGURE 4).[9]

Brain tissue monitoring has demonstrated the possible importance of increased inspired oxygen concentration on the reduction of secondary brain injury in head injury patients.[11] By increasing the fraction of inspired oxygen from 35% to 60% for three hours, and then from 60% to 100% for an additional three hours, significant increase in brain tissue oxygen, and decrease in brain tissue lactate and glucose (as measured by microdialysis) were accomplished (see FIGURE 5).[11] The argument is that the hyperoxygenation in the first few hours after severe head injury leads to significant improvement in aerobic metabolism in ischemic brain, and thus may improve outcome. Additional studies correlating brain tissue oxygen and clinical outcome, as well as measures of aerobic metabolism, such as adenosine triphosphate production, are necessary to establish the validity of hyperoxygenation (by increasing the fraction of inspired oxygen) as an early intervention for cerebral ischemia. Another application of the brain tissue monitoring technology may be in acute stroke patients, where hyperoxygenation could also prove beneficial.

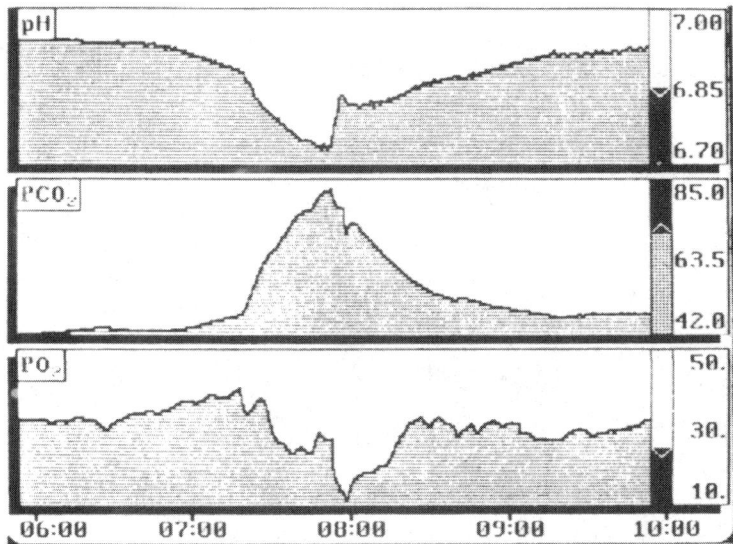

FIGURE 4. Graphs of brain PO_2, PCO_2, and pH measured during a generalized seizure in a patient with severe head injury that was confirmed by electroencephalography. The decrease in brain PO_2 is accompanied by an increase in brain PCO_2 and a decrease in brain pH. (Reprinted from Ref. 9 with permission from *Neurosurgery*.)

Continuous Local Brain Oxygen Monitoring—Near Infrared Spectroscopy

Near infrared spectroscopy (NIRS) permits monitoring of brain tissue oxygenation through the intact skull—a technique that shares with transcranial doppler CBF monitoring the advantage of non-invasiveness.[13,14] Since biological tissue is relatively transparent to light in the near infrared range (700–1000 nm), light in this wavelength range can be directed through the skull and similarly collected after

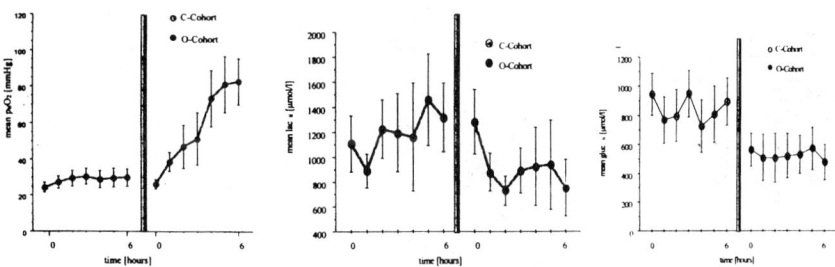

FIGURE 5. Graphs showing the time course of mean brain tissue PO_2 (**left**), mean brain tissue lactate (**center**), and mean brain tissue glucose (**right**), in 12 patients each in the control cohort (C-Cohort) and the O_2-treated cohort (O-Cohort). *Error bars* indicate the standard error of the mean. (Reprinted from Ref. 11 with permission from the *Journal of Neurosurgery*.)

passing through the brain. Fiber optic bundles (optodes) are used to transmit the light from the light source as well as to collect the light (see FIGURE 6). By placing opted arrays bilaterally, unilateral changes in hemoglobin saturation can be detected by noting a change in the ratio between the two sides; thus the relatively poor absolute

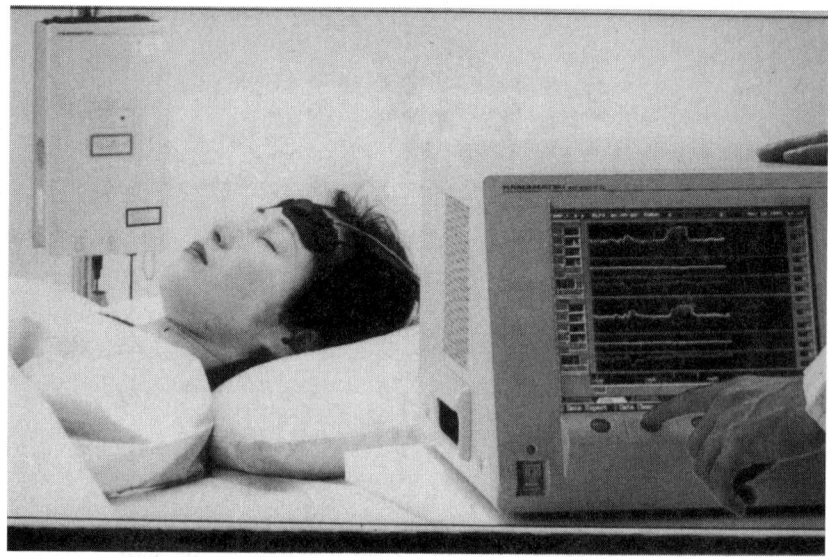

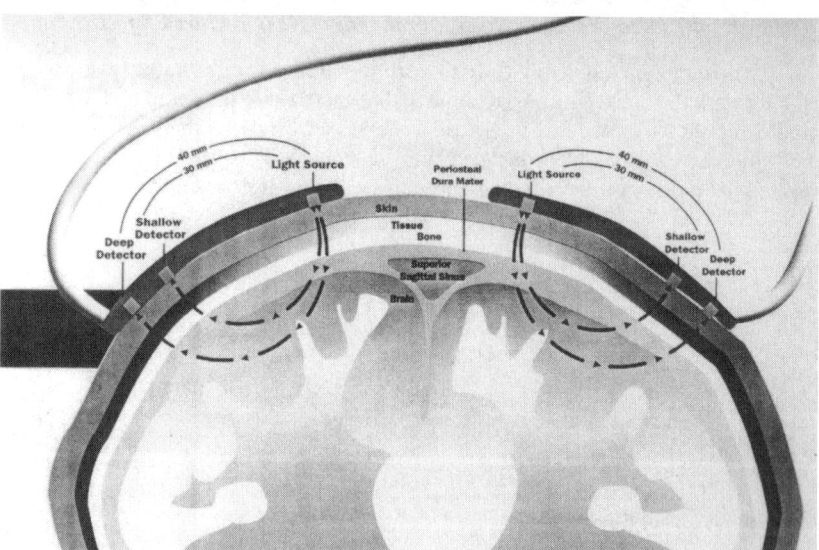

FIGURE 6. Typical NIRS monitoring set-up (**top**) and schematic of the optode arrays placed bilaterally with the shallow and deep detectors depicted (**bottom**). (Reprinted from with permission from Hamamatsu, Billerica, MA [**top**] and Somanetics, Troy, MI [**bottom**].)

values obtained by NIRS can be overcome, at least in part. By having more than one light receiving probe (one receiving probe two to four cm from the light source, the other one or two cm further from the light source), the contribution of extracranial blood flow can be subtracted (FIG. 6).

The potential applications of NIRS transcranial monitoring are considerable. One recent report suggests that NIRS may be superior to TCD in assessing patients with cerebral microangiopathy, a condition associated with aging, arterial hypertension,

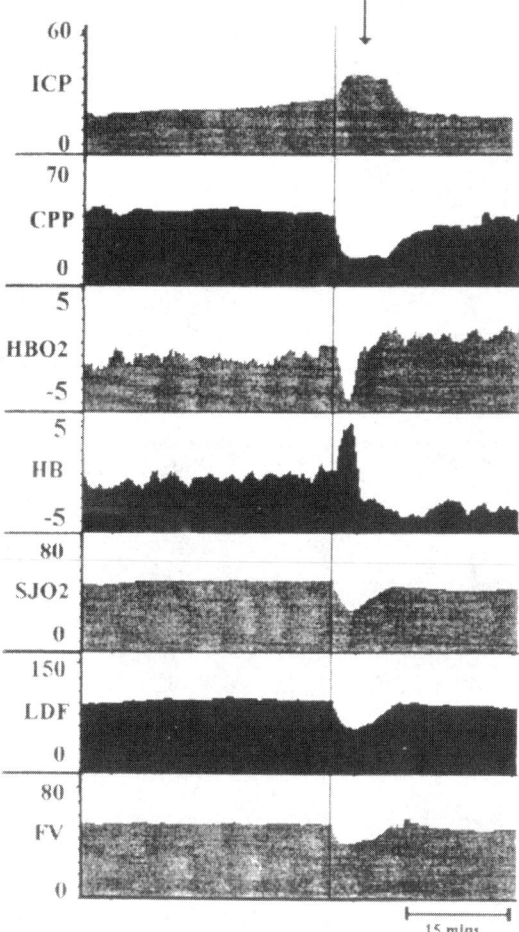

FIGURE 7. A spontaneous event recorded from the right side of the head in a 22-year-old man with a diffuse head injury. ICP, intracranial pressure (mm Hg); CPP, cerebral perfusion pressure (mm Hg); HBO_2, oxyhemoglobin; HB, deoxyhemoglobin; SJO_2, jugular venous blood oxygen saturation; LDF, laser Doppler cerebral blood flow; FV, right middle cerebral artery flow velocity (cm/sec). The temporal correlation of changes in oxy/deoxyhemoglobin with changes in the other parameters during this event lasting less than 15 min is clear. (Reprinted from Ref. 16 with permission from the *Journal of Neurosurgery*.)

110　ANNALS NEW YORK ACADEMY OF SCIENCES

lacunar infarction, and deep white matter demyelination.[15] Its main application, however, appears to be for neuromonitoring in the OR and the ICU (1) in patients who are undergoing operations with significant risk of reversible cerebral ischemia (e.g., carotid endarterectomy), and (2) in patients who have suffered severe head injury or stroke, and/or who may be at risk for intracranial hemorrhage.

FIGURE 7 records data obtained from multiple monitoring techniques during a spontaneous event in a 22-year-old man with a diffuse head injury.[16] The temporal correlation of changes in oxy/deoxyhemoglobin with changes in the other parameters during this event lasting less than 15 min is clear. Data such as these support the use of NIRS as a non-invasive, relatively inexpensive means of neuromonitoring.

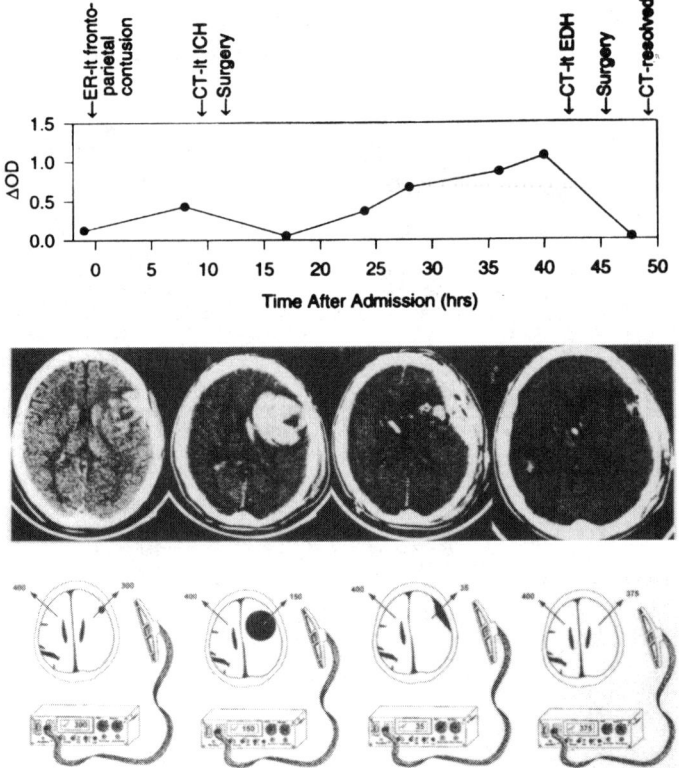

FIGURE 8. The patient suffered a severe closed head injury and underwent bilateral NIRS monitoring during the first 48 hours after admission. **Top,** time line of the events and surgical interventions (ΔOD is the difference in optical density between the right and left NIRS signals; lt, left). **Middle,** serial CT scans. **Bottom,** drawings of the NIRS examinations at the time of the CT scans. The inital left frontoparietal contusion "blossomed" into an intracerebral hematoma approximately eight hours later, and was detected first by an increase in the NIRS ΔOD. Approximately 40 hours after admission the NIRS, the ΔOD again increased, and the CT scan showed a new EDH. Following evacuation of the EDH the NIRS ΔOD normalized. (Reprinted from Ref. 17 with permission from the *Journal of Neurosurgery*.)

NIRS can also be helpful in the early detection of intracranial hematomas, including epidural hematomas (EDH), subdural hematomas (SDH), and intracerebral hematomas (ICH)—although ICH detection appears to be limited to those within two or three cm of the cortical surface due to the relatively superficial portion of the brain monitored by NIRS.[17] FIGURE 8 illustrates a series of events and the corresponding NIRS changes in a patient who suffered a severe closed head injury. An obvious limitation of NIRS monitoring is the small volume of brain tissue monitored by a single optode. The solution, application of multiple optodes (e.g., bilateral frontal, temporal, parietal, and occipital), should become reasonable when multichannel (four or more) NIRS monitoring becomes feasible.

The question has been raised, however, about the validity of NIRS monitoring to detect cerebral ischemia.[18,19] These studies on patients with severe head injuries document that changes in NIRS do not correlate well with changes in jugular venous blood oxygen saturation. When care is taken (1) in the application of the NIRS probes and (2) to avoid probe displacement, NIRS monitoring may actually be more sensitive than jugular venous saturation to ischemic events in head injured patients.[16] One potential problem with NIRS that can be easily avoided is that of normal or near normal NIRS brain oxygen saturation in dead or infarcted nonmetabolizing brain.[20] At least transiently, cerebral venous blood sequestered in capillaries may be misread by NIRS, since NIRS monitors the balance between local brain oxygen supply and demand. In dead or infarcted brain, both supply and demand are reduced—thus, NIRS may give an erroneously high reading.

Sound advice regarding the use of NIRS for neuromonitoring is given by Nemoto et al.:[20] "...examining the CT scan before placement of the sensor will avoid placing the sensor over subdural or epidural hematomas or CT-defined infarction. In contrast to jugular venous bulb saturation, NIRS for cerebral oxygenation provides information on oxygen availability at the tissue level in a given region of the brain. As with any other monitoring tool, the appropriate interpretation of the signal is of critical importance. It must be determined first whether it is artifactual, second, how it can explain the underlying pathology, and third, how it correlates with other physiologic variables."

CLINICAL MONITORING: BEYOND 2010

Although predictions are uncertain at best, there are two aspects of neuromonitoring in the OR and ICU that are likely to become increasingly important during the second decade of the new millennium. The first is the use of telemonitoring; the second is multiparameteric monitoring with automatic feedback (i.e., an "autopilot" OR and ICU).

Telemonitoring in the future will be an extension of the present day Holter monitor for continuous EKG monitoring. Sensors that are minimally invasive (e.g., the Codman Neurotrend), noninvasive but in contact (NIRS), or truly "remote" (i.e., no contact with the patient) will communicate wirelessly to a central interpretive station. The interpretive station may be in the same room (in the ICU setting) or around the world (where the interpretive station is prohibitively expensive, or expert human interpretation is necessary).

A greater step forward will be a truly self contained (or "smart") neuromonitoring capability, in which the multiparameter monitors provide feedback to the devices supporting the patient.[2,21] A simple example is an automated ventilatory machine which, by continuously sensing respiratory rate, blood gases, and so forth, adjusts the appropriate parameters (inspired oxygen content, respiratory rate, etc.) to maintain the patient as desired with respect to ventilatory status. Similar feedback devices could control fluid intake, as well as the administration of insulin, antibiotics, and anticonvulsants. As the pressure increases for both cost containment and reduction in management errors, such self contained sensor–effector devices for OR and ICU care will become commonplace. Just as automated, motorized rotating hospital beds have replaced some of the previous manual tasks of nurses, many of the monitoring/management activities of today in the OR and ICU will be taken over by devices. The physicians and nurses of the coming era will be concerned with the programming of monitoring, interventions, and supportive care activities—and their evaluation for efficacy—but much less in the actual carrying out of those health care interventions.

REFERENCES

1. LAM, A.M. & B.F. MATTA. 1996. Cerebral blood flow: transcranial doppler and microvascular doppler. *In* Intraoperative Neuroprotection. R.J. Andrews, Ed.: 217-247. Williams & Wilkins, Baltimore.
2. ANDREWS, R.J. 1999. Neuroprotective "agents" in surgery: secret "agent" man, or common "agent" machine? Ann N.Y. Acad. Sci. **890**: 59–72.
3. ANDREWS, R.J., J.R. BRINGAS & G. ALONZO. 1994. Cerebrospinal fluid pH and pCO_2 rapidly follow arterial blood pH and pCO_2 with changes in ventilation. Neurosurgery **34**: 466–470.
4. HOFFMAN, W.E., F.T. CHARBEL & G. EDELMAN. 1996. Brain tissue oxygen, carbon dioxide, and pH in neurosurgical patients at risk for ischemia. Anesth. Analg. 82: 582-586.
5. VAN SANTBRINK, H., A.I.R. MAAS & C.J.J. AVEZAAT. 1996. Continuous monitoring of partial pressure of brain tissue oxygen in patients with severe head injury. Neurosurgery **38**: 21–31.
6. CHARBEL, F.T., W.E. HOFFMAN, M. MISRA, *et al.* 1997. Cerebral interstitial tissue oxygen tenson, pH, HCO_3, CO_2. Surg. Neurol. **48**: 414–417.
7. DOPPENBERG, E.M.R., J.C. WATSON, W.C. BROADDUS, *et al.* 1997. Intraoperative monitoring of substrate delivery during aneurysm and hematoma surgery: initial experience in 16 patients. J. Neurosurg. **87**: 809–816.
8. HOFFMAN, W.E., F.T. CHARBEL, G. EDELMAN, *et al.* 1997. Brain tissue gases and pH during arteriovenous malformation resection. Neurosurgery **40**: 294–301.
9. ZAUNER, A., M.R. EGON, M.D. DOPPENBERG, *et al.* 1997. Continuous monitoring of cerebral substrate delivery and clearance: initial experience in 24 patients with severe acute brain injuries. Neurosurgery **41**: 1082–1093.
10. VALADKA, A.B., S.P. GOPINATH, C.F. CONTANT, *et al.* 1998. Relationship of brain tissue PO_2 to outcome after severe head injury. Crit. Care Med. **26**: 1576–1581.
11. MENZEL, M., E.M.R. DOPPENBERG, A. ZAUNER, *et al.* 1999. Increased inspired oxygen concentration as a factor in improved brain tissue oxygenation and tissue lactate levels after severe human head injury. J. Neurosurg. **91**: 1–10.
12. VAN DEN BRINK, W.A., H. VAN SANTBRINK, E.W. STEYERBERG, *et al.* 2000. Brain oxygen tension in severe head injury. Neurosurgery **46**: 868–878.
13. VILLRINGER, A. & B. CHANCE. 1997. Non-invasive optical spectroscopy and imaging of human brain function. Trends Neurosci. **20**: 435–442.

14. JOBSIS-VANDER VLIET, F.F. & P.D. JOBSIS. 1999. Biochemical and physiological basis of medical near-infrared spectroscopy. J. Biomed. Opt. **4**: 397–402.
15. TERBORG, C.F. GORA, C. WEILLER, et al. 2000. Reduced vasomotor reactivity in cerebral microangiopathy: a study with near-infrared spectroscopy and transcranial doppler sonography. Stroke **31**: 924–929.
16. KIRKPATRICK, P.J., P. SMIELEWSKI, M. CZOSNYKA, et al. 1995. Near-infrared spectroscopy use in patients with head injury. J. Neurosurg. **83**: 963–970.
17. GOPINATH, S.P., C.S. ROBERTSON, C.F. CONTANT, et al. 1995. Early detection of delayed traumatic intracranial hematomas using near-infrared spectroscopy. J. Neurosurg. **83**: 438–444.
18. LEWIS, S.B., J.A. MYBURGH, E.L. THORNTON, et al. 1996. Cerebral oxygenation monitoring by near-infrared spectroscopy is not clinically useful in patients with severe closed-head injury: a comparison with jugula venous bulb oximetry. Crit. Care Med. **24**: 1334–1338.
19. TER MINASSIAN, A.T., N. POIRIER, M. PIERROT et al. 1999. Correlation between cerebal oxygen saturation measured by near-infrared spectroscopy and jugular oxygen saturation in patients with severe closed head injury. Anesthesiology **91**: 985–990.
20. NEMOTO, E.M., H. YONAS & A. KASSAM. 2000. Clinical experience with cerebral oximetry in stroke and cardiac arrest. Crit. Care Med. **28**: 1052–1054.
21. ANDREWS, R.J., R. MAH, A. AGHEVLI, et al. 1999. Multimodality stereotactic brain tissue identification: the NASA Smart Probe project. Stereotact. Funct. Neurosurg. **73**: 1–8.

Neuroprotection for the New Millennium

Matchmaking Pharmacology and Technology

RUSSELL J. ANDREWS

NASA Ames Research Center, Moffett Field, California, U.S.A.

Department of Neurosurgery, Stanford University Medical Center, Stanford, California, U.S.A.

Division of Neurosurgery, Texas Tech University Health Sciences Center, El Paso, Texas, U.S.A.

ABSTRACT: A major theme of the 1990s in the pathophysiology of nervous system injury has been the multifactorial etiology of irreversible injury. Multiple causes imply multiple opportunities for therapeutic intervention—hence the abandonment of the "magic bullet" single pharmacologic agent for neuroprotection in favor of pharmacologic "cocktails". A second theme of the 1990s has been the progress in technology for neuroprotection, minimally- or non-invasive monitoring as well as treatment. Cardiac stenting has eliminated the need, in many cases, for open heart surgery; deep brain stimulation for Parkinson's disease has offered significant improvement in quality of life for many who had exhausted cocktail drug treatment for their disease. Deep brain stimulation of the subthalamic nucleus offers a novel treatment for Parkinson's disease where a technological advance may actually be an intervention with effects that are normally expected from pharmacologic agents. Rather than merely "jamming" the nervous system circuits involved in Parkinson's disease, deep brain stimulation of the subthalamic nucleus appears to improve the neurotransmitter imbalance that lies at the heart of Parkinson's disease. It may also slow the progression of the disease. Given the example of deep brain stimulation of the subthalamic nucleus for Parkinson's disease, in future one may expect other technological or "hardware" interventions to influence the programming or "software" of the nervous system's physiologic response in certain disease states.

KEYWORDS: Brain stimulation; Excitotoxic injury; Minimally invasive surgery; Movement disorders; Neuroprotection; Parkinson's disease; Subthalamic nucleus.

INTRODUCTION

Four years ago at the Third International Conference on Neuroprotective Agents (ICNA),[1] I presented evidence to support the concept that neuroprotection, in the specific instance of intraoperative neuroprotection, would require a pharmacologic "cocktail". A single agent "magic bullet" was simply naïve given the increasing complexity of nervous system injury, in terms of events occurring simultaneously as well

---

Address for correspondence: Russell J. Andrews, Division of Neurosurgery, Texas Tech University, Health Sciences Center, 4800 Alberta, El Paso, TX 79905, U.S.A., Voice: 915-545-6676; fax: 915-545-7584.

andrews_rj@compuserve.com

as sequentially. The failure in the early 1990s of a number of large clinical trials involving a single pharmacologic agent for the treatment of stroke and head injury led to several papers advocating a cocktail approach to pharmacologic neuroprotection.[2,3]

Two years ago at the Fourth ICNA,[4] my theme was technological advances overshadowing pharmacologic advances in the search for intraoperative neuroprotection. In recent years our ability to perform operations with efficacy and safety has improved, largely because of dramatic technological progress. Comparatively little progress has come from developments in pharmacologic neuroprotective agents used intraoperatively or perioperatively. Putative noxious agents—lurking for minutes to hours to days after injury to the nervous system—proliferate in daunting fashion. The ischemic cascade seemingly verifies, for the complexity of nervous system injury, Moore's Law for computer chips: that there is a doubling of processing speed every 18 months. The examples of coronary artery stenosis surgery and surgery for Parkinson's disease were used to illustrate the common "agent" of the computer outstripping the increasingly enigmatic or secret "agent" of nervous system pathophysiology.

For this, the inaugural ICNA of the new millennium, I carry the unfolding relation between pharmacology and technology one step further. With the advent of implantable electrostimulation devices to restore neurological function after stroke (e.g., a paralyzed limb or blindness),[5] the use of such devices for neuroprotection is feasible in certain situations. To provide continuity with the paper presented at the Fourth ICNA, I consider the example of Parkinson's disease.

PHARMACOLOGIC COCKTAIL NEUROPROTECTION—THE SOFTWARE

Multifactorial etiologies for various human pathophysiologic processes might be considered a pervasive theme of the 1990s. Consider the following quotations:

> "...the complex pathophysiology of...may require a multiple, simultaneous, or sequential drug-treatment approach. ...Therapy directed at only one of the processes involved in...may have only modest benefits at best... a series of single-agent trials might reject agents of limited efficacy in isolation that are effective in combination with other drugs."
> —Adams 1995 (Ref. 3)

> "...is not a single disease, but a very complex condition comprising a large variety of local and systemic...responses. Therefore, we propose that a combination of several therapies or multifunctional agents, directed at various phases of...might meet with more success than recent trials with single therapies."
> —Karima 1999 (Ref. 6)

> "Recent failures of so-called 'magic bullets' in randomized clinical trials suggest that multi-modality therapy may be required... If there are multiple mediators and factors involved in...then multiple agents may be needed."
> —Baue 1998 (Ref. 7)

> "...the greatest challenge of the future will lie in using new markers to direct patient treatment. It is unlikely that any single marker will have sufficient sensitivity or specificity to alter treatment choice with certainty. Therefore, future evaluations should focus on developing nomograms combining multiple markers..."
> —Reiter 1999 (Ref. 8)

"Treatment of patients with established...is still largely supportive and has made little impact on the patient mortality rate over the past 20 years. Future treatment strategies must focus on multimodality combination therapy aimed at specifically suppressing excessive activation of...while preserving...normal...defenses."
—Deitch 1999 (Ref. 9)

"...it may be important to test a cocktail of drugs with different therapeutic targets. ...rather than performing a limited number of costly clinical trials that test single agents in large numbers of unselected cases in search of small and incremental benefits, it may be more profitable to use the same dollars to perform a large number of clinical trials that test a larger number of agents and combination of agents..."
—Marsden 1998 (Ref. 10)

Although very similar in message, the above quotes describe quite different processes:
1. An editorial entitled: "Acute stroke treatment trials in the United States: rethinking strategies for success";
2. A review of sepsis entitled: "The molecular pathogenesis of endotoxic shock and organ failure";
3. A review of systemic inflammatory response and multiple organ failure;
4. An editorial considering methodologies for prostate cancer diagnosis;
5. A review entitled: "Prevention of multiple organ failure";
6. A review of Parkinson's disease entitled: "The causes of Parkinson's disease are being unraveled and rational neuroprotective therapy is close to reality".

However, beneath the appearance of disparate disease processes for stroke, septic shock and multiple organ failure, cancer, and a neurodegenerative process such as Parkinson's disease runs the common thread of host defense mechanisms being inappropriate—either inadequate (in the case of cancer) or excessive (stroke, neurodegenerative diseases, and sepsis/multiple organ failure).

Pharmacologic interventions can be considered the "software" that directs or modulates the host response to insult. The theme developed in the 1990s has been the necessity of multiple agents for effective therapy, given the complexity of the pathophysiological processes involved. The art and science of neuroprotection in the first decade of the new millennium will depend significantly on how well we can direct an increasing number of magic bullet drugs to their targets with as little collateral damage as possible.

TECHNOLOGICAL NEUROPROTECTION—THE HARDWARE

The potential technological advances benefitting neuroprotection are myriad:
1. Imaging techniques such as ultrasound (US), computed tomography (CT), functional magnetic resonance imaging (fMRI), MR spectroscopy (MRS), and image fusion (e.g., US, CT, and/or MRI).
2. Minimally invasive surgical techniques (e.g., endoscopic or laparoscopic surgery, and endovascular techniques such as stenting).

3. Robotic and other automated surgical techniques to minimize human error or human limitations (e.g., the need for direct vision and tactile feedback, and the drawback of physiologic tremor for precise surgery in ophthalmology and neurosurgery).

Another application of technology to neuroprotection and neurorestoration is the development of implantable devices to preserve, improve, or restore functionality. This can take the form of techniques to bridge or bypass the region of nervous system injury, for example neurotrophic electrodes placed into the motor cortex that allow a patient with the locked-in syndrome to communicate by altering brain electrical activity.[5] Another example is the use of stereotactic techniques to implant adrenal tissue precisely in the brain of a patient with Parkinson's disease. A third example, directly relevant to neuroprotection, is the use of chronic electrical stimulation as a neuroprotective agent, discussed here in the context of Parkinson's disease.

THE PATHOPHYSIOLOGY OF PARKINSON'S DISEASE

Parkinson's disease (PD) is a degenerative disorder of the nervous system that was first described by the English physician James Parkinson in 1817. It affects more than one million people in the North America, its incidence increases with increasing age, and the mortality rate among those affected with PD is two to five times that of age matched controls.[11]

The basic pathophysiology of PD is reasonably well understood, and its treatment has extensive roots in both pharmacologic and surgical interventions.[11,12] The pathological hallmark of PD is a loss of dopaminergic neurons in the substantia nigra compacta (SNc). Details of the effect of the loss of dopamine on the various nuclei involved in Parkinsonian movement disorders are illustrated in FIGURE 1, where (FIG. 1A) diagrams the normal situation and (FIG. 1B) the situation in PD. Note the separate direct and indirect pathways (the latter via the subthalamic nucleus [STN]) by which a decrease in dopamine can affect the circuit of nuclei involved in Parkinsonian movement disorders.

Among the etiologies of PD are idiopathic, postencephalitic (encephalitis lethargica or von Economo's disease), and drug induced. Drug induced PD (DIPD) occurs following the use of dopamine receptor blocking agents, although the list of pharmacologic agents that can cause DIPD includes—in addition to dopamine receptor blockers—the false neurotransmitter methyldopa, antiemetics (e.g., chlorpromazine), anticonvulsants (e.g., valproic acid), cholinomimetics (e.g., bethanecol), sympatholytics (e.g., reserpine), and calcium channel blockers (e.g., flunarizine).[13] Although DIPD is relatively common among those taking one of the offending drugs, it fortunately resolves in most cases on discontinuing use of the drug.

Idiopathic PD may begin before age 40, and unfortunately runs a relentlessly progressive course in most cases. The cardinal findings are tremor, bradykinesia, rigidity, and postural instability. The severe disability of PD led to many surgical explorations in the search for relief from the movement disorders, with stereotactic lesioning of the globus pallidus or thalamus proving to be of considerable benefit in the 1950s and 1960s.

The discovery that the dopamine precursor L-dopa, which readily crosses the blood–brain barrier, can ameliorate the dopamine deficit in PD resulted in a dramatic reduction in the number of surgical procedures for PD when L-dopa came into widespread clinical use in the late 1960s. For approximately 20 years after the late 1960s,

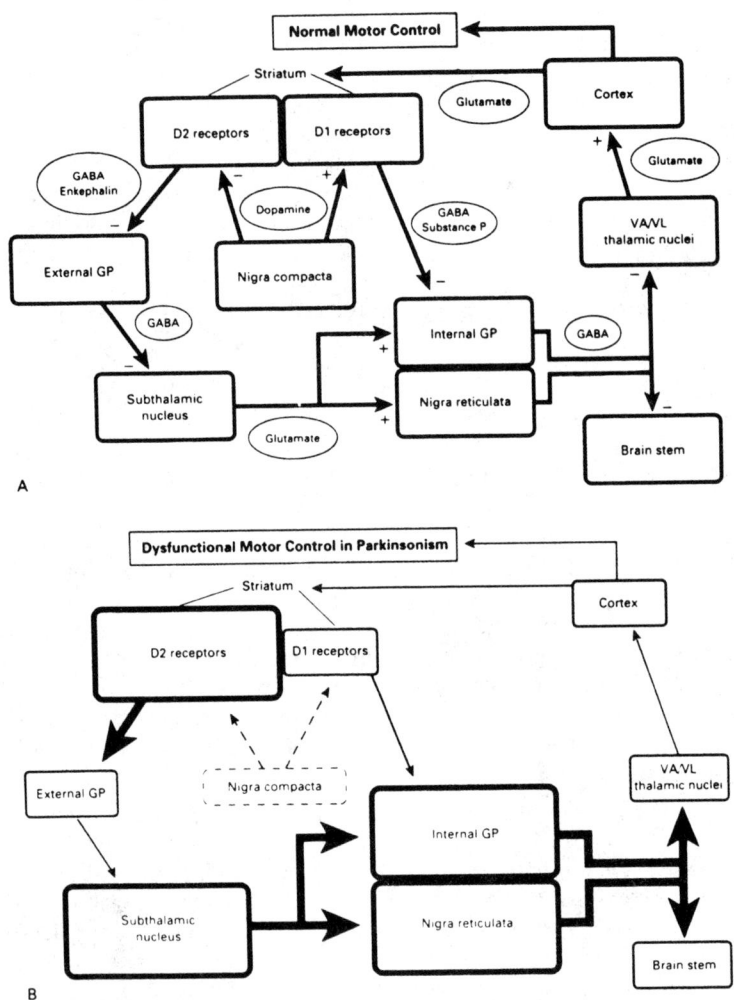

FIGURE 1. Model of the basal ganglia in persons with normal motor control (**A**) and Parkinson's disease (**B**). *Plus signs* indicate excitation; *minus signs* inhibition. *Arrow width* in (**B**) indicates the functional activity change in each pathway (change in neuronal firing rate) compared with normal activity. The size and outline of each box in (**B**) indicate the activity of the brain region compared with normal activity. Dashed lines and arrows indicate the dysfunctional nigrostriatal dopamine system in PD. The neurotransmitters used in each pathway are circled. VA/VL, ventral anterior and ventrolateral; GABA, g-aminobutyric acid. (Reprinted from Ref. 11 with permission from the *New England Journal of Medicine*.)

only a few major academic medical centers routinely performed stereotactic procedures for PD and other movement disorders.[14] L-dopa and other dopaminergic agents such as bromocriptine and pergolide, monoamine oxidase inhibitors such as deprenyl, and various anticholinergic agents all demonstrated varying degrees of success in treating PD.

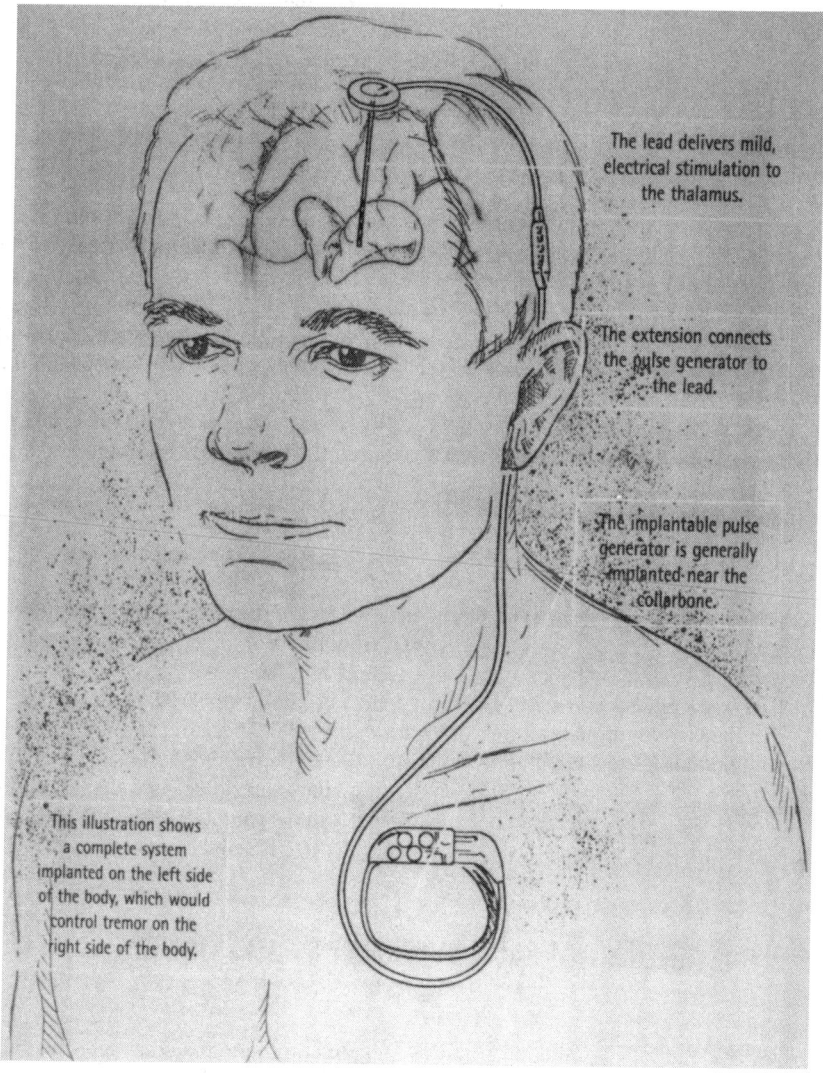

FIGURE 2. Schematic of the deep brain stimulator. Note the electrode placed into the thalamic or subthalamic region, the connecting wires placed subcutaneously, and the stimulator/battery placed subcutaneously inferior to the clavicle (Figure provided by Medtronic, Inc.)

These pharmacologic treatments for PD have several drawbacks, however. All of the agents appear to be symptomatic rather than curative in their efficacy. Side effects—notably nausea, confusion, and hallucinations—are common to all as well. Their benefit is not universal, in that only 1/2 to 2/3 of patients experience significant benefit. Bradykinesia and rigidity usually respond better than tremor. Perhaps most importantly, within two to five years of beginning L-dopa therapy (usually combined with the peripheral decarboxylase inhibitor carbidopa) the majority of patients experience a gradual but progressive decrease in efficacy. This decrease takes the form of either a diurnal variation in, or recurrence of, the Parkinsonian dyskinesias and rigidity (off stage)—alternating with periods of relative mobility (on stage). The failure of pharmacologic therapy to be an effective long term treatment for PD, together with improvements in neuroimaging (CT and MRI), stereotactic techniques for neurosurgery, and electrophysiologic recording, all served to rekindle interest in the surgical treatment of PD in the late 1980s.

The original surgical intervention used radiofrequency heating of a microelectrode to ablate either the ventrointermediate nucleus of the thalamus or the globus pallidus interna (GPi). In the late 1980s, Benabid in France pioneered the use of an implantable programmable stimulator to electrically "ablate" the region of interest in the thalamus or globus pallidus.[15] To effectively ablate, the stimulation must be relatively high frequency (100 to 200 Hz), with a voltage of 1 V to 3 V. The stimulator, derived from cardiac pacemakers, had been used during the early 1980s for epidural spinal cord stimulation—as well as by a few neurosurgeons in academic settings for deep brain stimulation (particularly for intractable chronic pain syndromes). In its present configuration, the microelectrode is stereotactically placed in the region of interest and connected by wires under the scalp and skin of the neck to the stimulator (a microprocessor and battery) placed—like a cardiac pacemaker—subcutaneously in the region of the clavicle (see FIGURE 2).

Extensive testing of deep brain stimulation (DBS) of the thalamus in particular—for Parkinsonian tremor as well as essential tremor—took place in Europe and the US during the late 1980s to mid 1990s. Food and Drug Administration (FDA) approval was granted in 1997 to market a device in the US for DBS of the thalamus as a treatment for contralateral tremor. In the early 1990s, Benabid and others found that stimulation of the subthalamic nucleus (STN) was effective in treating not only the tremor of PD, but the other symptoms (bradykinesia, rigidity) as well.[16] Presently the STN is the site of choice for most patients undergoing DBS for PD; the vast majority of patients benefit most from bilateral STN DBS placement.

SUBTHALAMIC NUCLEUS STIMULATION: NEUROPROTECTIVE MARRIAGE

The effect of STN stimulation (DBS)—where stimulation is considered equivalent to ablation or disruption of the normal electrochemical circuitry—is illustrated in FIGURE 3.[11] Note that one of the effects of PD is increased firing activity of the STN neurons, which in turn results in an excitatory effect on the SNc. STN DBS has an advantage over GPi DBS (or GPi ablation) in that it modulates the response of

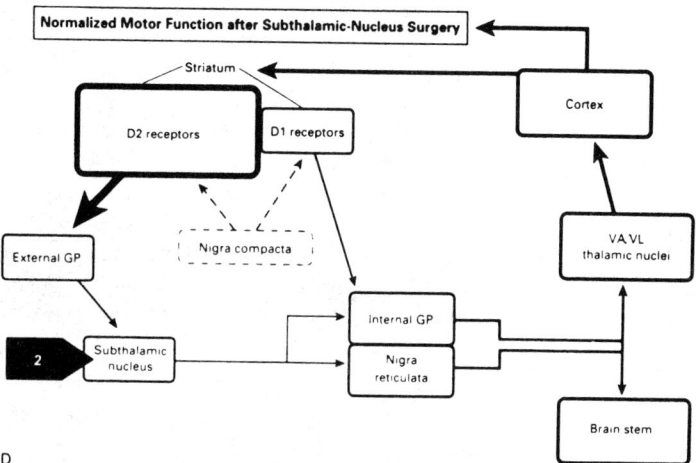

FIGURE 3. Model of the basal ganglia in persons with Parkinson's disease following surgical intervention in the subthalamic nucleus. Legend and abbreviations as in FIGURE 1. Note that reducing the excessive excitatory activity of the STN reduces the over-activity of both components of the output of the basal ganglia, the GPi and the SNr. (Reprinted from Ref. 11 with permission from the *New England Journal of Medicine*.)

both the GPi and the substantia nigra reticulata (SNr). This likely accounts for the greater effectiveness of STN DBS in most patients with PD.

It has recently been argued that STN DBS may actually be neuroprotective in addition to providing symptomatic relief for patients with PD.[17] This marriage of implanted hardware to alter the neurochemical processes that underlie the loss of dopaminergic neurons in the SNc warrants consideration in some detail. The following is in essence a summary of the evidence marshalled by Rodriguez et al.[17]

The STN is the central nucleus in the glutamatergic excitatory pathways in the basal ganglia, as shown in FIGURE 4.[17] The argument in favor of STN inhibition (e.g., by electrical stimulation) runs as follows:

1. SNc dopaminergic neurons can exhibit a bursting type discharge pattern which is associated with dopamine release.

2. STN excitation leads to increased SNc dopaminergic neuron bursting. Only a modest increase in STN firing is necessary—in Parkinsonian monkeys the STN discharge rate is only 29% greater than controls.

3. The STN-induced increase bursting in SNc dopaminergic neurons leads to stress on an already tenuous SNc, that is to say excitotoxic aggravation of the loss of SNc dopaminergic neurons.

The argument that excitotoxic injury is important—if not essential—to the neurodegeneration in Parkinson's disease is also readily summarized:

1. Glutamate mediates toxicity in SNc dopaminergic neurons. In MPTP treated monkeys, NMDA antagonists have been shown to be neuroprotective.

2. Glutamate induced toxicity is mediated in part by nitric oxide (NO). Agents that inhibit NO mediated toxicity are neuroprotective in animal models of PD.
3. A mitochondrial defect—mitochondrial complex 1 defect—may predispose SNc dopaminergic neurons to excitotoxic injury.[18] The defect leads to a loss of Mg^+ blockade of NMDA receptors, rendering the SNc dopaminergic neurons especially vulnerable by virtue of increased Ca^{2+} entry.
4. Decreasing the glutamatergic output of the STN by lesioning has been neuroprotective—in terms of SNc dopaminergic neurons—in animal models of PD.
5. There is experimental evidence that NO mediated toxicity is important in PD, and that SNc dopaminergic neurons may be especially vulnerable to NO, whereas other neurons may be protected (e.g., by increased levels of superoxide dismutase).

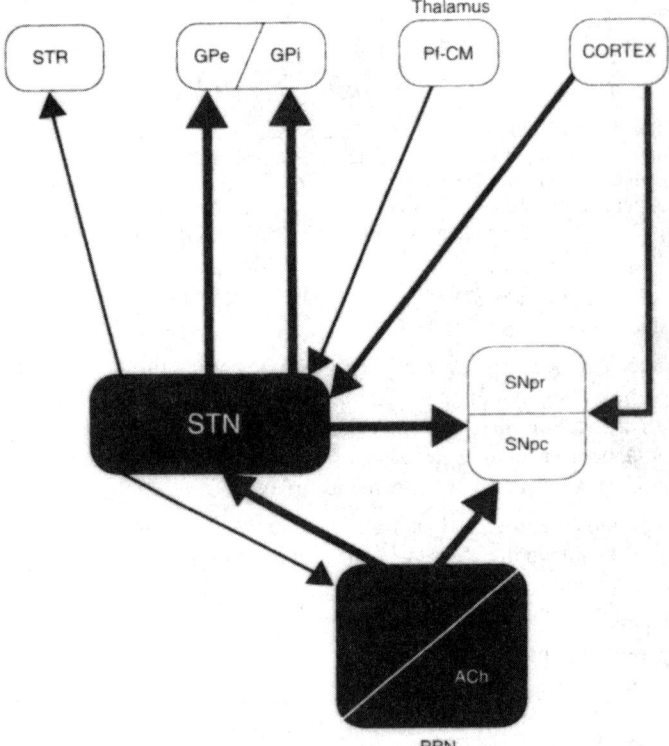

FIGURE 4. Schematic of the main glutamatergic pathways in the basal ganglia. The subthalamic nucleus (STN) has excitatory projections to the external and internal globus pallidus (GPe and GPi), the substantia nigra pars reticulata and pars compacta (SNpr and SNpc), the striatum (STR), and the pedunculopontine nucleus (PPN) in the brainstem. Glutamatergic projections to the STN arise from the cerebral cortex, the parafascicular and centromedian (Pf-CM) nuclei of the thalamus, and the PPN. Glutamatergic projections to the SNpc arise from the STN, the PPN, and the cerebral cortex. ACh, acetylcholine. (Reprinted from Ref. 17 with permission from the *Annals of Neurology*.)

Rodriguez *et al.* speculate that the following sequence of events can account for the neurodegeneration in PD:[17]

1. SNc dopaminergic neurons have an increased susceptibility to excitotoxic injury that leads to apotosis in at least some of these neurons. The predisposing factors, which may be either inherited and/or acquired, are:
   - an as yet undetermined vulnerability to oxidative stress (e.g., 5 in the previous paragraph)
   - the mitochondrial complex 1 defect in 3 in the previous paragraph

2. The loss of SNc dopaminergic neurons leads to decreased striatal dopamine, that in turn results in a decrease in inhibition of D_2 striatal neurons. This leads to increased inhibition of the GPe, that in turn disinhibits the STN.

3. Increased firing in the STN leads to glutamatergic excitation in SNc, as well as in SNr, GPi, and PPN.

4. The increased glutamate in SNc leads to further loss of dopaminergic neurons and, in turn, decreased striatal dopamine.

5. A "vicious cycle" is created in which the STN is further disinhibited and the SNc dopaminergic neurons exposed to further excitotoxic injury.

6. A some critical point—probably when 30 to 50% of dopaminergic neurons in the SNc are lost—the symptoms of PD become manifest.

7. Increased STN firing leads to glutaminergic excitotoxicity in SNr, GPi, and PPN—which may account for the late non-dopaminergic symptoms of PD—postural instability, speech impairment, and autonomic dysfunction.

The above discussion affords several prospects for the treatment of PD in addition to the use of levodopa, interventions that might be neuroprotective in that they break the vicious cycle of STN disinhibition and the resultant excitotoxic stress on the SNc dopaminergic neurons in particular. Most obvious is the use of dopamine agonists, as has been variably accepted for many years by those treating patients with PD. A recent article reporting that ropinirole, either alone or in conjunction with levodopa, is more effective than levodopa alone supports the use of certain dopamine agonists in the treatment of PD.[19] The issue from a neuroprotection standpoint is whether beginning such medications early in PD can slow the progression of the disease.

Another approach is to use pharmacologic agents that attack the excitotoxic cascade. There are various options: (1) inhibition of glutamate release (e.g., lamotrigine and riluzole); (2) NMDA receptor antagonists to decrease the excitotoxic effect of glutamate (the agents to date unfortunately have intolerable side effects); and/or (3) agents to oppose the effects of NO such as antioxidants. In time, this approach may benefit from a careful combination drug therapy—a rational pharmacologic cocktail. The PD drug treatment of choice may prove to be an "alloy".

Finally, there is deep brain stimulation of the STN. Although the mechanism by which stimulation of the STN at 100 to 200 Hz reduces the excitatory effect of the STN output is uncertain, its benefit on all the cardinal symptoms of PD is clear now that hundreds of PD patients have undergone STN DBS—some for well over five years.

The mechanism by which DBS affects the central nervous system is under debate. A simplistic explanation is that the stimulation jams the neural pathways, interrupting a circuit, and thus providing the desired clinical effect. However, recent research with brain slice preparations indicates that electrical stimulation in cortical gray matter activates axonal branches rather than cell bodies.[20] If so, the overall effect of DBS in regions that contain both axons passing through as well as cell bodies is uncertain. Additionally, the mechanism by which DBS decreases the firing rate in the STN, for example, is unclear: depolarization blockade, orthodromic excitation, and antidromic activation are all possibilities.[21] Another explanation for the therapeutic effect of DBS is stochastic resonance, where adding the regularity of high frequency DBS as noise to the signal of the STN neurons results in an improved signal-to-noise ratio.[22]

The future of DBS includes much uncharted and potentially productive territory. In addition to the hardware advances of more precise electrode placement and smaller electrodes (including microarrays), the benefits of varying the stimulation parameters beyond those of frequency, amplitude, and duration remain for the most part unexplored. For example, varying the frequency (perhaps in conjunction with varying the amplitude) may result in more effective inhibition. An intriguing possibility is that the neurons in a nucleus such as the STN may be taught by DBS to alter their firing rates or behavior.[23] Finally, DBS may see application in the treatment not only of disorders that might be considered extensions of its current applications in movement disorders (e.g., the motor and vocal tics of Gilles de la Tourette's syndrome[21]) but also in more novel situations such as epilepsy, obsessive compulsive disorders, and eating disorders.[23]

REFERENCES

1. ANDREWS, R.J. 1997. Neuroprotection in surgery: development of a pharmacologic cocktail for intraoperative use. Ann N.Y. Acad. Sci. **825**: 288–304.
2. HALLENBECK, J.M. & K.U. FRERICHS. 1992. Stroke therapy: it may be time for an integrative approach. Arch. Neurol. **50**: 768–770.
3. ADAMS, R.J., M. FISHER, A.J. FURLAN, et al. 1995. Acute stroke treatment trials in the United States. Stroke **26**: 2216–2218.
4. ANDREWS, R.J. 1999. Neuroprotective "agents" in surgery: secret "agent" man, or common "agent" machine?. Ann N.Y. Acad. Sci. **890**: 59–72.
5. KENNEDY, P.R. & R.A.E. BAKAY. 1998. Restoration of neural output from a paralyzed patient by a direct brain connection. NeuroReport **9**: 1707–1711.
6. KARIMA, R., S. MATSUMOTO, H. HIGASHI, et al. 1999. The molecular pathogenesis of endotoxic shock and organ failure. Molec. Med. Today **5**: 123–132.
7. BAUE, A.E., R. DURHAM & E. FAIST. 1998. Systemic inflammatory response syndrome (SIRS), multiple organ dysfunction syndrome (MODS), multiple organ failure (MOF): are we winning the battle? Shock **10**: 79–89.
8. REITER, R.E. 1999. Editorial: prostate cancer diagnostics - obstacles and opportunities. J. Urol. **162**: 2046–2047.
9. DEITCH, E.A. & E.R. GOODMAN. 1999. Prevention of multiple organ failure. Surg. Clin. North Am. **79**: 1471–1488.
10. MARSDEN, C.D. & C.W. OLANOW. 1998. The causes of Parkinson's disease are being unraveled and rational neuroprotective therapy is close to reality. Ann. Neurol. **44**: S189–S196.
11. LANG, A.E. & A.M. LOZANO. 1998. Parkinson's disease. N. Engl. J. Med. **339**: 1044–1053, 1130–1143.
12. GOETZ, C.G. 2000. Parkinson's disease. *In* Neurobase (CD-ROM). S. Gilman, Ed. Arbor, San Diego, CA.
13. DIEDERICH, N.J. & C.G. GOETZ. 1998. Drug-induced movement disorders. Neurol. Clin. North Am. **16**: 125–139.

14. GILDENBERG, P.L. 1987. Whatever happened to stereotactic surgery? Neurosurgery **20**: 983–987.
15. BENABID, A.L., P. POLLACK, A. LOUVEAU, et al. 1987. Combined (thalamotomy and stimulation) stereotactic surgery of the Vim thalamic nucleus for bilateral Parkinson's disease. Appl. Neurophysiol. **50**: 344–346.
16. LIMOUSIN, P., P. KRACK, P. POLLACK, et al. 1998. Electrical stimulation of the subthalamic nucleus in advanced Parkinson's disease. N. Engl. J. Med. **339**: 1105–1111.
17. RODRIGUEZ, M.C., J.A. OBESO & C.W. OLANOW. 1998. Subthalamic nucleus-mediated excitotoxicity in Parkinson's disease: a target for neuroprotection. Ann. Neurol. **44**: S175–S188.
18. MAREY-SEMPER, I., M. GELMAN & M. LEVI-STRAUSS. 1995. A selective toxicity toward cultured mesencephalic neurons is induced by the synergistic effects of energetic metabolism impairment and NMDA receptor activation. J. Neurosci. **15**: 5912–5918.
19. RASCOL, O., D.J. BROOKS, A.D. KORCZYN, et al. 2000. A five-year study of the incidence of dyskinesia in patients with early Parkinson's disease who were treated with ropinirole or levodopa. N. Engl. J. Med. **342**: 1484–1491.
20. NOWAK, L.G. & J. BULLIER. 1998. Axons, but not cell bodies, are activated by electrical stimulation in cortical gray matter. II. Evidence for selective inactivation of cell bodies and axon initial segments. Exp. Brain Res. **118**: 489–500.
21. GROSS, R.E. & A.M. LOZANO. 2000. Advances in neurostimulation for movement disorders. Neurol. Res. **22**: 247–258.
22. MONTGOMERY, E.B. & K.B. BAKER. 2000. Mechanisms of deep brain stimulation and future developments. Neurol. Res. **22**: 259–266.
23. BENABID, A.L., A. KOUDSIE, P. POLLACK, et al. 2000. Future prospects of brain stimulation. Neurol. Res. **22**: 237–246.

Neuroprotective Role of Neurophysiological Monitoring During Endovascular Procedures in the Spinal Cord

FRANCESCO SALA,[a] YASUNARI NIIMI,[b] ALEX BERENSTEIN,[b] AND VEDRAN DELETIS[c]

[a]*Department of Neurological Sciences and Vision, Section of Neurosurgery, Verona University Hospital, Italy*

[b]*Center for Endovascular Surgery, Beth Israel Medical Center, Singer Division, New York, New York, U.S.A.*

[c]*Division of Intraoperative Neurophysiology, Hyman Newman Institute for Neurology and Neurosurgery, Beth Israel Medical Center, New York, New York, U.S.A.*

ABSTRACT: The endovascular treatment of spinal vascular malformations places the spinal cord at risk for ischemia. When these procedures are performed using general anesthesia, the neurophysiological monitoring methods currently available provide the only means by which to assess the functional integrity of sensory and motor pathways. Neurophysiological monitoring allows a warning for the neuroradiologist of impending irreversible neurological damage so that action may be taken for the prompt restoration of adequate spinal cord perfusion. Muscle motor evoked potentials (mMEPs) better reflect spinal cord perfusion in the anterior spinal artery territory than do somatosensory evoked potentials (SEPs), although their use during spinal endovascular procedures remains anecdotal in the literature. In the study reported here we assessed: (1) the feasibility of intraoperative neurophysiological monitoring, (2) the role of provocative tests with Amytal and Xylocaine, and (3) the specific but complementary role played by SEPs and mMEPs, during endovascular embolization of spinal vascular malformations and tumors. The results suggest that: (1) neurophysiological monitoring is feasible during most endovascular procedures in the spine and spinal cord under general anesthesia, (2) provocative tests enhance the safety of the procedure, (3) mMEPs are more feasible than SEPs and more sensitive than SEPs to provocative tests. We strongly suggest the use of multimodal neurophysiological monitoring and provocative tests during the endovascular treatment of spinal and spinal cord vascular lesions.

KEYWORDS: Intraoperative neurophysiological monitoring; Motor evoked potentials; Spinal arteriovenous malformation; Spinal cord monitoring; Embolization; Provocative tests; Xylocaine; Amytal

Address for correspondence: Francesco Sala, M.D., Dept. of Neurological Sciences and Vision, Section of Neurosurgery, Verona University Hospital, Piazzale Stefani 1, 37126 Verona, Italy. Voice: 0039-045-8072695; fax: 0039-045-916790.
francescosala@yahoo.com

INTRODUCTION

Endovascular techniques are increasingly used in the treatment of hypervascularized lesions in the spinal cord and surrounding structures. Devascularization of spinal cord tumors, obliteration of intramedullary arteriovenous malformations, and occlusion of spinal or dural fistulas[1-6] can be achieved by injecting embolizing materials through a catheter selectively introduced into the vessels feeding the lesion. These procedures, however, put the spinal cord at risk for ischemia, which is mostly related to vasospasm or unrecognized obliteration of vessels feeding the normal spinal cord. Morbidity secondary to spinal angiography and embolization of vascular lesions in and around the cord is described elsewhere.[7-10]

Especially in cases of complex vascular malformations, the angiographic study and the following embolization can last for several hours and general anesthesia warrants less discomfort to the patient. In this setting, unless a so-called "wake-up" test is performed, the neurological status of the patient can only be assessed by using neurophysiological techniques throughout the procedure. The reliability of neurophysiological monitoring when compared to wake-up test has been established for spine surgery.[12,13] The goal of the neurophysiological monitoring in these cases is to detect spinal cord ischemia in time to allow for the restoration of adequate perfusion, thereby avoiding irreversible injury to the cord. Somatosensory evoked potentials (SEPs) have been used since the mid 1980s[14-16] during interventional procedures. At that time, techniques to elicit motor evoked potential under general anesthesia were not available and the functional integrity of motor tracts was indirectly determined through monitoring of SEPs. There was, in fact, clinical evidence that SEPs were sensitive to compromises in the anterior spinal artery circulation.[14] More recently, however, there have been reports warning about the reliability of SEPs in evaluating the integrity of descending motor tracts during spine and spinal cord surgery.[17,18] The unreliability of SEP monitoring in detecting cord injury has also been described during aortic surgery, suggesting the limited value of this technique also when the mechanism of injury is purely ischemic.[19,20]

Despite the advent of intraoperative motor evoked potentials (MEPs),[21-23] reports on their use during endovascular procedures in the spinal cord remain anecdotal.[24-28] Therefore, we retrospectively reviewed 110 consecutive endovascular procedures performed with neurophysiological monitoring under general anesthesia to assess the feasibility and reliability of motor evoked potentials. The complementary role of MEPs and SEPs was also evaluated.

MATERIAL AND METHODS

Patient Population

Between October 1996 and August 1999, we performed neurophysiological monitoring during 110 endovascular procedures in 85 patients who were treated for spine and/or spinal cord lesions. According to the expected risk of the procedure, the patients were divided in two groups. The high risk group (HRG) consisted of 26 patients, 7 harboring a perimedullary arteriovenous fistula and 19 with a spinal cord arteriovenous malformation. Overall, 41 procedures were performed in this group.

The low risk group (LRG) included 59 patients, 32 with a spine or spinal cord tumor and 27 patients with a dural arteriovenous fistula. Sixtynine procedures were performed on this group.

All procedures were performed under general anesthesia. In order to perform proper neurophysiological monitoring, continuous infusion of propofol (100–150 µg/kg/min) and fentanyl (1 µg/kg/h) with no halogenated agents were used throughout the procedures. Short acting muscle relaxants were given only for intubation.

Neurophysiological Monitoring

Neurophysiological monitoring consisted of SEPs, muscle motor evoked potentials (mMEPs) and the bulbocavernosus reflex (BCR). The Axon Sentinel-4 evoked potential system with modified software (AXON Systems, Inc., Hauppauge, NY) was used for both stimulation and recording.

SEPs were elicited by stimulation of the median nerve at the wrist (intensity 40 mA, duration 0.2 ms, repetition rate of 4.3 Hz) and the posterior tibial nerve at the ankle (intensity 40 mA, duration 0.2 ms, repetition rate of 4.3 Hz). Recordings were performed via corkscrew like electrodes inserted subcutaneously in the scalp (CS electrode, Neuromedical Inc., Herndon, VA) respectively at C3′/C4′-CZ′ (median nerve) and at CZ′- FZ (tibial nerve) according to the 10/20 International EEG system.

The mMEPs were elicited with transcranial electrical stimulation of the motor cortex using CS electrodes. Short trains up to seven square wave stimuli of 500 µs duration, interstimulus intervals of 4 ms and intensity up to 200 mA were applied at a repetition rate of 2 Hz through electrodes placed at C1 and C2 scalp sites, according to the International 10/20 EEG System. Muscle responses were recorded via needle electrodes inserted into the abductor pollicis brevis (APB) and hypothenar muscle for upper extremities and anterior tibial (TA) and abductor hallucis (ABH) muscles for the lower extremities, bilaterally. This technique has been used at our institution to monitor over 100 spinal cord tumor cases[29] and is described in detail elsewhere.[30]

In patients with a lesion at the lumbosacral segments of the spinal cord, BCR monitoring was added to the battery of neurophysiological tests.[31] This oligosynaptic reflex allows the assessment of the functional integrity of both the afferent and efferent fibers of the pudendal nerves together with the reflex center located in the gray matter at S2-S4 spinal segments. For stimulation of the dorsal penile nerve (pudendal afferents), two silver/silver chloride disc electrodes were placed on the dorsal aspect of the penis with the cathode proximal. In the female patients, the cathode was placed over the clitoris and the anode over the labia majora. Rectangular pulses of 0.2–0.5 ms duration were applied as a train of five stimuli (interstimulus intervals of 4 ms) at a repetition rate of 2.3 Hz. Stimulus intensities did not exceed 40mA. Recordings were made from the anal sphincter using two pairs of intramuscular teflon coated hooked wire electrodes inserted into the anal hemisphincters.

Endovascular Procedures and Provocative Tests

As far as endovascular procedures are concerned, we refer to the protocol we routinely use for the treatment of intramedullary arteriovenous malformations (AVMs) (see FIGURE 1). These represent the most complex lesions to treat as well as one of the most challenging to monitor from a neurophysiological perspective.

The first step of the procedure consists of a detailed and careful investigation of the vascular anatomy. This is necessary for successful treatment, independent of any additional neurophysiological support. The goal of the angiographic study is, therefore, the identification of the anterior spinal artery (ASA) and the posterior spinal arteries (PSAs) supplying the normal spinal cord below and above the AVM, as well as the identification of the AVM angioarchitecture. At this point the pedicle artery is catheterized, followed by a superselective catheterization of the radiculomedullary artery feeding the AVM. Once the catheter is in the embolizing position, and before any embolizing material is injected, we perform the provocative tests.

This sort of "Wada test" for the spinal cord[32] consists of the intra-arterial injection of short acting barbiturates (Amytal) and lidocaine (Xylocaine) through a microcatheter. Amytal blocks neuronal activity, whereas Xylocaine blocks axonal conduction. Accordingly, a positive Amytal or Xylocaine test (i.e., more than 50% decrease in SEP amplitude and/or mMEP disappearance) indicates that the vessel distal to the tip of the microcatheter supplies the functional gray or white matter of the spinal cord respectively (see FIGURE 2). Therefore, embolization is not performed with liquid embolic material from the same catheter position at which provocative

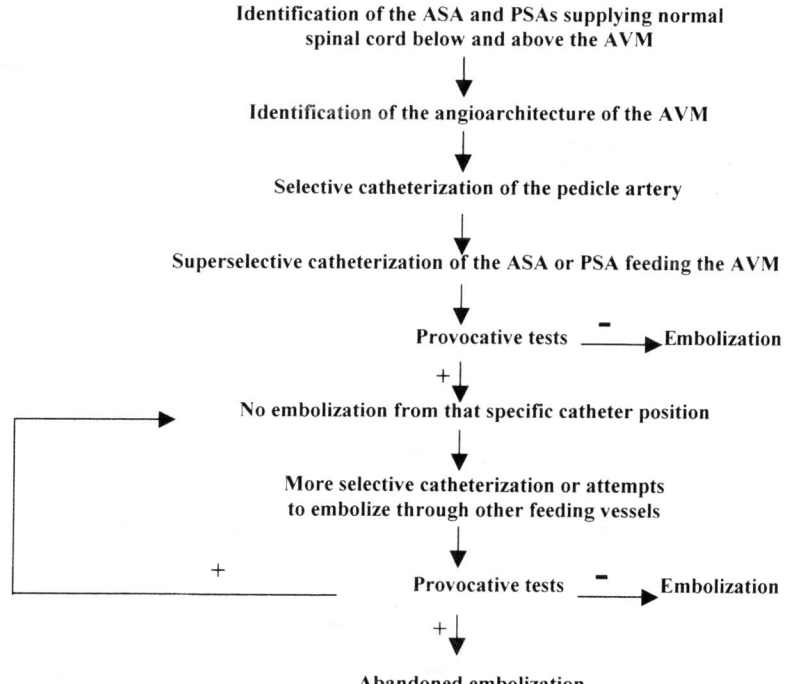

FIGURE 1. Protocol for endovascular embolization of spinal AVM using provocative tests.

tests are positive. In the case of positive provocative test, we attempt either to more selectively catheterize the vessel feeding the AVM or to approach the malformation from a different feeder. If provocative tests are repeatedly positive, embolization is abandoned.

Neurological outcome was evaluated at the end of the procedure, when the patient woke up from anesthesia. We consider a result to be false negative whenever a new

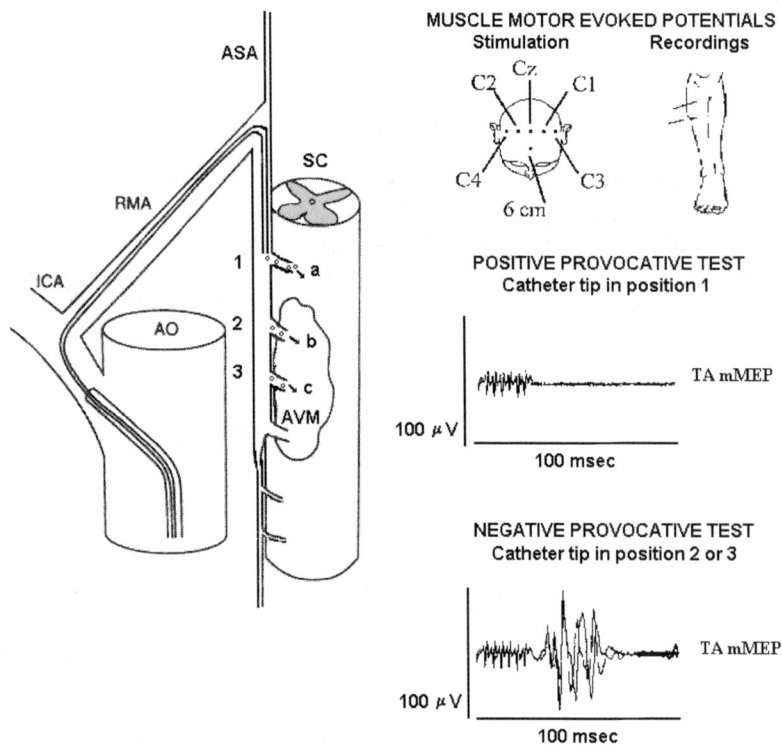

FIGURE 2. Left. Schematic illustration of the anatomic relationship between the aorta (*AO*), the intercostal artery (*ICA*) and the radiculomedullary artery (*RMA*) contributing to the anterior spinal artery axis (*ASA*). A spinal cord (*SC*) arteriovenous malformation (*AVM*) is represented. Xylocaine injected during a provocative test is schematically represented by *open circles*. (Modified from Touho *et al.*, used with permission from Elsevier). **Right.** Typical example of mMEP behavior after provocative test with Xylocaine injection at different catheter positions within the ASA. **Top right**. Muscle motor evoked potentials (mMEPs) are elicited through transcranial electrical stimulation and recorded from needle electrodes inserted in the anterior tibial muscle. **Middle right.** When the tip of the catheter is in position 1, the injected Xylocaine will flow through vessels feeding the normal spinal cord (*a*). The provocative test is positive: absence of the mMEP from the anterior tibial muscle (TA). **Bottom right.** When the tip of the catheter is more selectively advanced to position *2* or *3*, Xylocaine is injected only in vessels feeding the AVM (*b* and *c*) and, accordingly, the provocative test is negative: persistence of mMEP from the TA.

neurological deficit or a significant impairment of a preexisting deficit was not anticipated by permanent changes in either mMEPs, SEPs, or BCR at the end of the procedure.

Since we abandoned embolization when provocative tests were consistently positive, false positive results could not be tested.

RESULTS

Monitorability of evoked potentials, defined as the presence of a reliable response after the induction of anesthesia but before any interventional procedure, was 80% for SEPs, 85% for the BCR, and 92% for mMEPs.

Provocative tests were required by the endovascular surgeon in 23 out of 41 HRG procedures (56%), but in only eight out of 69 LRG procedures (12%). With the exception of two patients who had only a Xylocaine test before one of the embolizations they underwent during a multistage treatment, all other had provocative tests with both Amytal and Xylocaine. The results are presented in TABLES 1 and 2. Amytal test was positive in one out of 31 patients (3%); in this patient lower extremity mMEPs were lost after injection of the drug in the PSA. Nine out of 33 patients (27%) had a positive Xylocaine test: in seven patients mMEPs disappeared whereas in the other two patients SEPs decreased by more than 50% from baseline. In the seven patients where mMEPs were lost, Xylocaine was injected in the PSA in two cases, in the ASA territory in four cases, and in the posteroinferior cerebellar artery in one patient with a cervicomedullary malformation. SEPs decreased after injection of the drug, in the ASA in one patient and in the PSA in the other. Unexpectedly, there were no cases where both mMEPs and SEPs were simultaneously affected after the injection of either Amytal or Xylocaine in the ASA or PSA.

In terms of neurological outcome, no false negative results were observed. There was no neurological morbidity in the HRG. Paradoxically, two patients in the LRG woke up with a moderate and a severe paraparesis, which was not present before the procedure. In both of these patients, however, provocative tests were not done and neurophysiological monitoring could only document the disappearance of mMEPs from the lower extremities and anticipate the neurological motor deficit. On the other hand, there was no morbidity in either the LRG or in the HRG patients whenever provocative tests were performed.

TABLE 1. Results of provocative tests

	Number of Tests	Positive Tests	mMEPs Disappearance	SEPs Decrease Exceeding 50%	mMEPs Disappearance and SEPs Decrease Exceeding 50%
Amytal	31	1 (3%)	1	0	0
Xylocaine	33	9 (27%)	7	2	0

TABLE 2. Correlation between positive provocative tests and evoked potentials.

	Positive Tests	Vessel	Intraoperatively Changed EP
Amytal	1 (3%)	PSA	mMEPs
Xylocaine	9 (27%)	PSA	mMEPs
		PSA	mMEPs
		PSA	SEPs
		ASA	mMEPs
		ASA	mMEPs
		ASA	mMEPs
		ASA	mMEPs
		ASA	SEPs
		PICA	mMEPs

NOTE: EP, evoked potentials; PSA, posterior spinal artery; ASA, anterior spinal artery; PICA, posteroinferior cerebellar artery.

DISCUSSION

Neurophysiological monitoring offers a unique opportunity for the investigation of hemodynamic patterns in the spinal cord.[27,33] The complexity of these patterns in a healthy subject is due to the variability of spinal cord vascularization as well as to the anastomotic pial vessels bridging the anterior and posterior spinal artery compartments.[34,35] Such patterns are even more unpredictable in the presence of a spinal or spinal cord vascular lesion that introduces a rearrangement of the angioarchitecture. Consequently, the direction of blood flow in the ASA and PSA is variable and blood perfusion of the dorsal column does not always reflect that seen in the corticospinal tracts. This accounts for the unreliability of SEPs in assessing the functional integrity of the descending motor tracts.[17–20]

Despite the availability of techniques for intraoperative neurophysiological monitoring of MEPs under general anesthesia,[21–23] their use during endovascular procedures is anecdotal.[24-28]

In this study we reviewed intraoperative neurophysiological data from 110 endovascular procedures for AVM and tumors of the spine and spinal cord during a period of three years (1996–1999). The monitorablity of mMEPs was very high (92%) and exceeded that of SEPs and BCR, confirming that, under a proper anesthesiological regimen, transcranial electrical stimulation with a train of stimuli is highly effective in eliciting mMEPs from an impaired spinal cord.[29,30]

We routinely perform provocative tests before embolization. These tests have proven to be useful in enhancing the safety of endovascular obliteration of spinal vascular malformations.[36,37] Provocative tests are highly reliable, although they can overestimate the effects of embolization when the injection is not quite selective and the provocative drug spreads throughout the vascular network.[38]

In our experience, Xylocaine tests are more often positive than are Amytal tests. This may be explained by the different target of the two drugs. Xylocaine, at low doses, acts mainly on axonal conduction whereas Amytal switches off neuronal activities without suppressing axonal conduction.[39,40] Therefore, switching off the conduction in ascending or descending tracts will result in a positive test much more often than blocking the gray matter activity at a specific spinal level. The fact that these drugs test different pathways of the spinal cord implies that both should be used in every patient undergoing a spinal embolization.

The results of provocative tests in this series highlight the need for multimodal neurophysiological monitoring during these procedures. It is noteworthy that there were no cases where both SEPs and mMEPs were affected simultaneously after either Amytal or Xylocaine injection. This suggests that to monitor only SEPs or only mMEPs would expose the patient to the risk of neurological deficits. For example, in the seven patients where only mMEPs disappeared after Xylocaine injection, to monitor just SEPs would have resulted in a negative provocative test. Embolization would have been performed and most likely these patients would have suffered from new motor deficits. Although provocative tests affected mMEPs more than SEPs, we saw two cases of selective SEP loss and mMEP preservation after Xylocaine tests. Kalkman[41] described the possibility of selective sensory deficits with preserved motor function and its correlation with neurophysiological tests during removal of a spinal AVM. Thus, the use of mMEPs as a complement rather than as an alternative to SEPs monitoring must be emphasized.

It might be argued that neurophysiological tests can be chosen according to the vascular territory being dealt with. If a provocative test in the ASA territory is foreseen, mMEPs rather than SEPs should be used; if we expect to inject the drug in the PSA, SEPs may suffice. Unfortunately, strategies for embolization cannot be anticipated and mainly depend on the preliminary angiographic study of normal spinal cord vascularization and angioarchitecture of the malformation. Furthermore, our results showed that there is no correlation between the vascular compartment where the provocative test is performed (ASA or PSA) and the neurophysiological modality which is affected (SEPs or mMEPs). The unpredictability of provocative tests when Xylocaine is injected in the ASA and/or PSA supports the existence of vascular anastomoses and the variability of the spinal cord flow dynamic, the latter being even more complex in the presence of an AVM. The watershed area between the ASA and the PSA can also be shifted due to the existence of an AVM.

Touho et al.[37] described motor, but not sensory deficits, as a result of Xylocaine injection in the ASA. On the other hand, we have experienced one case where injection of Amytal/Xylocaine in the ASA caused loss of SEPs with mMEPs remaining unchanged. Therefore, we routinely monitor SEPs as well as mMEPs, even if the endovascular procedure is limited to the ASA territory.

Because any interventional procedure may acutely modify the local hemodynamic, it is critical to repeat provocative tests for both SEPs and MEPs before each embolization procedure.

There are situations where underlying pathology makes mMEPs and SEPs unmonitorable. Under these circumstances, setting up both monitoring modalities offers the opportunity to assess the functional integrity of at least one, should one be unmonitorable during the procedure. Permanent morbidity in this series was less then 3% and, surprisingly, higher in the LRG. In the HRG, where provocative tests were performed in 56% of the procedures, we never observed irreversible new or significantly exacerbated sensory or motor deficits immediately after the embolization. This is the result of our provocative test protocol, that selects patients who may benefit from the procedure without being exposed to the risk of ischemia. It might be argued that, because embolization is not performed whenever the tests are consistently positive, we did not evaluate the specificity of this protocol. However, we consider it unsafe and unethical to expose patients to the risk of embolization after a positive test. Since controlled trials are unlikely to occur, the ability to superselectively catheterize vessels feeding the vascular malformation is, therefore, of paramount importance.

Since morbidity was confined to two LRG patients on whom provocative tests were not performed, these tests should be considered also when a procedure is thought to be relatively safe. Hemodynamic patterns can be very unpredictable and the possibility of spinal cord ischemia even after embolization of a vertebral lesion has recently been described.[9]

CONCLUSIONS

In our experience, neurophysiological monitoring is feasible in the great majority of patients undergoing endovascular procedures for spine or spinal cord lesions. Although only mMEPs can reliably assess the functional integrity of motor tracts, they should be used as a complement rather than as an alternative to SEPs, whose specificity for ascending tracts cannot be ignored. The mMEPs turned out to be more sensitive than SEPs to provocative tests. This is noteworthy since, under general anesthesia, provocative tests with both Amytal and Xylocaine are mandatory in selecting those patients amenable to a safe embolization. Finally, neurophysiological monitoring offers a unique opportunity to investigate the hemodynamic of spinal cord vascular malformations and to integrate functional, anatomical, and clinical data.

REFERENCES

1. BRESLAU, J. & J.M. ESKRIDGE. 1995. Preoperative embolization of spinal tumors. J. Vasc. Interv. Radiol. **6**: 871–875.
2. BROADDUS, W.C., M.S. GRADY, J.B. DELASHAW, et al. 1990. Preoperative superselective arteriolar embolization: a new approach to enhance respectability of spinal tumors. Neurosurgery **27**: 755–759.
3. MEISEL, H.J., P. LASJAUNIAS & M. BROCK. 1995. Modern management of spinal cord vascular lesions. Minim. Invasive Neurosurg. **38**: 138–145.
4. NIIMI, Y., A. BERENSTEIN, A. SETTON, et al. 1997. Embolization of spinal dural arteriovenous fistulae: results and follow-up. Neurosurgery **40**: 675–683.
5. SMITH, T.P., L. GRAY, J.N. WEINSTEIN, et al. 1995. Preoperative transarterial embolization of spinal column neoplasms. J. Vasc. Interv. Radiol. **6**: 863–869.
6. VÁZQUEZ AÑÓN, V., C. BOTELLA, A. BELTRÁN et al. 1997. Preoperative embolization of solid cervicomedullary junction hemangioblastomas: report of two cases. Neuroradiology **39**: 86–89.
7. BERENSTEIN, A. & P. LASJAUNIAS. 1992. Surgical Neuroangiography: Endovascular Treatment of Spine and Spinal Cord Lesions. Vol. 5, 1–109, Springer Verlag. Berlin.

8. CLOFT, H.J., M.E. JENSEN, H.M. DO, et al. 1999. Spinal cord infarction complicating embolisation of vertebral metastasis: a result of masking of a spinal artery by a high-flow lesion. Interventional Neuroradiol. **5:** 61–65.
9. HALBACH, V.V., R.T. HIGASHIDA, C.F. DOWD et al. 1991. Management of vascular perforations that occur during neurointerventional procedures. AJNR **12:** 319–327.
10. MERLAND, J.J. & D. REIZINE. 1987. Treatment of arteriovenous spinal cord malformations. Semin. Intervent. Radiol. **4:** 281–290.
11. VAUZELLE, C., P. STAGNARA & P. JOUVNROUX. 1973. Functional monitoring of a spinal cord activity during spinal surgery. Clin. Orthop. **93:** 173.
12. BEN DAVID, B., P.D. TAYLOR, G.S. HALLER, et al. 1987. Posterior spinal fusion complicated by posterior column injury. A case report of a false-negative wake-up test. Spine **12:** 540–543.
13. PADBERG, A.M., T.J. WILSON HOLDEN, L.G. LENKE et al. 1998. Somatosensory- and motor-evoked potential monitoring without a wake-up test during idiopathic scoliosis surgery. An accepted standard of care. Spine **23:** 1392–1400.
14. BERENSTEIN, A., W. YOUNG, J. RANSOHOFF, et al. 1984. Somatosensory evoked potentials during spinal angiography and therapeutic transvascular embolization. J. Neurosurg. **60:** 777–785.
15. HACKE, W. 1989. Evoked potentials monitoring in interventional neuroradiology. In Neuromonitoring in Surgery. J. E. Desmedt, Ed.: 331–342. Elsevier, Amsterdam.
16. YOUNG, W. & D. MOLLIN. 1989. Intraoperative somatosensory evoked potentials monitoring of spinal surgery. In Neuromonitoring in Surgery. J. E. Desmedt, Ed.: 165–173. Elsevier, Amsterdam.
17. GINSBURG, H.H., A.G. SHETTER, & P.A. RAUDZENS. 1985. Postoperative paraplegia with preserved intraoperative somatosensory evoked potentials. J. Neurosurg. **63:** 296–300.
18. LESSER, R.P., P. RAUDZENS, H. LÜDERS et al. 1986. Postoperative neurological deficits may occur despite unchanged intraoperative somatosensory evoked potentials. Ann. Neurol. **19:** 22–25.
19. ZORNOW, M.H., M.R. GRAFE, C. TYBOR, et al. 1990. Preservation of evoked potentials in a case of anterior spinal artery syndrome. Electroencephal. Clin. Neurophys. **77:** 137–139.
20. TAKAKI, O. & F. OKUMURA. 1985. Application and limitation of somatosensory evoked potential monitoring during thoracic aortic aneurysm surgery: a case report. Anesthesiology **63:** 700–703.
21. TANIGUCHI, M., C. CEDZICH & J. SCHRAMM. 1993. Modification of cortical stimulation for motor evoked potentials under general anesthesia: technical description. Neurosurgery **32:** 219–226.
22. JONES, S.J., R. HARRISON, K.F. KOH et al. 1996. Motor evoked potential monitoring during spinal surgery: responses of distal limb muscles to transcranial cortical stimulation with pulse trains. Electroenceph. Clin. Neurophysiol. **100:** 375–383.
23. PECHSTEIN, U., C. CEDZICH, J. NADSTAWEK, et al. 1996. Transcranial high-frequency repetitive electrical stimulation of recording myogenic motor evoked potential with the patient under general anesthesia. Neurosurgery **39:** 335–344.
24. KATAYAMA, Y., T. TSUBOKAWA, T. HIRAYAMA, et al. 1991. Embolization of intramedullary spinal arteriovenous malformation fed by the anterior spinal artery with monitoring of the corticospinal motor evoked potential. Neurol. Med. Chir. **31:** 401–405.
25. ANDERSON, L.C., D.E. HEMLER, J.M. LUETHKE, et al. 1994. Transcranial magnetic evoked potentials used to monitor the spinal cord during neuroradiologic angiography of the spine. Spine **19:** 613–616.
26. KOTHBAUER, K., J.C. PRYOR, A. BERENSTEIN, et al. 1998. Motor evoked potentials predicting early recovery from paraparesis after embolisation of a spinal dural arteriovenous fistula. Interventional Neuroradiology **4:** 81–84.
27. SALA, F., Y. NIIMI, M.J. KRZAN, et al. 1999. Embolization of a spinal arteriovenous malformation: correlation between motor evoked potentials and angiographic findings: Technical Case Report. Neurosurgery **45:** 932–938.

28. SALA, F., Y. NIIMI, A. BERENSTEIN, et al. 2000. Role of multimodality intraoperative neurophysiological monitoring during embolization of a spinal cord arteriovenous malformation: a paradigmatic case. Intervent. Neuroradiol. **6:** 223–234.
29. KOTHBAUER, K., V. DELETIS & F.J. EPSTEIN. 1998. Motor evoked potential monitoring for intramedullary spinal cord tumor surgery: correlation of clinical and neurophysiological data in a series of 100 consecutive procedures. Neurosurg. Focus **4:** Article 1. <http://www.aans.org/journals/online j/may98/4-5-1>.
30. DELETIS, V. & K. KOTHBAUER. 1998. Intraoperative neurophysiology of the corticospinal tract. In Spinal Cord Monitoring. E. Stålberg, H.S. Sharma & Y. Olsson, Eds.: 421–444. Springer-Verlag, Vienna.
31. DELETIS, V., D.B. VODUSEK. 1997. Intraoperative recording of the bulbocavernosus reflex. Neurosurgery **40:** 88–93.
32. DOPPMAN, J.L., M. GIRTON & E.H. OLDFIELD. Spinal Wada Test. Radiology **161:** 319–321.
33. STECKER, M.M., P. MARCOTTE, R. HURST, et al. 1996. Spinal dural arteriovenous malformations. Intraoperative evoked potential evidence for pathophysiology: a case report. Spine **21:** 512–515.
34. DJINDJIAN, R. 1974. Angiography of the spinal cord. Surg. Neurol. **2:** 179–185.
35. KRAUSS, W.E. 1999. Vascular anatomy of the spinal cord. Neurosurg. Cl. N. Am. **10:** 9–15.
36. SADATO, A., W. TAKI, I. NAKAHARA, et al. 1994. Improved provocative test for the embolization of arteriovenous malformations: technical note. Neurol. Med. Chir. **34:** 187–190.
37. TOUHO, H., J. KARASAWA, H. OHNISHI, et al. 1994. Intravascular treatment of spinal arteriovenous malformations using a microcatheter: with special reference to serial Xylocaine tests and intravascular pressure monitoring. Surg. Neurol. **42:** 148–156.
38. KATSUTA, T., T. MORIOKA, K. HASUO, et al. 1993. Discrepancy between provocative test and clinical results following endovascular obliteration of spinal arteriovenous malformation. Surg. Neurol. **40:** 142–145.
39. TANAKA, K. & M. YAMASAKI. 1966. Blocking of cortical inhibitory synapses by intravenous lidocaine. Nature **209:** 207–208.
40. TERADA, T., K. NAKAI, M. NAKAI, et al. 1991. The differential action of Lidocaine and Amytal on the central nervous system as the provocative test drug. In Proceedings of 7th Annual Meeting of Japanese Society of Intravascular Neurosurgery, Mie, Japan. 95–101.
41. KALKMAN, C.J., J.C. DRUMMOND & S.U. HOI. 1994. Severe sensory deficits with preserved motor function after removal of a spinal arteriovenous malformation: correlation with simultaneously recorded somatosensory and motor evoked potentials. Anesth. Analg. **78:** 165–168.

The Role of Intraoperative Neurophysiology in the Protection or Documentation of Surgically Induced Injury to the Spinal Cord

VEDRAN DELETIS[a] AND FRANCESCO SALA[b]

[a]*Hyman Newman Institute for Neurology and Neurosurgery, Beth Israel Medical Center, New York, New York, U.S.A*

[b]*Department of Neurological Sciences and Vision, Section of Neurosurgery, Verona University Hospital, Italy*

ABSTRACT: Playing both neuroprotective and educational roles, introperative neurophysiology has become an intrinsic part of modern neurosurgery. In this article, we present evidence substantiating the neuroprotective role of intraoperative neurophysiology, specifically its capacity to help prevent injury to the corticospinal tracts and the dorsal columns during spinal cord injury.

KEYWORDS: Intraoperative neurophysiological monitoring; Motor evoked potentials; Spinal cord; Neuroprotection.

INTRODUCTION

Within the surgical community it is a common belief that certain neurosurgical interventions are intrinsically linked to postoperative neurological deficits. Examples include surgery for intramedullary spinal cord tumors (IMSCTs) and intrinsic brain stem tumors. For a long time, few surgeons dared to operate on the spinal cord or brain stem, or at least did not perform radical surgery. Therefore, most surgical interventions for IMSCTs and brain stem tumors were restricted to tumor biopsy.

The development and improvement of diagnostic tools (CT, MRI, ultraselective spinal angiography) and surgical techniques (operative microscope, bipolar coagulators, laser, contact laser) inspired a few surgical centers to start operating on IMSCTs. Since then, surgeons who operate on IMSCTs have been attempting to achieve gross total resection of these tumors, since this approach seems to result in tumor control or even complete remission in cases of low grade astrocytomas and ependymomas.

Regardless of how carefully the surgery to the spinal cord had been performed, serious neurological injuries were frequently the consequence of these surgeries. The most serious complications were paraplegia or, in cases of cervical spinal cord tumors, even quadriplegia.

Address for correspondence: Vedran Deletis, M.D., Ph.D., Hyman Newman Institute for Neuology & Neurosurgery, Beth Israel Medical Center, 170 East End Avenue, Room 311, New York, NY 10128, U.S.A. Voice: 212-870-9684; fax: 212-870-9690.
vdeletis@betheisraelny.org

The first attempts to prevent intraoperative injury to the spinal cord during surgical correction of scoliosis by using intraoperative neurophysiology began in the late 1970s. Intraoperative monitoring of the functional integrity of the spinal cord with somatosensory evoked potentials (SEPs) was the only neurophysiological tool available at that time.[1] Later, a similar approach was attempted during angiography and embolization of spinal vascular malformations.[2] These strategies have significantly decreased neurological injury to the spinal cord (from 20% to 2% in the cases of spinal angioembolization).

For a number of reasons, the initial enthusiasm generated by success with monitoring the functional integrity of the spinal cord by using SEPs did not generalize to surgery for IMSCTs. During surgery for scoliosis, and to a lesser extent during angioembolization of spinal cord vascular malformations, injury to the spinal cord is typically diffuse, affecting both ascending (sensory) and descending (motor) pathways. Therefore, monitoring spinal cord sensory pathways was sufficient to extrapolate data on the functional integrity of the motor pathways. Contrary to this, during surgery for IMSCTs, surgeons can selectively damage either the motor or sensory pathways. Therefore, monitoring only one of these pathways was not sufficient. Furthermore, at that time, a reliable method for intraoperative monitoring of the motor pathways did not exist. Advances in the method of intraoperatively eliciting and monitoring motor evoked potentials from the exposed spinal cord, and later from the limb muscles, enormously expanded the ability of intraoperative neurophysiology to prevent postoperative motor deficits.

Our experience with both techniques in surgeries performed on hundreds of patients with IMSCTs operated by a single surgeon, has been published elsewhere.[3,4] These two studies confirmed the validity of intraoperative monitoring of the functional integrity of the motor pathways by using motor evoked potentials.

METHODS OF INTRAOPERATIVE NEUROPHYSIOLOGY FOR MONITORING MOTOR PATHWAYS

There are two reliable methodologies for monitoring the motor pathways intraoperatively:
1. Transcranial electrical stimulation with a single stimulus, recording the D wave directly from the spinal cord (single pulse technique).
2. Transcranial electrical stimulation with a multipulse stimuli, recording muscle MEPs (mMEPs) from limb muscles (multipulse technique).

Transcranial Electrical Stimulation with a Single Stimulus and Recording of the D Wave Directly from the Spinal Cord (Single Pulse Technique)

The history of this method begins with experimental animal work by Patton and Amassian[5] who recorded the direct response of the pyramidal tract from pyramids of the medulla, after electrical stimulation of the exposed motor cortex in monkeys. Merton and Morton[6] demonstrated that the motor cortex can be electrically stimulated through the scalp, eliciting motor evoked potentials that can be recorded from the limb muscles (MEPs). The ability to intraoperatively monitor the functional integrity of the motor pathways was achieved by Boyd,[7] Burke,[8] Katayama,[9] Deletis,[10,11] and

Morota.[4] The schematics of this method are shown in FIGURE 1. Details of this method are described elsewhere.[3,10,11] Briefly, the motor cortex is stimulated transcranially with a stimulus of 0.5 ms duration and an intensity of up to 200 mA. The descending volleys of electrically activated corticospinal tract (CT) motoneurons are recorded in the vicinity of the spinal cord by placing a catheter electrode epi- or subdurally. Synchronized activity of the fast motoneurons of the CT is characterized by a D wave (representing direct activation of the CT neurons) and/or an I wave (representing transynaptical activation of the CT neurons). In the case of D waves, CT neurons are activated intracranially, distal to the cortical motoneuron cell body, and the electrical activity is recorded before the CT neurons synapse with the α motoneurons.

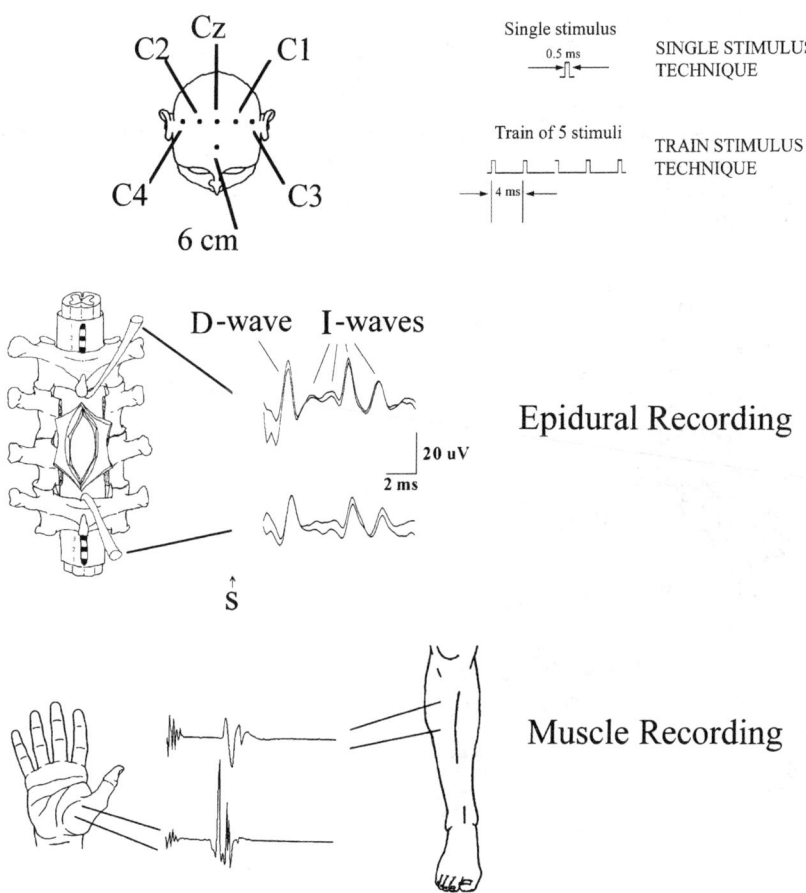

FIGURE 1. Transcranial electrical stimulation. Methodology for eliciting motor evoked potentials from the spinal cord (single stimulus technique) and from the limb muscles (multipulse technique). Transcranial stimuli were delivered through electrodes placed on the scalp according to the 10-20 EEG system (see text for explanation; modified with permission from Deletis and Kothbauer[11]).

Therefore, we can consider the D wave as a "neurogram" of the CT. The D wave exhibits very desirable features such as resistance to non-surgically induced changes (anesthesia, temperature, etc.) that are usually the main drawback of other intraoperative monitoring methods. Therefore, monitoring D waves is a suitable technique for judging the functional integrity of the CT.

Transcranial Electrical Stimulation with a Multipulse Stimuli and Recording Muscle MEPs from Limb Muscles (Multipulse Technique)

The position of the stimulating electrode is identical to that of the single pulse technique. Transcranially applied stimuli consist of a short train of five to nine stimuli, 4 ms apart. Recording MEPs from limb muscles is easily achieved with a needle or surface electrode (FIG. 1).

Temporal summation of descending activity to the CT at the level of the α motoneurons can overcome the suppressive influence of anesthetics. This accumulated activity brings α motoneurons to the firing threshold and generates mMEPs in patients under anesthesia.[12] Anesthetic regimens compatible with this techniques include Propofol (150 µg/kg/min) and Fentanyl (1 µg/kg/hour). Use of muscle relaxants should be restricted to a short acting type used only during the initial intubation of the patient.

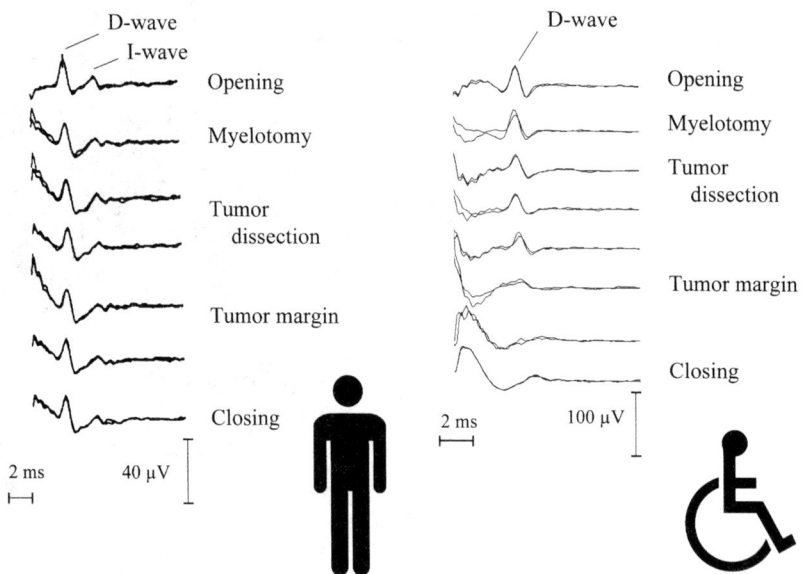

FIGURE 2. Left. D wave stable throughout an intramedullary spinal cord tumor removal. The presence of the D wave at the end of surgery predicts preserved motor function postoperatively. **Right.** Rapid loss of the D wave during intramedullary spinal cord tumor removal. In this instance, the patient became quadriparetic after surgery. (Note, when this patient was monitored no methods for eliciting muscle MEP intraoperatively were available; reproduced with permission from Deletis and Kothbauer[11]).

Critical Role of MEPs as Predictor for Postsurgical Motor Outcome

The D wave represents synchronous activity by electrically activated fast neurons of the CT. The preservation of the D wave at the end of brain or spinal cord surgery is a prerequisite for good motor outcome postoperatively. In FIGURE 2 we document the value of epidurally recorded MEPs. These figures demonstrate how the amplitude of the D wave is a crucial parameter for making prognosis of motor outcome after surgery. Unfortunately, no data exist concerning the minimum D wave amplitude that is sufficient to preserve motor function in man. From clinical experience we have learned that preservation of 50% of the amplitude of the D wave, when compared to the amplitude at the beginning of surgery, is sufficient for good motor outcome. If we assume that the amplitude of the D wave quantifies the number of fast fibers of the CT that are functional, it is theoretically possible to estimate the lowest number of CT axons needed to preserve good motor outcome. In cats, Blight et al.[13,14] showed that as much as 10% of surviving CT fibers is sufficient to support locomotion.

In the Case of mMEPs, Generation of These Potentials is a More Complex Process Than Those Underlying D Wave Generation

Together with preservation of function of fast motoneurons, generation of mMEPs require the preservation of other descending tracts of the spinal cord, including the propriospinal system. For purposes of clarity, we refer to all of the descending tracts, including the propriospinal systems, as the "supportive system" of the spinal cord. This supportive system is an important contributor in the generation of mMEPs. During surgery for IMSCTs, the supportive system can be selectively damaged whereas the CT is left intact. Acute damage to the supportive system prevents α motoneurons from reaching the firing threshold after electrical stimulation of the motor cortex. This kind of selective damage clinically results in postoperative transient paraplegia (TP) in the case of surgery for thoracic IMSCTs.

Neurophysiological features of TP include lost muscle MEPs with preserved or diminished amplitude of the D wave at the end of surgery (up to 50% of the initial amplitude[3,11]). Clinically, patients with TP are paraplegic immediately after surgery and gain complete recovery from the motor deficit within a few hours or days.

Due to the nature of spinal cord surgery, our experience has shown that, in most cases, the first warning sign of deteriorating function of the descending tract is a loss of mMEPs. At this point, the surgeon can continue with the surgery until the D wave

TABLE 1. Principles of motor evoked potential (MEP) interpretation

D Wave (at the end of surgery)	Muscle MEP[a] (at the end of surgery)	Motor Status (postoperatively)
Unchanged or 30–50% decrease	preserved	unchanged
Unchanged or 30–50% decrease	lost unilaterally or bilaterally	transient motor deficit
> 50% decrease	lost bilaterally	severe long term motor deficit

[a]In the tibial anterior muscle(s).

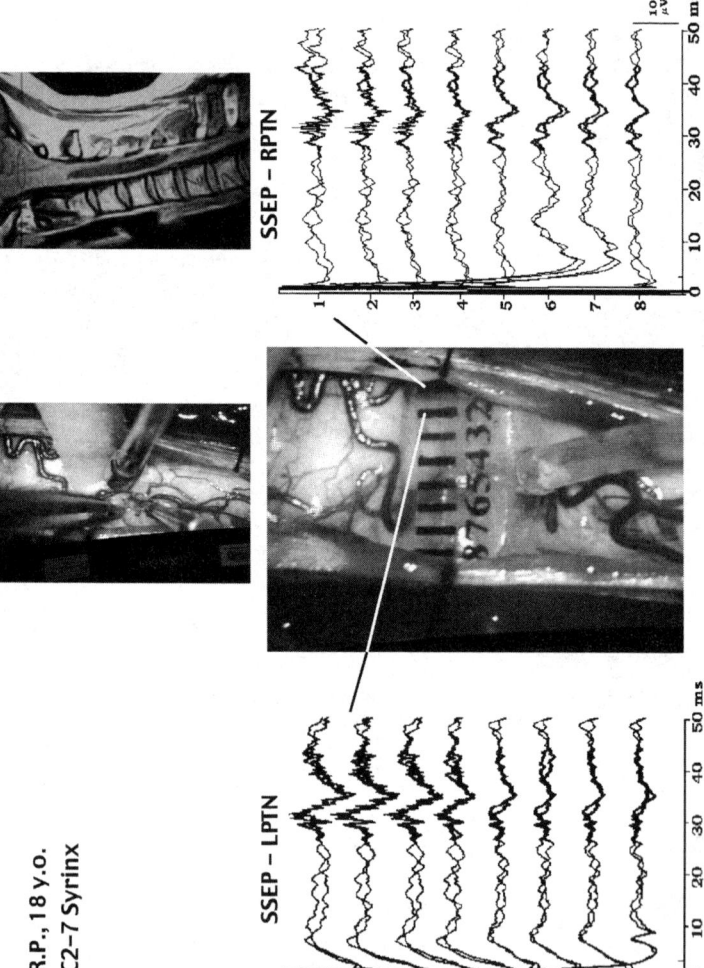

FIGURE 3. Dorsal column mapping in an 18 year old patient with a syringomyelic cyst between the C2 and C7 segments of the spinal cord. **Upper right.** MRI showing syrinx. **Lower middle.** placement of miniature electrode over surgically exposed dorsal column; *vertical bars* on the electrode represent the location of the underlying exposed electrode surfaces. SEPs after stimulation of the left and right tibial nerves showing maximum amplitude between electrode 1 and 2 (**lower left** and **right**). These data strongly indicate that both dorsal columns from the left and right lower extremities have been pushed to the extreme right side of the spinal cord. Using this data as a guideline, the surgeon performed the myelotomy through the left side of the spinal cord and inserted the shunt to drain the cyst (**upper middle**). Postoperatively, the patient did not suffer from a sensory deficit.

R.P., 18 y.o.
C2–7 Syrinx

starts to deteriorate, reaching a critical point at a decrease of 50% from the initial baseline. In all cases reaching this point, surgery must be terminated. Otherwise, serious motor deficits may result. In our experience, the guidance provided by information gained about the functional integrity of the descending tracts prevented any postoperative paraplegia from occurring in a series of 100 consecutive IMSCT patients operated on by a single surgeon.[3] TABLE 1 provides a summary of MEP monitoring that can be used for guidance and prognosis during surgery for IMSCTs.

Can We Preserve Dorsal Columns of the Spinal Cord from Being Damaged During Myelotomy?

The most common surgical approach to the IMSCT involves splitting the spinal cord between the dorsal columns (dorsal myelotomy). The surgeon attempts to perform the myelotomy on the midline between the right and left dorsal columns through the posterior median fissure of the spinal cord. Using this strategy, less lesioning to the dorsal columns of the spinal cord should occur. Unfortunately, this is not the case in most surgeries performed for IMSCTs. This strategy might fail for two reasons. First, the anatomical landmarks the surgeon uses to determine the posterior fissure are not always visible or reliable. Second, IMSCTs or other types of intramedullary pathology can grow asymmetrically. Therefore, the dorsal columns can be pushed to one side. In order to overcome these complications, we use neurophysiological mapping of the dorsal columns as a guide before proceeding with the myelotomy. This is accomplished by measuring the amplitude gradient of SEPs generated by dorsal columns of the spinal cord. SEPs are recorded by using a miniature electrode consisting of eight parallel stainless steel wires embedded in silicon, each wire with a diameter of 76 µm, an exposed length of 2 mm, and an interelectrode distance of 1 µm (see FIGURE 3).[15,16] The miniature electrode is placed over the surgically exposed dorsal columns before myelotomy. This unique electrode provides an ultraselective record of the amplitude gradient of SEPs directly from the dorsal column, after stimulation of the right and left tibial and/or median nerves (depending on the level of the tumor). Because the dorsal columns for the upper extremities lie laterally, and those of the lower extremities lie medially, it is possible to use the amplitude gradient of SEPs to determine the exact functional midline between right and left dorsal columns. Because of the physical features of the electrode, this can be done very precisely. The location of the maximal amplitude of recorded SEPs represents the electrode wire closest to the location of the dorsal columns (FIG. 3, wires are marked 1 and 2). As a result, the surgeon avoids cutting through the side of the spinal cord where the maximum amplitude SEPs are recorded (FIG. 3). The first experience using this method has helped us to preserve the dorsal columns from damage during the initial myelotomy. In order to prevent collateral damage from the surgical instrument used for the myelotomy, a contact laser with a cutting tip and a diameter of 200 µm has been used. Therefore, a very precise incision can be made.

CONCLUSION

After three decades of trial and error, intraoperative neurophysiology has become an integral part of neurosurgery. Intraoperative neurophysiology plays three roles:

prevention (neuroprotection), *documentation* of surgically induced neurological injury, and *education* for the young neurosurgeon gaining neurosurgical skills. The neuroprotective role of intraoperative neurophysiology has become especially important with the development and routine application of neurophysiological methods created specifically for intraoperative use.

REFERENCES

1. ENGLER, G.L., N.I. SPIELHOLZ, W.N. BERNHARD, *et al.* 1978. Somatosensory evoked potentials during Harrington instrumentation for scoliosis. J. Bone Joint Surg. **68**(4): 528–532.
2. BERENSTEIN, A., W. YOUNG, J. RANSOHOFF, *et al.* 1984. Somatosensory evoked potentials during spinal angiography and therapeutic transvascular embolization. J. Neurosurg. **60**: 777–785.
3. KOTHBAUER, K., V. DELETIS & F.J. EPSTEIN. 1998. Motor evoked potential monitoring for intramedullary spinal cord tumor surgery: Correlation of clinical and neurophysiological data in a series of 100 consecutive procedures. Neurosurg. Focus **4**: Article 1. <http://www.aans.org/journals/online_j/may98/4-5-1>.
4. MOROTA, N., V. DELETIS, V., S. CONSTANTINI, *et al.* 1997. The role of motor evoked potentials (MEPs) during surgery of intramedullary spinal cord tumors. Neurosurg. **41**: 1327–1336.
5. PATTON, H.D. & V.E. AMASSIAN. 1954. Single and multiple unit analysis of cortical state of pyramidal tract activation. J. Neurophysiol. **17**: 345–363.
6. MERTON, P.A. & M.H. MORTON. 1980. Stimulation of the cerebral cortex in the intact human subject. Nature. **285**: 227.
7. BOYD, S.G., J.C. ROTHWELL, J.M. COWAN, *et al.* 1986. A method of monitoring function in corticospinal pathways during scoliosis surgery with a note on motor conduction velocities. J. Neurol. Neurosurg. Psychiatry **49**: 251–257.
8. BURKE, D., R. HICKS, S.C. GANDEVIA, *et al.* 1993. Direct comparison of corticospinal volleys in human subject to transcranial magnetic and electrical stimulation. J. Physiol. (Lond.). **470**: 383–393
9. KATAYAMA, Y., T. TSUBOKAWA, S. MAIJIMA, *et al.* 1988. Corticospinal direct response in humans: identification of the motor cortex during intracranial surgery under general anesthesia. J. Neurol. Neurosurg. Psychiat. **51**: 50–59.
10. DELETIS, V. 1993. Intraoperative monitoring of the functional integrity of the motor pathways. *In* Advances in Neurology: Electrical and Magnetic Stimulation of the Brain. O. Devinsky, A. Beric & M. Dogali, Eds.: 201–214. Raven Press, New York.
11. DELETIS, V. & K. KOTHBAUER. 1998. Intraoperative neurophysiology of the corticospinal tract. *In* Spinal Cord Monitoring. E. Stalberg, H.S. Sharma & Y. Olsson, Eds.: 421–444. Springer-Verlag, Wien, New York.
12. TANIGUCHI, M., C. CEDZICH & J. SCHRAMM. 1993. Modification of cortical stimulation for motor evoked potentials under general anesthesia: technical description. Neurosurgery **32**(2): 219–226.
13. BLIGHT, A.R. 1983. Axonal physiology of chronic spinal cord injury in the cat: intracellular recording *in vitro*. Neuroscience **10**: 1471–1486.
14. BLIGHT, A.R. 1983. Cellular morphology of chronic spinal cord injury in the cat: analysis of myelinated axons by line sampling. Neuroscience **10**: 521–543.
15. KRZAN, M., V. DELETIS & V. ISGUM. 1997. Intraoperative neurophysiological mapping of dorsal columns. A new tool in the prevention of surgically induced sensory deficit? Joint Meeting of British Royal Society of Medicine and American Academy of Clinical Neurophysiology, London, June 1996. EEG Clin. Neurophys. **102**: 37P.
16. KRZAN, M.J., V. DELETIS & F.J. Epstein. 1998. Intraoperative Neurophysiological Mapping of the Dorsal Columns. *In* Neurophysiology—ECCN 98. E.V. Stalberg, A.W. De Weerd & J. Zidar, Eds.: 391–398. Monduzzi, Bologna, Italy.

Part IV: Neuroprotection in Neurosurgery

QUESTION FOR DR. SALA

From Dr. Maynard

What percentage of your patients underwent vasospasm and did it recur in the long term following the procedure?

ANSWER: Significant vasospasm during endovascular procedures is rare but depends on the size and tortuosity of the vessel catheterized. For spinal cord embolization, the incidence of significant vasospasm is slightly higher compared to the brain embolization because vessels are usually small and tortuous. We try to avoid spasm by using a floppy soft catheter and guidewire. For the last 100 cases of spinal cord embolization, we have had two to three cases of significant vasospasm that required us to abort or modify the procedure.

Have you followed your patients to determine long-term benefit, since endovascular interventions may cause intimal hyperplasia and decrease lumen size in blood vessels?

ANSWER: The long-term occlusive effect of embolization depends on the technique used. We try to use liquid adhesives to occlude the lesion, such as the fistula site or the nidus of a malformation. If this was achieved successfully, the occlusive effect can be considered permanent. If proximal feeder occlusion was performed without reaching the lesion itself, temporary improvement of symptoms may be obtained by decreasing flow and mass effect of the lesion, but recurrence of the symptoms may occur owing to development of collateral feeders. Also, if particles are used as an embolic agent, recanalization of the embolized vessel can occur. Therefore, we try to minimize the use of particles for spinal embolization.

QUESTIONS FOR DR. ANDREWS

From Dr. Paule

You mentioned that hyperoxygeneration may be beneficial in neuroprotection—yet in some instances (like retinopathy associated with prematurity) too much oxygen can be problematic—that is, oxidative stress—and therefore increase damage. How do we know when O_2 supplementation might be beneficial versus detrimental?

ANSWER: In all the instances considered (notably head or spinal cord injury, and stroke), there is a clear lack of oxygen at the tissue level. The interesting finding is that pushing the inspired O_2 very high appears to result in brain tissue O_2 levels that are significantly higher than expected based on dissolved oxygen in the blood. However, these brain tissue oxygen levels are still less than normal and thus probably not comparable to retinopathy of prematurity. Your point is well taken in that there is some evidence that long-term (greater than 24 h) exposure to 100% FiO_2 may be detrimental—not to the brain, but to the respiratory system (due to tracheal surface inflammation and increased pulmonary atelectasis)—as discussed briefly in Reference 11 in the article.

The comments in the previous paragraph may not apply in situations where neuroprotection is not based on ischemia/hypoxia (deficit of oxygen at the tissue level); for example, in Alzheimer's disease or Parkinson's disease (as discussed in my other presentation at this Conference). In these disorders it is unlikely that hyperoxygenation would be of significant benefit, and may be harmful—as you suggest.

From Dr. Maynard

How can surface pulse oximetry electrodes help us to monitor parenchymal oxygen tension subcortically or how effective is intraventricular monitoring for detecting oxygen tension, for example, in basal ganglia?

ANSWER: Near infrared spectroscopy (NIRS) for brain is not identical to transcutaneous pulse oximetry. Light in the near infrared portion of the spectrum (roughly 700 to 1000 nm) has the ability to penetrate tissue several centimeters, including the skull. By having more than one light-receiving probe or sensor, the absorption from the skull and extracranial tissues can be compensated for in comparison with light traversing the brain tissue itself. It is essential to stress that the values for brain tissue oxygenation obtained by NIRS are not particularly accurate in absolute terms, but are quite helpful in detecting changes in oxygenation. In this respect it is similar to certain techniques for cerebral blood flow monitoring, such as laser-Doppler.

Concerning intraventricular oxygen monitoring and basal ganglia hypoxia, it is unlikely that intraventircular monitoring would be of much help in a situation with focal ischemia of the basal ganglia such as a stroke. This is the same criticism leveled against jugular bulb oxygen tension monitoring—it measures the venous blood oxygen from much of one hemisphere and, thus, is poor at detecting focal ischemia. Clearly NIRS brain tissue oxygenation is only useful for the brain immediately beneath the optode (and probably only for a depth of a couple of centimeters)—it would thus be unlikely to detect basal ganglia focal ischemia any more successfully than intraventricular or jugular bulb monitoring.

From Dr. von Lubitz

Are you using "neural networks" as a specific, commercially available software package or as a "philosophical" approach in the process of data reduction?

Have you considered nanotechnology as a substitute in the development of essentially non-invasive site-specific physiological monitoring systems?

ANSWER: Concerning neural networks and "off the shelf" versus "philosophical"—the answer is "both." The NASA Ames Smart Systems Group is developing (proprietary) neural network/fuzzy logic software specifically for the integration of biometric data from multiple, dissimilar sensors in near real-time. Clearly there are similarities with commercially available software and, hopefully, advantageous differences.

Concerning nanotechnology—by some definitions this means less than 100 nm in size. Although we anticipate getting a probe with multiple microsensors down to a diameter of less than one mm, and possibly on the order of 0.1 mm, that is a far cry from true "nanoscale" intervention. Practically speaking, even in brain (not to mention other applications for which the NASA Smart Probe is being developed, such as breast cancer diagnosis and treatment) a probe 0.1 mm in diameter is quite minimally invasive—considering we are using it to diagnose and treat patients with brain tumors, disabling movement disorders, etc. Personally I see the next major step in non-invasive or minimally invasive monitoring, diagnosis, and treatment to be the expanded use of the vascular system to access tissues. One might develop the vascular system as the "intervention superhighway" for the brain and other organs—and unlike the Internet, there is no need to lay optical fiber or worry about bandwidth.

Neuroprotective Effects of Novel Cholinesterase Inhibitors Derived from Rasagiline as Potential Anti-Alzheimer Drugs

MARTA WEINSTOCK,[a] NATANJA KIRSCHBAUM-SLAGER,[a] PHILIP LAZAROVICI,[a] CORINA BEJAR,[a] MOUSSA B.H. YOUDIM,[b] AND SHAI SHOHAM[c]

[a]*Departments of Pharmacology, Hebrew University Faculty of Medicine, Jerusalem, Israel*

[b]*Technion-Faculty of Medicine, Eve Topf and NPF Centers, Haifa, Israel*

[c]*Herzog Hospital, Jerusalem, Israel*

ABSTRACT: TV3326, (*N*-propargyl-(3R)-aminoindan-5-yl-ethyl,methyl carbamate) was prepared in order to combine the neuroprotective effects of rasagiline, a selective inhibitor of monoamine oxidase (MAO)-B with the cholinesterase (ChE) inhibitory activity of rivastigmine as a potential treatment for Alzheimer's disease. The study reported here examined the neuropotective effects of TV3326 against various insults *in vitro* and *in vivo*. TV3326 caused a dose related (10–500 µM) reduction in death induced in NGF differentiated rat pheochromocytoma (PC12) cells by 3–4 hour exposure to oxygen–glucose deprivation. A single sc injection of TV3326 given five minutes after closed head injury in mice significantly reduced the cerebral edema, and accelerated the recovery of motor function and spatial memory several days later. Unilateral icv injection of streptozotocin (STZ) 1.5 mg in rats, caused specific damage to myelinated neurones in the fornix and corpus callosum accompanied by microgliosis. Three bilateral injections of STZ, 0.25 mg each, caused more widespread damage, and a marked impairment in spatial memory. Chronic oral treatment with TV3326 (75 µmols/kg) reduced the neuronal damage and microgliosis and almost completely prevented the memory impairment. The neuroprotective effect in PC12 cells may be due to a combination of ChE inhibition and antiapoptotic activity. The latter does not result from ChE inhibition. It is associated with the presence of the propargyl group, since it occurs with other propargylamines that do not inhibit MAO, but not with drugs that inhibit only ChE.

KEYWORDS: PC12 cells; Glucose–oxygen deprivation; Closed head injury; Streptozotocin memory deficits.

Address for correspondence: Professor Marta Weinstock, Leon & Mina Deutsch Professor of Psychopharmacology, Department of Pharmacology, School of Pharmacy, Hebrew University Medical Centre, Ein Kerem, Jerusalem 91120, Israel. Voice: 972-2-6758731; fax: 972-2-6758741.

martar@cc.huji.ac.il

INTRODUCTION

Although the cause of Alzheimer's disease (AD) has not yet been fully elucidated, indirect evidence has implicated oxidative stress in the etiology of the neurodegeneration by increasing glutamate release and inducing apoptosis.[1,2] The production of oxidative free radicals (ROS) is either a cause or a consequence of mitochondrial dysfunction. Complex IV activity has been reported to be decreased in the brains of patients with AD.[3,4] The initial neurodegeneration, whatever its source, is further exacerbated by the release of cytokines and the reactive oxygen species from activated microglia.[5,6]

In AD, nerve cell loss occurs principally in larger neurons of the superficial cortex and in terminal synapses of neurons arising in the nucleus basalis of Meynert. The degree of memory impairment correlates well with the loss of cholinergic transmission in the temporal lobe and other cortical brain regions.[7] Cholinesterase (ChE) inhibitors induce a modest improvement in memory and/or delay the rate of its decline.[8] However, there is little or no evidence that they can significantly reduce neuronal damage by interfering with the production or action of ROS, or by reducing microgliosis.

The monoamine oxidase (MAO)-B inhibitor, selegiline was reported to delay the progression of AD without significantly improving cognitive function.[9,10] In rats, selegiline suppresses the formation of the cytotoxic ·OH radical and protected nigrostriatal neurons from oxidative damage induced by MPTP.[11] It also decreases brain damage induced by hypoxia-ischemia, probably by increasing the activities of the protective enzymes, superoxide dismutase and catalase.[12] *In vitro*, selegiline protected cultured cells from the toxic effects of serum deprivation.[13] Selegiline is metabolized to desmethylselegiline, methamphetamine, and amphetamine. Although desmethylselegiline is responsible for the neuroprotective effect of the parent compound, the other metabolites appear to interfere with this activity[14] thereby compromising the effect of selegiline.

Rasagiline, *N*-propargyl-(3R)-aminoindan, is another selective inhibitor of MAO-B[15] that exerts neuroprotective actions against a variety of insults *in vitro* and *in vivo*. These actions are similar to those of selegiline,[16–19] but rasagiline is not metabolized to amphetamine or methamphetamine. TV3326, (*N*-propargyl-(3R)-aminoindan-5-yl-ethyl,methyl carbamate) was synthesized in an attempt to combine the neuroprotective properties of rasagiline with the ChE inhibiting activity of rivastigmine, a drug with proven efficacy in AD patients.[8] The introduction of a carbamate in the 6 position of rasagiline reduced by more than 1,000-fold, MAO-B inhibitory activity *in vitro* or after acute administration in mice and rats. However, after chronic oral administration to rats the compound inhibited MAO-A and B and ChE in the brain at similar doses, but showed very little effect on MAO in the intestine and liver.[20] This reduced the likelihood of tyramine potentiation and adverse cardiovascular reactions.

The rat pheochromocytoma cell line (PC12) has been widely used as a model of neuronal cell death induced by serum and nerve growth factor (NGF) deprivation,[21] toxins,[22] and ischemia.[17] Inhibitors of both ChE[23,24] and MAO-B[17,25] have been shown to protect PC12 cells against some of these insults. Rasagiline and rivastigmine significantly reduced brain edema and speeded the recovery of motor function and spatial memory after brain injury in mice.[18,26]

Unilateral intracerebroventricular (icv) injection of streptozotocin (STZ 1.5 mg) into middle-aged rats decreased glucose utilization in those brain areas that showed extensive morphological changes in AD patients.[27] Icv STZ also reduced choline acetyltransferase activity in these brain areas and altered levels of NGF in a manner consistent with cholinergic deafferentation.[28,29] Although it was suggested that STZ may act by interfering with insulin mediated glucose uptake,[30] it is possible that it produced its effects by inducing ROS and lipid peroxidation, as it does in peripheral tissues.[31,32] Parenteral administration of STZ also causes damage to myelin[33] associated with diabetes.

In this article, we report the neuroprotective effects of TV3326 against oxygen–glucose deprivation (OGD) in PC12 cell cultures and *in vivo* as seen by its reduction of the sequelæ of closed head injury in mice and neuronal damage caused in rats by icv injection of STZ.

MATERIALS AND METHODS

Drugs

TV3326 hemitartrate, TV3279 mesylate and TV3262 hydrochloride were supplied by Teva Pharmaceuticals Israel, Ltd. Rivastigmine hemitartrate was supplied by Novartis, Switzerland. Because of its chemical instability, streptozotocin (Sigma, St. Louis, MO, U.S.A.) was prepared freshly as a solution in artificial cerebrospinal fluid (CSF) immediately before injection into each rat. All doses or concentrations are expressed as μmoles/kg or μM of the appropriate salt.

Neuroprotective Effect of TV3326 Against Oxygen–Glucose Deprivation in Nerve Growth Factor Differentiated PC12 Cells

PC12 cells (1.5–2×10^5) per dish were grown in Dulbeco's modified Eagle's medium supplemented with 7% fetal calf and horse serum containing NGF (50 ng/ml) for 8–10 days and then subjected to OGD for 3–4 h in a specially constructed ischemic device as described elsewhere.[17] At the end of this period, glucose (4.5 mg/ml) was added and the cultures reoxygenated for an additional 18 h. Under these conditions, cell death determined by the leakage of lactic acid dehydrogenase (LDH) into the medium ranged between 30 and 50% of the total content. TV3326, its S-isomer, TV3279, or rivastigmine was added to the cultures 15 min before application of OGD in concentrations in the range 1–500 μM.

Cholinesterase Inhibition by Putative Neuroprotective Agents in PC12 Cells

ChE activity was measured by the method of Ellman *et al.*,[34] in PC12 cells grown in 25-ml culture bottles at a density of 100,000 cells/ml. The various experimental groups were incubated with the drugs (1–500 M) for 4 or 18 hours. They were then centrifuged at 1,000 rpm for 10 minutes and solubilized in 100 μl 0.1M phosphate buffer pH 8.0 containing 1% Triton. ChE inhibition was calculated by comparing the enzyme activity in cells taken from cultures with and without added drugs.

Cholinesterase Inhibition by TV3326 and TV3279 in Mouse Brain

Mice were injected sc. with saline, (1 ml/kg), TV3326 or TV3279 (50, 75, or 200 µmoles/kg). The mice were decapitated 60 and 120 min later, and the whole brain removed, rapidly weighed, and homogenized (100 mg/ml) in 0.1M phosphate buffer pH 8.0 containing 1% Triton. ChE activity was measured in 25 µl aliquots of enzyme homogenates.[34] Inhibition of ChE (percentage) induced by TV3326 and TV3279 was calculated by comparison with enzyme activity obtained from mice injected with saline under the same conditions.

Protective Effects of TV3326 Against the Sequelæ of Brain Trauma in Mice

The following experiments were performed in mice and rats according to the guidelines of the University Committee for Institutional Animal Care, based on those of the National Institutes of Health, U.S.A. Male mice of (40) Sabra Hebrew University strain, weighing 35–40 gm were trained in three daily trials for five days to find the escape platform in a Morris water maze (1 m diameter) that remained in the same place throughout the test. All the mice reached escape latencies ranging from 22–30 sec at the end of this period. They were then separated randomly into four equal groups. Moderate to severe closed head injury to the left hemisphere was induced in the mice under ether anesthesia by a weight-drop device as described elsewhere.[35] Saline (1 ml/kg), TV3326 or TV3279 (75 µmoles/kg or 200 µmoles/kg) were injected sc, five minutes after injury. Motor dysfunction and impairment of spatial memory was assessed for up to 14 days after injury, as described by Huang *et al.*[18] The maximal possible score for motor dysfunction one hour after injury was 14.

Protective Effect of TV3326 Against Neuronal Damage and Spatial Memory Deficits Induced in Rats by icv Injection of Streptozotocin (STZ)

The experiments were performed in order to determine the nature and extent of any morphological changes that are produced by icv STZ in rats, and to see if these changes could be prevented by TV3326. Ten 3–4-month old male Sprague-Dawley rats, weighing 250–300 g were given TV3326 (75 µmoles/kg) by gastric gavage, two hour before they were anesthetized with equithesin (1 ml/kg) and another 10 received 1 ml/kg water. The treatment with TV3326 or water was continued once daily for three weeks after injection of STZ. STZ 1.5 mg (prepared in artificial CSF immediately before injection) or CSF alone was injected once, unilaterally in a volume of 4 µl. The stereotaxic coordinates were −0.9 mm anterior, 1.8 mm lateral and −3.7 mm ventral from bregma. Forty days after the icv injection of STZ, the spatial memory of the rats was tested in the Morris water maze (1.4 m diam.) as described elsewhere.[36] Briefly, the rats were given two daily trials for five days in the maze with an intertrial interval of 15 min. The point of entry of the rat into the maze and the position of the escape platform were changed daily, but not between the first and second trials each day. The decrease in escape latency from day to day in trial 1 represents reference or long term memory, whereas that from trial 1 to trial 2, is consistent with working, or short term memory.[37] On the following day, the rats were deeply anesthetized with sodium pentobarbitone (60 mg/kg). The brains were fixed by transcardial perfusion, with phosphate buffered saline (0.02 M PBS, pH 7.4) containing heparin (5 U/ml) and 4% paraformaldehyde in 0.1 M PBS, pH 7.5 containing sucrose 4%. Brain sections, 30-μm thick, were cut and stained for potential axonal degeneration by means of silver impregnation.[38] In addition, reactive microglia were stained by means of an antibody to complement receptor 3.[39] Rating scales were constructed to evaluate the degree of pathology; one scale was based on the cresyl violet stain and depicted the damage to the hippocampus, fornix, and corpus callosum. The second scale was based on immunohistochemical staining of microglia and depicted the extent of microglial reactions in these structures. The scale for damage ranged from 0 for no damage to 5 for total loss of fornix, anterior hippocampus, and dorsal hippocampus. The rating scale for microglial reaction in the dorsal branch of fornix was also 0 for no change (microglia are evenly distributed with radial branching of dendrites) to 3 for a condition in which orientation of microglial fibers parallels orientation of axons together with an increase in the number of fibers over the entire dorsal fornix on that side (ipsilateral or contralateral).

In another experiment, the same total amount of STZ was given but injected bilaterally on day 1, 5, and 21 in three doses each of 0.25 mg/side. Control rats were given artificial CSF in the same way. TV3326 or water was given as described above for three weeks, starting two hour before the first STZ injection. Spatial memory was tested in the Morris water maze as described above, 40 days after the first STZ injection.

RESULTS

Neuroprotective Effect of Drugs Against Oxygen–Glucose Deprivation in Nerve Growth Factor (NGF) Differentiated PC12 Cells

Exposure of NGF differentiated PC12 cells to OGD for 3.5 h and reoxygenation resulted in death of about 45% of the cells compared to 8% under control conditions at 37°C (see TABLE 1). Rivastigmine reduced cytotoxicity by a similar amount (about 33%) at all concentrations tested, ranging from 1–500 μM. By contrast, TV3326 and TV3279 caused a dose related reduction in cell death that reached levels of 58% at a concentration of 100 μM, and more than 80% at 500 μM.

TABLE 1. Cell death (percent of total LDH) induced by OGD (3–4 h) in PC12 cells

Drug	Concentration (μM)				
	0	1	10	100	500
No insult	8.5 ± 1.5				
OGD	45.3 ± 3.5	—	—	—	—
Rivastigmine + OGD		29.0 ± 1.9*	30.2 ± 2.5*	27.2 ± 2.2*	26.8 ± 2.9*
TV3326 + OGD		43.9 ± 3.0	37.2 ± 4.9	18.9 ± 4.7*	9.4 ± 1.0**
TV3279 + OGD		40.0 ± 5.4	34.9 ± 4.8	19.1 ± 3.8*	8.9 ± 0.7**

Note: Significantly different from OGD alone, $^*p < 0.01$, $^{**}p < 0.001$.s

TABLE 2. Inhibition of cholinesterase by drugs (percent of control) after OGD (four hour) followed by reoxygenation (18 h)

Drug	Time of incub (h)	Concentration (µM)		
		1	10	100
Rivastigmine	4	34.8 ± 4.9	77.7 ± 2.2	96.0 ± 0.6
TV3326	4	0	9.5 ± 2.1	44.3 ± 4.0
	18	20.9 ± 8.1	38.0 ± 0.6	49.0 ± 5.6
TV3279	4	2.0 ± 3.9	12.5 ± 5.5	70.6 ± 8.7
	18	38.8 ± 6.0	52.9 ± 12.5	—

Inhibition by Drugs of ChE in PC12 Cells

The inhibition (percent of control) of ChE in PC12 cells incubated for 4 or 18 h with the drugs during their exposure to OGD is shown in TABLE 2. Although rivastigmine caused a significant inhibition of ChE after four hour incubation at a concentration of 1 µM, TV3326 and TV3279 were only effective at this concentration after 18 h incubation.

Inhibition by Drugs of ChE in Mouse Brain

The effect of TV3326 and TV3279 on brain ChE reached its peak two hour after sc injection. For TV3326 this was 33% at 75 µmoles/kg and 54% at 200 µmoles/kg. TV3279 (75 µmoles/kg) inhibited the enzyme by 49%. Neither isomer caused any

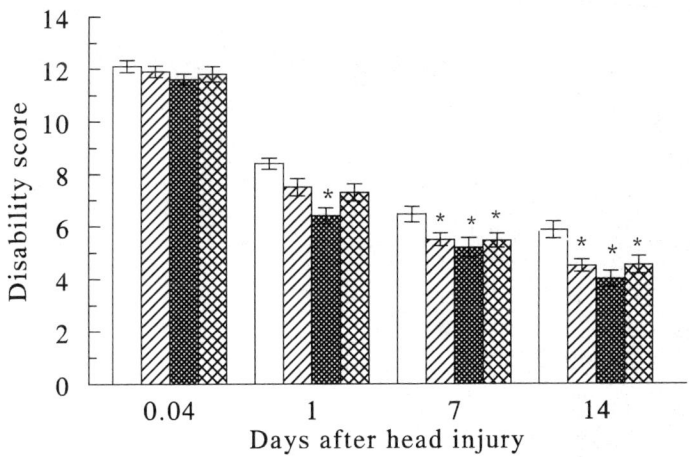

FIGURE 1. Effect of acute administration of TV3326 and TV3279 on the disturbance in motor function induced by closed head injury in mice. Significantly different from saline, *$p < 0.05$. Doses in µmoles/kg. ▢, saline; ▨, TV3326 (75); ▰, TV3326 (200); ▩, TV3279 (75).

inhibition of MAO-A or -B in the brain of mice after acute sc injection of 75 μmoles/kg (Gross, unpublished observations). TV3326 inhibited MAO-A by 17% and MAO-B by almost 50% at a dose of 200 μmoles/kg, whereas TV3279 was ineffective in doses below 500 μmoles/kg

Protective Effects of TV3326 Against the Sequelæ of Brain Trauma in Mice

All the mice showed a motor disability score of about 12 points, one hour after injury (see FIGURE 1). This gradually declined by about 50% in the saline treated animals over the next two weeks. TV3326 (200 μmoles/kg) reduced the motor disability score, even one day after injury, whereas 75 μmoles/kg of each isomer reduced this significantly from the seventh day onwards. Both isomers also significantly reduced escape latencies from the second day after injury at a dose of 75 μmoles/kg. The escape latencies of mice given the larger dose of TV3326 were significantly lower than those of saline-treated mice on day 4 (see FIGURE 2).

Effect of TV3326 Against Pathological Processes Induced in Rats by icv Injection of STZ

Two major types of pathological processes were observed after unilateral injection of STZ 1.5 mg: shrinkage of the dorsal hippocampus and adjacent fornix and gliosis in myelinated periventricular brain structures, including the corpus callosum

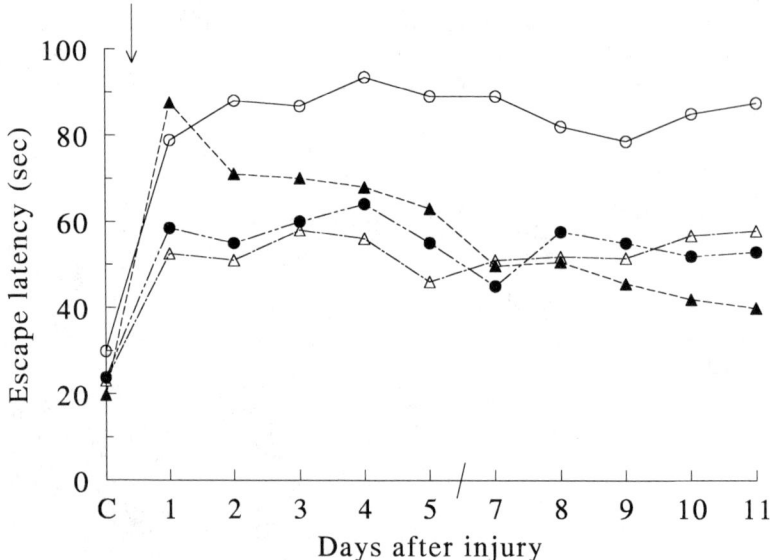

FIGURE 2. Effect of acute administration of TV3326 and TV3279 on spatial memory after closed head injury in mice. ○, saline; △, TV3279 (75 μmoles/kg); ●, TV3326 (75 μmoles/kg); ▲, TV3326, (200 μmoles/kg). Closed head injury was administered at the time indicated by *arrow*. Escape latencies of mice treated with TV3279 and TV3326, 75 μmoles/kg, were significantly lower than those of saline-treated mice from day two and those treated with TV3326, 200 μmoles/kg, from day four.

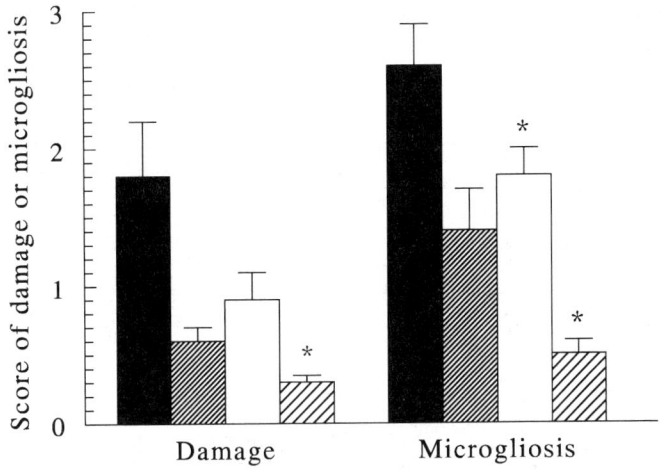

FIGURE 3. Effect of TV3326 75 µmoles/kg on neuronal damage and microgliosis induced in the dorsal fornix induced by a single unilateral icv injection of STZ in rats. Significantly different from water-treated rats, $*p < 0.05$. ■, STZ + water ipsilateral; ▨, STZ + water contralateral; ☐, STZ + TV 3326 ipsilateral; ▧, STZ + TV 3326 contralateral.

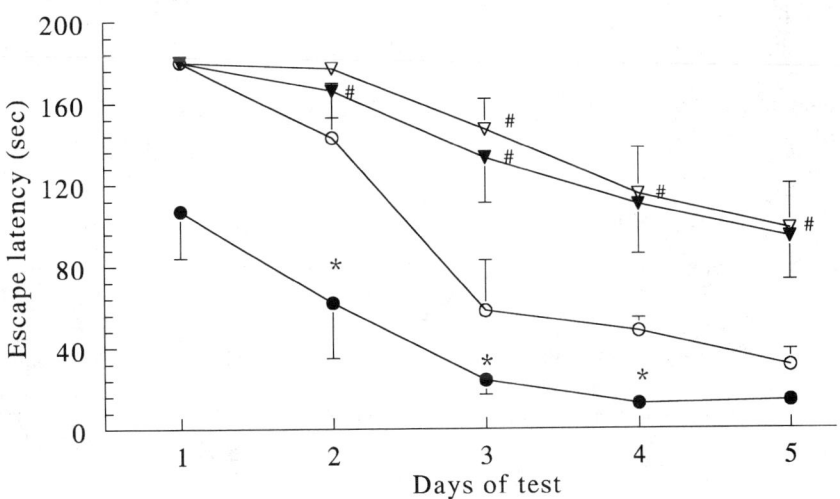

FIGURE 4. Effect of three bilateral icv injections of STZ on escape latencies of rats in the Morris water maze. CSF trial 1, ○; trial 2, ●; STZ trial 1, ▽; trial 2, ▼. Significantly different from artificial CSF, $\#p < 0.05$. Significantly different from trial 1, $*p < 0.05$.

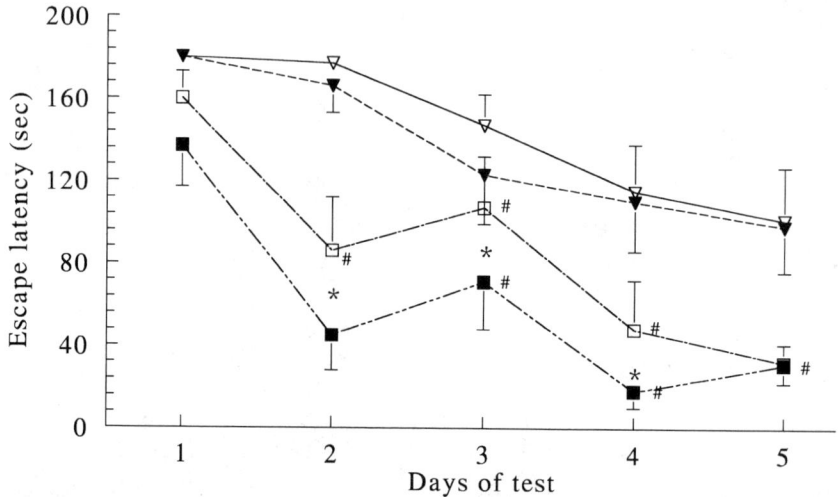

FIGURE 5. Reduction by TV3326 of the spatial memory impairment induced by three bilateral icv injections of STZ. STZ trial 1, ▽; trial 2, ▼; STZ + TV3326 (75 μmoles/kg) trial 1, □; trial 2, ■. Significantly different from STZ, #$p < 0.05$. Significantly different from trial 1, *$p < 0.05$.

and fornix. Damage to the fornix and anterior hippocampus was more severe ipsilateral than contralateral to the site of injection, both at the dorsal and hypothalamic branch of fornix. A change in the orientation of microglial fibers and an increase in their number occurred in the dorsal fornix, corpus callosum, and hypothalamic branch of the fornix. Chronic treatment with TV3326 reduced the degree of damage and microglial reaction in the dorsal fornix (see FIGURE 3) but not in the hypothalamic branch.

Effect of STZ and TV3326 Treatment on Spatial Memory in Rats

A single icv injection of STZ (1.5 mg) did not significantly affect spatial learning and memory in the Morris water maze compared to that in rats treated with icv CSF. However, when the same amount of STZ was given in three divided doses over a period of three weeks, both reference and working memory were significantly impaired 40 days after the first injection of STZ (see FIGURE 4). Chronic treatment with TV3326 (75 μmoles/kg) prevented the learning deficits and changes in spatial memory, restoring escape latencies to those seen in control rats injected with artificial CSF, bilaterally (see FIGURE 5).

DISCUSSION

A growing body of evidence has indicated that oxidative stress plays an important role in the progressive neuronal degeneration that occurs in AD.[40,41] Oxidative stress arises when the production of ROS exceeds the rate of consumption by cellular antioxidants, leading to peroxidation of lipids and cell damage.[42] The brain is particularly vulnerable to ROS because it contains a relatively high concentration of easily peroxidizable fatty acids[43] and a relatively low concentration of protective antioxidant enzymes.[44] In AD the damage occurs initially in cholinergic neurones in the basal forebrain and results in characteristic memory and cognitive impairment. It is, therefore, reasonable to suggest that a drug that can prevent the formation of ROS or accelerate their consumption by activating protective enzymes and also increase cholinergic transmission, should improve memory and delay the progression of the disease. When developing such a drug it is essential to show that it has protective effects against neuronal damage caused to cholinergic neurones in the brain, as well as in neuronal cultures.

In our study, TV3326, an aminoindan derivative with both carbamate and propargylamino groups, induced dose related protection against cell death caused by OGD in NGF differentiated PC12 cells. This could have resulted from blockade of ChE, since it also occurred with rivastigmine in concentrations that inhibited ChE by at least 40% in these cells. However, rivastigmine that lacks a propargyl amino group was only able to reduce cell death by about 33% at any concentration in the range 1–500 μM, whereas TV3326 and TV3279 reduced the number of dead cells by more than 80% at the higher concentrations. Such levels of neuroprotection against OGD were also seen with higher concentrations (1 μM or more) of rasagiline,[17] which has no ChE inhibitory activity. The mechanism of neuroprotection by ChE inhibitors and propargylamines is not yet established. However, other ChE inhibitors, tacrine, donepezil, and huperazine protected undifferentiated PC12 cells against oxidative stress resulting from exposure to β-amyloid fractions[23,45] or H_2O_2.[24] In this model, the effect of ChE inhibitors appeared to be mediated by activation of nicotinic receptors and involved an increase in the activity of catalase and glutathione peroxidase.[45]

Propargylamines may also prevent the reduction in the activity of these enzymes.[12] Furthermore, they have been shown to reduce apoptosis in response to oxidative stress or serum deprivation by increasing the synthesis of BCL-2 and by inhibiting activation of caspase 3.[19,46] These activities are shared by other propargylamine derivatives like the S-isomer of rasagiline, which does not block MAO in the concentrations used in the experiment (Naoi, personal communication).

The reduction by a single injection of TV3326 and TV3279 of the cerebral edema, induced by brain trauma and acceleration of the recovery of motor function and spatial memory are similar to that seen with another ChE inhibitor, rivastigmine.[26] It is likely that the drugs prevented the reduction in cholinergic transmission that has been shown to occur after brain trauma[47,48] since all the effects of rivastigmine were significantly reduced by cholinergic receptor blockade.[20,26]

In addition to the reported effect of icv STZ on cerebral glucose metabolism, our studies show that it causes ventricular enlargement and distinct white matter lesions in the fornix and corpus callosum, accompanied by activation of microglia. All these pathological changes are seen in the brains of AD patients and were reduced by TV3326. To our knowledge this is the first time that a detailed histological examination has been performed and the nature and extent of the morphological damage induced by icv STZ described. Damage to myelinated nerves in the periphery has also been reported following parenteral administration of STZ, in association with an increase in ROS, although this could have resulted from the ensuing hyperglycemia.[31] It is not known whether the damage to the fornix and corpus callosum results from oxidative stress or apoptosis, in a similar manner to that induced in pancreatic cells.[49] Examination of brain phospholipid metabolism three weeks after icv injection of STZ revealed significant losses in phosphatidyl ethanolamine, but none that are indicative of peroxidative damage.[50] If apoptosis is involved in the specific lesions induced by STZ and if it acts *in vivo* as it does in neuronal cell cultures, then this could explain how TV3326 causes protection.

In contrast to studies in older animals in which about 50% showed a deficit in spatial memory,[51] the limited damage caused by a single injection of STZ in young rats did not produce any memory impairment in the Morris water maze. In a more recent study, STZ was given in the same total amount bilaterally in three injections over a period of three weeks in rats aged one year. This produced memory deficits that were detected after 40 days.[52] We also obtained a marked impairment in working and reference memory after icv injection of STZ in young rats. Chronic treatment with TV3326 throughout the three weeks during which STZ was given almost completely prevented the occurrence of memory deficits when tested 19 days after the last administration.

The neuroprotective effects of TV3326 in cell cultures and animal models in which neurodegeneration occurs, together with its ability to inhibit ChE and MAO selectively in the brain, make it a potentially useful therapeutic agent for the treatment of AD.

ACKNOWLEDGMENTS

The authors gratefully acknowledge the support of Teva Pharmaceuticals Ltd., Israel for funding and for supplying the test compounds, and thank its personnel for helpful discussions during the course of this research.

REFERENCES

1. COYLE, J.T. & P. PUTTFARCKEN. 1993. Oxidative stress, glutamate, and neurodegeneraative disorders. Science **262:** 689–695.
2. BUTTKE, T.M. & P.A. SANDSTROM. 1994. Oxidative stress as a mediator of apoptosis. Immunol. Today **15:** 7–10.
3. KISH, S.J., C. BERGERON, A. RAJPUT, *et al.* 1992. Brain cytochrome oxidase activity in Alzheimer's disease. J. Neurochem. **59:** 776–779.
4. MUTISYA, E.M., A.C. BOWLING & M.F. BEAL. 1994. Cortical cytochrome oxidase is reduced in Alzheimer's disease. J. Neurochem. **63:** 2179–2184.
5. MCGEER, P.L., J. ROGERS & E.G. MCGEER. 1994. Neuroimmune mechanisms in Alzheimer's disease pathogenesis. Alzheimer's Disease and Associated Disorders **8:** 149–158.
6. COLTON C.A. & D.L. GILBERT. 1993. Microglia, an *in vivo* source of reactive oxygen species in the brain. Adv. Neurol. **59:** 321–326.
7. COYLE, J.T., D.L. PRICE & M.R. DELONG. 1983. Alzheimer's disease: a disorder of cortical cholinergic innervation. Science **219:** 1184–1190.
8. WEINSTOCK, M. 1999. Selectivity of cholinesterase inhibition: Clinical implications for the treatment of Alzheimer's disease. CNS Drugs **12:** 307–323.

9. SANO, M., C. ERNESTO, R.G. THOMAS, et al. 1997. A controlled trial of selegiline, alpha-tocopherol, or both as treatment for Alzheimer's disease. The Alzheimer's Disease Cooperative Study. New Eng. J. Med. **336:** 1216–1222.
10. FILIP, V. & E. KOLIBAS. 1999. Selegiline in the treatment of Alzheimer's disease: a long-term randomized placebo-controlled trial. Czech and Slovak Senile Dementia of Alzheimer Type Study Group. J. Psychiatry Neurosci. **24:** 234–243.
11. WU, R.-M., D.L. MURPHY & C.C. CHIUEH. 1996. Suppression of hydroxyl radical formation and protection of nigral neurons by l-deprenyl (selegiline). Ann. N.Y. Acad. Sci. **786:** 379–390.
12. KNOLLEMA, S., W. AUKEMA, H. HOM, et al. 1995. L-Deprenyl reduces brain damage in rats exposed to transient hypoxia-ischemia. Stroke **26:** 1883–1887.
13. MAGYAR, K., B. SZENDE, J. LENGYEL, et al. 1998. The neuroprotective and neuronal rescue effects of (−)deprenyl. J. Neural Transm. Suppl. **52:** 109–123.
14. TATTON, W.G. & R.M.E. CHALMERS-REDMAN. 1996. Modulation of gene expression rather than monoamine oxidase inhibition: (−)-deprenyl-related compounds in controlling neurodegeneration. Neurology **47**(Suppl. 3): S171–S183.
15. FINBERG, J.P.M, I. LAMENSDORF, J.W. COMMISSIONG & M.B.H. YOUDIM. 1996. Pharmacology and neuroprotective properties of rasagiline. J. Neural Transm. Suppl. **48:** 95–101.
16. FINBERG, J.P.M., T. TAKESHIMA, J.M. JOHNSTON & J.W. COMMISSIONG. 1998. Increased survival of dopaminergic neurons by rasagiline, a monoamine oxidase B inhibitor. Neuro-Report **9:** 703–707.
17. ABU-RAYA, S., E. BLAUGRUND, V. TREMBOVLER, et al. 1999. Rasagiline, a monoamine oxidase-B inhibitor, protects NGF-differentiated PC12 cells against oxygen-glucose deprivation. J. Neurosci. Res. **58:** 456–463.
18. HUANG, W., Y. CHEN, E. SHOHAMI & M. WEINSTOCK. 1999. Neuroprotective effect of rasagiline, a selective MAO-B inhibitor, against closed head injury in the mouse. Eur. J. Pharmacol. **366:** 127–135.
19. YOUDIM, M.B.H., J.S. WADIA & W.G. TATTON. 1999. Neuroprotective properties of the antiparkinson drug rasagiline and its optical S-isomer. Neurosci. Letts. Suppl. **54:** S45.
20. WEINSTOCK, M., T. GOREN & M.B.H. YOUDIM. 2000. Development of a novel neuroprotective drug (TV3326) for the treatment of Alzheimer's disease, with cholinesterase and monoamine oxidase inhibitory activities. Dev. Drug Res. **50:** 216–222.
21. MESNER, P.W., T.R. WINTERS & S.H. GREEN. 1992. Nerve growth factor withdrawal-induced cell death in neuronal PC12 cells resembles that in sympathetic neurons. J. Cell Biol. **119:** 1669–1680.
22. CALDERON, F.H., A. BONNEFONT, F.J. MUNOZ, et al. 1999. PC12 and neuro 2a cells have different susceptibilities to acetylcholinesterase-amyloid complexes, amyloid$_{25-35}$ fragment, glutamate, and hydrogen peroxide. J. Neurosci. Res. **56:** 620–631.
23. SVENSSON, A.L. & A. NORDBERG. 1998. Tacrine and donezepil attenuate the neurotoxic effect of A beta (25–35) in rat PC12 cells. Neuroreport **9:** 1519–1522.
24. XIAO, X.Q., J.W. YANG & X.C. TANG. 1999. Huperzine A protects rat pheochromocytoma cells against hydrogen peroxide-induced injury. Neurosci. Lett. **275:** 73–76.
25. TATTON, W.G., W.Y.L. JU, D.P. HOLLAND, et al. 1994. (−)-Deprenyl reduces PC12 cell apoptosis by inducing new protein synthesis. J. Neurochem. **63:** 1572–1575.
26. CHEN, Y., E. SHOHAMI, S. CONSTANTINI & M. WEINSTOCK. 1998 Rivastigmine, a brain-selective acetylcholinesterase inhibitor, ameliorates cognitive and motor deficits induced by closed-head injury in the mouse. J. Neurotrauma **15:** 231–237.
27. DUELLI, R., H. SCHRÖCK, W. KUSCHINSKY & S. HOYER. 1994. Intra- cerebroventricular injection of streptozotocin induces discrete local changes in cerebral glucose utilization in rats. Int. J. Dev. Neurosci. **12:** 737–743.
28. HELLWEG, R., R. NITSCH, C. HOCK, et al. 1992. Nerve growth factor and choline acetyltransferase activity levels in the rat brain following experimental impairment of cerebral glucose and energy metabolism. J. Neurosci. Res. **31:** 479–486.
29. BLOKLAND, A. & J. JOLLES. 1994. Behavioral and biochemical effects of an icv injection of streptozotocin in Old Lewis rats. Pharmacol. Biochem. Behav. **47:** 833–837.

30. MÜLLER, D., K. PLASCHKE & S. HOYER. 1995. Intracerebroventricular injection of streptozotocin: an animal model for sporadic Alzheimer's disease? *In* Alzheimer's and Parkinson's Diseases. I. Hanin, *et al.*, Eds.: 389–390. Plenum Press, New York.
31. BASTAR, I., S. SECKIN, M. UYSAL & G. AYKAC-TOKER. 1998. Effect of streptozotocin on glutathione and lipid peroxide levels in various tissues of rats. Res. Commun. Mol. Pathol. Pharmacol. **102:** 265–272.
32. ARAGNO, M., E. TAMAAGNO, V. GATTO, *et al.* 1999. Dehydroepiandrosterone protects tissues of streptozotocin-treated rats against oxidative stress. Free Rad. Biol. Med. **26:** 1467–1474.
33. RUSSELL, J.W., K.A. SULLIVAN, A.J. WINDEBANK, *et al.* 1999. Neurons undergo apoptosis in animal and cell culture models of diabetes. Neurobiol. Dis. **6:** 347–363.
34. ELLMAN, G.L., K.D. COURTNEY, F. ANDERS & R.M. FEATHERSTONE. 1961 A new and rapid colorimetric determination of acetylcholinesterase activity. Biochem. Pharmacol. **7:** 88–95.
35. CHEN, Y., S. CONSTANTINI, V. TREMBOVLER, *et al.* 1996. An experimental model of closed head injury in mice: Pathophysiology, histopathology and cognitive deficits. J. Neurotrauma **13:** 557–569.
36. BEJAR, C., R.H. WANG & M. WEINSTOCK. 1999. Effect of rivastigmine on scopolamine-induced memory impairment in rats. Eur. J. Pharmacol. **383:** 231–240.
37. MORRIS, R.G.M. 1983. An attempt to dissociate "spatial-mapping" and "working memory" theories of hippocampal function. *In* The Neurobiology of the Hippocampus. W. Siefert, Ed.: 405–432. Academic Press, London.
38. GALLYAS, F., F.H. GULDNER, G. ZOLTAY & R. WOLFF. 1990. Golgi-like demonstration of "dark" neurons with an argyrophilia. III method for experimental neuropathology. Acta Neuropathol. **79:** 620–628.
39. SHOHAM, S. & R.P. EBSTEIN. 1997. The distribution of beta amyloid precursor protein in rat cortex after kainate-induced seizures. Exp. Neurol. **147:** 361–376.
40. BENZI, G. & A. MORETTI. 1995. Are reactive oxygen species involved in Alzheimer's disease? Neurobiol. Aging **16:** 661–674.
41. MARKESBERRY, W.R. & J.M. CARNEY. 1999. Oxidative alterations in Alzheimer's disease. Brain Pathol. **9:** 133–146.
42. SMITH, M.A., L. SAYRE & G. PERRY. 1996. Is Alzheimer's a disease of oxidative stress? Alzheimer's Dis. Rev. **1:** 63–67.
43. DELEO, J.A., R.A. FLOYD & J.M. CARNEY. 1986. Increased *in vitro* lipid peroxidation of gerbil cerebral cortex as compared with rat. Neurosci. Lett. **67:** 63–67.
44. HALLIWEL, B. & J.M. GUTTERIDGE. 1984. Lipid peroxidation, oxygen radicals, cell damage, and antioxidant therapy. Lancet **1:** 1396–1397.
45. XIAO, X.Q., R. WANG, Y.F. HAN & X.C. TANG. 2000. Protective effects of huperizine A on -amyloid25-35 induced oxidative injury in rat pheochromocytoma cells. Neurosci. Lett. **286:** 155–158.
46. NOAI, M., W. MARUYAMA, M.B.H. YOUDIM, *et al.* 2001. Anti-apoptotic function of propargylamine series of monoamine oxidase inhibitors. J. Pharm. Pharmacol. In press.
47. LEONARD, J.R., D.O. MARIS & M.S. GRADY. 1994. fluid percussion injury causes loss of forebrain choline acetyltransferase and nerve growth factor receptor immunoreactive cells in the rat. J. Neurotrauma **11:** 379–392.
48. GORMAN, L.K., D.A. FU, D.A. HOVDA, *et al.* 1996. Effects of traumatic brain injury on the cholinergic system in the rat. J. Neurotrauma **13:** 457–463.
49. SAINI, K.S., C. THOMPSON, C.M. WINTERFORD, *et al.* 1996. Streptozotocin at low doses induces apoptosis and at high doses causes necrosis in a murine pancreatic beta cell line, INS-1. Biochem. Mol. Biol. Int. **39:** 1229–1239.
50. MÜLLER, D., R.M. NITSCH, R.J. WURTMAN & S. HOYER. 1998. Streptozotocin increases free fatty acids and decreases phospholipids in rat brain. J. Neural Transm. **105:** 1271–1281.
51. PRICKAERTS, J., T. FAHRIG & A. BLÖKLAND. 1999. Cognitive performance and biochemical markers in septum, hippocampus and striatum of rats after an icv injection of streptozotocin: a correlation analysis. Behav. Brain Res. **102:** 73–88.

52. LANNERT, H. & S. HOYER. 1998. Intracerebroventricular administration of streptozotocin causes long-term diminutions in learning and memory abilities and in cerebral energy metabolism in adult rats. Behav. Neurosci. **112:** 1199–1208.

The Action of Acetyl-L-Carnitine on the Neurotoxicity Evoked by Amyloid Fragments and Peroxide on Primary Rat Cortical Neurones

M.A. VIRMANI, V. CASO, A. SPADONI, S. ROSSI, F. RUSSO, AND F. GAETANI

Research and Development Department, Sigma-tau HealthScience s.p.a., Via Treviso 4, Pomezia, Italy

ABSTRACT: The amyloid β-peptides have been implicated in the excitotoxic mechanism of neuronal injury in the pathogenesis of Alzheimer's disease. In this paper we examine the effect of different amyloid fragments (β A1–40, A1–28, and A25–35), as well as potential neuroprotective compounds on rat cortical neuron viability. Exposure of neurones to β A25–35 or A1–40 at concentrations as low as 1 µg/ml inhibited, significantly, the MTT response and this level of inhibition was similar after 24-h or three-day exposure. Furthermore, the level of inhibition was not affected by the presence or absence of 5% horse serum in the medium. Preexposure (10 min) of neurones to ALC at concentrations of 0.1, 1, 5, and 10 mM attenuated the inhibition of the MTT response caused by β A25–35 (50 µg/ml) in serum free medium for 24 h. The treatment of cells with vitamin E (100 µM), catalase (4 mg/ml), NGF (0.1 and 10 ng/ml), or cycloheximide (0.1 µg/ml) significantly restored the MTT response that was inhibited by β A25–35. The mechanism for the protective actions of these compounds against β A25–35 toxicity is not clear but may involve free radical scavenger action and preservation of energy production, although other mechanisms, especially for ALC, such as a direct effect on A-β interaction with charged anionic phospholipids and/or stabilizing action on membranes, are also possible.

KEYWORDS: Acetyl-L-carnitine; Amyloid; Neurodegeneration; Neurotoxicity; Mitochondria; MTT response; Metabolic compromise; Antioxidant; Membrane; Fluidity; Permeability; Lipid peroxidation; Peroxidase; Amphiphilic; Zwitterion; Binding sites; A-β interaction; Charged anionic phospholipids; Membrane stabilization; Alzheimer's; Down's; Diabetes; Macular degeneration.

INTRODUCTION

According to the amyloid cascade hypothesis it is the amyloid β-peptides that are responsible for excitotoxic actions and neuronal injury. Amyloid β peptides deposit as plaques in vascular and parenchymal areas of Alzheimer's disease (AD) tissues and Down's syndrome patients. The accumulation of amyloid peptides may also be involved in the pathogenesis of other diseases, such as islet cell toxicity in type 2

Address for correspondence: Ashraf Virmani, Ph.D., Research & Development Department, Sigma-tau HealthScience s.p.a., Via Treviso 4, 00040 Pomezia, Italy. Voice: 0039 06 91619721; fax: 06 91612631.
ashraf.virmani@sigma-tau.it

diabetes and damage to the Brusch's membrane and retinal pigment epithelial cells in age related macular degeneration, in prion diseases, and other disorders.

The mechanism of amyloid β toxicity is still not clear, but is thought to involve mitochondrial inhibition, production of free radicals, changes in membrane characteristics leading to loss of Ca homeostasis, and apoptosis. The amyloid β-peptide may, therefore, be a key factor in the pathogenesis of various diseases, such as Alzheimer's, and the development of effective approaches to block this form of toxicity may be a preventative aid to stop or slow down disease progression. In this paper we examine the effect of different amyloid fragments (β A1–40, A1–28, and A25–35), as well as potential neuroprotective compounds on rat cortical neuron viability.

Under various pathological conditions it is the energy compromise due to loss of mitochondrial function that precedes the increased electron leakage, augmenting formation of activated oxygen species (superoxide and hydroxyl radical),[1–3] and in turn causing cell injury and death. In this study the effects of acetyl-L-carnitine (ALC) on the amyloid β toxicity were examined since it is a compound involved in mitochondrial energy producing pathways and has been reported have a neuroprotective action in various conditions of metabolic stress; for example, ischemia, hypoxia, aging, alcohol, MPTP, AZT, and diabetes,[4–8] as well as in traumatic injury (nerve transection and degeneration of axotomised motoneurons.[9,10] Under hypoxic conditions, ALC prevents the increase in L-type voltage sensitive calcium channels[11] and even reverses these effects.[12] It can also affect the expression of receptors; that is, it prevents the decrease in NMDA receptors,[13] dopamine D2 receptors in hippocampus, and muscarinic M2 receptors in neostriatum[14] that normally occurs with age.

Furthermore, ALC is metabolized intracellularly to produce L-carnitine (L-C), which plays a central role in fatty acid metabolism.[15] L-C has also been reported to exert protective actions. It prevents loss of mitochondrial function in isolated liver mitochondria,[16] prevents lipid peroxidation damage by hypobaric hypoxia in rat brain cells,[17] and protects against mitochondrial inhibitor induced toxicity in neurones.[7,18] The exact mechanism for these actions is not yet known, but it has been suggested that L-C and ALC may exert a direct effect on the membrane. They may prevent cell damage by stabilizing the membrane against free radical damage, and also prevent mitochondrial injury, thus increasing energy production and decreasing the leakage of free radicals and Ca^{2+}.[7,18]

In this paper we report that ALC can attenuate amyloid β-peptide related neurotoxicity in rat cortical cells. This work has been reported, in part, in abstract form.[19]

METHOD

Cell Culture

The primary cells of rat fetal brain, approximately 16–19 days of development, were prepared by dissecting out the cortex and dispersing the cells from the intact tissue using the enzyme collagenase.[20] The separated brain cells were than cultured in 24-well plates (Corning) that had been precoated with poly-D-lysine to aid the attachment of the cells. The culture media was MEM modified medium containing

10% fetal calf serum (Flow Labs) and 5% horse serum (Gibco). Neurotoxicity protocol was carried out using two-week-old neurones (more than 14 days *in vitro* [DIV]) in 24-well plates. These were washed twice and incubated in serum free culture medium. The neuroprotective ability of the compounds, acetyl-L-carnitine (Sigma-tau labs, Pomezia, Italy), glutathione, vitamin E, nerve growth factor (NGF), catalase, and cycloheximide (Sigma, St. Louis, MO, USA) was assessed by incubating them for 10 min prior to, and during insult, with various amyloid peptides (β A1–40, A1–28, and A25–35; Sigma, St. Louis, MO, U.S.A.). The substances were present throughout the 24-h incubation period. The serum free medium alone did not trigger cell death during this incubation period in our experiments. In long term exposure experiments (three days) the neurones were incubated together with the test substances in a medium containing 5% horse serum. In one experiment the cells were preincubated with ALC from DIV1 with fresh substance and medium added every fourth day until the time of experiment.

Cell Viability Assessment

Neuron viability was assessed by aspirating 200 μl of incubation media from the wells and measuring the release of lactate dehydrogenase (LDH). This gave an index for the loss of cell integrity. The level of LDH activity was determined using the assay system reported elsewhere (Roche optimized one-vial test) using the COBAS automatic analyzer. In the same experiments, cell viability was also assessed by following the conversion of 3-[4,5-dimethylthiazol]-2,5-diphenyltetrazolium bromide (MTT) (Sigma) to a blue formazan product by the live cells as reported elsewhere.[20] Briefly, the cells were washed twice with serum free medium before addition of 50 μl of MTT (5 mg/ml), and the reaction stopped after 20 min incubation at 37°C. Absorbance was measured at 540 nm.

RESULTS

Effect of Acetyl-L-carnitine (ALC) Against the Neurotoxic Action of Amyloid β A25–35

In the present experiment, 1×10^6 primary rat cortical cells were cultured until 14 DIV, at which time they were well differentiated and forming good connections (see FIGURE 1). At this time they were exposed to amyloid fragments and the resulting toxicity assessed.

Exposure of neurones to β A1–40 (see FIGURE 2), β A1–28 (see FIGURE 3), or β A25–35 (see FIGURE 4) or at concentrations as low as 1 μg/ml significantly inhibited the MTT response, as shown elsewhere,[21] and this level of inhibition was similar in percentage terms after 24-h or three-day exposure. The level of aggregation of the β-amyloid in solution is thought to be related to the potential neurotoxic ability.[22,23] We found that the A1–40 fragment formed aggregates almost immediately, as seen in the light microscope, whereas A1–28 and A25–35 did not. However, after 24 h, aggregates of A25–35 were also observed.

A1–40 Effects on Neuron Viability

After 24 h in serum free medium, A1–40 at 1 µg/ml caused 30% inhibition and 47% at 25 µg/ml (FIG. 2 A). LDH only increased at the higher concentration of A1–40 25 µg/ml (FIG. 2 B). The same batch of cells exposed to A1–40 for three days in serum gave the same percentage decrease in MTT response. Thus, the level of MTT response inhibition was not affected by the time of exposure and the presence or absence of 5% horse serum in the medium.

A1–28 Effects on Neuron Viability

The MTT response of the cortical neurones reduced only slightly after 24 h serum free exposure to A1–28, even at 25 µg/ml (7% inhibition) (FIG. 3 A). The level of LDH was not affected (FIG. 3 B). Exposure for three days in the presence of serum reduced the MTT response by 16% at 5 µg/ml, but LDH release was not increased, indeed it was reduced by 12%, suggesting that there were no major toxic actions by A1–28.

A25–35 Effects on Neuron Viability

Acute exposure (24 h) in absence of serum led to 33% inhibition of the MTT response, and an increase in LDH levels by 258% at 25 µg/ml A25–35 (FIG. 4). Chronic exposure for three days in the presence of serum did not augment the MTT response, which remained inhibited by 33%, but the increase in LDH was not seen, possibly due to interference by the 5% horse serum in the culture medium.

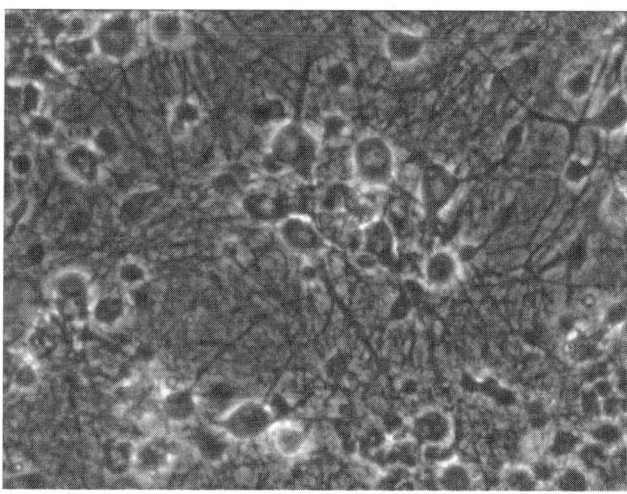

FIGURE 1. Primary rat cortical cells taken from rat embryos at 16–19 days of development, dissociated, and cultured at a density of 1×10^6 cells per well until 14 DIV (see METHODS) at which time they were well differentiated and forming good connections. At this time they were exposed to amyloid fragments and the resulting toxicity assessed.

Effect of Acetyl-L-carnitine on A25–35 Toxicity on Cortical Neuron Viability

The treatment of neurones with ALC (added 10 min before the A25–35) at concentrations of 0.1, 1, 5, and 10 mM counteracted inhibition of the MTT response caused by β A25–35 (50 µg/ml) in serum free medium for 24 h (see FIGURE 5). This effect was already significant at a concentration of 0.1 mM ALC ($p < 0.05$). The levels of LDH were not affected by the A25–35 in this experiment.

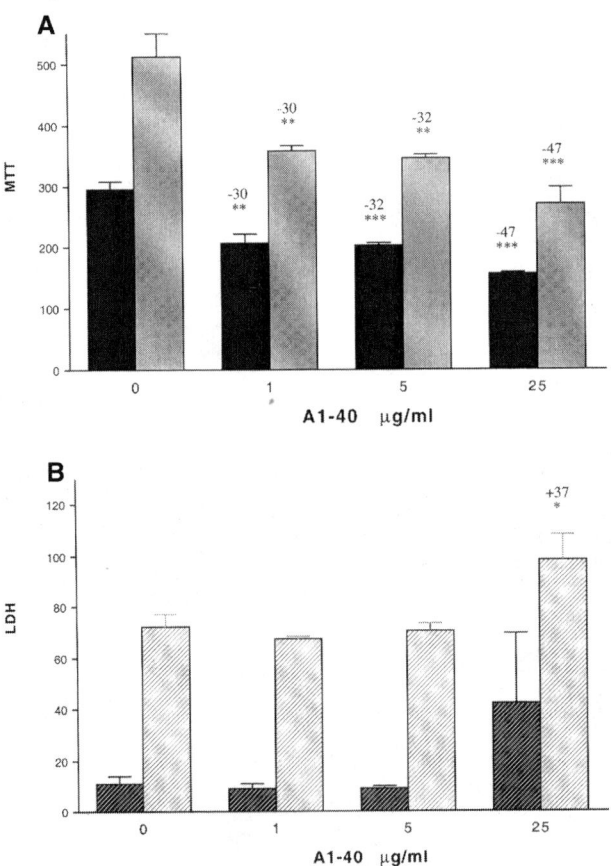

FIGURE 2. Exposure of two-week old primary rat cortical to amyloid β fragments, A1–40 for 24 h in serum free medium, or three days in serum containing medium, caused a 47% at 25 µg/ml of the MTT response. The level of LDH was only increased by the higher concentration of A1–40 25 µg/ml. Each point is the mean ±SD of the response from three wells. The significance was calculated using the Student's t-test versus the control with $*p < 0.05$, $**p < 0.01$, and $***p < 0.001$; $n = 3$. **A:** ■, A1–40 w/o serum, 24 h; ▫, A1–40 + serum, three days. **B:** ▨, A1–40 w/o serum, 24 h; ▫, A1–40 + serum, 3 day.

Long term exposure of neurones from DIV1 to ALC (0.01, 01, and 1 mM) showed that the subsequent toxicity, in terms of MTT response inhibition by amyloid A25–35 (5 µg/ml) was completely prevented by the higher concentration of 1 mM ALC (see FIGURE 6 A). The levels of LDH were not affected (FIG. 6 B).

Possible Protective Effects Against A25–35 Toxicity by Other Agents: Glutathione, Vitamin E, NGF, Catalase, and Cycloheximide

The preexposure of neurones to the antioxidant glutathione (0.1 or 0.5 mM) did not significantly alter the inhibition of the MTT response by β A25–35 (25 µg/ml) (see TABLE 1) and the LDH levels remained normal under all conditions (results not shown). The inhibition by β A25–35 of MTT response was significantly restored by

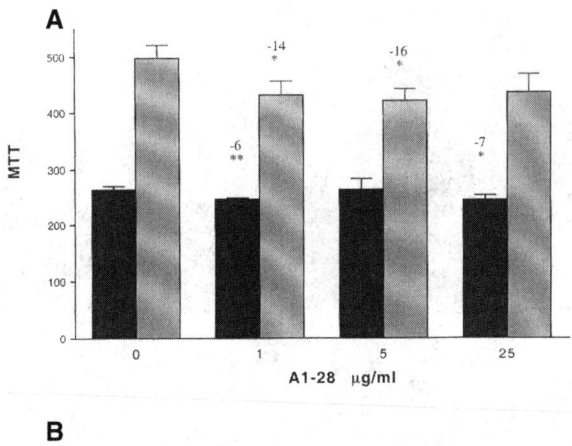

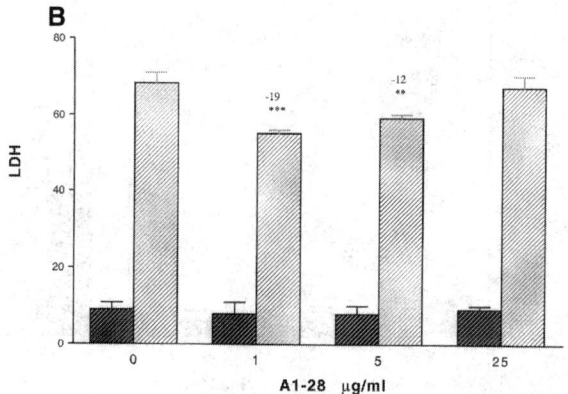

FIGURE 3. Exposure of two-week old primary rat cortical to amyloid β fragments, A1–28 for 24 h in serum free medium or three days in serum containing medium caused a small inhibition of the MTT response. The level of LDH was not affected or was slightly reduced. **A**: ■, A1–28 w/o serum, 24 h; ▨, A1–28 + serum, three days. **B**: ▨, A1–28 w/o serum, 24 h; ▨, A1–28 + serum, three days. X SD, $n = 3$. $*p < 0.05$, $**p < 0.01$, and $***p < 0.001$ versus control.

the treatment of cells with either vitamin E (100 μM) ($p < 0.05$), catalase (4 mg/ml) ($p < 0.001$), NGF (0.1 and 10 ng/ml) ($p < 0.05$), or cycloheximide (0.1 μg/ml) ($p < 0.05$) (TABLE 1).

Possible Protective Effects Against Peroxidase (H2O2) Toxicity by ALC, Glutathione, and Catalase

The toxicity to peroxidase (H2O2 0.2 mM) applied to the neurones for one hour was evident after 24 h as an inhibition of the MTT response (see FIGURE 7 A) and an increase in the LDH levels (FIG. 7 B). Both of these parameters were significantly

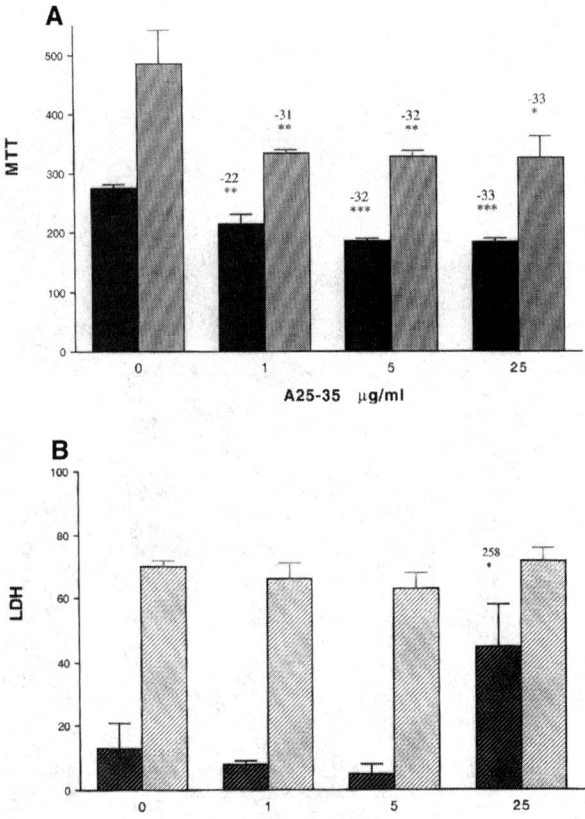

FIGURE 4. Exposure of two-week old primary rat cortical to amyloid β fragments, A25–35 for 24 h in serum free medium or three days in serum containing medium caused a 33% inhibition at 25 μg/ml of the MTT response. The level of LDH was evident in serum free conditions at the higher concentration of A25–35 25 μg/ml. **A**: ■, A1–28 w/o serum, 24 h; ▨, A1–28 + serum, three days. **B**: ▨, A25–35 w/o serum, 24 h; ▨, A25–35 + serum, three days. X SD, $n = 3$. *$p < 0.05$, **$p < 0.01$, and ***$p < 0.001$ versus control.

reduced by the addition of either glutathione (0.5 mM) or catalase (40 µg/ml or 400 µg/ml). The protective effects of ALC were also seen in this form of toxicity, but only on the MTT response ($p < 0.05$), the LDH levels were not affected (see FIGURE 8).

DISCUSSION

Effects of Amyloid β on Neuron Viability

Fragments of amyloid beta-peptide truncated at the N-terminus are present in AD brain early in the course of Alzheimer's disease. It is proposed that these species contribute to the pathogenesis of AD. It has been shown previously that β A1–42 and

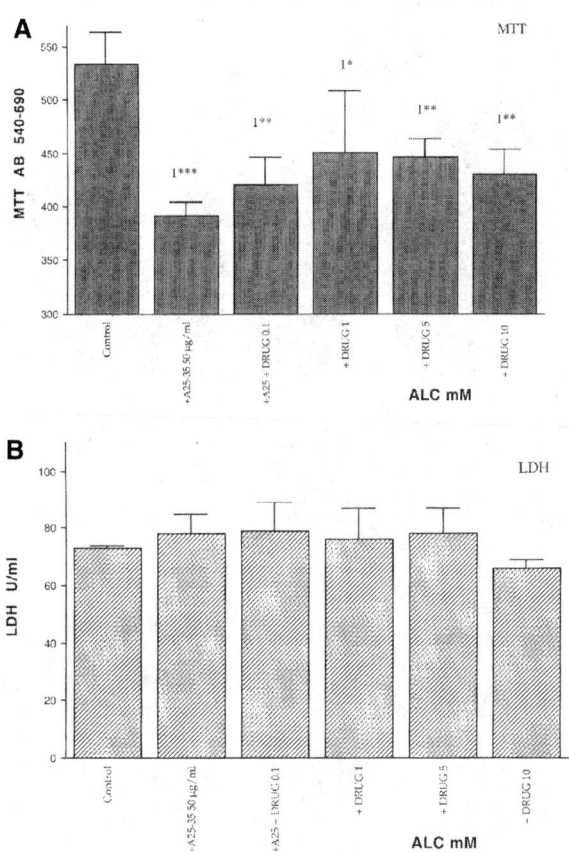

FIGURE 5. Effect of acetyl-L-carnitine on amyloid β A25–35 evoked toxicity in rat cortical neurones. The cotreatment of rat cortical neurones with ALC (0.1–10 mM) counteracted inhibition of the MTT response caused by the presence of β A25–35 (50 µg/ml). The levels of LDH were not affected. X SD, $n = 4$. $*p < 0.05$, $**p < 0.01$, and $***p < 0.001$ versus control.

β A25–35 are toxic to neurones and this was also seen in the experiments described here. The main effect seen was the inhibition of the MTT response, with maximum effects at 25 µg/ml; 47% inhibition by A1–42 and 33% by A25–35. An increase in the release of LDH also occurred at these concentrations suggesting actual cell membrane breakdown. The effects of β A1–28 were much smaller, with a maximum of 16% inhibition of the MTT response at 5 µg/ml after three days of exposure. Studies have shown that the active core of amyloid β is located within the A25–35 sequence but the flanking peptide regions can also influence, not only the folding and aggregation properties of the amyloid β fragments, but also their neurotoxic potential.[24]

The main effects of A25–35 in these experiments were on inhibition of the MTT response elicited by the cultured rat cortical neurones, indeed, on most occasions, the effects on LDH release were not evident. This has been seen by various other

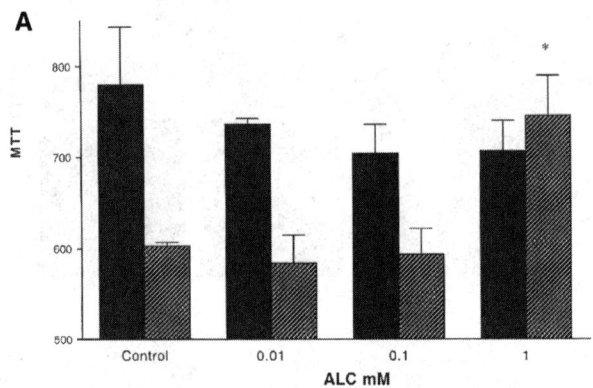

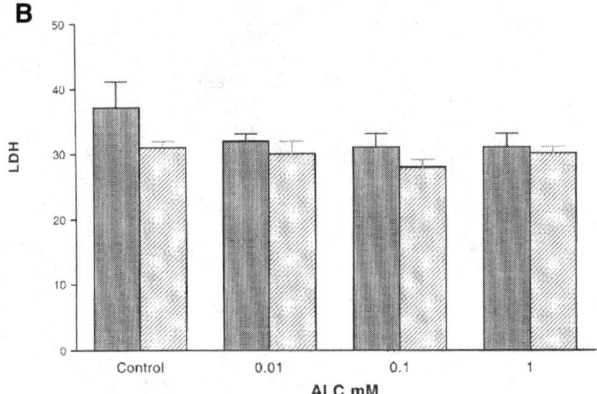

FIGURE 6. The long term treatment of rat cortical neurones (DIV 1 to 14) with acetyl-L-carnitine at concentrations of 0.001, 0.1, and 1 mM prevented inhibition of the MTT response at the highest concentration (1 mM) of the amyloid β A25–35 inhibition after 24 h exposure. The levels of LDH were not affected. **A.** MTT, ■, long ALC treated; ▨, long term ALC + A25–35 5 µg/ml. **B.** ▣, LDH, long ALC treated; ▣, long term ALC + A25–35 5 µg/ml.

investigators. Thus, in differentiated mouse neuroblastoma N1E–115 cells or rat hippocampal neurones, A25–25 caused loss of trypan blue exclusion but fluorescein diacetate staining or release of LDH were not affected.[25] These authors concluded that amyloid β might induce trypan blue adsorption on the cell membrane causing an artifact in measuring viability. In another study on astroglial cells in culture, a significant inhibition of the MTT response by amyloid β was seen, but trypan blue uptake and LDH release were not affected.[26] Indeed, various groups have now shown that amyloid enhances the MTT formazan exocytosis and that the formazan crystals might be toxic *per se* in PC12 and neuronal cells.[27,28] However, the *in vivo*

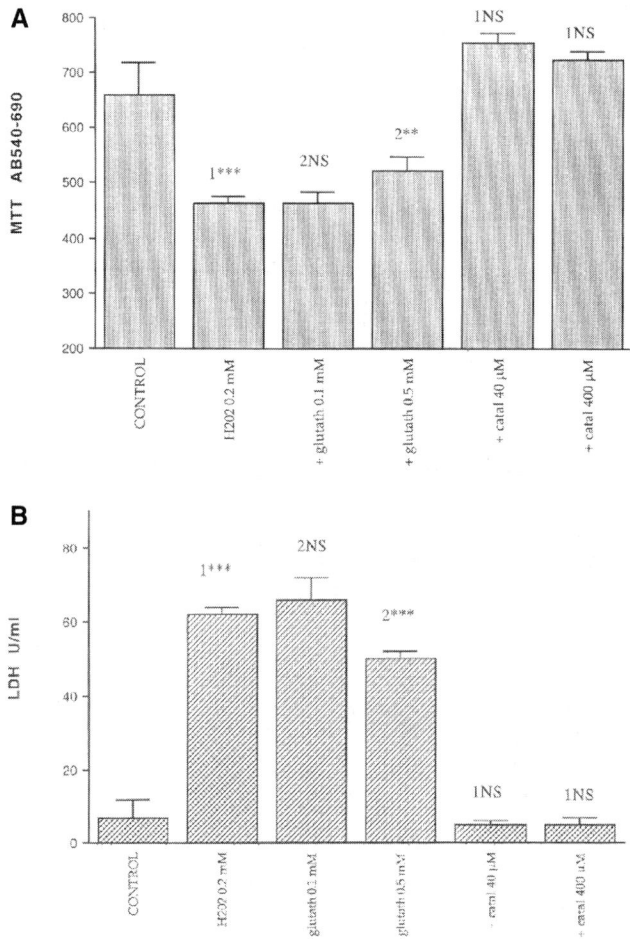

FIGURE 7. Protective effect of glutathione and catalase on H_2O_2 evoked (0.2 mM, one hour) toxicity in rat cortical neurones as seen by the MTT response and LDH release 24 h later. X, SD, $n = 4$. 1, versus control; 2, versus H_2O_2 0.2 mM.

degeneration activity of β-amyloids is well correlated with their having β-structure and activity to suppress the MTT reduction activity.[29] Thus, part of the inhibition of the MTT response was found to be related to the amyloid β induced, oxyradical mediated impairment of glucose transport and mitochondrial function in rat neocortical synaptosomes.[30] This effect was antagonized by antioxidants suggesting that amyloid β beta induced membrane lipid peroxidation. Thus, it is becoming evident that amyloids, as well as other compounds like cholesterol, lysophosphatidic acid, sterol sex hormones, and chloroquine[31,32] are able to reduce MTT reduction by increasing exocytosis and that this occurs at nanomolar concentrations, whereas at micromolar concentrations and time scales of less than one hour of the MTT assay, it is the actual inhibition of the MTT reduction that prevails.[31]

Mechanism of Protection by ALC Against A25–35 Toxicity: Possible Mitochondrial Action

The cotreatment of rat cortical neurones with ALC (0.1–10 mM) counteracted inhibition of the MTT response caused by the presence of β A25–35 (50 μg/ml) and long term exposure of neurones form DIV1 to ALC (1 mM) completely prevented the subsequent toxicity to amyloid A25–35 (5 μg/ml). A protective effect of ALC against A25–35 in hippocampal cultures has also been reported[33] and its analogue, acetyl-L-carnitine arginine amide was reported to prevent A25–35 induced neurotoxicity in cerebellar granule cells.[34]

Metabolic hypofunction is a common finding in a number of neurodegenerative diseases, including AD.[35] Defects in mitochondrial oxidative metabolism, in particular decreased activity of cytochrome c oxidase, have been demonstrated in AD. The incubation of isolated rat brain mitochondria with A25–35 caused a rapid, dose dependent decrease in the activity of complex IV.[36] It has been suggested that the primary target of β-amyloids is the suppression of mitochondrial succinate dehydrogenase, a

TABLE 1. Effect of glutathione, catalase, vitamin E, NGF and cycloheximide on amyloid β A25–35 evoked (25 μg/ml) toxicity in rat cortical neurones

Treatment	Percent Inhibition of MTT versus untreated control	Significance versus β A25–35
β A25–35 (25 μg/ml)	18.4	
+ glutathione (0.1 mM)	17.8	ns
+ glutathione (0.5 mM)	15.8	ns
+ catalase (4 mg/ml)	1.1	$p < 0.001$
+ vitamin E (100 μM)	11.4	$p < 0.05$
+ NGF (0.1 ng/ml)	13.6	$p < 0.05$
+ NGF (10 ng/ml)	10.9	$p < 0.05$
+ cycloheximide (0.1 μg/ml)	10.8	$p < 0.05$

The MTT response was measured following 24 h culture in serum free medium. Each percentage value is the mean of the MTT response from three wells compared to that of the untreated control. The significance was calculated using the Student *t*-test versus the β A25–35 treated cells.

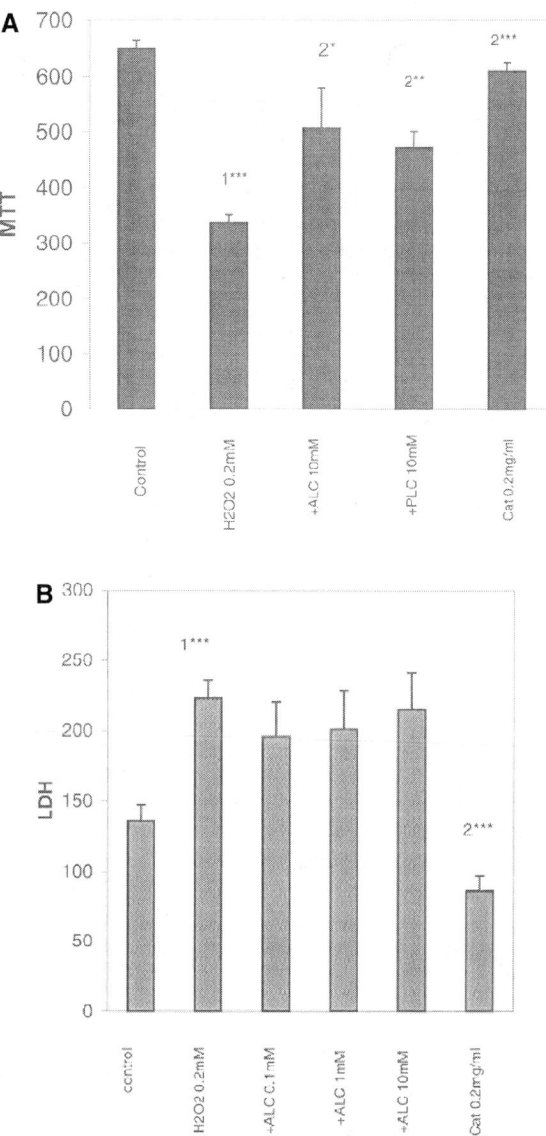

FIGURE 8. A small evidence of the protective effect of acetyl-L-carnitine on the H_2O_2 evoked (0.2 mM, one hour) toxicity in rat cortical neurones as seen by the MTT response, but not on LDH release.

constituent of both Krebs cycle and complex II of the mitochondrial respiratory chain, and the resulting neuronal damage occurs due to the brains large dependence on mitochondrial energy production.[37] The action of ALC could be at the level of mitochondria, thereby protecting against the loss of energy and formation of free radicals.

We have previously shown that the toxicity of 3-nitropropionic acid (3-NPA), an inhibitor of succinate dehydrogenase, is prevented by ALC in cultured cortical neurones.[7] A protective action by L-C against 3-NPA-evoked toxicity has also been demonstrated.[18] Another possibility is that the there is an intracellular shortage of either L-C or ALC, especially under conditions of energy compromise, as reported elsewhere for AZT-myopathy.[38] Impairment of L-C uptake by dysfunctioning mitochondria and a shift to the glycolytic pathway was also reported.[39] Thus, additional L-C or ALC would probably overcome the mitochondrial insufficiency and enable the cell to mobilize energy via beta oxidation. The free L-C levels are reported to be significantly lower in plasma in patients with AD.[40] Other compounds that augment cellular energy levels have also been shown to exhibit protective actions. Toxicity in embryonic hippocampal neurons exposed to A25–35 for 48 h was partially prevented by creatine.[41] Ketone bodies also protect neurons in culture.[42] Part of the toxicity of the A25–35 and 1–40 is related to the depletion of ATP levels, but they also induce an accumulation of reactive oxygen species (ROS). The amyloid β-peptide generates free radicals in a metal catalyzed reaction inducing neuronal cell death by a ROS mediated process. These free radicals damage neuronal membrane lipids, proteins, and nucleic acids. Studies suggest that amyloid β produces hydrogen peroxide (H_2O_2) by a mechanism that involves the reduction of metal ions, Fe(III) or Cu(II), setting up conditions for Fenton type chemistry.[43] This process that leads to inhibition of mitochondrial function and of glycolysis is prevented by antioxidants—for example, vitamin E, idebenone, superoxide dismutase, GSH ethyl ester, and Ginkgo biloba extract (EGb 761).[44–47] In our studies, vitamin E, but not glutathione, was able to protect against A25–35 mediated toxicity, as is reported elsewhere.[48] Part of the toxicity seen in these studies is probably related to the apoptosis mechanism, since the protein synthesis inhibitor cycloheximide and NGF prevented neuronal death evoked by A25–35 as shown elsewhere.[49,50] Protection by ALC against peroxide damage, part of the protective effect ALC, could also be related to the attenuation of free radical formation and escape from the dysfunctioning mitochondria. In the present study ALC (1–10 mM) protected cultured rat cortical neurones from toxicity following direct exposure to H2O2 (0.2 mM, 1 h). A reduction in intracellular peroxide formation following exposure of B12 cells to amyloid protein stimulated peroxide toxicity by ALC has also been reported.[51]

Possible Inhibition of A-β Fibril Formation and Protection of Membrane by ALC

Since L-C and ALC are amphiphilic molecules they could interact directly with the charges on the membrane phospholipids, glycolipids, and proteins.[52] Both L-C and ALC have two binding sites, one recognizing the trimethylamino group and the second by the carboxylic group (SEE FIGURE 9). Thus ALC may interact directly with the amphipathic binding site on the β-amyloid peptides precluding the assembly into fibrils or it could affect the membrane negative surface potential directly, thereby interfering with A-β toxicity to the membrane. Thus, ALC like other reported quaternary ammonium compounds e.g. *N*-hexadedecyl-*N*-methylpiperidinium

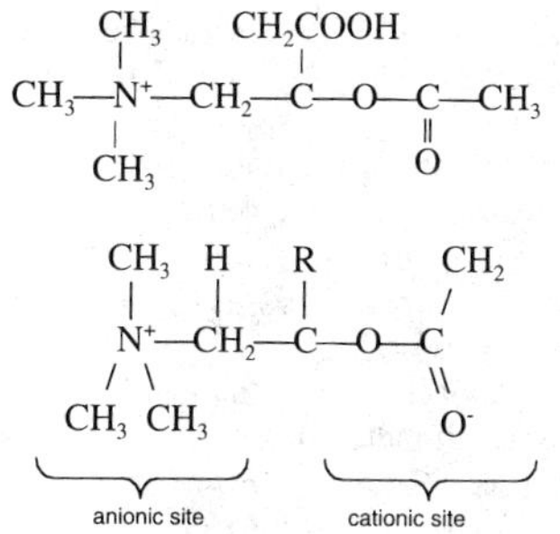

$$CH_3-\overset{\underset{|}{CH_3}}{\underset{|}{N^+}}-CH_2-\overset{\underset{|}{CH_2COOH}}{C}-O-\overset{O}{\underset{\|}{C}}-CH_3$$

$$\underset{\underbrace{}_{\text{anionic site}}}{\overset{CH_3}{\underset{CH_3}{{}_/{N^+}\backslash}}}-\overset{H}{\underset{|}{CH_2}}-\overset{R}{\underset{|}{C}}-O-\underset{\underbrace{}_{\text{cationic site}}}{\overset{CH_2}{\underset{O^-}{{}^/C\backslash\backslash}}}$$

FIGURE 9. Structure of acetyl-L-carnitine showing its zwitterionic nature and the two potential binding sites, one recognizes the trimethylamino group and the other the carboxylic group.

bromide,[53] could inhibit A-β polymerization or interferes with the A-β interaction with charged anionic phospholipids[54,55] but these hypothesis remain to be tested. The possibility that ALC may affect the membrane directly and in some way make it more resistant to damage has been proposed by various groups.[7,53,56,57] Part of this process may also involve changes in membrane fluidity thereby affecting the membrane phase transition behavior.[7] ALC has been reported to increase fluidity in rat brain microsomes and liposomes.[58] The β-amyloid peptide and several of the β-fragments alter the neuronal membrane lipid environment by changing fluidity and inducing free radical lipid peroxidation.[59] They have been reported to decrease the fluidity of human cortex membranes in a concentration dependent fashion.[60] The mechanism for ALC protective actions against A-β is not clear, but it may prevent cell damage by stabilizing the membrane against amyloid action and thereby prevent free radical damage and mitochondrial injury, thus preserving energy production and membrane functions.

ACKNOWLEDGMENTS

A special thanks to the participants of the Fifth International Conference on Neuroprotective Agents, (Lake Tahoe, California) for all their helpful comments and suggestions and to Miss A. Intilangelo for help in preparation of the manuscript.

REFERENCES

1. ALOI-TOTARO, E.M., M. VARRICCHIO, M. VERZA, et al. 1987. Mitochondrial damage, environmental factors and aging: acetyl-L-carnitine as a protective factor. Quad. Med. Chir. **3:** 182–186.
2. BONDY, S.C. 1995. The relation of oxidative stress and hyperexcitation to neurological disease. Proc. Soc. Exp. Biol. Med. **208:** 337–345.
3. FISKUM, G., A.N. MURPHY & M.F. BEAL. 1999. Mitochondria in neurodegeneration: acute ischemia and chronic neurodegenerative diseases. J. Cereb. Blood Flow Metab. **19:** 351–369.
4. HARIK, S.I. & M.A. HRITZ. 1993. Effect of acetyl-L-carnitine on 1-methyl-4-phenyl-1,2,3,6-tetrahydropyridine (MPTP) neurotoxicity. Biochem. Pharmacol. **45:** 2170–2172.

5. FISKUM, G., Y. LIU, K.M. MYERS, et al. 1992. Prevention of post-ischemic molecular alterations and neurological injury by acetyl-L-carnitine. Soc. Neurosci. Abst. **18:** 1453.
6. IDO, Y., et al. 1994. Neural dysfunction and metabolic imbalances in diabetic rats. Prevention by acetyl-L-carnitine. Diabetes **43:** 1469–1477.
7. VIRMANI, M.A., R. BISELLI, A. SPADONI, et al. 1995. Protective actions of L-carnitine and acetyl-L-carnitine on the neurotoxicity evoked by mitochondrial uncopuling or inhibitors. Pharmacol. Res. **32:** 383–389.
8. SONERU, I.L., et al. 1997. Acetyl-L-carnitine effects on nerve conduction and glycemic regulation in experimental diabetes. Endocr. Res. **23:** 27–36.
9. MANFRIDI, A., G.L. FORLONI, E. ARRIGONI-MARTELLI, et al. 1992. Culture of dorsal root ganglion neurons from aged rats. Effects of acetyl-L-carnitine and NGF. Int. J. Dev. Neurosci. **10:** 321–329.
10. FERNANDEZ, E., R. PALLINI, L. LAURETTI, et al. 1997. Motonuclear changes after cranial nerve injury and regeneration. Arch. Ital. Biol. **135:** 343–351.
11. HOEHNER, P.J., T.J. BLANCK, R. ROY, et al. 1992. Alteration of voltage-dependent calcium channels in canine brain during global ischemia and reperfusion. J. Cereb. Blood Flow Metab. **12:** 418–424.
12. PERRY, D.C., H. WEI, R.E. ROSENTHAL, et al. 1994. Autoradiographic analysis of L- and N-type voltage-dependent calcium channel binding in canine brain after global cerebral ischemia/reperfusion. Brain Res. **657:** 65–72.
13. FIORE, L. & L. RAMPELLO. 1989. L-Acetylcarnitine attenuates the age dependent decrease of NMDA sensitive glutamate receptors in rat hippocampus. Acta Neurol. (Napoli) **11:** 346–350.
14. SERSHEN, H., L.G. HARSING, M. BANAY-SCHWARTZ, et al. 1991. Effect of acetyl-L-carnitine on the dopaminergic system in aging brain. J. Neurosci. Res. **30:** 555–559.
15. Bremer, J. 1983. Carnitine: metabolism and function. Physiol. Rev. **63:** 1420–1480.
16. DI LISA, F., V. BOBYLEVA-GUARRIERO, P. JOCELYN, et al. 1985. Stabilising action of carnitine on energy linked processes in rat liver mitochondria. Biochem. Biophys. Res. Commun. **131:** 968–973.
17. KOUDELOVA, J., J. MOUREK, Z. DRAHOTA, et al. 1994. Protective effect of carnitine on lipoperoxide formation in rat brain. Physiol. Res. **43:** 387–389.
18. BINIENDA, Z., J.R. JOHNSON, A.A. TYLER-HASHEMI, et al. 1999. Protective effect of L-carnitine in the neurotoxicity induced by the mitochondrial inhibitor 3-nitropropionic acid (3-NPA). Ann. N.Y. Acad. Sci. **890:** 173–178.
19. VIRMANI, M.A. & V. CASO. 2000. Action of acetyl-L-carnitine on the neurotoxicity evoked by amyloid fragments on primary rat cortical neurones. Fifth International Conference on Neuroprotective Agents. Clinical and Experimental Aspects. September 17–21, Lake Tahoe, California.
20. VIRMANI, M.A., N. CORSICO & E. ARRIGONI-MARTELLI. 1992. Tests for evaluating the potential toxic effects of compounds on the neuroendocrine system. Review. Euro. Bull. Drug Res. **1:** 53–62.
21. SHEARMAN, M.S., S.R. HAWTIN & V.J. TAILOR. 1995. The intracellular component of cellular 3-(4,5-dimethylthiazol-2-yl)-2, 5-diphenyltetrazolium bromide (MTT) reduction is specifically inhibited by beta-amyloid peptides. J. Neurochem. **65:** 218–227.
22. HOWLETT, D.R., K.H. JENNINGS, D.C. LEE, et al. 1995. Aggregation state and neurotoxic properties of Alzheimer beta-amyloid peptide. Neurodegeneration **4:** 23–32.
23. HARKANY, T., T. HORTOBAGYI, M. SASVARI, et al. 1999. Neuroprotective approaches in experimental models of beta-amyloid neurotoxicity: relevance to Alzheimer's disease. Prog. Neuropsychopharmacol. Biol. Psych. **23:** 963–1008.
24. PIKE, C.J., A.J. WALENCEWICZ-WASSERMAN, J. KOSMOSKI, et al. 1995. Structure-activity analyses of beta-amyloid peptides: contributions of the beta 25–35 region to aggregation and neurotoxicity. J. Neurochem. **64:** 253–265.
25. HU, J. & E.E. EL-FAKAHANY. 1994. An artifact associated with using trypan blue exclusion to measure effects of amyloid beta on neuron viability. Life Sci. **55:** 1009–1016.

26. PATEL, A.J., S. GUNASEKERA, A. JEN, et al. 1996. Beta-amyloid-mediated inhibition of redox activity (MTT reduction) is not an indicator of astroglial degeneration. Neuroreport **7:** 2026–2030.
27. HERTEL, C., E. TERZI, N. HAUSER, et al. 1997. Inhibition of the electrostatic interaction between beta-amyloid peptide and membranes prevents beta-amyloid-induced toxicity. Proc. Natl. Acad. Sci. U.S.A. **94:** 9412–9416.
28. LIU, Y. & D. SCHUBERT. 1998. Steroid hormones block amyloid fibril-induced 3-(4,5-dimethylthiazol-2-yl)-2,5-diphenyltetrazolium bromide (MTT) formazan exocytosis: relationship to neurotoxicity. J. Neurochem. **71:** 2322–2329.
29. KANEKO, I., T. KUBO & K. MORIMOTO. 2000. Neurotoxicity of beta-amyloid. Nippon Yakurigaku Zasshi **115:** 67–77.
30. KELLER, J.N., Z. PANG, J.W. GEDDES, et al. 1997. Impairment of glucose and glutamate transport and induction of mitochondrial oxidative stress and dysfunction in synaptosomes by amyloid beta-peptide: role of the lipid peroxidation product 4-hydroxynonenal. J. Neurochem. **69:** 273–284.
31. ABE, K. & H. SAITO. 1999. Both oxidative stress-dependent and independent effects of amyloid beta protein are detected by 3-(4,5-dimethylthiazol-2-yl)-2, 5-diphenyltetrazolium bromide (MTT) reduction assay. Brain Res. **830:** 146–154.
32. ISOBE, I., M. MICHIKAWA & K. YANAGISAWA. 1999. Enhancement of MTT, a tetrazolium salt, exocytosis by amyloid beta-protein and chloroquine in cultured rat astrocytes. Neurosci. Lett. **266:** 129–132.
33. FORLONI, G., N. ANGERETTI & S. SMIROLDO. 1994. Neuroprotective activity of acetyl-L-carnitine: studies in vitro. J. Neurosci. Res. **37:** 92–96.
34. SCORZIELLO, A., O. MEUCCI, M. CALVANI, et al. 1997. Acetyl-L-carnitine arginine amide prevents beta 25-35-induced neurotoxicity in cerebellar granule cells. Neurochem. Res. **22:** 257–265.
35. ALBERS, D.S. & M.F. BEAL. 2000. Mitochondrial dysfunction and oxidative stress in aging and neurodegenerative disease. J. Neural. Transm. Suppl. **59:** 133–154.
36. CANEVARI, L., J.B. CLARK & T.E. BATES. 1999. Beta-amyloid fragment 25-35 selectively decreases complex IV activity in isolated mitochondria. FEBS Lett. **457:** 131–134.
37. KANEKO, I., N . YAMADA, Y. SAKURABA, et al. 1995. Suppression of mitochondrial succinate dehydrogenase, a primary target of beta-amyloid, and its derivative racemized at Ser residue. J. Neurochem. **65:** 2585–2593.
38. DALAKAS, M.C., M.E. LEON-MONZON, I. BERNARDINI, et al. 1994. Zidovudine-induced mitochondrial myopathy is associated with muscle carnitine deficiency and lipid storage. Ann. Neurol. **35:** 482–487.
39. SEMINO-MORA, M.C., M.E. LEON-MONZON, M.C. DALAKAS. 1994. The effect of L-carnitine on the AZT-induced destruction of human myotubes. Part II: treatment with L-carnitine improves the AZT-induced changes and prevents further destruction. Lab Invest. **71:** 773–781.
40. RUBIO, J.C., F. DE BUSTOS, J.A. MOLINA, et al. 1998. Cerebrospinal fluid carnitine levels in patients with Alzheimer's disease. J. Neurol. Sci. **155:** 192–195.
41. BREWER, G.J. & T.W. WALLIMANN. 2000. Protective effect of the energy precursor creatine against toxicity of glutamate and beta-amyloid in rat hippocampal neurons. J. Neurochem. **74:** 1968–1978.
42. KASHIWAYA, Y., T. TAKESHIMA, N. MORI, et al. 2000. D-beta-hydroxybutyrate protects neurons in models of Alzheimer's and Parkinson's disease. Proc. Natl. Acad. Sci. U.S.A. **97:** 5440–5444.
43. HUANG, X., C.S. ATWOOD, M.A. HARTSHORN, et al. 1999. The A beta peptide of Alzheimer's disease directly produces hydrogen peroxide through metal ion reduction. Biochemistry **38:** 7609–7616.
44. SUO, Z., C. FANG, F. CRAWFORD, et al. 1997 . Superoxide free radical and intracellular calcium mediate A beta(1-42) induced endothelial toxicity. Brain Res. **762:** 144–152.
45. PEREIRA, C., M.S. SANTOS, C. OLIVEIRA, et al. 1999. Involvement of oxidative stress on the impairment of energy metabolism induced by A beta peptides on PC12 cells: protection by antioxidants. Neurobiol. Dis. **6:** 209–219.

46. BUTTERFIELD, D.A., T. KOPPAL, R. SUBRAMANIAM, et al. 1999. Vitamin E as an antioxidant/free radical scavenger against amyloid beta-peptide-induced oxidative stress in neocortical synaptosomal membranes and hippocampal neurons in culture: insights into Alzheimer's disease. Rev. Neurosci. **10:** 141–149.
47. BASTIANETTO, S., C. RAMASSAMY, S. DORE, et al. 2000. The Ginkgo biloba extract (EGb 761) protects hippocampal neurons against cell death induced by beta-amyloid. Eur. J. Neurosci. **12:** 1882–1890.
48. HUANG, H.M., H.C. OU & S.J. HSIEH. 2000. Antioxidants prevent amyloid peptide-induced apoptosis and alteration of calcium homeostasis in cultured cortical neurons. Life Sci. **66:** 1879–1892.
49. FURUKAWA, K., S. ESTUS, W. FU, et al. 1997. Neuroprotective action of cycloheximide involves induction of bcl-2 and antioxidant pathways. J. Cell Biol. **136:** 1137–1149.
50. DORE, S., S. BASTIANETTO, S. KAR, et al. 1999. Protective and rescuing abilities of IGF-I and some putative free radical scavengers against beta-amyloid-inducing toxicity in neurons. Ann. N.Y. Acad. Sci. **890:** 356–364.
51. BEHL, C., J.B. DAVIS, R. LESLEY, et al. 1994. Hydrogen peroxide mediates amyloid beta protein toxicity. Cell **77:** 817–827.
52. VIRMANI, M.A., S. ROSSI, R. CONTI, et al. 1996. Structural, metabolic and ionic requirements for the uptake of L-carnitine by primary rat cortical cells. Pharmacol. Res. **33:** 19–27.
53. WOOD, S.J., L. MACKENZIE, B. MALEEFF, et al. 1996. Selective inhibition of Abeta fibril formation. J. Biol. Chem. **271:** 4086–4092.
54. CHAUHAN, A., I. RAY & V.P. CHAUHAN. 2000. Interaction of amyloid beta-protein with anionic phospholipids: possible involvement of Lys28 and C-terminus aliphatic amino acids. Neurochem. Res. **25:** 423–429.
55. VARGAS, J., J.M. ALARCON & E. ROJAS. 2000. Displacement currents associated with the insertion of alzheimer disease amyloid beta-peptide into planar bilayer membranes. Biophys. J. **79:** 934–944.
56. PASTORINO, J.G., J.W. SNYDER & A. SERRONI. 1993. Cyclosporin and carnitine prevent the anoxic death of cultured hepatocytes by inhibiting the mitochondrial permeability transition. J. Biol. Chem. **268:** 13791–13798.
57. FRITZ, I.B. & E. ARRIGONI-MARTELLI. 1993. Sites of action of carnitine and its derivatives on the cardiovascular system. Interactions with membranes. Trends Pharmacol. Sci. **14:** 355–360.
58. ARIENTI, G., M.T. RAMACCI, A. MACCARI, et al. 1992. Acetyl-L-carnitine influences the fluidity of brain microsomes and of liposomes made of rat brain microsomal lipid extracts. Neurochem. Res. **17:** 671–675.
59. AVDULOV, N.A., S.V. CHOCHINA, U. IGBAVBOA, et al. 1997. Amyloid beta-peptides increase annular and bulk fluidity and induce lipid peroxidation in brain synaptic plasma membranes. J. Neurochem. **68:** 2086–2091.
60. MULLER, W.E., G.P. ECKERT, K. SCHEUER, et al. 1998. Effects of beta-amyloid peptides on the fluidity of membranes from frontal and parietal lobes of human brain. High potencies of A beta 1–42 and A beta 1–43. Amyloid **5:** 10–15.

Post-Stroke Dementia

Nootropic Drug Modulation of Neuronal Nicotinic Acetylcholine Receptors

XILONG ZHAO, JAY Z. YEH, AND TOSHIO NARAHASHI

Department of Molecular Pharmacology and Biological Chemistry, Northwestern University Medical School, Chicago, Illinois, U.S.A.

ABSTRACT: Nefiracetam is a new pyrrolidone nootropic drug that is being developed for clinical use in the treatment of post-stroke vascular-type and Alzheimer's-type dementia. Among a few neuroreceptors that have been identified as potential targets of nootropics, neuronal nicotinic acetylcholine receptors (nnAChRs) are deemed the most important since they are related to learning, memory, and Alzheimer's disease dementia. We have recently found potent stimulating action of nefiracetam on nnAChRs. Rat cortical neurons in long-term primary culture expressed nnAChRs. Whole-cell patch clamp experiments revealed two types of currents induced by ACh, α-bungarotoxin (α-BuTX)-sensitive, rapidly desensitizing, α7-type currents and α-BuTX-insensitive, slowly desensitizing, α4β2-type currents. Although α7-type currents were only weakly inhibited by nefiracetam, α4β2-type currents were potently and efficaciously potentiated by nefiracetam. Nefiracetam at 0.1 nM reversibly potentiated ACh-induced currents to 200–300% of control. Very high concentrations (about 10 μM) also potentiated these currents, but to a lesser extent, indicative of the bell-shaped dose–response relationship known to occur for nefiracetam, even in animal behavior experiments. Three specific inhibitors of each of PKA and PKC did not prevent nefiracetam from potentiating ACh-induced currents, indicating that these protein kinases are not involved in nefiracetam action. Pretreatment with pertussis toxin did not alter nefiracetam potentiation, indicating G_i/G_o proteins are not involved. Pretreatment with cholera toxin did abolish nefiracetam potentiation. Thus, nefiracetam potentiation is mediated via G_s proteins. In conclusion, nefiracetam stimulates α4β2-type nnAChRs via G_s proteins at nanomolar concentrations. The potentiation of α4β2-type nnAChRs is thought to be at least partially responsible for cognitive enhancing action.

KEYWORDS: Stroke; Dementia; Alzheimer; Nootropic drug; Neuronal nicotinic acetylecholine receptor.

INTRODUCTION

Nootropic drugs are becoming a hot subject of investigation. At least four drugs, all of which are anticholinesterases, have been approved by FDA in the treatment of Alzheimer's patients and patients with post-stroke dementia. Some oxopyrrolidine acetic acid (racetam) derivatives are being developed into clinical use for

---

Address for correspondence: Dr. Toshio Narahashi, Department of Molecular Pharmacology and Biological Chemistry, Northwestern University Medical School, 303 East Chicago Avenue, Chicago, IL 60611, U.S.A. Voice: 312-503-8284; fax: 312-503-1700.

tna597@northwestern.edu

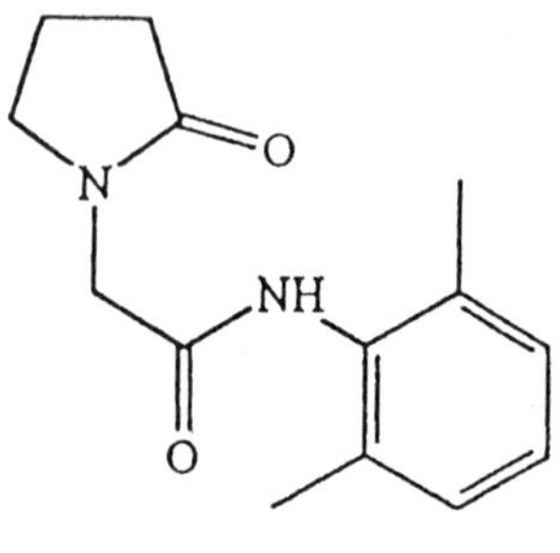

FIGURE 1. Structure of nefiracetam (DM-9384), $C_{14} H_{18} N_2 O_2$.

patients with dementia. One of these derivatives is nefiracetam (see FIGURE 1, DM-9384), N-(2,6-dimethylphenyl)-2-(2-oxo-1-pyrrolidinyl) acetamide, which is undergoing an extensive phase II and phase III test following a variety of animal behavior experiments.[1]

The mechanism of action of nefiracetam at the cellular and molecular level has been studied by several investigators. Nefiracetam augmented high voltage-gated calcium channels at micromolar concentrations (about 1 µM) via interactions with G proteins.[2–4] It also modulated the activity of the GABAergic system.[5] $GABA_A$ receptor currents were either potentiated or inhibited by nefiracetam, depending on the GABA concentration, via protein kinase A (PKA) and G proteins. Neuronal nicotinic acetylcholine receptors (nnAChRs) in PC12 cells were affected by nefiracetam in a manner similar to that seen in $GABA_A$ receptors.[6]

The importance of the cholinergic system in nefiracetam action has also been emphasized by Nishizaki and his associates. Nefiracetam at 10–100 nM caused a short-term depression of ACh-induced currents in *Torpedo* nicotinic AChRs expressed in *Xenopus* oocytes due to activation of G protein–regulated PKA activity, causing a long-term potentiation at higher concentrations (1–10 µM) due to activation of Ca^{2+}-dependent protein kinase C (PKC).[7] Nefiracetam also potentiated the activity of human α4β2 and α7 AChRs expressed in *Xenopus* oocytes,[8] and it induced LTP-like facilitation of hippocampal synaptic transmission.[9]

Since cholinergic activity in the brain of Alzheimer's patients and patients with other forms of dementia is known to be downregulated, stimulation of this activity is thought to improve the learning and memory of those patients. Alzheimer's drugs with anticholinesterase activity work through stimulation of the cholinergic system. Thus, we recently launched an extensive study of nefiracetam interactions with the nnAChRs *native* to brain neurons.[10] It is important to emphasize that receptors/channels recombinantly expressed in various cells do not necessarily respond to drugs in the same manner as they do in native neurons that possess the same receptors/channels.[11–14]

MATERIALS AND METHODS

Rat cortical neurons in primary culture for 4–7 weeks[15] were used as material. Currents evoked by ACh were recorded by the whole-cell patch clamp technique at room temperature (21–22°C). The external solution contained (in mM): 150 NaCl,

5 KCl, 2.5 $CaCl_2$, 1 $MgCl_2$, 15 HEPES acid, 10 HEPES sodium, and 10 D-glucose. Tetrodotoxin 100 nM was added to eliminate voltage-gated sodium channel currents. Atropine sulfate 20 nM was added to block the muscarinic AChR currents. The pH was 7.3 and the osmolarity was adjusted to 300 mOsm with D-glucose. The internal pipette solution contained (in mM): 140 Cs gluconate, 2 $MgCl_2$, 1 $CaCl_2$, 11 EGTA, 10 HEPES acid, 2 ATP-Mg^{2+}, and 0.2 GTP-Na^+. The pH was adjusted to 7.3 with CsOH, and the osmolarity was adjusted to 290–300 mOsm with D-glucose.

Two methods for drug application were used: one was application via a U-tube and the other was perfusion through the bath. The ACh solution was applied through a fast U-tube application system[16] controlled by computer-operated magnetic valves. The external solution surrounding the cell could be completely changed with the ACh solution within 30–40 ms. Test drugs were added to the external solution and continuously perfused to the recording chamber via a glass syringe and Teflon tube.

RESULTS

Rat cortical neurons in long-term primary culture are endowed with two types of nnAChRs:[17] one generates α-bungarotoxin (α-BuTX)-sensitive, fast desensitizing currents in response to ACh application; the other generates α-BuTX-insensitive, slowly desensitizing currents (see FIGURE 2). The former AChRs have a low ACh affinity with an EC_{50} of about 300 µM and are composed primarily of α7-type receptors, whereas the latter AChRs have a high ACh affinity with an EC_{50} of about 3 µM and comprise primarily α4 and α2 receptor subunits (referred to as α4β2 type).

Nefiracetam exerted a weak and irreversible inhibitory action on the α7-type AChRs.[10] After correction for run down of receptor responses that amounted to 3.8% in 20 min, nefiracetam inhibitions at 1, 10, and 100 µM were estimated to be 2.8, 8.7, and 20.1%, respectively. By contrast, the α4β2 type currents were potently and efficaciously potentiated by nefiracetam.[10] An example of such an experiment is shown in FIGURE 3. The experiment was performed in the presence of 25 nM α-BuTX to block α7-type nnAChRs. ACh 10 µM was applied via a U tube for 0.5 sec at one-minute intervals. After applying 1 nM nefiracetam to the bath following several control currents evoked by ACh, the current amplitude was gradually increased and attained a steady state that was maintained at a constant level for 45 min. Washing out nefiracetam caused a gradual recovery of ACh currents toward the control level.

The effects of nefiracetam in a variety of experiments including animal behavioral tests are known to show a bell-shaped dose–response curve indicating an optimal dose to produce a maximal effect.[1] This was also the case for the nefiracetam potentiation of nnAChR responses: the mean percentages of current increase induced by nefiracetam at 0.1, 1, 10, 1,000, and 10,000 nM were 11, 125, 75, 68, and 35%, respectively.[10] The mechanism that underlies the bell-shaped dose–response relationship remains to be determined.

Drug-induced potentiation of currents is often accompanied by a shift of the agonist dose–response curve in the direction of lower agonist concentrations. The ACh EC_{50} was indeed decreased by 10 nM nefiracetam from the control value of 2.0 µM to 1.2 µM. However, such a shift does not explain the large increases in current amplitude observed at very high concentrations of ACh (100–1,000 µM) that gave a saturating response.

A α-BuTX-SC ACh Dose-Response

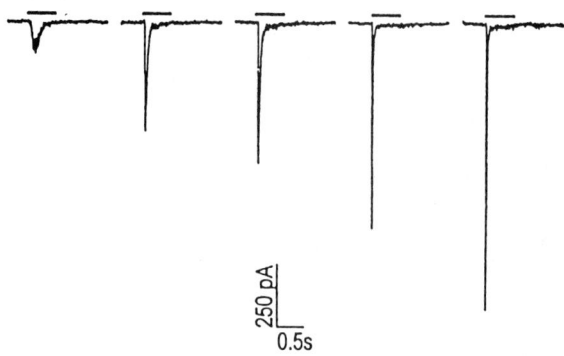

B α-BuTX-IC ACh Dose-Response

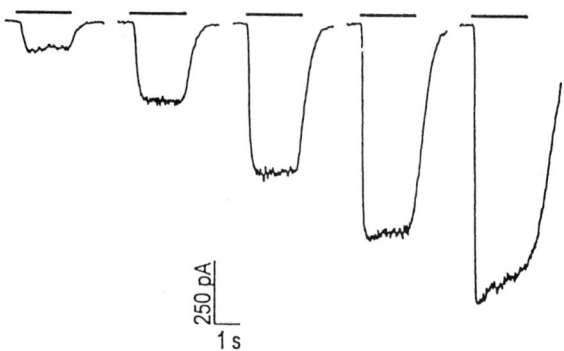

FIGURE 2. ACh induces α-BuTX-sensitive, rapidly desensitizing currents (**A**) and α-BuTX-insensitive, slowly desensitizing currents (**B**) in rat cortical neurons (from Narahashi et al.[26]).

As briefly described in the INTRODUCTION, nefiracetam modulates various receptors and channels via G proteins and/or protein kinases. This possibility was explored for the α4β2-type nnAChRs.[10] Preapplication of the PKA-specific inhibitors, H-89, peptide 5–24, or KT 5720 did not prevent nefiracetam from exerting its potentiating action on the α4β2-type receptor currents. Similarly, the PKC-specific inhibitors peptide 19–36, calphostin C, or chelerythrine failed to change nefiracetam potentiation. Pretreatment of neurons with pertussis toxin for 24 h did not prevent nefiracetam from potentiating the α4β2-type receptor currents. However, pretreatment with cholera toxin for 24 h did prevent nefiracetam action in potentiating ACh-induced currents. Thus, it was concluded that nefiracetam potentiated the α4β2-type receptor currents via G_s proteins, not G_i/G_o proteins, PKA, or PKC.

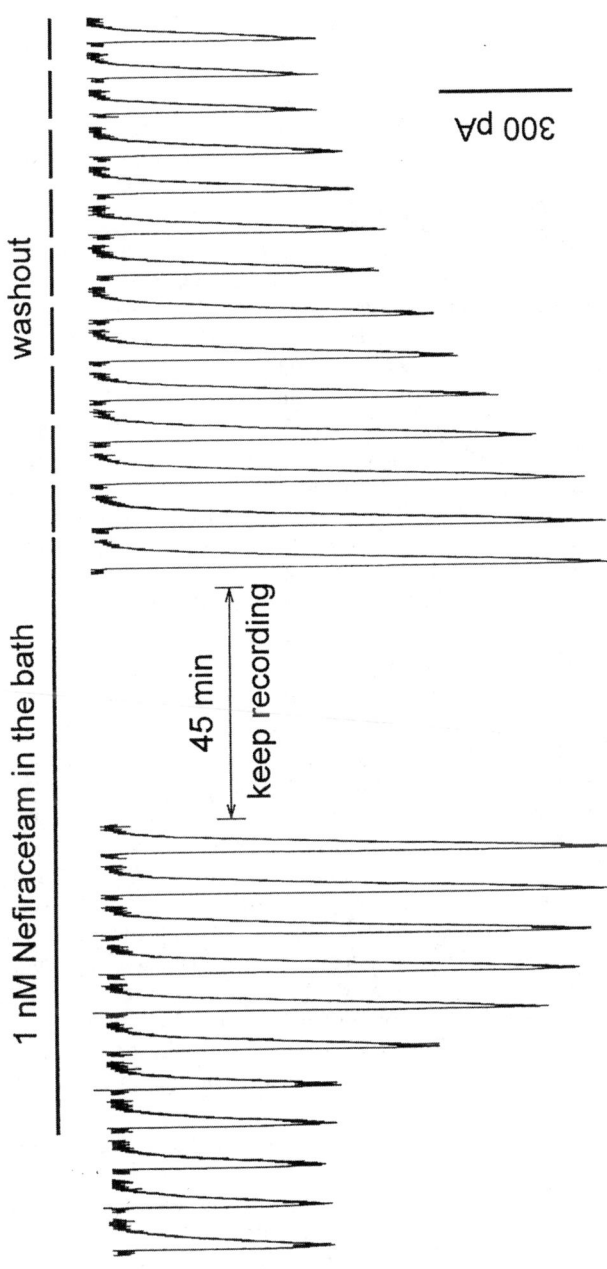

FIGURE 3. Nefiracetam 1 nM reversibly potentiates α4β2-type nnAChR currents evoked by ACh pulses (10 μM, 0.5 sec) applied at intervals of 1 min in a rat cortical neuron. Holding potential −70 mV.

DISCUSSION

In the study reported here, nefiracetam potently augmented α4β2-type currents while weakly suppressing α7-type currents evoked by ACh in rat cortical neurons. There is general agreement that the nicotinic system plays a major role in higher cognitive functions,[18] and the loss of brain nnAChRs is a neurochemical hallmark of Alzheimer's disease.[18,19] Compared to the α4β2 subunits, reductions in the α7 subunits appear less extensive in the cortex and hippocampus of Alzheimer's patients.[20,21] Thus, potentiation of α4β2-type nnAChRs may play an important role in the therapeutic action of nefiracetam.

Our observations in rat native neurons are different from those in *Xenopus* oocytes expressing the rat α4β2 or α7 nnAChRs. Nishizaki *et al.*[8] reported that nefiracetam at 100 nM potentiated not only α4β2 but also α7 receptor currents. This discrepancy on the effects of nefiracetam on α7 nnAChRs may be due to the difference in expression system. The folding, assembly, and subcellular localization of cloned nnAChR subunits are critically dependent upon host cells.[11,14] These properties of nnAChRs are known to be influenced by the choice of heterologous expression system.[12,13] We also observed that nefiracetam had no potentiating effect on cloned human α4β2 nnAChRs expressed in HEK cells.[10] Therefore, the cloned nnAChRs expressed either transiently in *Xenopus* oocytes or stably in mammalian cell lines may not accurately represent native nnAChRs in neurons. Another factor that complicates the comparison is the source of nnAChR subunits. We used human α4β2 subunits expressed in HEK cells whereas Nishizaki *et al.*[8] used rat α4β2 subunits expressed in *Xenopus* oocytes.

It is well established that protein phosphorylation plays an important role in various neuroreceptors.[22,23] Previous studies indicated that nefiracetam modulation of $GABA_A$ receptors,[5] nnAChRs,[6–8] and high voltage-gated calcium channels[2,4] occurred via protein kinases and G proteins. However, these results are not necessarily consistent. In PC12 cells expressing the α3-type nAChRs, the PKA inhibitor KT 5720 and the G_i/G_o protein inhibitor pertussis toxin, but not the PKC, inhibitor abolished the nefiracetam stimulation of nnAChRs.[6] In *Xenopus* oocytes expressing *Torpedo* nAChRs or rat nnAChRs (α4β2 or α7), PKC inhibitors, but not PKA inhibitors, blocked the nefiracetam potentiation of these nAChRs.[7,8] Previous studies also showed that nefiracetam modulated $GABA_A$ receptor currents in rat dorsal root ganglion neurons[5] and L-type Ca^{2+} channels in neuroblastoma × glioma hybrid (NG108-15) cells[2] by interacting with a PKA pathway. Our experiments with native cortical neuron nAChRs indicated that neither PKA nor PKC inhibition prevented nefiracetam potentiation. The difference between the data in the literature and our results may be due to the receptor subunit composition, the host cells, and/or the modulators applied.

Pertussis toxin is a selective, irreversible blocker of G_i/G_o. Pertussis toxin catalyzes ADP ribosylation of α-subunits of G_i and G_o proteins, prevents the G_i heterotrimers from interacting with receptors, and blocks their coupling and activation.[24] Since the $G_{i\alpha}$ remains in the GDP-bound and inactive state, it becomes unable to inactivate adenylyl cyclase. In contrast, cholera toxin catalyzes ADP ribosylation of $G_{s\alpha}$, which in turn activates adenylyl cyclase, resulting in an increase in the level of cAMP.[25] cAMP activates cAMP-dependent protein kinases including PKA.

Although pretreatment with pertussis toxin did not prevent nefiracetam from potentiating α4β2 type currents, cholera toxin did prevent nefiracetam potentiation. It is possible that the potentiating effect of nefiracetam is not seen because the receptors are already maximally modulated after cholera toxin treatment. However, this possibility was excluded by the fact that the ACh currents with and without cholera toxin treatment were not significantly different from each other. The lack of effect of PKA and PKC inhibitors on nefiracetam potentiating action suggests that G_s proteins may regulate the activity of the receptor via membrane delimited pathways.

In conclusion, the nootropic drug nefiracetam potently enhances the α4β2-type nnAChR response of rat cortical neurons in long-term primary culture. In contrast, nefiracetam does not potentiate, but slightly inhibits the α7-type currents. The potentiation of α4β2-type nnAChR can be blocked by a G_s protein modulator but not by a G_i/G_o protein inhibitor, nor PKA and PKC inhibitors. These results indicate that α4β2-type nnAChR is an important site of action of nefiracetam and that G_s proteins may play a crucial role in the nefiracetam potentiation.

ACKNOWLEDGMENTS

We thank Nayla Hasan for technical assistance and Julia Irizarry for secretarial assistance. This work was supported by a grant from the National Institutes of Health NS 14144, USA, and Daiichi Pharmaceutical Company, Tokyo, Japan.

REFERENCES

1. NABESHIMA, T. 1994. Ameliorating effects of nefiracetam (DM-9384) on brain dysfunction. Drugs Today **30**: 357–379.
2. YOSHII, M. & S. WATABE. 1994. Enhancement of neuronal calcium channel currents by the nootropic agent, nefiracetam (DM-9384), in NG108-15 cells. Brain Res. **642**: 123–131.
3. YOSHII, M., S. WATABE, T. SAKURAI & T. SHIOTANI. 1997. Cellular mechanism of underlying cognition-enhancing actions of nefiracetam (DM-9384). Behav. Brain Res. **83**: 185–188.
4. YOSHII, M., S. WATABE, Y. MURASHIMA, et al. 2000. Cellular mechanism of action of cognitive enhancers: effects of nefiracetam on neuronal Ca^{2+} channels. Alzheimer Dis. Assoc. Disord. **14**(Suppl 1): S95–S102.
5. HUANG, C.S., J.Y. MA, W. MARSZALEC & T. NARAHASHI. 1996. Effects of the nootropic drug nefiracetam on the GABAA receptor-channel complex in dorsal root ganglion neurons. Neuropharmacology **35**: 1251–1261.
6. OYAIZU, M. & T. NARAHASHI. 1999. Modulation of the neuronal nicotinic acetylcholine receptor-channel by the nootropic drug nefiracetam. Brain Res. **822**: 72–79.
7. NISHIZAKI, T., T. MATSUOKA, T. NOMURA, et al. 1998. Nefiracetam modulates acetylcholine receptor currents via two different signal transduction pathways. Mol. Pharmacol. **53**: 1–5.
8. NISHIZAKI, T., T. MATSUOKA, T. NOMURA, et al. 2000. Presynaptic nicotinic acetylcholine receptors as a functional target of nefiracetam in inducing a long-lasting facilitation of hippocampal neurotransmission. Alzheimer Dis. Assoc. Disord. **14**(Suppl 1): S82–S94.
9. NISHIZAKI, T, T. MATSUOKA, T. NOMURA, et al. 1999. A 'long-term-potentiation-like' facilitation of hippocampal synaptic transmission induced by the nootropic nefiracetam. Brain Res. **826**: 281–288.

10. ZHAO, X., A. KURYATOV, J.M. LINDSTROM, *et al.* 2001. Nootropic drug modulation of neuronal nicotinic acetylcholine receptors in rat cortical neuorns. Mol. Pharmacol. **59:** 674–683.
11. COOPER, S.T. & N.S. MILLAR. 1997. Host cell-specific folding and assembly of the neuronal nicotinic acetylcholine receptor α7 subunit. J. Neurochem. **68:** 2140–2151.
12. LEWIS, T.M., P.C. HARKNESS, L.G. SIVILOTTI, *et al.* 1997. The ion channel properties of a rat neuronal nicotinic receptor are dependent on the host cell type. J. Physiol. (Lond.) **505:** 299–306.
13. SIVILOTTI, L.G., D.K. MCNEIL, T.M. LEWIS, *et al.* 1997. Recombinant nicotinic receptors, expressed in *Xenopus* oocytes, do not resemble native rat sympathetic ganglion receptors in single-channel behavior. J. Physiol. (Lond.) **500:** 123–138.
14. SWEILEH, W., K. WENBERG, J. XU, *et al.* 2000. Multistep expression and assembly of neuronal nicotinic receptors is both host-cell- and receptor-subtype-dependent. Mol. Brain Res. **75:** 293–302.
15. MARSZALEC, W. & T. NARAHASHI. 1993. Use-dependent pentobarbital block of kainate and quisqualate currents. Brain Res. **608:** 7–15
16. FENWICK, E.M., A. MARTY & E. NEHER. 1982. A patch-clamp study of bovine chromaffin cells and of their sensitivity to acetylcholine. J. Physiol. **331:** 577–597.
17. AISTRUP, G.L., W. MARSZALEC & T. NARAHASHI. 1999. Ethanol modulation of nicotinic acetylcholine receptor currents in cultured cortical neurons. Mol. Pharmacol. **55:** 39–49.
18. VIDAL, C. & J.P. CHANGEUX. 1996. Neuronal nicotinic acetylcholine receptors in the brain. News Physiol. Sci. **11:** 202–208.
19. WOODRUFF-PAK, D.S. & R.M. HINCHLIFFE. 1997. Mecamylamine- or scopolamine-induced learning impairment: ameliorated by nefiracetam. Psychopharmacology **131:** 130–139.
20. MARTIN-RUIZ, C.M., J.A. COURT, E. MOLNAR, *et al.* 1999. α4 but not α3 and α7 nicotinic acetylcholine receptor subunits are lost from the temporal cortex in Alzheimer's disease. J. Neurochem. **73:** 1635–1640.
21. BURGHAUS, L., U. SCHUTZ, U. KREMPEL, *et al.* 2000. Quantitative assessment of nicotinic acetylcholine receptor proteins in the cerebral cortex of Alzheimer patients. Mol. Brain Res. **76:** 385–388.
22. HUGANIR, R.L. & P. GREENGARD. 1990. Regulation of neurotransmitter receptor desensitization by protein phosphorylation. Neuron **5:** 555–567.
23. HOFFMAN, P.W., A. RAVINDRAN & R.L. HUGANIR. 1994. Role of phosphorylation in desensitization of acetylcholine receptors expressed in *Xenopus* oocytes. J. Neurosci. **14:** 4185–4195.
24. KATAKA, T., G.M. BOKOCH, J.K. NORTHUP, *et al.* 1984. The inhibitory guanine nucleotide-binding regulatory component of adenylate cyclase: properties and functions of the purified protein. J. Biol. Chem. **259:** 3568–3577.
25. GILL, D.M. & R. MEREN. 1978. ADP-ribosylation of membrane proteins catalyzed by cholera toxin: basis of the activation of adenylate cyclase. Proc. Natl. Acad. Sci. U.S.A. **75:** 3050–3054.
26. NARAHASHI, T., G.L. AISTRUP, W. MARSZALEC & K. NAGATA. 1999. Neuronal nicotinic acetylcholine receptors: a new target site of ethanol. Neurochem. Internat. **35:** 131–141.

Part V: Neuroprotection in Alzheimer's and Related Diseases

QUESTIONS FOR DR. WEINSTOCK

From Dr. Slikker

Is there a role of hypothermia in the neuroprotective action of rasagiline?

ANSWER: I do not think so, because rasagiline did not lower body temperature in mice or rats in doses up to 10 mg/kg, but neuroprotection against head injury was achieved at a dose of 0.2 mg/kg.

From Dr. Palmer

Do you know the development status of TV 3326?

ANSWER: Teva, the company developing the drug, is currently preparing to perform chronic toxicity tests in rats and monkeys prior to phase I clinical studies.

From Dr. Banik

Elegant presentation: the question is that the neuroprotection you have shown *in vitro* with your drugs, how do you know these protected cells are functionally active?

ANSWER: We do not know that they are functionally active but we have shown that after treatment with TV3326 more of them are viable as indicated by the MTT stain and they contain more ATP than those which were not treated with the drug but exposed to oxygen–glucose deprivation.

You mentioned that streptozotocin treatment caused olegodendrocyte death. I wonder whether you found any myelin loss or demyelination. Also, do you have any information whether Schwann cells die of diabetic neuropathy, where there is PNS demyelination.

ANSWER: By immunohistochemical staining we showed that intracerebroventricular injection of streptozotocin caused a loss of myelin around the third ventricle at the level of the hypothalamus and at the paraventricular nucleus of the thalamus. I do not know anything specifically about Schwann cells, but streptozotocin causes a loss of myelin in peripheral nerves that can be prevented by the antidiabetic drugs troglitazone and gliclazide, irrespective of their effects on blood glucose.

From Dr. Manev

Could you elaborate more on the MAO inhibitory effect of TV 3326?

ANSWER: TV3326 is a relatively weak inhibitor of MAO-A and B from rat brain *in vitro*, with IC_{50} of more than 300 μM. A single sc injection or oral dose of 350 μmoles/kg in rats or mice only inhibits the brain enzyme by 50%. However, after chronic oral daily administration for two weeks, in rats at a dose of 75 µmoles/kg, brain MAO-A and B are inhibited by more than 70%, but the enzymes in the intestine and liver, by less than 20%. After two months of daily oral dosing the degree of enzyme inhibition in the brain exceeded 85% with no further inhibition in the intestine and liver.

QUESTIONS FOR DR. VIRMANI

From Dr. Lin

The pathophysiological concentration of Aβ protein on Alzheimer's patients is quite lower than the dose you have used.

Aggregated Aβ may decrease the dose required to induce cell death.

The degree of aggregated Aβ may be checked by congo red.

ANSWER: The overall concentration of amyloid β proteins may be lower in the brain, but they do form plaques and local concentrations may be higher, approaching the micromolar range that we used in our study. We did some preliminary investigation on the level of aggregation and level of toxicity and found that A1–40 did form aggregates almost immediately and that, therefore, the toxicity we observed *in vitro* was due to the aggregated form of the A1–40. The level of aggregation of A25-35 did vary and maybe some of the differences in toxicity, especially to short term exposure (less than 24 h) are related to differences in aggregation, but this has to be studied in more detail.

From Dr. Skaper

β-Amyloid has been described as a cause of MTT extrusion from cells and to cause a false positive response for vitality. Could the difference between MTT and LDH responses to β-amyloid reflect this extrusion phenomenon?

ANSWER: This is an important point since it is becoming evident that amyloids, as well as other compounds like cholesterol, lysophosphatidic acid, sterol sex hormones, and chloroquine, are able to reduce MTT by increasing exocytosis, but this phenomenon occurs at nanomolar concentrations, whereas at micromolar concentrations and time scales of less than one hour of the MTT assay, as in our study, it is the actual inhibition of the MTT reduction that is thought to prevail.

From Dr. Abbracchio

β-amyloid toxicity has been shown to depend on an aberrant reentry of neurons into the cell cycles, which is, hence, followed by abortive mitosis. Have you considered the possibility that acetyl-L-carnitine may interfere with such a mechanism (e.g., by interfering with cyclon synthesis and/or by inhibiting abortive mitosis)?

ANSWER: This is an interesting mechanism for the amyloid β protein toxicity, which can be the target of compounds like acetyl-L-carnitine. We have not yet looked at this mechanism but will certainly investigate its role further in future.

From Dr. Binienda

L-Carnitine–acetyl-L-carnitine shuttle was suggested to mediate acetylcholine synthesis by transfer of the acyl moieties from mitochondria to cytoplasm. Would it be feasible to include ACh measurements in your model?

ANSWER: Both L-carnitine and acetyl-L-carnitine have been reported to enhance acetylcholine production, probably by facilitating transfer of acetyl groups across mitochondrial membranes, thus regulating the availability in the cytoplasm of acetyl-CoA, a substrate of choline acetyltransferase in cholinergic neurons. Indeed L-carnitne has been reported to stimulate acetylcholine synthesis by up to 18% in cortex cells from adult rats. The levels of acetylcholine can be easily measured in our model of cultured cortical cells.

Antioxidant and Antiaging Activity of *N*-Acetylserotonin and Melatonin in the *in Vivo* Models

G. OXENKRUG,[a] P. REQUINTINA,[a] AND S. BACHURIN[b]

[a]*Pineal Research Laboratory, Department of Psychiatry,
St. Elizabeth's Medical Center/Tufts University, Boston, Massachusetts, U.S.A.*

[b]*Institute of Physiologically Active Compounds, Russian Academy of Science,
Chernogolovka 142432, Russia*

ABSTRACT: It is generally accepted that antioxidant properties of melatonin significantly contribute to its antiaging effect. Antioxidant effects of *N*-acetylserotonin (NAS), a melatonin precursor and metabolite, might predict its antiaging action as well. The antiaging effect of NAS was studied in female retired breeders and male C3H mice. Both NAS and melatonin administered with drinking water prolonged life span in male animals by about 20% versus control animals ($p < 0.01$) but did not affect the life span of female mice. Antioxidative activity was evaluated by determining the malonaldehyde + 4-hydroxynonenal (MDA + 4-HNE) and cellular glutathion peroxidase (GPx) levels in male, 11-month-old, C57Bl/6J mice with very limited (if any) capacity to convert pineal NAS into melatonin. NAS increased the antioxidant capacity of kidney. Both NAS and melatonin (four weeks daily i.p. injections) increased the antioxidant capacity of brain as demonstrated by decreased MDA + 4-HNE and increased GPx levels. NAS-treated C57Bl/6J mice experienced a weight loss of 9%, whereas the saline and melatonin groups only 3%. NAS- and melatonin-treated animals had healthy and luxuriant fur coats with some gray fur in the melatonin group; animals in the saline group had large areas of baldness. This study demonstrates, for the first time, the antiaging effect of NAS. This effect needs to be confirmed in animals with impaired capacity to convert NAS into melatonin.

KEYWORDS: Antioxidant activity; Antiaging activity; *N*-acetylserotonin; Melatonin.

INTRODUCTION

According to the free radical theory of aging, reactive oxygen species initiate degradative processes that contribute to the development of aging.[1] The increased vulnerability of aging organisms to oxidative stress implies that antioxidants might exert antiaging effects. In this vein, the antioxidative properties of melatonin suggest that it might exhibit an antiaging effect.[2] Indeed, melatonin prolongs the life span of rodents.[3] However, the antiaging effect of melatonin was found to vary with strain,

---

Address for correspondence: Prof. G.F. Oxenkrug, M.D., Ph.D., Pineal Research Laboratory, Department of Psychiatry, St. Elizabeth's Medical Center, QN-3P, 736 Cambridge St., Boston, MA, 02135, U.S.A. Voice: 617-789-2925; fax: 617-789-2066.
goxenkrug@cchcs.org

gender, age at the initiation of melatonin administration, and other factors. Thus, melatonin increases the life span of female BALB and C57Bl mice of both genders, whereas it shortens the life span of female C3H mice.[4] We are not aware of any studies assessing the melatonin effect on the life span of male C3H mice. Therefore, the first aim of this study was to compare the effects of melatonin on the life span of female and male C3H mice.

Apart from melatonin, several related pineal constituents also poses antioxidant properties.[5–8] One of them is N-acetylserotonin (NAS), the melatonin precursor and metabolite.[9,10] Although the antioxidant capacities of NAS suggest that it may show antiaging effects, there are no reports evaluating NAS effect on longevity. Thus, the second aim of the study reported here was to evaluate NAS effect on the life span of male and female C3H mice.

The antioxidant effects of melatonin and, especially, NAS were tested either against prooxidant agents or in *in vitro* models. The third aim of this study was to evaluate the antioxidant effects of melatonin and NAS in the *in vivo* models under naturally occurring prooxidant conditions, that is, aged mice.

METHODS AND PROCEDURE

Experiment #1: Effects of Melatonin and NAS on the Life Span of C3H Mice

Mice of C3H strain were purchased from Harlan Sprague-Dawley, Inc. (Indianapolis, IN). Male mice were four-weeks of age, female mice were retired breeders (approximately eight-months old). Animals were housed five per cage under a 12/12 light/dark cycle with lights on/off at 0600/1800. NAS and melatonin were administered with drinking water at 2.5 mg/kg/day. Drinking solutions were prepared from melatonin and NAS stock solutions of 10 mg/ml in 1% Tween-20. Melatonin and NAS content in drinking water were adjusted according to body weight and amount of water intake. Control animals were given the solvent solution. Each group consisted of 20 animals.

Experiment #2: Effects of Melatonin and NAS on the Antioxidant Capacities of Aged C57Bl/6J Mice

To differentiate between the effects of melatonin and NAS, we used mice of C57Bl/6J strain, since this strain does not have enzymes for pineal melatonin biosynthesis from serotonin.[11,12] Therefore, the effects of NAS in C57Bl/6J mice could be studied without interference from melatonin converted from NAS and/or from the endogenously produced NAS. We used 11-month old mice since, having passed the midpoint of their life span, these animals were at a stage that is known to show age-associated decline in their antioxidant capacities.[13]

Male C57Bl/6J mice (Harlan Sprague-Dawley, Inc., Indianapolis, IN) were housed five per cage under 12/12 light/dark cycle with lights on/off at 0600/1800. NAS (20 mg/kg) and melatonin (1 mg/kg) were dissolved in 1% Tween-20 saline solution and administered daily, i.p., at 1200H for four weeks. Control animals were injected with the solvent solution. Each group consisted of eight animals.

Antioxidant activity was evaluated by measuring the malonaldehyde and 4-hydroxynonenal (MDA + 4-HNE) levels in tissues, expressed in micromol/mg protein, and from the cellular glutathion peroxidase (GPx) level expressed as mU/mg protein, using the Bioxytech LPO-586 kit and Biotech GPX-340 kit, respectively, (Oxis International, Portland, OR). Protein was determined by the Lowry method (Sigma Chem. Co., St. Louis, MO). Twenty-four hours after the last injection, animals were decapitated; brains, kidneys and livers were immediately removed, frozen in dry ice, and stored at −70°C until assayed. In addition, changes in body weight and quality of fur were noted as supplemental outcome measures. The results were obtained as mean values ± SEM and analyzed using one-way ANOVA and the Student's t-test. The level of significance was $p < 0.05$.

RESULTS

Experiment #1

Neither NAS nor melatonin affected the life span of female C3H mice (see FIGURE 1). However, melatonin did not cause premature death as is described elsewhere.[4] On the other hand, both NAS and melatonin prolonged the lifespan in male C3H mice by approximately 20% when compared to control animals ($p < 0.01$) (see FIGURES 2 and 3).

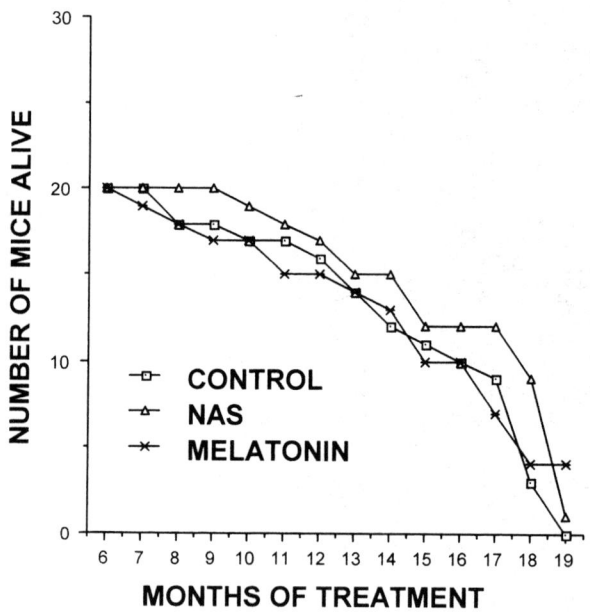

FIGURE 1. Effect of *N*-acetylserotonin and melatonin on life span in female C3H mice. NAS and melatonin were administered with drinking water at 2.5 mg/kg/day to male (starting at four weeks of age) and female (starting at about eight months of age) mice. Control animals were given the solvent solution. Each group consisted of 20 animals.

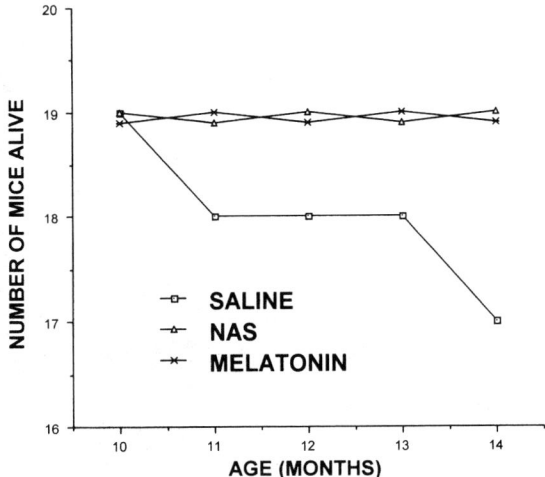

FIGURE 2. Effect of *N*-acetylserotonin and melatonin on life span in male C3H mice. NAS and melatonin were administered with drinking water at 2.5 mg/kg/day to male mice (starting at four weeks of age). Control animals were given the solvent solution. Each group consisted of 20 animals.

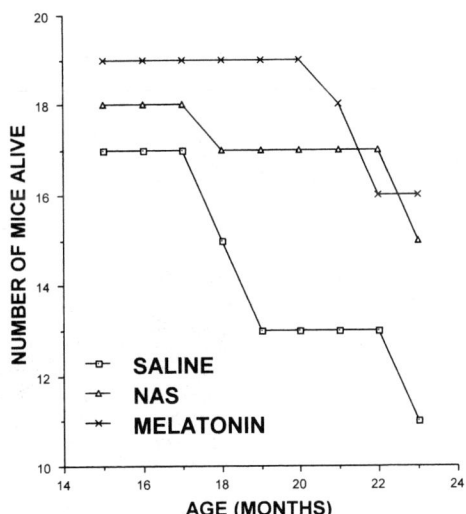

FIGURE 3. Effect of *N*-acetylserotonin and melatonin on life span in male C3H mice. NAS and melatonin were administered with drinking water at 2.5 mg/kg/day to male mice (starting at four weeks of age). Control animals were given the solvent solution. Each group consisted of 20 animals.

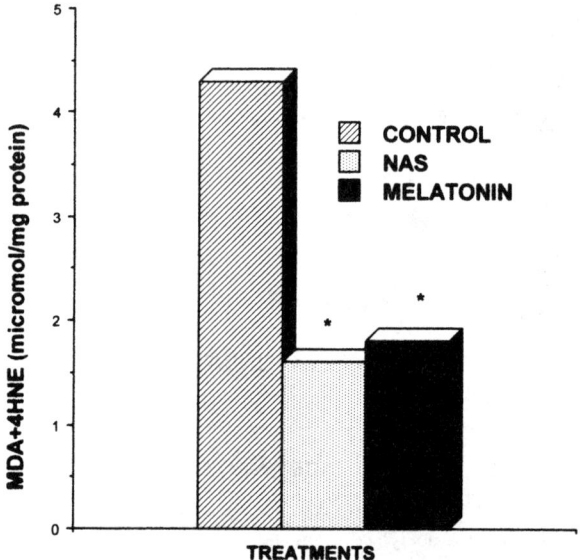

FIGURE 4. Effect of *N*-acetylserotonin and melatonin on malonaldehyde and 4-hydroxynonenal (MDA + 4-HNE) levels in C57BL/6J mouse brain tissue. NAS (20 mg/kg) and melatonin (1 mg/kg) were administered daily, i.p., at 1200 h for four weeks. Control animals were injected with the solvent solution. Each group consisted of eight animals, $*p < 0.01$ (ANOVA and Student's *t*-test).

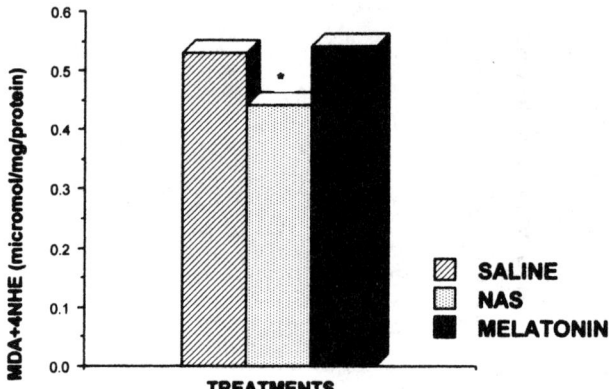

FIGURE 5. Effect of *N*-acetylserotonin and melatonin on malonaldehyde and 4-hydroxynonenal (MDA + 4-HNE) levels in C57BL/6J male mouse kidney. NAS (20 mg/kg) and melatonin (1 mg/kg) were administered daily, i.p., at 1200 h for four weeks starting at 11 months of age. Control animals were injected with the solvent solution. Each group consisted of eight animals, $*p < 0.03$ (ANOVA and Student's *t*-test).

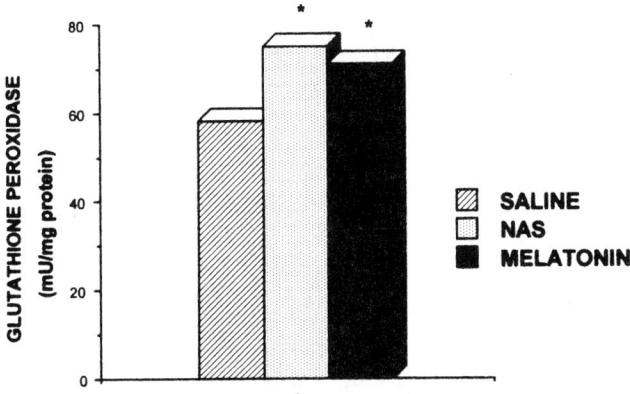

FIGURE 6. Effect of N-acetylserotonin and melatonin on cellular glutathion peroxidase levels in brain tissues of C57BL/6J male mice. NAS (20 mg/kg) and melatonin (1 mg/kg) were administered daily, i.p., at 1200 h for four weeks starting at 11 months of age. Control animals were injected with the solvent solution. Each group consisted of eight animals, $*p < 0.01$ (ANOVA and Student's t-test).

Experiment #2

There was an almost three-fold reduction in the MDA + 4-HNE formation in the brains of both NAS and melatonin treated mice, in comparison with the control group ($p < 0.01$), showing increased antioxidant capacity in the brain of NAS and melatonin treated animals (see FIGURE 4). Only the NAS treated animals showed

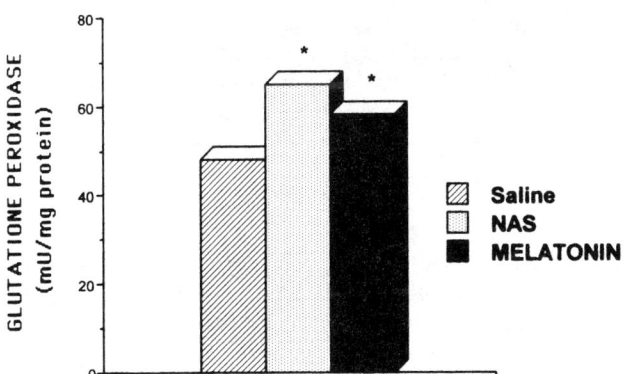

FIGURE 7. Effect of N-acetylserotonin and melatonin on cellular glutathion peroxidase levels in kidney tissues of vC57BL/6J male mice. NAS (20 mg/kg) and melatonin (1 mg/kg) were administered daily, i.p., at 1200 H for four weeks starting at 11 months of age. Control animals were injected with the solvent solution. Each group consisted of eight animals, $*p < 0.01$ (ANOVA and Student's t-test).

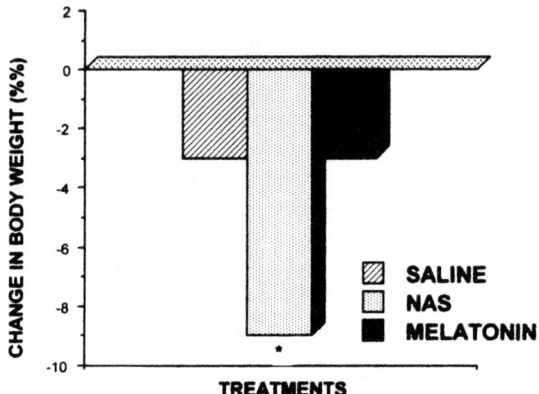

FIGURE 8. Effect of *N*-acetylserotonin and melatonin on body weight of C57BL/6J male mice. NAS (20 mg/kg) and melatonin (1 mg/kg) were administered daily, i.p., at 1200 h for four weeks starting at 11 months of age. Control animals were injected with the solvent solution. Data show the difference between body weights at the beginning and the end of a four-week drug administration period. Each group consisted of eight animals, *$p < 0.03$ (ANOVA and Student's *t*-test).

decrease in the MDA + 4HNE levels in the kidneys ($p < 0.03$) (see FIGURE 5). NAS and melatonin treated mice showed increased levels of cellular GRx in both brains and kidneys ($p < 0.01$) (see FIGURES 6 and 7). No difference between the groups studied was found in the liver (data not shown).

The NAS-treated animals revealed a weight loss of 9% ($p < 0.01$) from their respective baselines whereas the control and melatonin-treated groups had weight loss of only 3% (non-significant) (see FIGURE 8). Additional outcome measures noted: NAS-treated animals had the healthiest and most luxuriant fur coat among the groups; the animals in the control group had large areas of baldness; and, although the melatonin-treated animals did not have areas of baldness, the fur had more graying compared to the NAS-treated animals (see FIGURE 9).

DISCUSSION

The study we report shows, for the first time the antiaging effect of melatonin in male C3H mice, confirming previous similar data on BALB and C57Bl/6J strains.[3,4] The lack of the antiaging effect of melatonin (and NAS) in the female C3H mice may suggest gender differences in response to antiaging medication. However, the earlier commencement of the administration of pineal hormones in male (at one month of age) than in female (about eight months of age) mice might be also accountable for the difference.

The present study is the first demonstration of the antiaging effect of NAS. Considering that NAS is a biosynthetic precursor of melatonin, one might suggest that the antiaging effect of NAS depends on its conversion to melatonin. The pineals of

FIGURE 9. Effect of four weeks' administration of *N*-acetylserotonin and melatonin on C57BL/6J male mice. Note: NAS-treated animals with the healthiest and most luxuriant fur coat; the large areas of baldness in the control group; the coat of melatonin-treated mice had more graying fur than the NAS-treated animals.

the C3H mice do capable of converting NAS to melatonin.[14] However, this capacity is very limited in comparison to rat pineals. Efforts in our laboratory to induce the biosynthesis of melatonin in the pineals of C3H mice have been unsuccessful. The administration of NAS and selective monoamine oxidase inhibitor, clorgyline, agents known to increase the biosynthesis of melatonin in the rat pineals,[15,16] failed to induce melatonin biosynthesis in the pineals of C3H mice. NAS (30 mg/kg) administered to mice at the beginning of the dark cycle did not increase the melatonin content in the pineal (0.12 ± 0.02 ng/pineal for NAS group versus 0.09 ± 0.02 ng/pineal for control, 30 minutes after injection). Clorgyline (2.5 mg/kg) administered to light-primed mice also failed to induce melatonin biosynthesis in the pineal (0.05 ± 0.01 ng/pineal for clorgyline group versus 0.05 ± 0.01 for control group, 90 minutes after injection) (unpublished results). In agreement with the previous reports on antioxidant effects of NAS in the *in vitro* models[6–8] our study demonstrated the antioxidant effects of NAS proper in the brain and kidney tissues of the aged C57Bl/6J mice. Since the antioxidant effect might contribute to the antiaging action, one might suggest that NAS indeed has its own antiaging effect. Further evaluation of the NAS effect on life span conducted in mice unable to convert NAS into melatonin (i.e., C57Bl/6J strain) could help to differentiate between NAS proper and melatonin-mediated effects of NAS on aging.

REFERENCES

1. HARMAN, D. 1981. The aging process. Proc. Natl. Acad. Sci. U.S.A. **78:** 7124–7128.
2. REITER, R.J., M.I. PABLOS, T.T. AGAPITO & J.M. GUERRERO. 1996. Melatonin in the context of the free radical theory of aging. Ann. N.Y. Acad. Sci. **786:** 362–378.
3. PIERPAOLI, W. & G.J.M. MAESTRONI. 1987. Melatonin: a principal neuroimmunoregulatory and anti-stress hormone: its antiaging effects. Immunol. Lett. **16:** 355–362.
4. PIERPAOLI, W., A. DALL'ARA, E. PEDRINIS & W. REGELSON. 1991. The effects of melatonin and pineal grafting on the survival of older mice. Ann. N.Y Acad. Sci. **621:** 291–313.
5. SIU, A.W., R.J. REITER & C.H. TO. 1999. Pineal indoleamines and vitamin E reduce nitric oxide-induced lipid peroxidation in rat retinal homogenates. J. Pineal Res. **27:** 122–128.
6. LONGONI, B., W.A. PRYOR & P. MARCHIAFAVA. 1997. Inhibition of lipid peroxidation by N-acetylserotonin and its role in retinal physiology. Biochem. Biophys. Res. Comm. **233:** 778–780.
7. LEZOUALE'H, F., M. SPARAPANI & C. BEHL. 1998. N-acetyl-serotonin (normelatonin) and melatonin protect neurons against oxidative challenges and supress the activity of the transcription factor NF-kB. J. Pineal Res. **24:** 168–178.
8. BACHURIN, S., G. OXENKRUG, N. LERMONTOVA, *et al.* 1999. N-acetyl-serotonin, melatonin and their derivatives improve cognition and protect against β-amyloid-induced neurotoxicity. Ann. N.Y. Acad. Sci. **890:** 155–166.
9. LEONE, R.M. & R.E. SILMAN. 1984. Melatonin can be differentially metabolized in the rat to produce N-acetyl-serotonin in addition to 6-hydroxy-melatonin. Endocrinology **114:** 1825–1832.
10. YOUNG, I.M., R.M. LEONE, P. FRABCIS, *et al.* 1985. Melatonin is metabolized to N-acetyl-serotonin and 6-hydroxy-melatonin in man. J. Clin. Endocrinol. Metab. **60:** 114–119.
11. EBIHARA, S., T. MARKS, D.J. HUDSON & M. MENAKER. 1986. Genetic control of melatonin synthesis in the pineal gland of mouse. Science **231:** 491–493.

12. ROSEBOOM, P.A., M.A. ARYAN NAMBOODIRI, D.B. ZIMONJIC, *et al.* 1998. Natural melatonin "knockdown" in C57Bl/6J mice: rare mechanism truncates serotonin N-acetyltransferase. Molec. Brain Res. **63:** 189–197.
13. IMRE, J., J.H. FIRBAS & R.C. NOBLE. 2000. Reduced lipid peroxidation capacity and desaturation as biochemical markers of aging. Arch. Gerontol. Geriat. **31:** 5–12.
14. CONTI, A. & G.J.M. MAESTRONI. 1998. Endogenous melatonin in serum, pineal gland and bone marrow in inbred and outbred mice. Thymus and Pineal gland in Neuroimmunoendocrinology, Swieradow Zdrog 22B.
15. OXENKRUG, G.F. 1991. The acute effect of monoamine oxidase inhibitors on serotonin conversion to melatonin. *In* 5-hydroxytryptamine in Psychiatry: A Spectrum of Ideas. M. Sandler, A. Coppen & S. Harnett, Eds.: 98–109. Oxford University Press.
16. OXENKRUG, G.F. & P.J. REQUINTINA. 1994. Stimulation of rat pineal melatonin biosynthesis by N-acetylserotonin. Intern. J. Neurosci. **77:** 237–241.

Free Radical–Mediated Molecular Damage

Mechanisms for the Protective Actions of Melatonin in the Central Nervous System

RUSSEL J. REITER, DARIO ACUÑA-CASTROVIEJO, DUN-XIAN TAN, AND SUSANNE BURKHARDT

Department of Cellular and Structural Biology, The University of Texas Health Science Center at San Antonio, San Antonio, Texas, U.S.A.

ABSTRACT: This review briefly summarizes the multiple actions by which melatonin reduces the damaging effects of free radicals and reactive oxygen and nitrogen species. It is well documented that melatonin protects macromolecules from oxidative damage in all subcellular compartments. This is consistent with the protection by melatonin of lipids and proteins, as well as both nuclear and mitochondrial DNA. Melatonin achieves this widespread protection by means of its ubiquitous actions as a direct free radical scavenger and an indirect antioxidant. Thus, melatonin directly scavenges a variety of free radicals and reactive species including the hydroxyl radical, hydrogen peroxide, singlet oxygen, nitric oxide, peroxynitrite anion, and peroxynitrous acid. Furthermore, melatonin stimulates a number of antioxidative enzymes including superoxide dismutase, glutathione peroxidase, glutathione reductase, and catalase. Additionally, melatonin experimentally enhances intracellular glutathione (another important antioxidant) levels by stimulating the rate-limiting enzyme in its synthesis, γ-glutamylcysteine synthase. Melatonin also inhibits the proxidative enzymes nitric oxide synthase and lipoxygenase. Finally, there is evidence that melatonin stabilizes cellular membranes, thereby probably helping them resist oxidative damage. Most recently, melatonin has been shown to increase the efficiency of the electron transport chain and, as a consequence, to reduce election leakage and the generation of free radicals. These multiple actions make melatonin a potentially useful agent in the treatment of neurological disorders that have oxidative damage as part of their etiological basis.

KEYWORDS: Melatonin; Oxidative stress; Antioxidant; Free radicals; Neuroprotective mitochondria.

INTRODUCTION

Molecular abuse by free radicals and reactive oxygen species is common in the central nervous system and the resulting damage to neurons and glia is believed to contribute to a variety of neurodegenerative conditions.[1-5] Remarkably, the brain is

Address for correspondence: Russel J. Reiter, Department of Cellular and Structural Biology, Mail Code 7762, The University of Texas Health Science Center At San Antonio, 7703 Floyd Curl Drive, San Antonio, TX 78229-3900, U.S.A. Voice: 210/567-3859; fax: 210/567-6948.
Reiter@uthscsa.edu

readily plundered by reactive species for several reasons, including its high utilization of O_2 (the source of many damaging species) and its relatively weak antioxidative defense system. Since neurodegenerative changes and dementia are major and debilitating features of aging, curtailing the actions of free radicals and other reactants in the brain is of great importance in potentially alleviating or forestalling cognitive and mental decline in older individuals.

Agents that neutralize free radicals and reduce the molecular destruction that they mediate are referred to as antioxidants. There are a variety of endogenous and exogenous antioxidants that reduce free radical damage and preserve neuronal and glial integrity. Best known are the vitamin antioxidants, such as, tocopherol (vitamin E), ascorbic acid (vitamin C), and β-carotene. These agents directly interact with reactive species to neutralize them before they inflict damage. Additionally, a variety of antioxidative enzymes, for example, the superoxide dismutases (SOD), glutathione peroxidases (GPx), and catalase (CAT) metabolically convert reactive species to innocuous agents. The relative importance of each antioxidant varies with the organ, the cell, and the subcellular compartment that is under study since most antioxidants vary in terms of their concentrations in tissues and are limited in their subcellular distribution.

Melatonin, a secretory product of the vertebrate pineal gland, was recently shown to possess free radical scavenging activity.[6–8] Although the discovery of melatonin as a free radical scavenger and antioxidant occurred less than a decade ago,[9] there are already hundreds of reports that have documented its ability to reduce oxidative damage, not only in the brain, but in every organ where it has been tested.[10–12] Melatonin seems to be especially effective in restraining free radical damage in the brain.[5,13] This high efficacy may relate to the ease with which melatonin crosses the blood–brain barrier and to the fact that melatonin levels in the cerebrospinal fluid are orders of magnitude higher than in the blood.

This brief report summarizes the multitude of actions by which melatonin neutralizes reactive oxygen and nitrogen species and its effects in limiting their generation. The findings have obvious implications in neurobiological protection since a variety of neural disorders have free radical damage as a contributing factor. More extensive reviews of the actions of melatonin in reducing the severity of experimental neurological damage have been published.[2,5,13–15]

MELATONIN LEVELS IN BODILY FLUIDS AND TISSUES

Judgments in reference to tissue levels of melatonin are often made exclusively on the basis of blood levels of the indole. Blood concentrations of melatonin are typically in the mid-pM to low-nM range. It has become obvious, however, that levels of melatonin in other bodily fluids are not dictated by those in the serum. For example, it was recently noted that concentrations of melatonin in bile are several orders of magnitude in excess of those measured in blood.[16] Furthermore, there is evidence that biliary melatonin functions as an antioxidant to inhibit gallstone formation, a process believed to be mediated by free radicals.[17] The report of unusually high levels of melatonin in biliary fluid was followed shortly by one that confirmed earlier observations,[18,19] that similarly high concentrations of melatonin are found in ventricular cerebrospinal fluid (CSF).[20] Thus, the assumption that blood concentrations determine the levels of this indole in other bodily fluids is clearly erroneous.

As with fluid levels, cellular concentrations of melatonin are not necessarily uniform from one tissue to the next and there are certainly some cells that have very significantly higher levels than would occur if blood levels of melatonin were in equilibrium with those within cells. Thus, the measurement of melatonin in the gut[21–23] and bone narrow[24,25] has shown highly elevated concentrations of the indole; furthermore, removal of the pineal, an important source of circulating melatonin, does not markedly diminish melatonin concentrations in these tissues.[24,26] These results imply that melatonin is produced in a variety of organs other than the pineal gland. The evidence supporting this implication has recently been reviewed[27] and, in the case of bone marrow, Conti et al.[25] feel that the melatonin is produced locally. Consistent with this possibility are the observations that the mRNAs for the two enzymes that convert serotonin to melatonin are found in a large variety of cells, albeit in small amounts.[28] However, neither the associated activities of these enzymes nor the actual production of melatonin by these cells has yet been documented.

The uptake of melatonin by brain cells was shown by Menendez-Pelaez et al.[29] Shortly after the peripheral administration of melatonin, the indole appears in brain cells in high concentrations. Also, within cells melatonin may not be uniformly distributed. Thus, melatonin concentrations in cell nuclei may exceed those in the cytosol[29,30] and mitochondria have also been reported to contain exceptionally high levels of melatonin.[31] Indeed, recent evidence suggests mitochondria may be a site of action of melatonin.[32] Its differential distribution within cells and its higher levels in cells relative to blood melatonin concentrations may depend on the presence of binding proteins for the indole in certain cells and subcellular compartments.[33]

MELATONIN IN THE DETOXIFICATION OF REACTIVE SPECIES AND FREE RADICALS

A host of endogenously generated toxic agents damage and kill neurons and glia in the CNS; this mutilation is believed to account for at least some of the neural loss in age related neurodegenerative diseases.[1–5] Many of these toxic molecules are derived from the partial reduction of oxygen (O_2) (see FIGURE 1); these are referred to as reactive oxygen species (ROS) and, if they have an unpaired electron in their valence orbital, oxygen free radicals. Besides being derived from O_2, there are also reactive molecules that are nitrogen based, that is, reactive nitrogen species (RNS). Both these classes of toxic agents are illustrated in FIGURE 1.

The most devastatingly reactive free radical is generally considered to be the free radical that derives from the three electron reduction of O_2, namely, the hydroxyl radical (·OH). This species is unrelenting in its attack on molecules near where it is formed. The resulting mangled and defiled molecules—that may be proteins, lipid, DNA, RNA, or others—cause dysfunction in the affected cells and, in the worst case scenario, they kill cells via either apoptosis or necrosis.

Melatonin, like a variety of other antioxidants, is a highly effective scavenger of the ·OH. This observation, first reported in 1993, has been repeatedly confirmed using a large variety of techniques (see TABLE 1) that collectively, unequivocally document the ·OH scavenging activity of melatonin. The rate constants (that varied from 1.2×10^{10} to 0.6×10^{11} $M^{-1}s^{-1}$ in the reports summarized in TABLE 1) for the scavenging ·OH by melatonin are similar to those of other well known antioxidants.[50] The efficacy of melatonin in detoxifying the ·OH certainly contributes to its ability to reduce oxidative damage in the CNS, but precisely what percentage of melatonin's protection is a consequence of the melatonin–·OH interaction remains unknown. Besides neutralizing the ·OH in pure chemical systems, melatonin has been shown to act in this manner *in vivo*.[36,42] The product that is formed when melatonin detoxifies the ·OH *in vivo* is cyclic 3-hydroxymelatonin (see FIGURE 2), a metabolite that is excreted in the urine.[42]

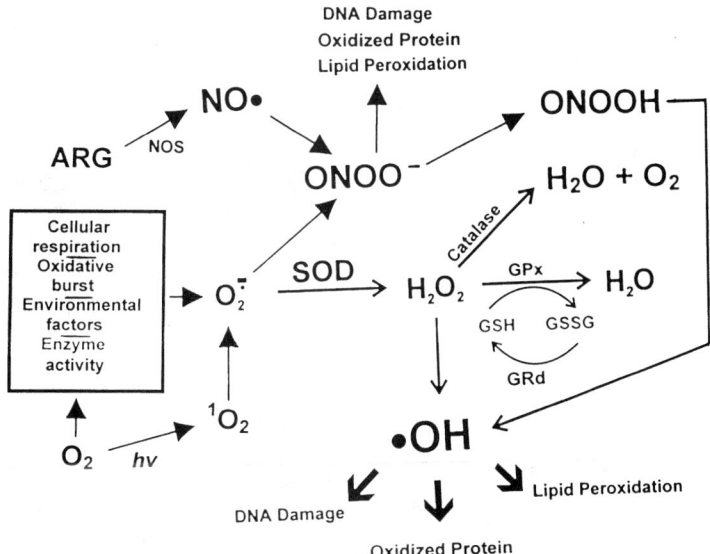

FIGURE 1. The partial reduction of oxygen (O_2) in cells generates a variety of reactive species and free radicals including the superoxide anion radical ($O_2^{-\cdot}$), hydrogen peroxide (H_2O_2) and the hydroxyl radical (·OH). Additionally, the addition of energy (hv) to O_2 produces another toxic species, singlet O_2 ($^1O_2$). The dismutation of $O_2^{-\cdot}$ to H_2O_2 usually requires one of a family of antioxidative enzymes, the superoxide dismutases (SOD). H_2O_2 is catalytically removed from cells by the enzymes catalase and glutathione peroxidase (GPx). Catalase activity is very low in most parts of the CNS and, therefore, it is generally not considered an important antioxidative enzyme in the brain. GPx uses H_2O_2 as a substrate in the reduction of glutathione (GSH) to its disulfide form (GSSG), which is metabolized back to GSH by glutathione reductase (GRd). Besides these oxygen-based toxicants, several nitrogen-based species can also be generated intracellularly. Although nitric oxide (NO·), which is catalytically formed from arginine (ARG) by nitric oxide synthase (NOS), has obvious beneficial effects, it also is toxic under some circumstances, for example, ischemia/reperfusion injury. NO· couples with $O_2^{-\cdot}$ to produce the highly reactive peroxynitrite anion (ONOO-). Besides its inherent toxicity, $ONOO^-$ degrades into another reactive intermediate, peroxynitrous acid (ONOOH). Finally, ONOOH may degrade to form the ·OH or a similarly reactive species.

TABLE 1. Publications that have documented the scavenging of the ·OH by melatonin[a]

Reference	Source of ·OH	Method of measurement
Tan et al.[9]	Photolysis of H_2O_2	Spin trapping and ESR
Poeggeler et al.[34]	Fenton reagents	Kinetic competition with ABTS[b]
Poeggeler et al.[35]	Fenton reagents	Reduction in melatonin
Li et al.[36]	Ischemia/reperfusion	Production of 2,3-dihydroxybenzoate
Matuszek et al.[37]	Fenton reagents	Spin trapping and ESR[c]
Susa et al.[38]	Chromium and H_2O_2	Spin trapping and ESR
Pähkla et al.[39]	Fenton reagents	Kinetic competition with terephthalic acid
Stasica et al.[40]	Pulse radiolysis of water	Absorption spectra of indolyl radical
Stasica et al.[41]	Pulse radiolysis of water	Absorption spectra of indolyl radical
Tan et al.[42]	Ionizing radiation	Formation of cyclic 3-hydroxymelatonin
Chyan et al.[43]	Ultraviolet photolysis of H_2O_2	Kinetic competition with ABTS or DMPO[d]
Mahal et al.[44]	Pulse radiolysis of water	Absorption spectra of indolyl radical
Bandyopadhyay et al.[45]	Cu^{2+}/ascorbate system	Kinetic study with methanesulfonic acid
Oosthuizen et al.[46]	Morsalt/H_2O_2 and nitrilotriacetic acid/$FeCl_3$/H_2O_2	Change in luminol chemiluminescence
Bromme et al.[47]	Glutathione/alloxan/Fe^{2+} system	Spin trapping and ESR
Kaneko et al.[48]	Ischemia/reperfusion	Production of 2,3-dihydrobenzoate
Khaldy et al.[49]	Iron-catalyzed oxidation of dopamine	Production of 2,3-dihydroxybenzoate

[a]This interaction has been verified using a variety of methods which are normally employed to confirm the actions of free radical scavengers. The reports are listed in the order of their appearance in the literature.
[b]ABTS, 2,2′-azuno-bis(c-ethylbenz-thiazoline-6-sulfonic acid).
[c]ESR, electron spin resonance spectroscopy.
[d]DMPO, 5,5-dimethyl-pyrroine-N-oxide.

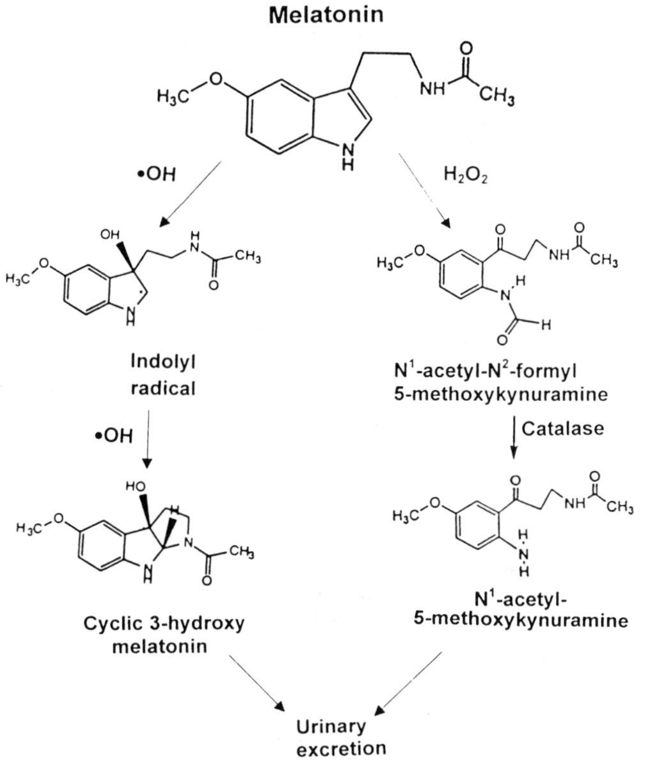

FIGURE 2. Pathways by which melatonin scavenges the ·OH and H_2O_2; also shown are the products that result from these interactions. Melatonin actually scavenges two ·OH to produce cyclic 3-hydroxymelatonin, the intermediate being the indolyl radical. When melatonin scavenges H_2O_2 the product is N'-acetyl-N^2-formyl-5-methoxykynuramine.

Although antioxidants that scavenge the ·OH are obviously important, preventing this free radical from being formed would avoid the possibility of it escaping detoxification, and mangling an adjacent molecule. As is illustrated in FIGURE 1, there are two enzymes, GPx and CAT, that catalyze the enzymatic degradation of H_2O_2. That melatonin may also interact with the H_2O_2 was first suggested by Zang and coworkers,[51] although the evidence they presented was not compelling. Subsequently, however, a more definitive documentation of this interaction was offered.[52] To verify melatonin's scavenging of H_2O_2, Tan et al.[52] measured both the depletion of H_2O_2 and melatonin and the accumulation of a product identified (by HPLC, GC-MS, and proton and carbon nuclear resonance) as N^1-acetyl-N^2-formyl-5-methoxykynuramine (AFMK) (FIG. 2). The rate constant for this reaction is 2.3×10^6 $M^{-1}s^{-1}$. AFMK itself is an effective scavenger. Thus, if the secondary and possibly the tertiary metabolites are equally as effective as melatonin in detoxifying reactive species, we feel that this cascade of reactions may make melatonin an especially important neutralizer of reactive toxicants.

The $O_2^{-\cdot}$ is not especially toxic although it obviously can be converted to more highly toxic species (FIG. 1). Although melatonin may not directly scavenge the $O_2^{-\cdot}$,[53,54] it may secondarily do so.[6]

$^1O_2$ is a high energy form of O_2 with significant potential to damage macromolecules. This activated form of O_2 has been implicated in pathologies that are aggravated by high intensity light. Melatonin is believed to scavenge $^1O_2$,[34,51] which is consistent with the ability of indole to reduce neural cytotoxicity of rose bengal, a photosensitive dye, in cerebellar cells exposed to bright light.[55]

NO·, a nitrogen-centered molecule, is widely produced in the brain and exhibits toxicity in some neurological disorders; for example, ischemia/reperfusion injury. In a cell-free system, melatonin scavenged NO· more effectively than either N-acetylserotonin or 5-hydroxytryptophan.[56] There have been no tests of this interaction in vivo.

NO· couples with $O_2^{-\cdot}$ to produce a nitrogen-centered toxicant referred to as the peroxynitrite anion ($ONOO^-$). This molecule is sufficiently reactive as to damage the major categories of macromolecules (i.e., lipids, proteins, and DNA) within cells (FIG. 1). Gilad and coworkers[57] were the first to show that melatonin scavenged this reactant in vitro, and subsequent studies suggested melatonin also had this action in vivo since it prevented the toxicity of agents believed to generate the $ONOO^-$.[58,59] In these studies, the authors found that melatonin reduced the amount of immunocytochemically identified nitrotyrosine, a product found when $ONOO^-$ nitrates tyrosine. Substantial support for the $ONOO^-$/melatonin interaction was recently provided by Blanchard et al.[6] They showed, in a phosphate buffered solution, that melatonin reacts with the $ONOO^-$ causing nitrosation and oxidation of the pyrrole nitrogen, leading to the formation of 1-nitrosomelatonin and 1-hydroxymelatonin. The kinetics of these biotransformations were strictly dependent on the decay of $ONOO^-$.

That melatonin reacts with nitrogen-centered reactants was also demonstrated in the elegant studies of Zhang and colleagues.[60,61] Again, using a pure chemical assay, they found that melatonin was capable of scavenging $ONOO^-$, but at physiological pH the indole was more reactive toward peroxynitrous acid (ONOOH) (FIG. 1) or an activated form of that agent, namely, *ONOOH. They also succeeded in identifying the product that was formed from this interaction. Interestingly it is 6-hydroymelatonin; this is the same metabolite produced in the liver when melatonin is enzymatically degraded.

That melatonin is capable of detoxifying reactive oxygen and nitrogen species is certainly supported by the evidence summarized herein. Although the bulk of these studies have been performed in vitro,[62] when similar experiments were conducted in vivo (e.g., in the case of ·OH), the outcomes were similar.[36,42,45,63] Thus, it is reasonable to conclude that melatonin plays an important direct scavenging role in defense against oxygen and nitrogen based toxicants.

Another highly destructive radical that is generated during the peroxidation of polyunsaturated fatty acids is the peroxyl radical (LOO·). Initial reports claiming melatonin was more effective than vitamin E in neutralizing LOO·[64] are not supported by the outcome of more recent studies.[65] Thus, whereas melatonin is a powerful inhibitor of in vivo lipid peroxidation in the brain[5,13,66] and elsewhere,[10,67,68] it probably does so by scavenging the initiating radicals, such as, ·OH and $ONOO^-$, rather than functioning as a chain-breaking antioxidant in the peroxidative cascade.

THE EFFECTS OF MELATONIN ON ENZYMES RELATED TO OXIDATIVE STRESS

Enzymes play an important role in maintaining steady state levels of ROS. The major catalytic factors that metabolically remove reactive species before they plunder nearby molecules include SOD, GPx, and CAT (see FIGURE 3). As well as using H_2O_2 (and other hydroperoxides) as substrates in a reaction that oxidizes GSH to GSSG, GPx also functions as a peroxynitrite reductase. Once GSSG is generated by the action of GPx it is converted back to GSH by the antioxidative enzyme, GRd. When the mRNAs and/or the activities of SOD, GPx, CAT, and GRd were measured after the *in vivo* administration of melatonin, each was found to be elevated.[69–71] Additionally, in the brain and in other organs as well, there are night time increases in the activities of GPx and GRd; these rises were shown to depend on the concurrent nocturnal increase in endogenous melatonin levels.[72,73] Furthermore, inhibiting the night time increase in melatonin also severely blunted rises in the activities of the antioxidative enzymes, indicating that the rhythms in the activities of these key antioxidative enzymes are driven by endogenous melatonin levels.

GSH itself is present in very high concentrations (in the mM range) in most cells and is a critical molecule in defense against oxidative stress. In a recent report, Urata and colleagues[74] found that the rate limiting enzyme in GSH synthesis, that is, γ-glutamylcysteine synthase, is stimulated by melatonin as are the levels of GSH; maintaining GSH levels in cells may be an important action by which melatonin assists cells in coping with oxidative terrorism.

There are also proxidative enzymes whose activity leads to the generation of free radicals (FIG. 3). At least two of these (lipoxygenase[75] and nitric oxide synthase[76,77]) have been shown to be inhibited by melatonin, both at physiological and pharmacological levels. Reduction in the activities of these enzymes would function in reducing oxidative stress by lowering the production of reactive toxicants.

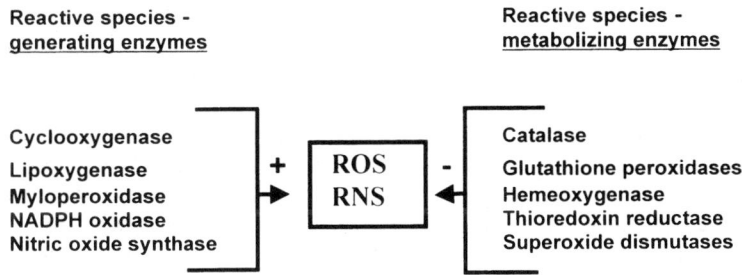

FIGURE 3. Enzymes that either cause (+) the generation of toxic reactive oxygen (ROS) or reactive nitrogen species (RNS) or that remove (−) these agents from cells by metabolizing them to innocuous products. Steady state levels of these toxicants are controlled by these antioxidative (metabolizing) and proxidative (generating) enzymes. Several antioxidative enzymes are stimulated by melatonin whereas some of the proxidative enzymes are inhibited by the indole.

MELATONIN AND MEMBRANE FLUIDITY

The membranes of cells are normally maintained in a viscous state. When the fluidity of the membranes is altered, for example, if they become more rigid, the processes that depend on an optimal state of fluidity, such as on channel and receptor functions, are accordingly altered. In advanced age it is generally known that cell membranes become increasingly rigid, thereby negatively impacting on the cells.

One of the factors, although there are others, that cause cell membranes to become more rigid is the peroxidation of the lipids they contain. Garcia et al.[78,79] have shown that membrane rigidity is prevented when cells are exposed to agents that normally cause lipid peroxidation. In most cases the reduction of lipid breakdown correlated with the loss of membrane fluidity, although this was not always the case. Melatonin was shown to prevent both lipid breakdown and the increased rigidity of the membrane. It is possible that melatonin has effects that optimize the fluidity of cellular membranes independent of the state of the lipids they contain. This feature could assist melatonin in helping cells resist free radical damage. This is an area of research that warrants more intensive investigation.

MELATONIN AND MITOCHONDRIAL PHYSIOLOGY

The mitochondrion, an organelle of eukaryotic cells that originated from bacterial endosymbionts, is mainly responsible for energy generation through oxidative phosphorylation, yielding ATP required for cell function.[80] As a consequence of the oxygen consumption by the respiratory chain, mitochondria produce large amounts of $O_2^{-\cdot}$ and H_2O_2. The formation of these reactive species must be controlled in order to avoid their detrimental effects on the electron transport chain (ETC). Oxidative stress leads to respiratory chain dysfunction and, in some instances, may increase permeability transition inducing cell death. Apoptosis and mitochondrial permeability transition are two regulated and closely related events. When mitochondrial dysfunction becomes irreversible, permeability transition pores are opened releasing proapoptotic factors, such as cytochrome c and apoptosis-inducing factor (AIF), into the cytosol. These agents activate proapoptotic nuclear signaling through the activation of caspases.[81] Thus, free radicals, mitochondrial dysfunction, and ATP production are closely related to apoptosis and cell death.

Under ATP demands by the cell, increased activity of oxidative phosphorylation parallels increased oxygen consumption and ROS generation. To avoid oxidative damage to the ETC, mitochondria remove $O_2^{-\cdot}$ via SOD and H_2O_2 through glutathione redox cycling; mitochondria, however, lack CAT. The consequence is a decrease in reduced GSH, which must be recycled to maintain mitochondrial protection.[82]

The antioxidant activity of melatonin and its potentiation of the redox enzymes make this molecule unique in terms of its antioxidant activity. These features of melatonin, and its ubiquitous distribution within cells, suggest a role in mitochondria, where the production of ROS is higher than in any other organelle. *In vivo* experiments have shown that melatonin administration increases the activity of the oxidative mitochondrial complexes I and IV in liver and brain and counteracts the toxicity of ruthenium red on the ETC.[83]

To further characterize these effects, a series of *in vitro* studies were performed using mitochondria experiencing oxidative stress due to exposure to *t*-butyl-hydroperoxide (*t*-BHP). Nanomolar concentrations of melatonin significantly increased the GSH content and the activities of both GPx and GRd in control mitochondria; these effects were accompanied by significant increases in the activities of complexes I and IV of the mitochondrial ETC. After incubation with *t*-BHP virtually all GSH was oxidized and the activities of GPx and GRd were totally inhibited. The presence of melatonin reversed the oxidative damage induced by *t*-BHP, and the mitochondria recovered their normal enzyme activities. Interestingly, neither vitamin E nor vitamin C was able to prevent mitochondrial oxidative damage at doses up to 1 mM, 100–1000 times those of melatonin.[84]

From these results, several conclusions may be drawn. Melatonin counteracts the excess of ROS produced during mitochondrial activity and scavenges those that may damage the ETC (see FIGURE 4). Melatonin also maintains high levels (more than 85%) of mitochondrial GSH, reducing the requirement for these organelles to take up GSH from the cytosol. Thus, the indoleamine preserves the ability of mitochondria to produce the required ATP for cell function. In addition to maintaining mitochondrial GSH homeostasis, melatonin stabilizes GSH levels sufficient to prevent transition pores from opening. These effects reflect a protective role of melatonin on cellular physiology. Due to the relation between damaged mitochondria, aging and ROS, these effects of melatonin may also help to explain the antiapoptotic and antiaging properties of melatonin.

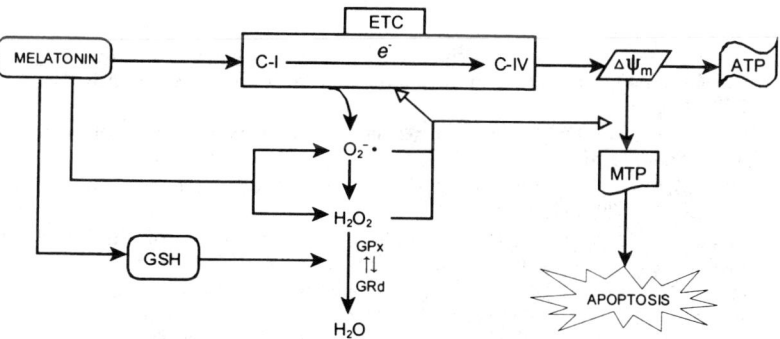

FIGURE 4. Effects of melatonin on mitochondrial homeostasis. Melatonin increases the activity of complex I (C-I) and IV (C-IV) of the electron transport chain (ETC), which in turn maintains mitochondrial membrane potential ($\Delta\psi_m$) and ATP synthesis. O_2^- and H_2O_2 generated in the mitochondria are scavenged by SOD and GPx, respectively. These ROS may damage the ETC, decrease the $\Delta\psi_m$, and induce mitochondrial transition permeability (MTP). This in turn leads to release of proapoptotic signals. Melatonin acts by increasing the activity of C-I and C-IV of the ETC, thereby improving ATP synthesis. Additionally, melatonin scavenges ROS directly and indirectly by increasing the activity of antioxidative enzymes.

CONCLUDING REMARKS

The evidence summarized herein indicates melatonin has at its disposal a myriad of functions by means of which it restrains oxidative damage in the brain. Besides its direct scavenging actions, it indirectly reduces molecular mangling and malfunction by promoting the removal of toxic and potentially toxic agents by stimulating antioxidative enzymes and inhibiting proxidative enzymes. As an additional contribution to the repertoire by which melatonin indirectly limits oxidative destruction, it may stabilize phospholipid interactions in cellular membranes and limit free radical generation at the mitochondrial level by increasing the efficiency of the ETC. Finally, there may be other beneficial actions of melatonin that remain to be uncovered.

It is certainly well documented that melatonin protects a variety of macromolecules from the abuse normally inflicted by reactive oxygen or reactive nitrogen species. Thus, although its suppressive effects on lipid peroxidation are most frequently described,[10,85–88] it seems equally effective in reducing DNA[10,86,89–91] and protein[92–94] destruction.

Considering the ease with which melatonin crosses the blood–brain barrier and its ready uptake by both neurons and glia, the findings described in this report clearly indicate that melatonin may be an effective agent for the treatment of neurological disorders and conditions that involve free radical damage. In fact, it has been tested in a wide variety of experimental conditions that involve toxic damage to neurons or glia *in vitro*[14,95] or the CNS *in vivo*.[3,5,13,15,36,88] Finally, melatonin seems to lack significant toxicity[96–98] and preliminary evidence in humans suggests that it may be protective against neurodegenerative disorders.[99]

REFERENCES

1. YOUDIM, M.B.H., D. BEN-SHACBAR, G. ESHEL, *et al.* 1993. The neurotoxicity of iron and nitric oxide: relevance to the etiology of Parkinson's disease. Adv. Neurol. **60:** 259–266.
2. REITER, R.J. 1995. Oxidative processes and antioxidative defense mechanisms in the aging brain. FASEB J. **9:** 526–533.
3. SMITH, M.A., G. PERRY, P.L. RICHEY, *et al.* 1996. Oxidative damage in Alzheimer's. Nature **382:** 120–121.
4. DALKARA, T. & M.A. MOSCHOWITZ. 1994. The complex role of nitric oxide in the pathophysiology of focal cerebral ischemia. Brain Pathol. **4:** 49–57.
5. REITER, R.J. 1998. Oxidative damage in the central nervous system: Protection by melatonin. Prog. Neurobiol. **56:** 359–384.
6. HARDELAND, R., R.J. REITER, B. POEGGELER & D.X. TAN. 1993. The significance of the metabolism of the neurohormone melatonin: antioxidative protection and formation of bioactive substances. Neurosci. Biobehav. Rev. **17:** 347–357.
7. BLANCHARD, B., D. POMPON & C. DUCROCQ. 2000. Nitrosation of melatonin by nitric oxide and peroxynitrite. J. Pineal Res. **29:** 184–192.
8. REITER, R.J. 2000. Melatonin: lowering the high price of free radicals. News Physiol. Sci. **15:** 256–250.
9. TAN, D.X., L.D. CHEN, B. POEGGELER, *et al.* 1993. Melatonin: a potent, endogenous hydroxyl radical scavenger. Endocrine J. **1:** 57–60.
10. REITER, R.J., L. TANG, J.J. GARCIA & A. MUÑOZ-HOYOS. 1997. Pharmacological actions of melatonin in oxygen radical pathophysiology. Life Sci. **60:** 2255–2271.

11. TAN, D.X., L.C. MANCHESTER, R.J. REITER, et al. 2000. Significance of melatonin in antioxidant defense system: reactions and products. Biol. Signals Recept. **9:** 137–159.
12. REITER, R.J., D.X. TAN, W. QI, et al. 2000. Pharmacological and physiology of melatonin in the reduction of oxidative stress in vivo. Biol. Signals Recept. **9:** 160–171.
13. REITER, R.J., J.J. GARCIA & J. PIE. 1998. Oxidative toxicity in models of neurodegeneration: responses to melatonin. Rest. Neurol. Neurosci. **12:** 135–142.
14. PAPPOLLA, M.A., Y.J. CHYAN, B. POEGGELER, et al. 2000. An assessment of the antioxidant and antiamyloidogenic properties of melatonin: implications for Alzheimer's disease. J. Neural Transm. **107:** 203–231.
15. REITER, R.J., J. CABRERA, R.M. SAINZ, et al. 1999. Melatonin as a pharmacological agent against neuronal loss in experimental models of Huntington's disease, Alzheimer's disease and Parkinsonism. Ann. N.Y. Acad. Sci. **890:** 471–484.
16. TAN, D.X., L.C. MANCHESTER, W. QI, et al. 1999. High physiological levels of melatonin in the bile of mammals. Life Sci. **65:** 2523–2529.
17. SHIESH, S.C., C.Y. CHEN, X.Z. LIN, et al. 2000. Melatonin prevents pigment gallstone formation induced by bile duct ligation in guinea pigs. Hepathology **32:** 455–460.
18. HEDLUND, L., M.M. LISCHKO, M.D. ROLLAG & G.D. NISWENDER. 1977. Melatonin: daily cycle in plasma and cerebrospinal fluid. Science **195:** 686–687.
19. KANEMATSU, N., Y. MORI, S. HAYASHI & K. HOSHINO. 1989. Presence of a distinct 24-hour melatonin rhythm in the ventricular spinal fluid of the goat. J. Pineal Res. **7:** 143–152.
20. SKINNER, D.C. & B. MALPAUX. 1999. High melatonin concentrations in third ventricular cerebrospinal fluid are not due to Galen vein blood recirculating through the choroid plexus. Endocrinology **140:** 4399–4405.
21. HUETHER, G., B. POEGGELER, A. REINER & A. GEORGE. 1992. Effect of tryptophan administration on circulating melatonin levels in chicks and rats: Evidence for stimulation of melatonin synthesis and release in the gastrointestinal tract. Life Sci. **51:** 945–953.
22. HUETHER, G. 1993. The contribution of extrapineal sites of melatonin synthesis to circulating melatonin levels in higher vertebrates. Experientia **48:** 665–670.
23. SAARELA, S., M. VUORI, E. ELORANTA & O. VAKKURI. 1997. Melatonin, a candidate signalling molecule for energy sparing. Ornis Fenn. **76:** 231–235.
24. TAN, D.X., L.C. MANCHESTER, R.J. REITER, et al. 1999. Identification of highly elevated levels of melatonin in bone marrow: Its origin and significance. Biochim. Biophys. Acta **1472:** 206–214.
25. CONTI, A., S. CONICONI, E. HERTEN, et al. 2000. Evidence for melatonin synthesis in mouse and human bone marrow. J. Pineal Res. **28:** 193–202.
26. BUBENIK, G.A. 1980. Localization of melatonin in the digestive tract of the rat. Effect of maturation, diurnal variation, melatonin treatment and pinealectomy. Horm. Res. **6:** 313–323.
27. KVETNOY, I. 1999. Extrapineal melatonin: Location and role within diffuse neuroendocrine system. Histochem. J. **31:** 1–12.
28. STEFULJ, J., M. HOERTNER, M. GHOSH, et al. 2001. Gene expression of the key enzymes of melatonin synthesis in extrapineal tissues of the rat. J. Pineal Res. **30:** 243–247.
29. MENENDEZ-PELAEZ, A., B. POEGGELER, R.J. REITER, et al. 1993. Nuclear localization of melatonin in different mammalian tissues: immunocytochemical and radioimmunoassay evidence. J. Cell. Biochem. **51:** 373–382.
30. MENENDEZ-PELAEZ, A. & R.J. REITER. 1993. Distribution of melatonin in mammalian tissues: the relative importance of nuclear verses cytosolic localization. J. Pineal Res. **15:** 59–69.
31. MARTIN, M., M. MACIAS, G. ESCAMES, et al. 2000. Melatonin but not vitamins C and E maintain glutathione homeostasis in t-butyl hydroperoxide-included mitochondrial oxidative stress. FASEB J. **14:** 1677–1679.
32. ACUÑA-CASTROVIEJO, D., M. MARTIN, M. MACIAS, et al. 2001. Melatonin, mitochondria and cellular bioenergetics. J. Pineal Res. **30:** 65–74.
33. LAHIRI, D.K. 1999. Melatonin affects metabolism of the β-amyloid precursor protein in different cell types. J. Pineal Res. **26:** 137–146.

34. POEGGELER, B., S. SAARELA, R.J. REITER, *et al.* 1994. Melatonin—a highly potent endogenous radical scavenger and electron donor: new aspects of oxidation chemistry of this indole assessed *in vitro*. Ann. N.Y. Acad. Sci. **738**: 419–420.
35. POEGGELER, B., R.J. REITER, R. HARDELAND, *et al.* 1996. Melatonin and structurally related endogenous indoles act as potent electron donors and free radical scavengers *in vitro*. Redox Rept. **2**: 179–184.
36. LI, X.J., L.M. ZHANG, J. GU, *et al.* 1997. Melatonin decreases production of hydroxyl radical during ischemia-reperfusion. Acta Pharmacol. Sinica **18**: 394–396.
37. MATUSZEK, Z., K.J. RESZKA & C.F. CHIGNELL. 1997. Reaction of melatonin and related indoles with hydroxyl radicals: ESR and spin trapping investigations. Free Rad. Biol. Med. **23**: 267–272.
38. SUSA, N., S. UENO, Y. FURUKAWA, *et al.* 1997. Potent protective effect of melatonin on chromium(VI) induced DNA single strand breaks, cytotoxicity, and lipid peroxidation in primary cultures of rat hepatocytes. Toxicol. Appl. Pharmacol. **144**: 373–384.
39. PÄHKLA, R., M. ZILMER, T. KULLISAR & L. RÄGO. 1998. Comparison of the antioxidant activity of melatonin and pinoline *in vitro*. J. Pineal Res. **24**: 96–101.
40. STASICA, P., P. ULANSKI & J.M. ROSIAK. 1998. Melatonin as a hydroxyl radical scavenger. J. Pineal Res. **25**: 65–66.
41. STASICA, P., P. ULANSKI & J.M. ROSIAK. 1998. Reaction of melatonin with radicals in deoxygenated aqueous solutions. J. Radioanal. Nucl. Chem. **232**: 107–113.
42. TAN, D.X., L.C. MANCHESTER, R.J. REITER, *et al.* 1998. A novel melatonin metabolite, cyclic 3-hydroxymelatonin: a biomarker of *in vivo* hydroxyl radical generation. Biochem. Biophys. Res. Commun. **253**: 614–620.
43. CHYAN, Y.J., B. POEGGELER, R.A. OMAR, *et al.* 1999. Potent neuroprotective properties against Alzheimer β-amyloid by an endogenous melatonin-related indole structure, indole-3-propionic acid. J. Biol. Chem. **274**: 21937–21942.
44. MAHAL, H.S., H.S. SHARMA & T. MUKHERJEE. 1998. Antioxidant properties of melatonin: a pulse radiolysis study. Free Radical Biol. Med. **26**: 557–565.
45. BANDYOPADHYAY, D., K. BISWAS, U. BANDYOPADHYAY, *et al.* 2000. Melatonin protects against stress-induced gastric lesions by scavenging the hydroxyl radical. J. Pineal Res. **25**: 143–151.
46. OOSTHUIZEN, M.M.J. & D. GREYLING. 1999. Antioxidants suitable for use with chemiluminescence to identify oxyradical species. Redox Rept. **4**: 277–290.
47. BRÖMME, H.J., W. MÖRKE, E. PESCHKE, *et al.* 2000. Scavenging effect of melatonin on hydroxyl radical generated by alloxan. J. Pineal Res. **29**: 201–208.
48. KANEKO, S., K. OKUMURA, Y. DUMAGUCHI, *et al.* 2000. Melatonin scavenges hydroxyl radical and protects isolated rat hearts from ischemia reperfusion injury. Life Sci. **67**: 101–112.
49. KHALDY, H., G. ESCAMES, J. LEON, *et al.* 2000. Comparative effects of melatonin, L-deprenyl, Trolox and ascorbate in the suppression of hydroxyl radical formation during dopamine autoxidation *in vivo*. J. Pineal Res. **29**: 100–107.
50. REITER, R.J., D.X. TAN, L.C. MANCHESTER & W. QI. 2001. Biochemical reactivity of melatonin with reactive oxygen and nitrogen species: a review of the evidence. Cell Biochem. Biophys. In press.
51. ZANG, L.Y., G. COSMA, H. GARDNER & V. VALLYNATHAN. 1998. Scavenging of reactive oxygen species by melatonin. Biochim. Biophys. Acta **1425**: 469–477.
52. TAN, D.X., L.C. MANCHESTER, R.J. REITER, *et al.* 2000. Melatonin directly scavenges hydrogen peroxide: A potentially new metabolic pathway of melatonin biotransformation. Free Radical Biol. Med. **29**: 1177–1185.
53. MARSHALL, K.A., R.J. REITER, B. POEGGELER, *et al.* 1996. Evaluation of the antioxidant activity of melatonin in vivo. Free Radical Biol. Med. **21**: 307–315.
54. CHAN, T.Y. & P.L. TANG. 1996. Characterization of the antioxidant effects of melatonin and related indoleamines *in vitro*. J. Pineal Res. **20**: 187–191.
55. CAGNOLI, C.M., C. ATABAY, E. KHARLAMOVA & H. MANEV. 1995. Melatonin protects neurons from singlet oxygen-induced apoptosis. J. Pineal Res. **18**: 222–226.
56. NODA, Y., A. MORI, R. LIBURTY & L. PACKER. 1999. Melatonin and its precursors scavenge nitric oxide. J. Pineal Res. **27**: 159–163.

57. GILAD, E., S. CUZZOCREA, B. ZINGARELLI, et al. 1997. Melatonin as a scavenger of peroxynitrite. Life Sci. **60:** PL169–PL174.
58. CUZZOCREA, S., D.X. TAN, G. COSTANTINO, et al. 1999. The protective role of melatonin in carrageen-induced pleurisy in rat. FASEB J. **13:** 1930–1938.
59. EL-SOKKARY, G.H., R.J. REITER, S. CUZZOCREA, et al. 1999. Role of melatonin in reduction of lipid peroxidation and peroxynitrite formation in non-septic shock induced by zymosan. Shock **12:** 402–408.
60. ZHANG, H., G.L. SQUADRITO & W.A. PRYOR. 1998. The reaction of melatonin with peroxynitrite: formation of melatonin radical cation and absence of stable nitrated products. Biochem. Biophys. Res. Commun. **251:** 83–87.
61. ZHANG, H., G.L. SQUADRITO, R. UPPI & W.A. PRYOR. 1999. Reaction of peroxynitrite with melatonin: A mechanistic study. Chem. Res. Toxicol. **12:** 526–534.
62. LIVREA, M.A., L. TESORIERE, D.X. TAN & R.J. REITER. 2001. Radical and reactive intermediate scavenging properties of melatonin in pure chemical systems. *In* Handbook of Antioxidants. E. Cardenas & L. Packer, Eds.: Marcell Dekker, New York. In press.
63. QI, W., R.J. REITER, D.X. TAN, et al. 2000. Increased levels of oxidatively damaged DNA induced by chromium (III) and H_2O_2: Protection by melatonin and related molecules. J. Pineal Res. **29:** 54–61.
64. PIERI, C., F. MORONI, M. MARRA, et al. 1995. Melatonin as an efficient antioxidant. Arch. Gerontol. Geriatrics **20:** 159–165.
65. ANTUNES, F., L.R.C. BARCLAY, K.U. INGOLD, et al. 1999. On the antioxidant activity of melatonin. Free Radical Biol. Med. **26:** 117–128.
66. CUZZOCREA, S., G. COSTANTINO, E. GITTO, et al. 2000. Protective effects of melatonin in ischemic brain injury. J. Pineal Res. **29:** 217–227.
67. OHTA, Y., M. KONGO, E. SASAKI, et al. 2000. Protective effect of melatonin against α-naphthlisocyanate-induced liver injury in rats. J. Pineal Res. **29:** 15–23.
68. SAILAJA DEVI, M.M., Y. SURESH & U.N. DAS. 2000. Preservation of the antioxidant status in chemically induced diabetes mellitus by melatonin. J. Pineal Res. **29:** 108–115.
69. PABLOS, M.I., M.T. AGAPITO, R. GUTIERREZ, et al. 1995. Melatonin stimulates the activity of the detoxifying enzyme glutathione peroxidase in several tissues of chicks. J. Pineal Res. **19:** 111–115.
70. ANTOLIN, I., C. RODRIQUEZ, R.M. SAINZ, et al. 1996. Neurohormone melatonin prevents cell damage: effect on gene expression for antioxidative enzymes. FASEB J. **10:** 882–890.
71. MONTILLA LOPEZ, P., I. TUNEZ-FINAÑA, C. MUÑOZ DE AGUEDA, et al. 2000. Protective effect of melatonin against oxidative stress induced by ligature of extra-hepatic biliary duct in rats: comparison with the effect of S-adenosyl-L-methionine. J. Pineal Res. **28:** 143–149.
72. PABLOS, M.I., R.J. REITER, G.G. ORTIZ, et al. 1998. Rhythms of glutathione peroxidase and glutathione reductase in brain of chick and their inhibition by light. Neurochem. Int. **32:** 69–75.
73. ALBARRAN, M.T., S. LOPEZ-BURILLO, M.I. PABLOS, et al. 2001. Endogenous rhythms of melatonin, total antioxidant status and superoxide dismutase activity in several tissues of chick and their inhibition by light. J. Pineal Res. **30:** 227–233.
74. URATA, Y.S., S. HONMA, S. GOTO, et al. 1999. Melatonin induces γ-glutamylcysteine synthase mediated by activator protein-1 in human vascular endothelial cells. Free Rad. Biol. Med. **27:** 838–847.
75. CARLBERG, C. & I. WIESENBERG. 1995. The orphan receptor family RZR/ROR, melatonin and lipoxygenase: an unexpected relationship. J. Pineal Res. **18:** 171–178.
76. POZO, D., R.J. REITER, J.M. CALVO & J.M. GUERRERO. 1997. Inhibition of cerebellar nitric oxide synthase and cyclic GMP production by melatonin via complex formation with calmodulin. J. Cell. Biochem. **65:** 430–432.
77. GILAD, E., H.R. WONG, B. ZINGARELLI, et al. 1998. Melatonin inhibits expression of the inducible isoform of nitric oxide synthase in murine macrophages: role in inhibition of NF-κB activation. FASEB J. **12:** 685–693.

78. GARCIA, J.J., R.J. REITER, J.M. GUERRERO, et al. 1997. Melatonin prevents changes in microsomal membrane fluidity during induced lipid peroxidation. FEBS Lett. **408:** 297–300.
79. GARCIA, J.J., R.J. REITER, J. PIE, et al. 1999. Role or pinoline and melatonin in stabilizing hepatic microsomal membranes against oxidative stress. J. Bioenerg. Biomembr. **31:** 609–616.
80. SKULACHEV, P.C. 1999. Mitochondrial physiology and pathology: concepts of programmed death of organelles, cells and organisms. Mol. Aspects Med. **20:** 139–184.
81. SUSIN, S.A., N. ZAMZAMI & G. KROEMER. 1998. Mitochondria as regulators of apoptosis: doubt no more. Biochim. Biophys. Acta **1366:** 151–165.
82. NICHOLLS, D.G. & S.L. BUDD. 2000. Mitochondria and neuronal survival. Physiol. Rev. **80:** 315–360.
83. MARTIN, M., M. MACIAS, G. ESCAMES, et al. 2000. Melatonin-induced increased activity of the respiratory chain complexes I and IV can prevent mitochondrial damage induced by ruthenium red in vivo. J. Pineal Res. **28:** 242–248.
84. MARTIN, M., M. MACIAS, G. ESCAMES, et al. 2000. Melatonin but not vitamins C and E maintains glutathione homeostasis in t-butyl hydroperoxide-induced mitochondrial oxidative stress. FASEB J. **14:** 1677–1679.
85. LIVREA, M.A., L. TESORIERE, D. ARPA & M. MORREALE. 1997. Reaction of melatonin with lipoperoxyl radicals in phospholipid biolayers. Free Rad. Biol. Med. **23:** 706–711.
86. REITER, R.J., D.X. TAN, S.J. KIM & W. QI. 1998. Melatonin as a pharmacological agent against oxidative damage to lipids and DNA. Proc. West. Pharmacol. Soc. **41:** 229–236.
87. CUZZOCREA, S., G. COSTANTINO, E. MAZZON, et al. 2000. Beneficial effects of melatonin in a rat model of splanchnic artery occlusion and reperfusion. J. Pineal Res. **28:** 52–63.
88. ACUÑA-CASTROVIEJO, D., G. ESCAMES, M. MACIAS, et al. 1995. Cell protective role of melatonin in the brain. J. Pineal Res. **19:** 57–63.
89. KIM, S.J., R.J. REITER, M.V. ROUVIER-GARAY, et al. 1998. Nitropropane-induced lipid peroxidation and DNA damage: Antitoxic effects of melatonin. Toxicology **130:** 183–190.
90. MELCHIORRI, D., G.G. ORTIZ, R.J. REITER, et al. 1998. Melatonin reduces paraquat-induced genotoxicity in mice. Toxicol. Lett. **95:** 103–108.
91. QI, W., R.J. REITER, D.X. TAN, et al. 2000. Increased levels of oxidatively damaged DNA induced by chromium (III) and H_2O_2: protection by melatonin and related indoles. J. Pineal Res. **29:** 54–61.
92. COTO-MONTES, A. & R. HARDELAND. 1999. Antioxidative effects of melatonin in Drosophila melanogaster: antagonization of damage induced by the inhibition of catalase. J. Pineal Res. **27:** 154–158.
93. KIM, S.J., R.J. REITER, W. QI, et al. 2000. Melatonin prevents oxidative damage to protein and lipid induced by ascorbate-Fe^{3+}-EDTA: comparison with glutathione and α-tocopherol. Neuroendocrinol. Lett. **21:** 269–276.
94. CERAULO, L., M. FERRUGIA, L. TESORIERE, et al. 1999. Interactions of melatonin with membrane models: portioning of melatonin in AOT and lecithin reversed micelles. J. Pineal Res. **26:** 108–112.
95. BOLOGNAN, C.V., M. YAMAMOTO, N. TAKEI, et al. 2000. Gial cell survival is enhanced during melatonin-induced neuroprotection against cerebral edema. FASEB J. **14:** 1307–1317.
96. SEABRA, M. DE L.V., M. BIGNOTTO, L.R. PINTO, JR. & S. TUFIK. 2000. Randomized, double blind clinical trial, controlled with placebo, of the toxicology of chronic melatonin treatment. J. Pineal Res. **29:** 193–200.
97. JAN, J.E., D. HAMILTON, N. SEWARD, et al. 2000. Clinical trials of controlled-release melatonin in children with sleep-wake cycle disorders. J. Pineal Res. **29:** 34–39.
98. JAHNKE, G., M. MARR, C. MYERS, et al. 1999. Maternal and developmental toxicity evaluation of melatonin administration orally to pregnant Sprague-Dawley rats. Toxicol. Res. **50:** 271–279.

99. BRUSCO, L.I., M. MARQUEZ & D.P. CARDINALI. 1998. Monozygotic twins with Alzheimer's disease treated with melatonin: case report. J. Pineal Res. **25:** 260–263.

Part VI: Endogenous Neuroprotectants—Melatonin

QUESTION FOR DR. OXENKRUG

From Dr. Manev

Would you comment on a possibility that NAS, acting via 5-HT receptors, reduces food intake and thus increases life span?

ANSWER: It certainly might be one of the mechanisms of the antiaging effects of NAS. We, unfortunately, did not assess the food intake in our experiments. There are some other possible explanations of the body weight reduction in NAS-treated rats, such as the increase of locomotor activity. It would be very interesting to evaluate these and other factors in the future experiments.

QUESTIONS FOR DR. REITER

From Dr. Scallet

TMT produces greater damage for a given dose in older animals; an effect which is blocked by caloric restriction. Might this be related to the decreased melatonin in aging?

ANSWER: Certainly the loss of melatonin with age could exaggerate TMT damage, especially the free radical component of the damage. Loss of melatonin during aging (or after pinealectomy in younger animals) exaggerates the destruction induced by free radical process (Manev, *et al.*, FASEB J. **10:** 1546, 1996; Kilic, *et al.*, J. Cerebr. Blood Flow Metab. **19:** 511, 1999).

From Dr. Maynard

Are there melatonin transporters? There must be an explanation for high levels in mitochondria and CSF relative to serum levels, yet melatonin crosses all biological membranes.

ANSWER: No melatonin transporters have been defined. The only explanation currently available for different concentrations of melatonin in different subcellular compartments may relate to binding proteins for melatonin. These have been described and if their concentrations differ between subcellular compartments, melatonin levels may vary accordingly.

From Dr. Auer

What is the effect of caloric restriction, which alters the rate of aging, on melatonin production and melatonin rhythms throughout life?

ANSWER: Melatonin production is normally reduced in advanced age (Reiter, Bio Essays **14:** 169, 1992). Food restriction defers the loss of melatonin during aging (Stokkan, *et al.*, Brain Res. **545:** 66, 1991). Whether the higher melatonin levels (preserved by food restriction) are beneficial in the older animals remains unknown.

From Dr. Youdim

I am a little concerned about the usefulness of antioxidants such as melatonin as neuroprotectants in neurodegenerative disease. I say this because brain autopsy studies have shown that neither Vit. E or Vit. C is altered in Parkinson's disease or Alzheimer's disease. Thus, would giving more antioxidant to patients be useful? Is it possible that melatonin's action is related to its compartmentalization in different brain areas, and that its protective action in hippocampus and not striatum supports this.

Is the methoxy group on melatonin necessary for antioxidant neuroprotection activities?

ANSWER: Melatonin acts very differently from classic antioxidants such as vitamins C and E. In models of Alzheimer's and Parkinson's disease, melatonin effectively reduces the neural damage that occurs in these situations. How or whether this translates into a beneficial effect in humans is unknown although there is a modicum of data to suggest that it may (Brusco, *et al.*, J. Pineal Res. **25:** 260, 1998).

The methoxy group at position 5 on the indole nucleus is significant for the hydroxyl radical-scavenging capacity of melatonin (Tan, *et al.*, Endocrine J. **1:** 57, 1993).

QUESTIONS FOR DRS. REITER AND OXENKRUG

From Dr. Slikker

Dietary restriction has been show to extend life span in animal models. Are your NAS or melatonin effects seen on (animal) mouse life span influenced by the reduced weight gain (dietary restriction equivalent) that you showed?

ANSWER: (Oxenkrug) It would be very tempting to suggest that NAS/melatonin treatments are the equivalents of caloric restriction! It is noteworthy that some researchers explained the antiaging effect of caloric restriction by the increase of antioxidant activity. Thus, it might be that caloric restriction and NAS/melatonin treatments share the same mechanism of action.

Does melatonin decrease body temperature? *In vitro* data clearly demonstrate that hypothermia will enhance cell survival. In your rat study, did hypothermia play a role in enhanced life span?

ANSWER: (Oxenkrug) We have not measured body temperature in our experiments. Dr. Morton, who, a few years ago, was a fellow in Professor Reiter's laboratory, reported that both melatonin and NAS decrease body temperature. That, as you suggest, might contribute to NAS/melatonin and aging effects.

(Reiter) There have been recent publications indicating that melatonin reduces body weight of old animals. This could be important in terms of an effect on aging.

Melatonin has a minor inhibitory (transient) effect on body temperature. Pharmacological doses of melatonin usually cause less than 0.5°C drop in body temperature. I do not feel that this transient reduction would be consequential in terms of longevity of an animal.

Neuroprotective and Cognition-Enhancing Properties of MK-801 Flexible Analogs

Structure–Activity Relationships

SERGEY BACHURIN, SERGEY TKACHENKO, IGOR BASKIN, NADEGDA LERMONTOVA, TATYANA MUKHINA, LYUDMILA PETROVA, ANATOLIY USTINOV, ALEXEY PROSHIN, VLADIMIR GRIGORIEV, NIKOLAY LUKOYANOV, VLADIMIR PALYULIN, AND NIKOLAY ZEFIROV

Institute of Physiologically Active Compounds RAS, 142432, Chernogolovka, Russia

ABSTRACT: Neuroprotective and biobehavioral properties of a series of novel open chain MK-801 analogs, as well as their structure–activity relationships have been investigated. Three groups of compounds were synthesized: monobenzylamino, benzhydrylamino, and dibenzylamino (DBA) analogs of MK-801. It was revealed that DBA analogs exhibit pronounced glutamate-induced calcium uptake blocking properties and anti-NMDA activity. The hit compound of DBA series, NT-1505, was investigated for its ability to improve cognition functions in animal model of Alzheimer's disease type dementia, simulated by treating animals with cholinotoxin AF64A. The results from an active avoidance test and a Morris water maze test showed that experimental animals, treated additionally with NT-1505, exhibited much better learning ability and memory than the control group (AF64A treated) and close to that of the vehicle group of animals (treated with physiological solution). Study of NT-1505 influence on locomotor activity revealed that it is characterized by a spectrum of behavioral activity radically different from that of MK-801, and in contrast to the latter one does not produce any psychotomimetic side effects in the therapeutically significant dose interval. The computed docking of MK-801 and its flexible analogs on the NMDA receptor elucidated the crucial role of the hydrogen bond formed between these compounds and the asparagine residue for magnesium binding in the NMDA receptor. It was suggested that strong hydrophobic interaction between MK-801 and the hydrophobic pocket in the NMDA receptor–channel complex determines much higher irreversibility of this adduct compared to the intermediates formed between this site and Mg ions or flexible DBA derivatives, which might explain the absence of PCP-like side effects of the latter compounds.

KEYWORDS: Neuroprotection; MK-801; NMDA-receptor; Antagonists; Docking; Cognition enhancer.

INTRODUCTION

It is generally recognized that specific blockade of calcium ion influx via hyperactivated glutamate receptor–channel complexes (GluR), in particular NMDA

Address for correspondence: S.O. Bachurin, Institute of Physiologically Active Compounds RAS, 142432, Chernogolovka, Russia. Voice/fax: 7(095) 913-21-13.
bachurin@ipac.ac.ru

subtype receptors, can provide efficient protection against diverse group of neurological disorders related to glutamate excitotoxicity, such as brain ischemia and Alzheimer's disease (AD). In the latter, it is considered that NMDA receptor antagonists may provide efficient protection of neurons against neurotoxic action of beta-amyloid peptide (Aβ), realized, in part, by potentiating the excitotoxic effects of endogenous glutamate.[1] On the other hand, it is assumed that glutamate receptor agonists (in particular, AMPA subtype receptor agonists) can exhibit strong cognitive enhancing effects due to activation of glutamatergic neurotransmission that is otherwise reduced in the course of AD.[2] Such dualism in the properties of glutamatergic compounds explains the interest to these substances as promising neuroprotectors and cognition enhancers. Previously it had already been revealed that the well known noncompetitive NMDA receptor antagonist, MK-801, shows strong neuroprotective properties in cell culture experiments. However, the therapeutic potential of MK-801 is diminished by side remarkable psychotomimetic effects associated with its high affinity interaction with the intrachannel binding site (ICS) of the NMDA receptor.[3] In contrast to MK-801, the low affinity open channel blockers, such as amantadine, memantine, or ARL 15896, often show better therapeutic indices because of their rapid blocking kinetics and strong voltage dependency.[4,5] The goal of the study reported here was to provide: (1) focused synthesis of flexible analogs of MK-801, (2) a computational estimate of the binding of MK-801 and its flexible analogs to the NMDA receptor by the molecular docking method, and (3) a study of neuroprotective and behavioral (in particular, cognition-enhancing) properties of series of novel open chain MK-801 analogs whose side effects were expected to be minimized due to less rigid interaction of ICS with flexible molecules relative to MK-801.

METHODOLOGY OF DIRECTED SEARCH FOR NEUROPROTECTORS AND COGNITION ENHANCERS

To screen and study the action of potential neuroprotectors and cognition enhancers among wide spectrum of GluR ligands a battery of *in vitro* and *in vivo* tests and models was used in the present study. This set of tests includes:

1. Inhibition of total and specific glutamate-stimulated calcium ion (Ca^{2+}) uptake in rat brain synaptosomes;

2. Study of the ability of the compound to prevent the development of NMDA-induced convulsion and lethality in experimental animals—as a measure of anti-NMDA activity.

3. Animal behavioral study of the cognition-enhancing properties for the most promising compounds in an active avoidance test and in a water maze test, after artificial cholinergic neurodegeneration in rat brain induced by a specific synthetic cholinotoxin, ethylcholine aziridinium ion (AF64A).

As a first step the ability of each compound to inhibit glutamate-induced Ca^{2+} uptake (in rat brain synaptosomes) and to manifest anti-NMDA activity was estimated. This permitted us to limit the total number of potentially promising structures and to create a uniform date bases for QSAR analysis and docking. In the second stage of our research, the most promising compounds ("hit compounds") were studied for their neuroprotective and cognition enhancing properties in animal models of AD-type degeneration.

TOTALLY OPEN CHAIN ANALOGS OF MK-801 STRATEGY FOR TEMPLATE GENERATION AND SYNTHESIS

Disclosure of the MK-801 bicycle leads to the generation of two types of compounds: (1) open chain analogs as a result of breaking one cycle (see FIGURE 1); and (2) totally open chain analogs as a result of breaking two cycles (see FIGURE 2). Open chain analogs include two groups of compounds: the isoquinoline series and the isoindoline series (FIG. 1). The molecular shapes of these molecules are very similar to MK-801, and some of the compounds display properties of high affinity open channel blockers and associated side effects, such as isoquinoline FR-115427.[6]

Four templates result from the further opening of isoquinoline and/or isoindoline cycles to form totally open chain types of MK-801 analogs: monobenzylamine, benzhydrylamine, phenethylamine, and dibenzylamine series (FIG. 2). These compounds are very similar, but essentially more flexible molecules, relative to MK-801. Focusing libraries contained 100–200 substances were synthesized in each series, forming a total of about 600 compounds. The functional variation[7] in each series included changes "around" the template (introduction of substituents to aryl, and/or alkyl, and/or nitrogen, and bioisosteric substitution of aryls in the template) but not involving a change of template (e.g., transformation of the template into novel cycles).

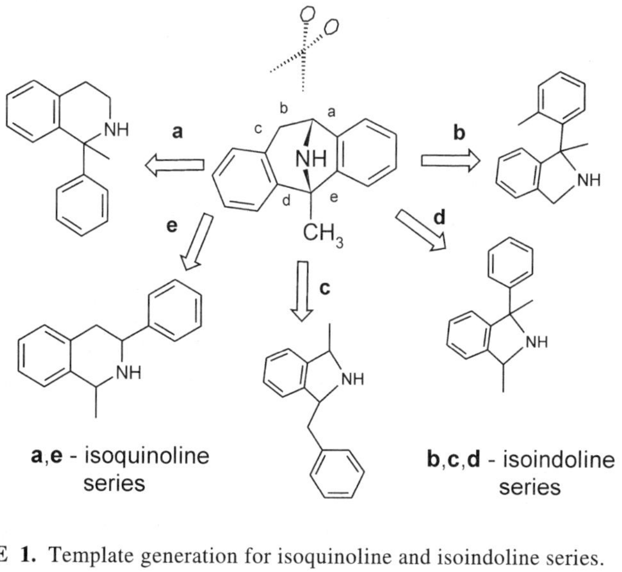

FIGURE 1. Template generation for isoquinoline and isoindoline series.

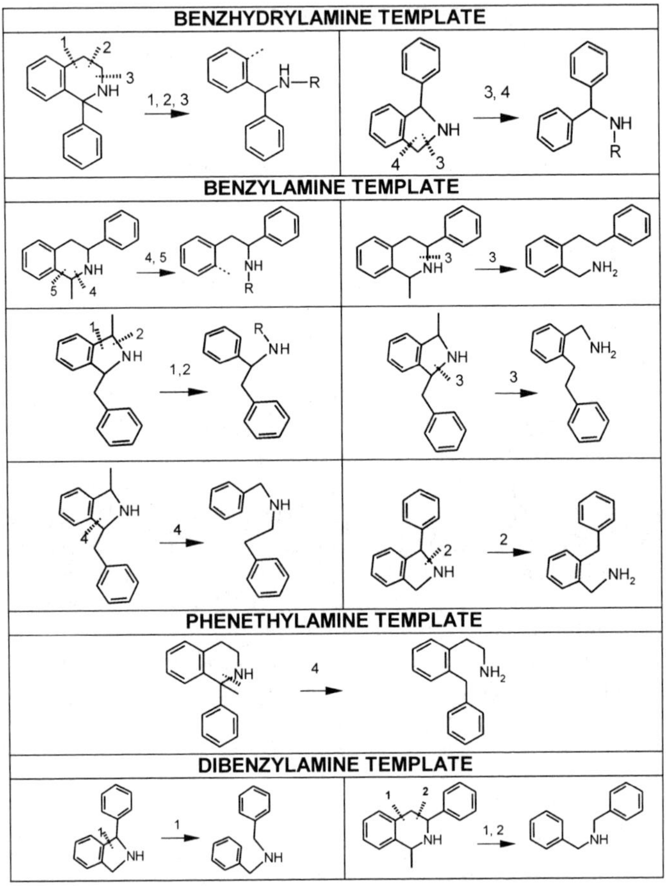

FIGURE 2. Opening of isoquinoline and/or isoindoline cycles leads to the generation of total open-chain templates (benzhydrylamines, benzylamines, and dibenzylamines).

MOLECULAR DOCKING

Homology Modeling

Amino acid sequences of human GLUR1 (Swiss Protein Data Bank accession number P42261), GLUR2 (P42262), GLUR3 (P42263), GLUR4 (P48058), GLUR5 (P39086), GLUR6 (Q13002), GLUR7 (Q13003), KA1 (Q16099), KA2 (Q16478), NR2A (Q12879), NR2B (Q13224), NR2C (Q14957), NR2D (O15399), and NR1 (P35437) subunits were multiply aligned with the amino acid sequence of the K$^+$ channel from *Streptomyces lividans*[8] (KcsA K$^+$ channel, Brookhaven Protein Data Bank[9] accession number 1BL8) using the program Clustal X.[10–12] The BLOSUM 30 homology matrix was used to evaluate amino acid similarity. This sequence alignment was used as the basis for homology modeling using the Biopolymer module in SYBYL 6.5.[13]

The X-ray structure of the KcsA K^+ channel was used as a template for building a NR1/NR2B/NR1/NR2B tetramer of the NMDA receptor. Chains A and C from the K^+ channel were "mutated" to NR2B subunits, whereas chains B and D were mutated to NR1 subunits in accord with the following procedure. All aligned non-identical amino acids were mutated to each other by changing their side chains while keeping main chain atoms unchanged and fixed at their original positions in space. Internal torsion angle values in side chains of new amino acids were chosen to be those most appropriate for a given amino acid in a given secondary structure state (which remains unchanged by mutating residues). All insertions and deletions were handled by defining "loops" containing all residues being inserted and several neighboring atoms to the positions of insertion or deletion were followed by searching for them in the Brookhaven Protein Data Bank. Subsequently, that search results were visually inspected, and main chain atoms of those *loops* for which (1) the geometry of the main chain atoms neighboring the deletion or insertion positions underwent smaller changes, (2) the amino acid sequence possessed the strongest homology to the required sequence, and (3) no unfavorable spatial hindrance between atoms belonging to the *loop* and to the remaining part of the protein could be expected, were inserted in the protein model and followed by adding appropriate side chains in accord with the aforementioned procedure. Next, all proline residues and side chains of all other residues were "fixed" by finding new values for several torsion angles so as to remove the spatial overlap between atoms. This was followed by adding hydrogen atoms that were capable of forming hydrogen bonds. The protein was then subjected to 5,000 steps of constrained energy minimization, with all main chain atoms fixed in space using the Tripos force field, as implemented in the SYBYL package. This was followed by 5,000 steps of constrained minimization without fixing the main chain atoms, but keeping the mutual spatial arrangement of several oxygen atoms at the narrowest part of the pore unchanged so as to preserve the requirements for effective ion conduction.

Docking Ligands

All ligands were manually docked to the putative binding site followed by 500 steps of energy minimization for the entire protein–ligand complex. Interactive docking was carried out by means of the Sybyl 6.5 molecular-modeling package running on a SGI Octane workstation.

Sequence Alignment

In order to avoid chance effects and increase validity of the alignment used for predicting the three-dimensional structure of the NMDA receptor channel, we included all members if the ionotropic glutamate receptor family subunits into multiple alignment of amino acid residues, along with the amino acid sequence of the K^+-channel. Since all sequences have different numbering for the residues, we introduced a uniform numbering for them, the rows are labeled in the multiple alignment diagram with the abbreviation C.N. The mere existence of such multiple alignment implies that the general architecture of channel pores in subunits of all ionotropic receptors can actually be expected to be the same. Thus, the part in each sequence that is generally associated with transmembrane segment M1 corresponds to C.N. 3–21, the segment M2 has C.N. 51–71, and the amino acids in transmembrane segment M3 have C.N. 82–100. Four "fingerprint" amino acid residues (that correspond to the ion selectivity filter of the KcsA K^+ channel) in the nearest part of the channel pore have C.N. 67–70. The most important among them is generally believed to be the amino acid with C.N. 67 (Q/R/N site). For the case of AMPA receptors this residue can undergo RNA editing,[14] which controls the Ca^{2+} permeability of these receptors. In the NMDA receptor, this amino acid in a NR2 subunit (Asn614 in NR2A, Asn615 in NR2B, Asn612 in NR2C, and Asn642 in NR2D) has been shown to take part in the voltage-dependent Mg^{2+} blockage of the receptor channel.[15] Homological residue Asn616 in NR1 has been shown to control the Ca^{2+} permeability[15] of the NMDA receptor channel, as well as being involved in the voltage-dependent blockage by compounds believed to bind to the PCP blocking site in the receptor pore.[16] This residue, along with Trp611 and Ala645 in the NR1 subunit are supposed to form the PCP binding site.[16]

Architecture of the NMDA Receptor Channel

A three-dimensional model of the NMDA receptor channel, built in accordance with the procedure considered above, is depicted in FIGURE 3. The receptor channel in this model is a heterotetramer with two-fold symmetry about a central pore. Each of its subunits consists of two transmembrane α-helices, segments M1 and M3, connected by a pore region, segment M2, that consists of a turret, pore helix, and selectivity filter. Four subunits are arranged in the heterotetramer such that one transmembrane helix (segment M3) faces the central pore and the other (segment M1) faces the lipid membrane. The transmembrane helices M3 are tilted with respect to the membrane, so that the subunits open like the petals of a flower. The M3 helices are attached to the pore region near the intracellular surface of the membrane. This region contains the selectivity filter of the channel.

The selectivity filter is formed by four amino acids with C.N. 67–70. Sixteen main chain carbonyl oxygen atoms of these residues face the pore and take part in stabilizing dehydrated cations moving through it. In addition to the main chain atoms, the oxygen atoms of the asparagine side chains (C.N. 67) located near the entrance to the narrow pore formed by the selectivity filter amino acids may play an important role in dehydrating and stabilizing metal cations (since it controls the Mg^{2+} blockage for the case of a NR2 subunit of the NMDA receptor and Ca^{2+} permeability for the cases of the AMPA and the NR1 subunit of the NMDA receptor).

Two important features of the cavity inside the channel should be considered in connection with their possible pharmacological implications. First, the walls of the pore are formed by hydrophobic amino acids. And second, there is a narrow hydrophobic pocket between segments M2 and M3 just above the selectivity filter. The shape of this pocket is nearly complementary to the shape of the most potent phencyclidine binding site ligands, such as PCP and MK-801, and in the vicinity of this pocket there is the important Asn616 residue that controls the Ca^{2+} permeability of the channel. The space location of this amino acid permits easy formation of a strong hydrogen bond with PCP, MK-801, and their analogs. Thus, all important features of

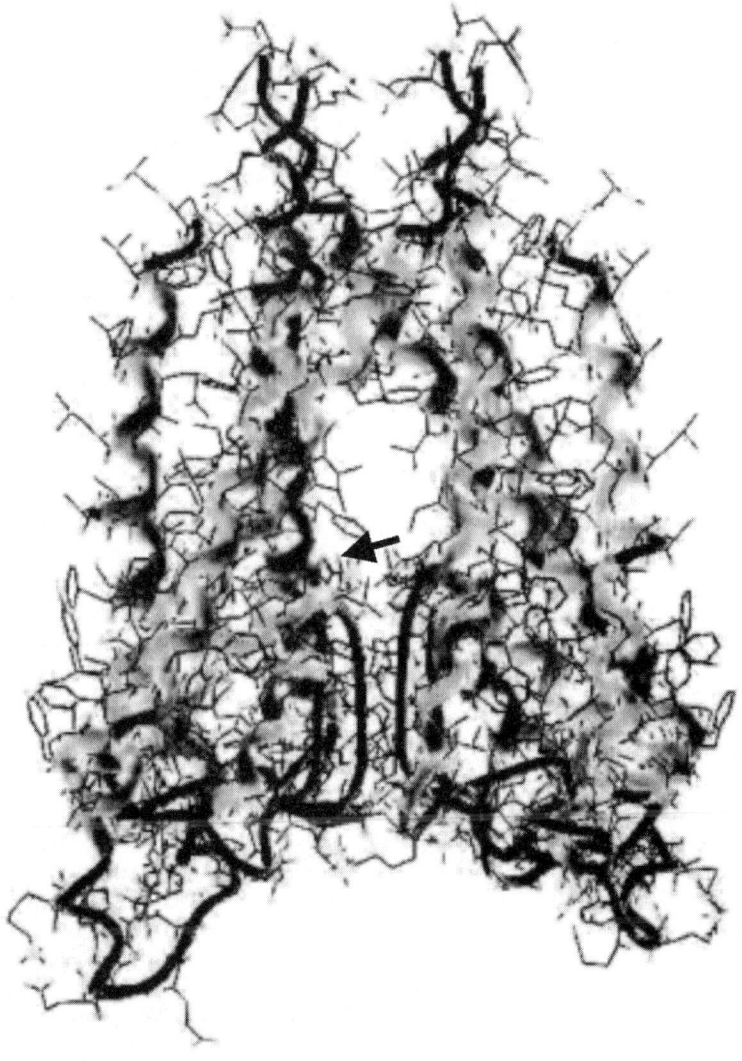

FIGURE 3. General view of the NMDA receptor channel modeled by homology (the arrow indicates the approximate location of MK-801-binding site).

the phencyclidine binding site pharmacophore, two hydrophobic areas (one formed by the pocket and one by the surface of the M3 helix facing the central part of the pore) arranged nearly perpendicularly to each other and the hydrogen bond acceptor part located near the corner of this triangle, are present here. We suggest that this part of the channel contains the phencyclidine binding site.

MATERIALS AND METHODS

Estimation of Calcium-Blocking Activity

Interaction between compounds and glutamate-dependent Ca^{2+} uptake system was studied on newborn (8–11 days old) rat brain synaptosomal P_2-fraction. Specific Ca^{2+} uptake was measured in the presence of the compound tested as a variation between $^{45}Ca^{2+}$ content in glutamate stimulated synaptosomes and $^{45}Ca^{2+}$ content in unstimulated synaptosomes according to the equation:

$$K = [(Ca4 - Ca3)/(Ca2 - Ca1)] \times 100\%$$

Where Ca1 is the Ca^{2+} influx in a blank experiment (without glutamate and tested compounds); Ca2 is the Ca^{2+} influx in the presence of glutamate only (glutamate-induced Ca^{2+} influx); Ca3 is the Ca^{2+} influx in the presence of tested compound (without glutamate); Ca4 is the Ca^{2+} influx in the presence of both glutamate and the tested compound.

Estimation of Anti-NMDA Activity

Adult non-pedigree male mice weighing 20–24 g were used. The compounds tested were administered i.p. 30 min prior to administration of NMDA. NMDA was i.c.v. injected in doses of 0.1 µg in 1.0 µl of distilled water (pH = 7.0); this induced clonic seizures in 100% of mice. The anti-NMDA activity was determined as a dose of the compound under i.p. administration needed to prevent NMDA-induced seizures in 50% of mice. ED_{50} values were calculated from the linear portions of sigmoidal curves. The χ-test was used to compare ED_{50} values for different treatments.

Animal Model of AD Type Dementia

Neuronal atrophy related to the loss of cholinergic markers in some areas of human brain is known to be the most typical pathological feature of AD.[17] It was shown previously that i.c.v. injection of AF64A leads to degeneration of cholinergic neurons in the brain of experimental animals and reproduces some specific features of chronic cholinergic hypofunction, along with learning and memory impairments analogous to those observed in AD.[18,19] In our research we used a cholinotoxin-induced animal model of AD type neurodegeneration to study cognition-enhancing properties of the compounds in a Morris water maze test and in an active avoidance test.

Experimental Design

Compound Injections

The experiments were performed on male Wistar rats (200–230 g weight). AF64A was prepared from commercial samples of AF64 (RBI) according to a method described elsewhere[20] and diluted with artificial cerebrospinal fluid (CSF). Before the surgery, rats were anesthetized with ether. Freshly prepared AF64A solution in a dose of 3 nmol/3 µl was injected into each of lateral cerebral ventricles of experimental rats. Control rats received 3 µl CSF. One day after the surgery AF64A injected rats received orally 2 mg/kg of Aricept (Pfizer, USA) or Memantine (RBI,

Memantine HCl) or NT1505 as a suspension in 1% starch. The compounds were injected over a period of 23 days. Ten days after the first injections, animal training in the pool was started. During the training the compounds were injected 1–1.5 hours after the daily learning trials (to study their effect on long-term memory).

Morris Water Maze Test

Experiments were performed in a circular black swimming pool 180 cm in diameter. The surface of a removable platform (10 cm) was located 2–3 cm below the water level, so that the platform was invisible to the swimming rat. All experiments were detected by means of a video camera connected with a computer and S-VHS recorder.

The training procedures aimed at finding the hidden platform were started 10 days after the first injections, with all animals receiving two escape trials a day from two different positions. On the day 11, the daily training was interrupted and a retention test was performed only on days 14 and 26. On days 5, 11, 14, and 26 after commencing the training, the movements of the rats during the test were recorded and then analyzed with the original computerized system "Behavioral Vision".[21,22] Four parameters were analyzed: *escape latency* (the time in seconds required for the rat to find the hidden platform), *depression index* (characterizing immobility of rats in the pool and calculated as a ratio between time that the rat spent with a speed less than a threshold value to the time of swimming with the higher speed; the threshold was set experimentally at 10 cm/s), *straightness index* (characterizing the track directionality and calculated as a ratio between trajectory length and distance from the track start to the platform center, i.e., deviation of the trajectory from the straight path–vector from start position to platform), and *strategy index* (calculated as trajectory length within the non-entering zone, as a percentage of the full trajectory of animal during the trial).

Statistical Analysis

The results were analyzed by ANOVA followed by post hoc LSD comparisons using program package STATISTICA ver. 5.0. The level of significance was preset at $p < 0.05$. Data are presented as mean ± S.E.M.

Active Avoidance Test

The test procedure is described elsewhere.[23] Briefly it is as the follows: two weeks after the injection of AF64A or CSF, active avoidance training was conducted in a two-chamber shuttle box. Each chamber was 30 cm long and 20 cm wide, with a guillotine door (7 × 7 cm) between them, and a grid floor of steel rods 1 cm apart. A 12-W light bulb was fixed in the ceiling on each side. Before training, the animals were given a five-minute familiarization period in the darkened shuttle box with the central door open. The conditioned stimulus was a five-second light followed by the unconditioned stimulus, a 1 mA shock was delivered to the grid in the lit chamber. During the presentation of both stimuli, the door was open. After the rat either escaped or avoided the shock by crossing through to the other chamber, the door closed and a new trial sequence began. The intertrial interval was 30–60 sec. The time taken to escape or to avoid the shock (response latency) was measured. The avoidance during the conditioned stimulus (5 sec) was considered a conditioned avoidance response (CAR). Training procedure consisted of 35 trials. Data from the last 15 trials (learning test) were collected and analyzed in five-trial blocks, as percentage of CAR summarized over each block. From this was calculated the mean percentage of CAR in three blocks. Twenty-five further trials were given 24 h later (retention test). In this case, data were analyzed as the mean percentage of CAR in first two blocks of five trials. Statistical significance was tested using an unpaired Student's t-test. $p < 0.05$ was taken as indicative of statistical significance.

RESULTS AND DISCUSSION

The results obtained at the first stage of investigation revealed the group of compounds that manifested strong anti-NMDA and calcium-blocking activity. Examples

TABLE 1. Calcium-blocking property (IC_{50}) and anti-NMDA activity (ED_{50}) of several totally open chain analogs of MK-801 in the monobenzylamine and dibenzylamine series

Structure	IC_{50} (μM)	ED_{50} (mg/kg)	Structure	IC_{50} (μM)	ED_{50} (mg/kg)
NT-1505	15	10	LS-4206	11	30
NT-3101	6	16	NT-5001	18	35
[structure]	16	100	*[structure]*	11	65
[structure]	100	>100	*[structure]*	100	55
[structure]	20	25	*[structure]*	100	62
[structure]	31	52	*[structure]*	100	70
[structure]	56	10	*[structure]*	107	43

of anti-calcium and anti-NMDA activities are presented in TABLE 1. The most interesting and promising results have been obtained in a series of newly synthesized monobenzylamine and dibenzylamine derivatives. Structure–activity analysis in these series of compounds revealed that their biological effects depend appreciably on the substituents in the benzene rings. The introduction of a bulky *ortho*-group into the each benzene rings leads to the increase of inhibition effect. The bioisosteric substitution of the benzene ring on the heteroaryl moiety (furyl, pyridyl, etc.) leads mainly to the appreciable stimulation of uptake. The hit compound in both series (NT-1505 and NT-3101) demonstrated a strong calcium-blocking property (IC_{50} = 15.0 and 6.0 µM, respectively) and anti-NMDA activity (ED_{50} = 10 and 16 mg/kg). According to these results the most promising group of compounds, in particular, compound NT-1505, were selected for an expanded biobehavioral study. In parallel, the comparative docking study of NT-1505 and MK-801 binding with the NMDA receptor was performed.

Binding of Ligands to the Channel

The three-dimensional structure of the complex between MK-801 and the putative PCP-binding site of the NMDA receptor is depicted in FIGURE 4. MK-801 forms two hydrogen bonds with the receptor, one with Asn616, and the other with the main chain carbonyl oxygen atom belonging to Leu615. In addition, MK-801 binds with one of its benzene rings deeply into the hydrophobic narrow pocket and the other sticks to the hydrophobic M3 helix. The side-chain amide group of Asn 616 also forms two hydrogen bonds: one between its oxygen atom and protonated nitrogen of MK-801, and one between its amide nitrogen and the oxygen atoms belonging to the main chain of the same residue. As a result, the oxygen atom belonging to the side

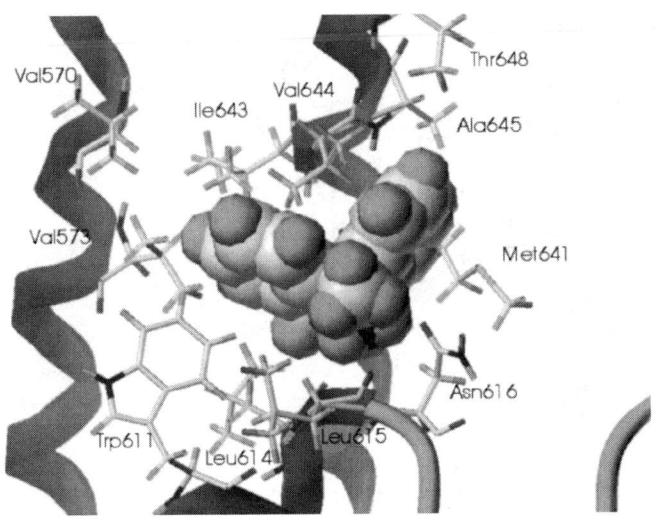

FIGURE 4. Predicted binding mode of MK-801 to the putative phencyclidine site inside the NMDA receptor channel.

chain of Asn616 and the oxygen atom belonging to its main chain can no further stabilize cations inside the narrow pore formed by the selectivity filter amino acids.

The flexible MK-801 analog NT-1505 was docked to the putative PCP binding site in a conformation that mimics the spatial structure of MK-801. NT-1505 appeared to be able to bind to the same hydrophobic pocket as MK-801 in the same mode (see FIGURE 5). However, it binds to the receptor with only one hydrogen bond, the opposite site is more exposed to the channel pore, and the binding involves freezing of torsion degrees of freedom. Thus, one can suppose more favorable condition for dissociation of NT-1505 upon channel closure.

Behavioral Properties of NT-1505

A preliminary examination of the cognition-enhancing properties of each compound and its ability to improve learning and memory from a toxin-induced animal model of dementia, studied in the active avoidance test. The results presented in FIGURE 6 indicate that rats treated with cholinotoxin AF64A demonstrate significant reduction in learning ability and pronounced memory impairment when compared to the vehicle group of animals treated with CSF (approximately 50% from the control level of CAR). Oral administration of NT1505 causes the recovery of the cognitive functions that deteriorate after the injection of neurotoxin AF64A, both in the learning test (the learning trials follow training on the same day) and in the retention test (the trials were performed the day after the training set). The compound produced statistically significant memory-enhancing effects at all doses, in the interval 0.5 to 5.0 mg/kg (orally) that lasted for at least 10 days after the learning trial.

A comparative long-term study of cognition enhancing properties of compound NT-1505 and widely used anti-Alzheimer medicines, Aricept and Memantine, was

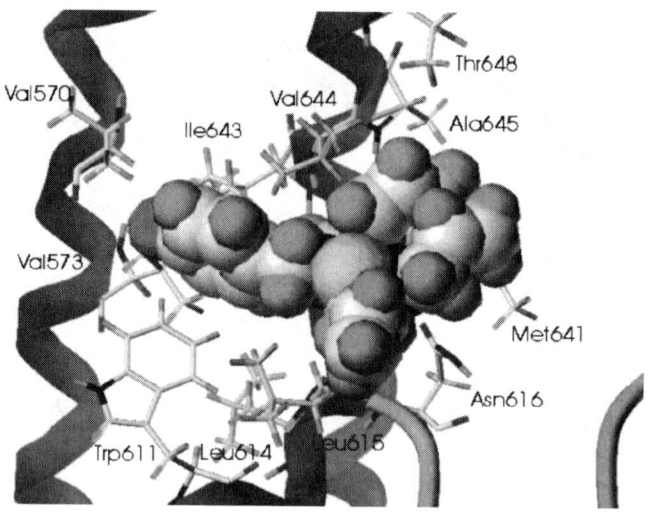

FIGURE 5. Predicted binding mode of NT-1505 to the putative phencyclidine site inside the NMDA receptor channel.

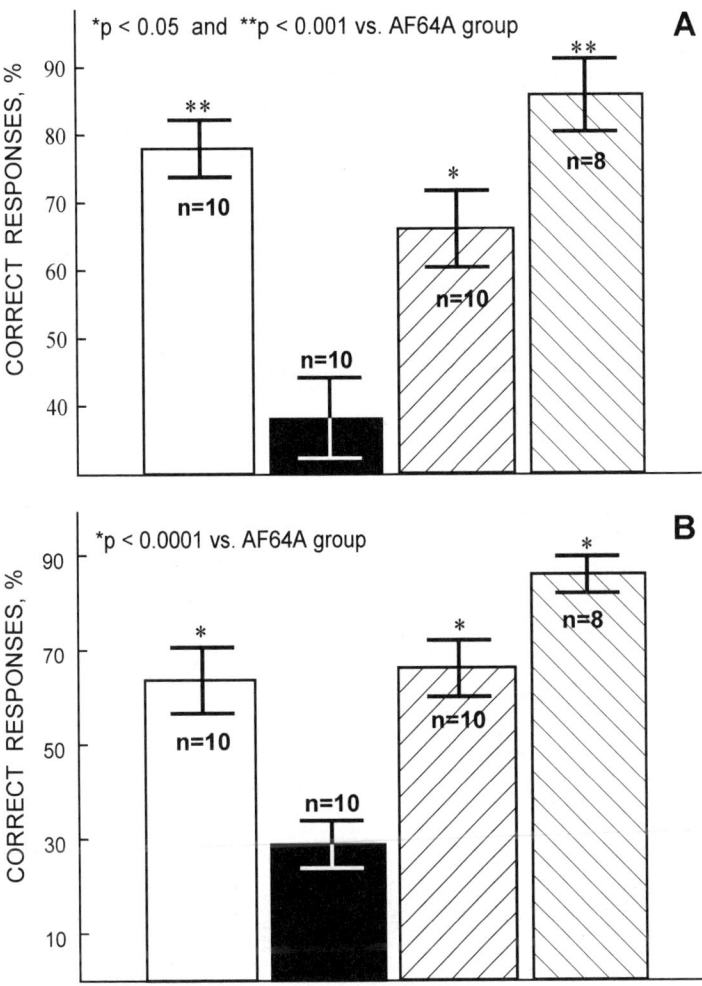

FIGURE 6. Cognition-enhancing effect of NT1505 (n is the number of animals). **A.** Learning test. **B.** Retention test (the day after the learning trials). ☐, control; ■, AF64A; ▨, AF64A+NT-1505, 1mg/kg; ▨, AF64A+NT-1505, 5mg/kg.

performed in a variant of the Morris water maze test in rats treated with toxin AF64A.

The superposition of swimming trajectories (start position 1) of all groups of rats in search of the hidden platform in the Morris water pool is presented in FIGURE 7. The results of computer estimated indices of escape latency and track directionality (straightness index) are presented in FIGURE 8.

Despite rather large mean escape latency observed on day 5 of the training, Aricept significantly ($p = 0.048$) decreased the time and improved the trajectory in the search of the platform, Memantine demonstrated a tendency to decrease the search time in comparison with AF64A-treated rats.

On day 11 of training, the difference between control and AF64A-injected groups in the escape latency was almost statistically significant ($p = 0.077$). The AF64A treated rats receiving Aricept and NT-1505 escaped to the platform in almost the same time and this was significantly different from the AF64A group. Two-day interruption of the training with continued injections of the compounds led to a small increase in the search time for all the rats; however, for the rats receiving Aricept, this was significantly ($p = 0.035$) less than that of AF64A-injected group.

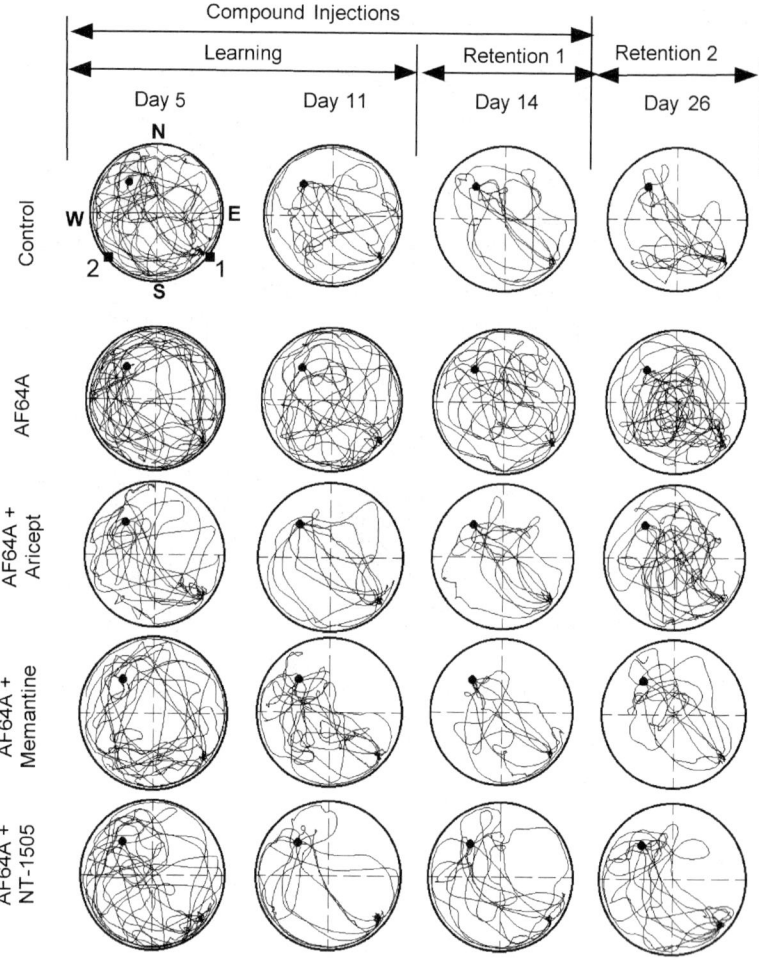

FIGURE 7. Swimming trajectories of rats groups receiving different compounds. ●, hidden platform position; ■, start position.

The best values of the straightness index on day 14 was obtained for Aricept and Memantine. The rats receiving Memantine and NT-1505 had rather good values for the strategy index on day 14 (data not shown).

An eleven-day break (day 26) in the training procedure after completing the compound injections led to a considerable deterioration in the spatial orientation of rats

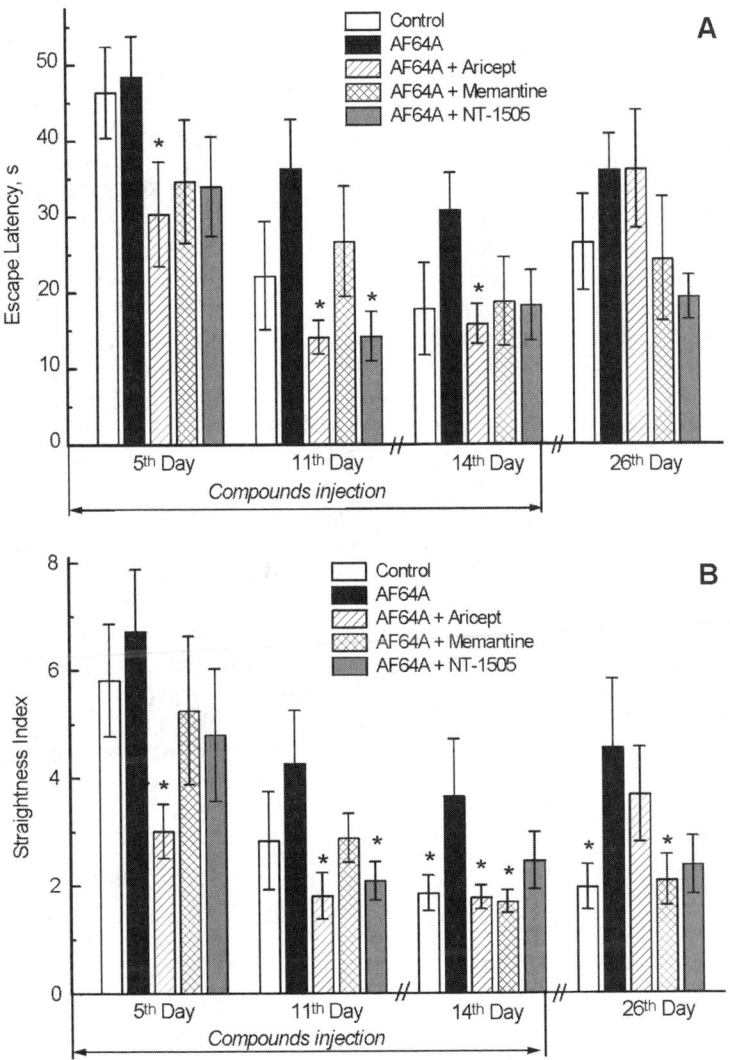

FIGURE 8. Escape latency (**A**) and track directionality relative to platform position (**B**). Start position 1. Control, $n = 10$; AF64A, $n = 10$; AF64A+Aricept, $n = 10$; AF64A+Memantine, $n = 8$ (days 5 and 26), $n = 9$ (days 11 and 14); n is the number of animals; $*p < 0.05$ versus AF64A group (ANOVA followed by post hoc comparisons).

receiving Aricept, and a slight decrease for the other groups, except for NT-1505-injected rats for which the ability to find the platform remained almost the same as on day 14. The the most statistically significant differences among the groups, relative to AF64A-injected rats, were obtained for the straightness index.

The depression was slightly pronounced during all experiments. However, on day 26, some rats receiving Memantine and control rats swam reluctantly, demonstrating stress or less motivation (data not shown) that could led to an increase in the escape latency.

Summarizing the results obtained at this stage, it was noted that Aricept, Memantine, and NT-1505 demonstrated statistically significant improvement in the learning ability of the experimental animals so that it became equal to the results for the control group, or even exceeded it. A significant imperfection of Aricept is its symptomatic effect, namely, cancellation of injections (results obtained on day 26) leads to a decrease in the cognitive functions of rats down to the level of the untreated group. Memantine and NT-1505 demonstrated a prolonged effect, observed after completing their injections, that allowed us to suppose that they influence regenerative and/or compensatory processes in brain. All the above data permit us to conclude that NT-1505 improves learning of rats with partial chronic deprivation of cholinergic functions in the task of finding a hidden platform, and demonstrates a long-term effect on their retention ability after finishing training and injection of the compound. The efficiency of its influence on spatial learning in the chronic experiment is almost the same as for Aricept. However, NT-1505 causes a more prolonged effect on memory functions than Aricept, which let us assume its effect on regenerative and/or compensatory processes in brain.

To reveal possible similarities and differences between NT-1505 and MK-801, a comparative study of the effects of the compounds at various doses on locomotor activity of rats was performed. It was found that MK-801 in doses of 0.1 mg/kg and higher produces a significant psychogenic effect in animals manifesting in stereotypic behavior, ataxia, and hyperactivity (see TABLE 2). In contrast to MK-801, agent NT-1505 did not potentiate stereotypy, ataxia, or behavioral activity; that is, it did not exhibit any side effects typical of MK-801. It is important to note that the absence of MK-801-like side effects is observed over wide interval of doses, from 0.5 to 50 mg/kg, which completely covers the interval of its neurological activity

TABLE 2. Side effects of MK-801 and NT-1505

MK-801				NT-1505			
ED_{50} (Anti-NMDA) = 0.2 mg/kg				ED_{50} (Anti-NMDA) = 10 mg/kg			
Dose, mg/kg	Stereotypic movement	Ataxy	Activity	Dose, mg/kg	Stereotypic movement	Ataxy	Activity
0.1	NO	NO	↑	0.5	NO	NO	NO
0.2	↑	↑	↑↑	1	NO	NO	NO
0.4	↑	↑↑	↑↑↑	5	NO	NO	NO
0.8	↑↑↑	↑↑↑	↑↑↑	10	NO	NO	NO
1.0	↑↑↑↑	↑↑↑↑	↑↑↑↑	50	NO	NO	↓

(according to previously published results, NT-1505 shows anti-NMDA activity in doses 10–50 mg/kg and exhibits cognition-enhancing properties at doses of 0.5–5 mg/kg). The results obtained permit us to conclude that NT-1505 possesses strong cognition-enhancing properties, manifested in a toxin-induced animal model of AD-type degeneration and characterized by a spectrum of behavioral activity radically different from that of MK-801. In particular, it does not produce any psychotomimetic side effects over the therapeutically significant dose interval.

ACKNOWLEDGMENTS

This work was supported in part by the Russian Foundation for Basic Research (RFBR Grant 00-04-48398 and Grant 98-04-48616) and Russian Federal Program "Integration" (Grant 312).

REFERENCES

1. MATTSON, M.P., B. CHENG, D. DAVIS, et al. 1992. β-Amyloid peptodes destabilize calcium homeostasis and render human cortical neurons vulnerable to excitotoxicity. J. Neurosci. **12:** 376–389.
2. GREENAMYRE, J.T., J.B. PENNEY, C.J. D'AMATO & A.B. YOUNG. 1987. Dementia of the Alzheimer's type: changes in hippocampal L-[^{3}H]glutamate binding. J. Neurochem. **48:** 543–551.
3. VANDERSCHUREN, L.J.M.J., A.N.M. SCHOFFELMEER, A.H. MULDER & T.J. DE VRIES. 1998. Trend Dizocilpine (MK801): use or abuse? Pharm. Sci. **19:** 79–81.
4. BLANPIED, T.A., F.A. BOECKMAN, E. AIZENMAN & J.W. JOHNSON. 1997. Trapping channel block of NMDA-activated responses by amantadine and memantine. J. Neurophysiol. **77:** 309–323.
5. PALMER, G.C., J.A. MILLER, E.F. CREGAN, et al. 1997. Low-affinity NMDA receptor antagonists. The neuroprotective potential of ARL 15896AR. Ann. N.Y. Acad. Sci. **825:** 220–231.
6. PARSONS, C., W. DANYSZ & G. QUACK. 1998. Glutamate in CNS disorders as a target for drug development. Drug News Perspect. **11:** 523–579.
7. KATRITZKY, A., J. KIELY, N. HEBERT & C. CHASSAING. 2000. Definition of Templates within Combinatorial Libraries. J. Comb. Chem. **2:** 2–5.
8. DOYLE, D., J. CABRAL, R. PFUETZNER, et al. 1998. The structure of the potassium channel: molecular basis of K^+ conduction and selectivity. Science **280:** 69–77.
9. ABOLA, E., F. BERNSTEIN, S. BRYANT, et al. 1987. Protein data bank. In Crystallographic Databases—Information Content, Software Systems, Scientific Applications. F.H. Allen, G. Bergerhoff & R. Sievers, Eds.: 107–132. Data Commission of the International Union of Crystallography, Bonn, Cambridge, Chester.
10. HIGGINS, D. & P. SHARP. 1989. Fast and sensitive multiple sequence alignments on a microcomputer. CABIOS **5:** 151–153.
11. HIGGINS, D., A. BLEASBY & R. FUCHS. 1991. CLUSTAL V: improved softwarefor multiple sequence alignment. CABIOS **8:** 189–191.
12. THOMPSON, J.D., D.G. HIGGINS & T.J. GIBSON. 1994. CLUSTAL W: improving the sensitivity of progressive multiple sequence alignment through sequence weighting, positions-specific gap penalties and weight matrix choice. Nucleic Acids Res. **22:** 4673–4680.
13. SYBYL 6.5; Tripos Assoc.: St. Louis, MO.
14. SOMMER, B., M. KOEHLER, R. SPRENGEL & P.H. SEEBURG. 1991. RNA editing in brain controls a determinant of ion flow in glutamate-gated channels. Cell **67:** 11–19.

15. BURNASHEV, N., R. SCHOEPFER, H. MONYER, et al. 1992. Control by asparagine residues of calcium permeability and magnesium blockage in the NMDA receptor. Science **257**: 1415–1419.
16. FERRER-MONTIEL, A.V., W. SUN & M. MONTAL. 1995. Molecular design of the N-methyl-D-aspartate binding site for phencyclidine and dizolcipine. Proc. Natl. Acad. Sci. U.S.A. **92**: 8021–8025.
17. MESULAM, M.M. 1996. The system level organization of cholinergic innervation in the human cerebral cortex and its alterations in Alzheimer's disease. Prog. Brain Res. **109**: 285–297.
18. HANIN, I. 1996. The AF64A model of cholinergic hypofunction: an update. Life Sci. **58**: 1955–1964.
19. WALSH, T. & K. OPELLO. 1994. The use of AF64A to model Alzheimer disease. *In* Toxin-Induced Models of Neurological Disorders. 259–337. Plenum Press, New York, London.
20. FISHER, A., C. MANTIONE., D. ABRAHAM & I. HANIN. 1982. Long-term central cholinergic hypofunction induced in mice by ethylcholine aziridinium ion (AF64F) *in vivo*. J. Pharm. Exp. Ther. **22**: 140–145.
21. MUKHINA, T.V., N.N. LERMONTOVA & S.O. BACHURIN. 2000. Automated system for tracking and behavioral analysis. Presented at the 3rd International Conference on Methods and Techniques, Nijmegen, The Netherlands, 15–18 August.
22. MUKHINA, T.V. & S.O. BACHURIN. 2000. Software-hardware complex behavioral vision ("BVision"). Russian certificate for registration of the software for PC. No. 2000610678. Date of registration, July 25, 2000.
23. LERMONTOVA, N., N. LUKOYANOV, T. SERKOVA, et al. 1998. Effect of tacrine on deficits in acute avoidance performance induced by AF64A in rats. Molec. Chem. Neuropathol. **33**: 51–61.

Effects of Chronic NMDA Receptor and Fast Sodium Channel Blockade during Development on the Acquisition of Visual Discriminations and Learning Abilities in Rhesus Monkeys

MERLE G. PAULE,[a] E. JON POPKE,[a] RICHARD R. ALLEN,[b] EDWIN PEARSON,[c] AND TIM HAMMOND[c]

[a]*Behavioral Toxicology Laboratory, Division of Neurotoxicology, National Center for Toxicological Research, Jefferson, Arkansas, U.S.A.*

[b]*Peak Statistical Services, 5691 Northwood Drive, Evergreen, Colorado, U.S.A.*

[c]*Safety Assessment, AstraZeneca Charnwood, Bakewell Road, Loughborough, Leicester, England*

KEYWORDS: Dizocilpine; Chronic toxicity; Behavioral toxicity; Operant behavior.

The effects of MK-801 (a relatively selective NMDA receptor antagonist) and remacemide (an NMDA receptor antagonist that also blocks fast sodium channels) on the acquisition and performance of several complex, food-reinforced operant behaviors was assessed in juvenile female rhesus monkeys. Color and position discrimination, short-term memory, learning, and motivation were modeled using conditioned position responding, delayed matching-to-sample, incremental repeated acquisition, and progressive ratio tasks, respectively. Low and high doses of both drugs (0.1 and 1.0 mg/kg/day for MK-801 and 20 and 50 mg/kg/day for remacemide) were administered by oral gavage immediately after daily (M-F) behavioral assessments and at the same time of day on Saturday and Sunday for 18 months. Neither drug had any significant effects on acquisition of short-term memory task performance. Additionally, neither drug decreased response rate in any task at any dose or time, suggesting that there were no adverse effects of treatment on aspects of motivation or motoric capabilities. Low doses of either compound had no significant effects on any of the behaviors monitored. Developmental exposure to high doses of either MK-801 or remacemide significantly delayed—but did not prevent—acquisition of color and position discrimination. Developmental exposure to high doses of remacemide significantly disrupted acquisition of learning task performance and this effect persisted during subsequent three-month periods of dose reduction and cessation, respectively. These data suggest that chronic blockade of fast sodium channels (perhaps in conjunction with NMDA receptor blockade) during development has long-term consequences for the development of important brain functions.

Address for correspondence: Merle G. Paule, Ph.D., Head, Behavioral Toxicology Laboratory, Division of Neurotoxicology, HFT-132, National Center for Toxicological Research, 3900 NCTR Road, Jefferson, AR 72079-9502, U.S.A. Voice: 870-543-7147; fax: 870-543-7720.
mpaule@nctr.fda.gov

Synaptic Deprivation and Age-Related Vulnerability to Hypoxic-Ischemic Neuronal Injury

A Hypothesis

ANN M. MARINI,[a] JOHN CHOI,[a,b] AND ROBERT LABUTTA[a,b]

[a]*Departments of Neurology and Neuroscience,
Uniformed Services University of the Health Sciences, Bethesda, Maryland, U.S.A.*

[b]*Department of Neurology, Walter Reed Army Medical Center, Washington, D.C., U.S.A.*

ABSTRACT: Advanced age is associated with physiological changes, such as cerebral autoregulation dysfunction, atrial fibrillation, reduced cerebral blood flow, elevated blood pressure, and other changes. Stroke-related dementia is associated with brain loss principally due to strokes, and neuropathological examination of the brains of old people shows a direct correlation between the extent of brain loss and dementia. However, the exact mechanism of the age related vulnerability to hypoxic-ischemic neuronal injury remains unknown. The majority of synapses in the brain use excitatory amino acids as their neurotransmitter. Glutamate, a major endogenous excitatory amino acid required for normal physiological excitation, is also involved in the pathophysiology of hypoxic-ischemic neuronal injury. The *N*-methyl-D-aspartate (NMDA) glutamate receptor subtype plays a major role in mediating hypoxic-ischemic neuronal injury. NMDA receptors also mediate adaptive responses important for synaptic plasticity. This report explores the possible role of synaptic activity as a protective mechanism against neuronal cell death. Specifically, the role of NMDA receptors in neuronal plasticity by upregulating a survival pathway is discussed. Loss of a neuronal population that uses glutamate as its neurotransmitter leads to a loss of activity on the postsynaptic neurons or synaptic deprivation. Deprivation of excitatory amino acids on the postsynaptic neurons results in the failure of activity-dependent induced intrinsic survival pathways induced by NMDA receptors. The loss of neuroprotective intrinsic survival pathways increases the vulnerability of these neurons to more hypoxic-ischemic neuronal damage. Since cerebral infarction is also age related, this hypothesis provides a plausible explanation of how we become more vulnerable to hypoxic-ischemic neuronal injury as a function of age.

KEYWORDS: Synaptic deprivation; Age related vulnerability; Hypoxic-ischemic neuronal injury; NMDA receptors; Plasticity; Intrinsic survival pathways; Excitatory amino acids.

Address for correspondence: Ann M. Marini, Departments of Neurology and Neuroscience, Uniformed Services University of the Health Sciences, Bethesda, MD 20814, U.S.A.

BACKGROUND

Age is the major unmodifiable risk factor for hypoxic and ischemic neuronal injury. The stroke risk doubles with each decade past the age of 55 and two-thirds of all cerebral infarctions occur in people over the age of 65. Stroke related dementia and Alzheimer's disease also increase exponentially with age.[1] Stroke related dementia or vascular dementia is a clinical syndrome defined as acquired cognitive impairment due to ischemic neuronal injury. Another age related stroke risk factor is atrial fibrillation and the incidence of atrial fibrillation also doubles with every decade of life starting at age 55.[2] Atrial fibrillation is an independent risk factor for poor cognition[3-5] and may also have a role in accelerating Alzheimer's disease expression.[6]

Age related changes in cerebral autoregulation and cerebral blood flow may increase neuronal vulnerability to hypoxia and ischemia. Cerebral autoregulation dysfunction may result in borderzone and end-arterial ischemia in the presence of labile blood pressure. The presence of white matter lesions, an age related phenomenon, is associated with borderzone infarctions.[7] Furthermore, atherosclerosis may increase susceptibility to borderzone infarction as observed in carotid artery disease and internal borderzone infarcts.[8] Cerebral blood flow decreases with age.[9,10] This reduction of blood flow may enhance neuronal vulnerability by eliminating any reserve during hypoxic or low flow states.

STROKE RISK FACTORS AND COGNITIVE IMPAIRMENT

Detailed neuropathological examinations of the brains of old people reveals a direct relationship between the extent of brain loss and dementia.[11] Since this published report and the advent of CT and MRI scans, accumulated evidence indicates that stroke related dementia is associated with a higher number of infarcts,[12-14] more neuronal substance loss in the dominant hemisphere, deep frontal strokes,[13] and the total extent of cerebral brain loss.[15]

Risk factors associated with cognitive impairment due to vascular dementia include hypertension,[12] cigarette smoking,[16] diabetes mellitus,[13] hypercholesterolemia,[17] and heart attack.[16] Recent evidence suggests that there is a significant association between hypertension, cerebral atrophy, left hemispheric atrophy, and an increased incidence of white matter hyperintensities.[18] In another study, the volume of white matter hyperintensities, left parietal infarcts, and the number of thalamic lacunæ were important for the development of dementia.[19] Many studies suggest that cerebrovascular disease amplifies the clinical expression of Alzheimer's disease.[20-21] No differences in the amount of neuritic plaques or neurofibrillary tangles were noted on neuropathology but the presence of vascular lesions affected the severity of dementia.[22] Small vessel ischemic disease appears to have the most significant effect.[20,23] In addition, amyloid angiopathy may occur in associated with Alzheimer's disease. The consequent damage to the media and adventitia of the microvasculature[24] may promote additional brain injury via accelerated atherosclerosis and intracerebral hemorrhages. Genetic factors also play a role in stroke and dementia but the exact underlying mechanisms of how these factors are involved with aberrant protein production, loss of brain substance, and cerebral infarctions remain to be elucidated. One genetic component, apolipoprotein

E-4 (Apo E-4), was associated with cognitive impairment, brain atrophy, and infarct like lesions on MRI.[25]

The effect of neurodegenerative diseases on memory has lead to several studies on the temporal lobe. Several investigators have found that medial temporal lobe atrophy is strongly associated with dementia[26–28] and appears to be independent of ApoE-4.[29] The hippocampus has the highest concentration of glucocorticoid receptors in the brain. People with Cushing's syndrome, a disorder associated with hypercortisolemia, experience neuropsychiatric disturbances.[30,31] Congenital adrenal hyperplasia studies have suggested possibly lower intelligence and increased learning disabilities due to early androgen exposure.[32,33] On the cellular level, glucocorticoids enhance oxidative stress induced cell death in hippocampal neurons *in vitro*[34–36] and are associated with hippocampal dendritic atrophy.[37] Thus, increased glucocorticoids induced by stress disorders may be another mechanism for increasing the vulnerability of neurons to injury and death.

HYPOTHESIS

Dementia is a multifactorial disease involving many levels of dysfunction. Alzheimer's and vascular dementia share many features mentioned above and ultimately result in the loss of neurons. Neuronal loss from discrete and strategic brain regions, such as the hippocampus or thalamus, produces memory loss or the inability to form new memory. We hypothesize that there are other consequences of neuronal loss and the resultant presynaptic deprivation. The remaining postsynaptic neurons may have a loss of plasticity and adaptability that renders them more susceptible to hypoxic/ischemic insult as age increases.

NEURONAL DEVELOPMENT

Undifferentiated neurons migrate to their designated brain region where they differentiate into mature neurons. The differentiation process requires many steps but one of the central features of differentiation is the formation of processes called dendrites and axons. Numerous dendrites bring information into the cell body and the single axon sends information away from the cell body. Neurons communicate with one another through these processes. A presynaptic bouton is at the terminal end and is separated from the postsynaptic neurons by a synaptic cleft. Neurotransmitters are released and bind to postsynaptic sites or receptors. The binding activates receptors with resultant electrical, biochemical and molecular changes within the postsynaptic neuron.

EXCITATORY AMINO ACID RECEPTOR ACTIVATION

There are many receptors on neurons that mediate intracellular events specific to each individual stimulus. The vast majority of synapses in the central nervous system use excitatory amino acids as their neurotransmitters. Excitatory amino acid receptors not only mediate normal synaptic transmission but also participate in the modification of synaptic connections during development and changes in the efficacy of synaptic transmission. In addition to participating in synaptic plasticity through physiological excitation, excitatory amino acids can cause neurodegeneration via excessive activation of their receptors.

EXCITATORY AMINO ACIDS AND RECEPTOR SUBTYPES

There are two general classes of excitatory amino acid receptors called ionotropic and metabotropic. The ionotropic receptors include the N-methyl-D-aspartate (NMDA) and α-amino-3-hydroxy-5-methyl-4-isoxazolepropionic acid (AMPA), kainate, and AP-4 receptors. There are three groups of metabotropic glutamate receptors (Group I–III) that either stimulate inositol phosphate metabolism (38) or are negatively coupled to adenylate cyclase.[39] Glutamate is the endogenous neurotransmitter required for normal physiological excitation in the central nervous system and communicates with other neurons by binding to all of the glutamate receptor subtypes. Glutamate is also involved in the pathophysiology of hypoxic and ischemic neuronal damage and although this neuropathological process can be mediated by any of the excitatory amino acid receptors, the NMDA glutamate receptor subtype plays a major role.[40,41] The neuronal damage incurred by hypoxia or ischemia is delayed, since it has been shown that neurons appear morphologically normal for about one day followed by neurodegeneration.[42] In addition to the evidence linking excitatory amino acid receptors to hypoxic and ischemic neuronal injury, evidence has accumulated to suggest that the adrenal stress hormones glucocorticoids increase neuronal vulnerability to hypoxic and ischemic neuronal damage.[43] Stress levels of glucocorticoids induced a four-fold increase in the local extracellular concentration of glutamate in the hippocampus.[44] It has also been shown that hippocampal injury is dependent, in part, on NMDA receptor activation.[45] Moreover, a significant number of patients with depression and dementia in later life were found to have an increase in the glucocorticoid cortisol level.[46] These results suggest that multiple factors may be involved in age related brain substance loss. Thus, one of the consequences of degeneration of specific neuronal populations is the loss of presynaptic input that the neuronal population communicated with along the synaptic pathway.

SYNAPTIC PLASTICITY AND EXCITATORY AMINO ACIDS

Synaptic plasticity may be defined as the ability of synapses to modify their synaptic strength in response to activity. For example, high frequency synaptic stimulation results in long term potentiation in the hippocampus that is thought to be the basis of learning and memory and mediated by NMDA receptors.[47] Activation of NMDA receptors in this model allows for short term and long term changes in neurons and results in activation of pre- and postsynaptic mechanisms to produce persistent changes in synaptic strength. Recent studies have now provided direct evidence for activity dependent changes in synaptic structure. These are required for synaptic plasticity and may underlie memory formation or other plastic responses. For example, it has been shown that the NMDA receptor subunits NR1 and

NR2A-2C are enriched in the postsynaptic density, a region that plays a major role in neurotrophin mediated synaptic plasticity.[48] Enhanced phosphorylation of NR1 by specific kinases resulted in non-covalent association with activated protein kinase Cγ and the major postsynaptic density protein suggesting that NMDA receptors are regulated by phosphorylation.[48] Neurotrophins such as brain derived neurotrophic factor (BDNF) can also modulate the phosphorylation state of NR2B suggesting another mechanism of synaptic plasticity regulation.[49] Recently, it has also been shown that physiologic synaptic activity increases a novel protein called Arc or activity regulated, cytoskeleton associated protein.[50] This protein is enriched in brain and found in the subplasmalemma cortex of the neuronal soma and dendrites. The protein is induced under normal physiological synaptic activity and by NMDA receptor dependent synaptic stimuli in association with long term potentiation.[50]

Although synaptic plasticity in the form of long-term potentiation has been known to exist in the hippocampus and the neocortex, it was not known to occur in a teleological older part of the brain, the cerebellum. Recently, however, long term potentiation has been demonstrated in depolarized cerebellar granule cells *in vivo* and is mediated by NMDA and metabotropic glutamate receptors.[51] Intrinsic excitability potentiation in the deep nuclear neurons of the cerebellum has also been observed.[52] Mossy fiber high frequency stimulation modulated granule cell excitability where stronger inputs resulted in potentiation to increased excitatory postsynaptic potentials and spike generation.[53] Thus, the cerebellum also appears to be a site for synaptic plasticity in the form of long term potentiation.

Future research will no doubt unravel the purpose of long term potentiation in the cerebellum. Further knowledge of NMDA receptor diversity will likely reveal more similarities between the cerebellum and hippocampus. Currently, the similarities between NMDA receptors in the cerebellum and hippocampus include synaptic plasticity and neurodegeneration. Despite these similarities, the cerebellum is less vulnerable to neurodegenerative disorders compared to the hippocampus. The mechanism(s) leading to the selective vulnerability remain unclear.

NMDA RECEPTOR MEDIATED NEUROPROTECTION IN CULTURED CEREBELLAR GRANULE CELLS

We used cultured rat cerebellar granule cells as an *in vitro* model for NMDA receptor mediated excitotoxicity and neuroprotection. Since the initial observation that the glutamate uptake blocker, DL-threo-3-hydroxyaspartic acid (THA), protected cultured rat cerebellar granule cells from hypoglycemia induced neuronal cell death,[54] we have characterized this intrinsic survival pathway mediated by subtoxic concentrations of NMDA. The neuroprotective effect of NMDA is activity dependent since NMDA receptors must be activated by subtoxic concentrations of NMDA in order for NMDA to mediate neuroprotection. We have shown recently that one of the molecular mechanisms responsible for NMDA neuroprotection involves a member of the neurotrophin family, brain derived neurotrophic factor (BDNF). The neurotrophin family are a family of trophic factors related by primary amino acid sequence homology and whose members include brain derived neurotrophic factor, nerve growth factor (NGF), neurotrophin-3 (NT-3), and NT-4/5.[55–60] The biological activity of neurotrophins depends upon the activation of high affinity receptors (Trk), a family of structurally related receptors that have similar intrinsic protein–tyrosine kinase activity[61] but different ligand binding properties. TrkA, TrkB, TrkC are, respectively, the receptors for NGF, BDNF, and NT-3.[62–64] The binding of specific neurotrophins to their cognate receptor results in a signal transduction cascade, with Trk phosphorylation serving as the primary event in the initiation of the signaling leading to the upregulation of several genes and their products as a later event. Neurotrophins have also been shown to attenuate excitotoxic cell death suggesting that activation of Trk receptors by enhanced phosphorylation may participate in protecting neurons against excitotoxic cell death. Brain derived neurotrophic factor has been shown to function in an autocrine manner in the peripheral nervous system.[65] Because it had been shown that BDNF could protect cerebellar granule cells *in vitro* from NMDA receptor mediated excitotoxicity,[66] we hypothesized that BDNF may play an important role in modulating glutamate mediated excitotoxicity through activation of NMDA receptors. We found that NMDA exerted trophic activity by activating neurotrophin signaling in that BDNF is the only neurotrophin that protected cerebellar granule cells against glutamate excitotoxicity. Importantly, we showed that NMDA elicited a time dependent increase in BDNF protein in the culture medium and TrkB tyrosine phosphorylation, suggesting that one of the mechanisms by which NMDA protects neurons against glutamate toxicity is by activation of TrkB receptors mediated by the release of BDNF into the culture medium. Although BDNF protects cultured granule cells against the excitotoxic effects of glutamate, BDNF does not have the same profile as observed with NMDA in that BDNF requires longer exposure times to elicit the same level of protection. Thus, the activity dependent autocrine loop, activation of TrkB receptors by BDNF, serves to enhance and strengthen the neuroprotective effect mediated by the activation of NMDA receptors because blocking the activation of TrkB receptors with the fusion protein TrkB-IgG or the Trk signal transduction pathway with K252a completely blocks NMDA neuroprotection.[67] These results also indicate that NMDA receptors regulate the release of BDNF where a rapid release of BDNF was observed within two minutes; sustained levels of BDNF were observed even after a three-hour incubation with NMDA. Within this later time frame (three hours), we observed an increase in BDNF mRNA levels that corresponds to enhanced transcription since previous results showed that pretreatment with actinomycin D abolished NMDA neuroprotection.[68] Thus, the neuroprotective effect of NMDA requires two temporally distinct responses: an early release of BDNF protein and a later increased BDNF synthesis and release. In this way, the autocrine loop can be maintained and perpetuated for as long as NMDA receptors are activated.

As indicated above, BDNF modulates both acute and long term synaptic function through NMDA receptor phosphorylation/dephosphorylation and the major postsynaptic density protein.[48,49] Our observations suggest that activity dependent neurotrophin signaling via an autocrine loop may play an important role in synaptic plasticity, neuronal growth, and development. These intrinsic survival pathways may play an important role in synaptic plasticity to promote neuronal survival. In addition, pretreatment of the cultured neurons with NMDA can be viewed as preconditioning, perhaps analogous to the tolerance observed in brain *in vivo* following repeated ischemia, as well as other insult models.[69] To this end, we have undertaken studies to further define other molecular mechanisms involved in NMDA receptor mediated neuroprotection.

THE ROLE OF IMMEDIATE EARLY GENES IN NMDA RECEPTOR MEDIATED NEUROPROTECTION

Long term responses in neurons involved in activity dependent processes require synthesis of new mRNA and proteins. We have shown that NMDA neuroprotection results in the rapid release of BDNF protein into the culture medium that was identical to recombinant BDNF. NMDA increased intracellular BDNF and extracellular concentrations of BDNF, such that by three hours BDNF levels in the culture medium increased more than three-fold. Enhanced release of BDNF by NMDA is also associated with a later synthesis of BDNF mRNA that is correlated with the increased intracellular levels of BDNF protein.[67] Growth factor stimulation induces a rapid and transient expression of a set of genes, termed immediate early genes (IEG), that encode for transcription factors as well as other molecules believed to regulate long term cellular responses.[70] We showed previously that pretreatment of the cultured neurons with cycloheximide, a known protein synthesis inhibitor, blocked the neuroprotective activity of NMDA. These results suggest that transcription factors play a role in NMDA neuroprotection. We sought to determine IEGs, other than c-fos and jun that are known to be induced by glutamate, that may be involved in NMDA receptor mediated neuroprotection.

The nuclear factor κB (NF-κB) family of dimeric transcription factors regulates genes involved in many different functions including cell survival and cell death. Also known as the Rel family of transcription factors, the transcriptionally active forms of NF-κB are made up from combinations of the five monomeric polypeptides, p50, p65, p52, and RelB.[71] Previous studies demonstrated that glutamate activates NF-κB in neurons.[72,73]

NF-κB is activated in neurons following injury to the central nervous system, but its role is unclear. For example, it has been shown that when NF-κB activation is blocked in cultured rat cerebellar granule cells in the presence of an excitotoxic concentration of glutamate, neuronal survival is observed.[74] It has also been shown that suppression of NF-κB protects hippocampal cultures against oxidative damage[75] but induces nerve growth factor resistant apoptosis in PC12 cells.[76] In contrast, mouse fibroblasts deficient in the p65 subunit of NF-κB are susceptible to the toxic effects of tumor necrosis factor-κ, whereas cells containing the p65 subunit were unaffected by tumor necrosis factor-α.[77] Thus, it appears that NF-κB can mediate either cell death or survival depending upon the stimulus and cell type.

We first determined the time– and concentration–course of neuroprotection by NMDA in cultured rat cerebellar granule cells. Pretreatment of cultured cerebellar granule cells with a maximum neuroprotective concentration of NMDA (100 μM) for variable times showed a time dependent increase in neuroprotection against glutamate induced cell death (see FIGURE 1). Pretreatment of neuronal cultures with NMDA for six hours protected all of the vulnerable neurons against glutamate mediated cell death. To further characterize the neuroprotective properties of NMDA, we examined different subtoxic concentrations of NMDA against glutamate neurotoxicity (100 μM,

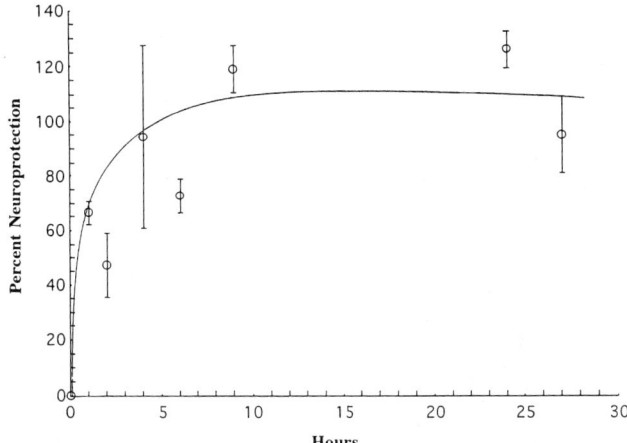

FIGURE 1. *N*-methyl-D-aspartate neuroprotection is time dependent. Rat cerebellar neurons were pretreated with NMDA (100 μM) for the indicated times followed by removal and replacement of the medium with sister culture medium. An excitotoxic concentration of glutamate (100 μM) was added and neuronal viability was assessed 24 hours later with fluorescein diacetate. Data are expressed as mean ± SEM of two separate experiments ($n = 3$).

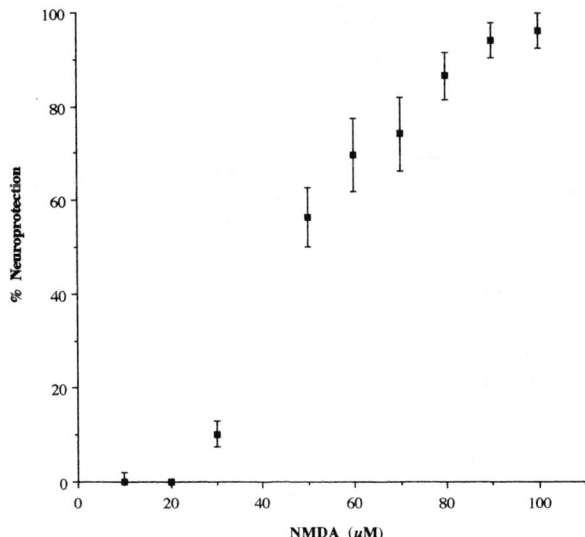

FIGURE 2. *N*-methyl-D-aspartate neuroprotection is concentration-dependent. Rat cerebellar neurons on day eight *in vitro* were pretreated with the indicated NMDA concentration for six hours followed by removal and replacement of the culture medium. Glutamate (100 μM) was added and neuronal viability was assessed 24 hours later with fluorescein diacetate. Data are expressed as mean ± SEM of two separate experiments ($n = 3$).

see FIGURE 2). Nearly 100% neuroprotection was achieved when cultures were pretreated with 100 μM NMDA (FIG. 2). The concentration response curve is consistent with the hypothesis that NMDA receptors require a requisite level of stimulation to mediate neuroprotection, whereas the time dependent nature of the neuroprotective activity of NMDA suggest that gene expression is required to achieve maximal neuroprotection. We have shown that BDNF mRNA begins to increase significantly within three hours of NMDA receptor activation.[67] The time dependent nature of the transcriptional activation of BDNF mRNA suggests that transcription factors may be involved prior to the observed increase in BDNF mRNA.

We showed previously that the protein synthesis inhibitor cycloheximide blocked the neuroprotective effect of NMDA supporting a role for transcription factors. In addition, glutamate receptor agonists have been shown to activate NF-κB. Thus, we tested the hypothesis that NF-κB played an important role in the neuroprotective activity of NMDA. We tested whether NF-κB played a role in the neuroprotective activity of NMDA using a double stranded DNA oligonucleotide based on a candidate site we observed within the 5′ flanking region of exon III from the rat BDNF gene. Once this target DNA is internalized, it binds activated NF-κB located in the cytoplasm, preventing it from entering the nucleus and binding to NF-κB binding motifs. Pretreatment of the neurons with this target DNA in the presence of a maximum neuroprotective concentration of NMDA followed by an excitotoxic concentration of glutamate completely blocked NMDA neuroprotection (see FIGURE 3). We also prepared a target DNA composed of the same nucleotides but in random order as a control (scramble oligonucleotide). We observed no toxicity when neurons were exposed to either the NF-κB target DNA alone (FIG. 3D) or to the scramble oligonucleotide (results not shown). These results suggest that NF-κB is activated by NMDA and show for the first time that the neuroprotective activity of NMDA requires activation of NF-κB to block glutamate-mediated excitotoxicity.

REVERSAL AND PREVENTATIVE STRATEGIES AGAINST BRAIN SUBSTANCE LOSS

Stroke may be a preventable neurological disease. The two types of risk factors for stroke are controllable or uncontrollable. Controllable risk factors include hypertension, hypercholesterolemia, atrial fibrillation and coronary artery disease, smoking, history of a stroke or transient ischemic attack (TIA), excessive alcohol, and obesity. It is of utmost importance for all providers in the health care system to be aggressive in controlling these risk factors. Similarly, people who have these modifiable risk factors should ensure commitment to risk factor reduction and compliance with current medical therapy. The importance of preventing stroke by risk factor modification is underscored by neurologic disability that accounts for over forty billion dollars per year in the United States. Neurologic disability is the loss of brain function, which is to receive, associate, and process information from the external milieu. The billions of neurons and intricate neurite network within the brain impart this enormous function. Examples are cited in the following paragraph that are used to emphasize the ability of the brain to adapt, and remodel based upon the information it receives via synaptic input and activity.

Animals exposed to an enriched environment stimulates neurogenesis in the dentate gyrus of the hippocampus as well as increasing the number of granule cells and size of the granule cell layer.[78] In an adult mouse model, simple running also enhanced the number of neurons and the formation of new neurons.[79] Older mice also had the capacity for neurogenesis in the dentate gyrus when exposed to an enriched environment.[80] Even after the withdrawal of the enriched environment neurogenesis

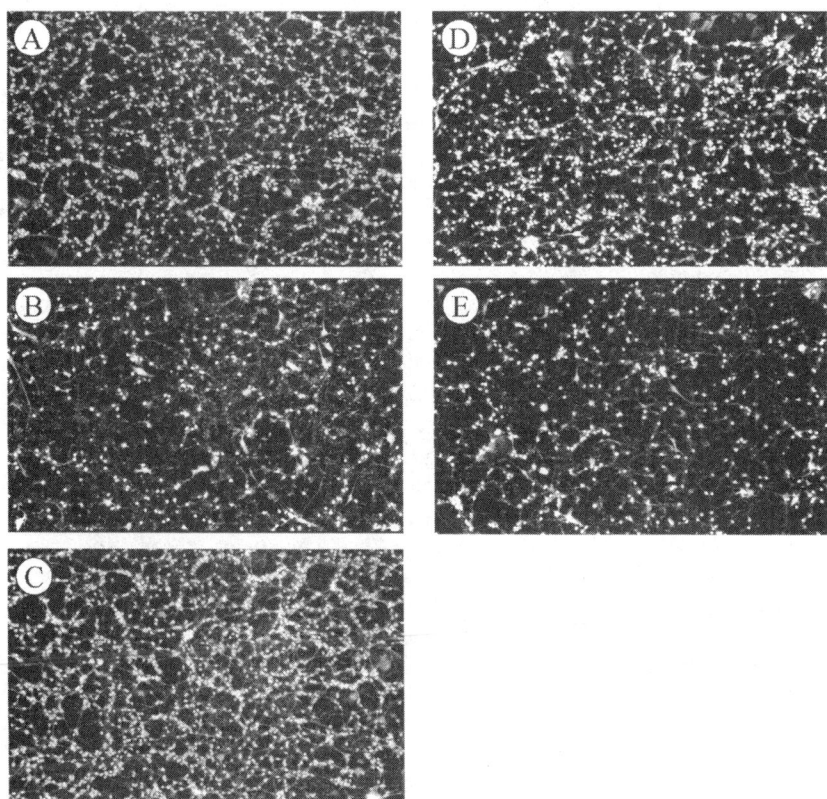

FIGURE 3. The NF-κB decoy oligonucleotide blocks the neuroprotective activity of N-methyl-D-aspartate in cultured neurons. Granule cell neurons (day 7 *in vitro*) were treated with the NF-κB double stranded oligonucleotide for 24 hours, followed by the addition of NMDA (100 μM) for six hours and neuronal viability was assessed 24 hours after the addition of an excitotoxic concentration of glutamate (100 μM) (**E**). Cultured neurons (day 8 *in vitro*) were either treated with glutamate (100 μM) (**B**) or pretreated with a maximum neuroprotective concentration of NMDA (100 μM) for six hours followed by the addition of glutamate (100 μM) (**C**) or treatment with the double stranded decoy oligonucleotide for 24 hours (**D**). Note that pretreatment of the cultured neurons with the decoy oligonucleotide completely blocked the neuroprotective activity of NMDA (100 μM) against glutamate excitotoxicity (compare **E** with **C**) and was comparable to glutamate alone (compare **E** with **B**). Also note that NMDA also blocked glutamate excitotoxicity (compare **C** with **B**) and the NF-κB decoy oligonucleotide did not affect neuronal viability (compare **D** with **A**).

continued.[81] These results suggest the capability of the brain in late life to make adjustments based upon new information received from the enriched environment.

It has been known for a long time that children who undergo hemispherectomy before the age of five can completely or nearly completely recover from their neurologic deficits including language. It was long believed that only young brain had this remarkable ability to remodel. However, observations of adult patients with stroke have also shown that a wide range of recovery exists spanning minimal to extensive recovery.[82–87] Defining the remodeling in human brain after stroke is being undertaken using functional imaging studies.[88–91]

Our own work has shown that activity dependent activation of NMDA receptors results in the release of the neurotrophic factor, BDNF, which in turns binds to TrkB receptors leading to a neuroprotective state that protects all of the vulnerable neurons against the excitotoxic effects of glutamate acting on NMDA receptors. Moreover, activation of NMDA receptors activates nuclear factor κB, a dimeric transcription factor involved in diverse functions of the immune system, cell proliferation, growth factor regulation, and apoptosis. Nuclear factor κB translocates to the nucleus where it binds to κB elements of DNA. Since a double stranded oligonucleotide prepared from the 5' flanking region of Exon III of the BDNF gene blocks the neuroprotective activity of NMDA, perhaps NF-κB plays an important role in protecting neurons against the excitotoxic actions of glutamate.

Neuronal responses are inextricably linked to synaptic activity by neurotransmitters released from nearby neurons. Thus, release of neurotransmitters from presynaptic neurons results in binding to and activating cognate receptors on the postsynaptic neuron that in turn depolarizes and elicits signals as well as other mechanisms to perform a whole host of functions. If neurons are deprived of synaptic activity by neighboring neurons either because of acute neurodegeneration, such as damage from hypoxic and ischemic neuronal injury, or neuronal loss from chronic neurodegenerative diseases such as Alzheimer's disease, then synaptic activity to the postsynaptic neuron from that subset of neurons is permanently lost. The significance of lost input into the postsynaptic neuron means an end to all communication from those neighboring neurons. Thus, a part of the postsynaptic neuron becomes quiescent. For example, assuming that excitatory amino acids mediate similar neuroprotective mechanisms *in vivo,* the synaptic activity through release of glutamate acting on NMDA receptors maintains a continuous neuroprotective or survival state through the release of BDNF via an autocrine loop; other types of neuroprotective mechanisms may also exist *in vivo*. If glutamatergic neurons that activate NMDA receptors on postsynaptic neurons are destroyed, the postsynaptic neuron now becomes vulnerable to hypoxic and ischemic neuronal injury because the intracellular postsynaptic intrinsic neuronal survival pathway can no longer be activated. In this case, glutamate acts as a trophic factor. Similarly, it has been shown that estradiol induces formation of new dentritic spines and synapses in the hippocampus and enhances the sensitivity of hippocampal pyramidal cells to NMDA receptor mediated input.[92] These results suggest that estradiol activates an intraneuronal pathway(s) that lead to an increase number of dendritic spines that then increases sensitivity to NMDA receptor mediated input. Moreover, an estrogen antagonist was shown to completely block the formation of spines on CA1 pyramidal neurons.[93] Perhaps, the loss of estrogens in postmenopausal women results in the loss of these dendritic spines and sensitivity to NMDA receptor mediated synaptic input resulting in enhanced neuronal vulnerability to hypoxic and ischemic neuronal injury. Changes in or loss of synaptic activity may also alter the composition of NMDA receptor channels. It has been shown that molecular diversity of NMDA receptor channels exist and that multiple subunits with distinct distributions, properties and regulation, implies that NMDA receptor channels are heterogeneous in their properties dictated by their location in specific brain regions and developmental stage.[94]

In summary, synaptic activity is required for normal neuronal function. Synaptic input from all of presynaptic neurons communicate this information to postsynaptic neurons. Changes in the local neuronal environment, resulting from death of a specific population of neurons, results in synaptic deprivation of the postsynaptic neuron. Once the postsynaptic neuron is deprived of this communication, changes in the postsynaptic neuron result. These changes may be permanent resulting in altered physiology and manifesting the loss of synaptic input by the inability of postsynaptic neurons to adapt, that is, loss of synaptic plasticity. More deleterious results such as the inability to activate intrinsic survival pathways may lead to an enhanced susceptibility to hypoxic and ischemic neuronal damage. The importance of intrinsic survival pathways *in vivo* dictates the fundamental concept that every neuron is important and loss of neurons disrupt brain networking and function of the brain.

REFERENCES

1. Rocco, W.A., A. Hofman, C. Brayne, *et al.* 1991. The prevalence of vascular dementia in Europe: facts and fragments from 1980–1990 studies. Ann. Neurol. **30:** 817–824.
2. Benjamin, E.J., D. Levy, S.M. Vaziri, *et al.* 1994. Independent risk factors for atrial fibrillation in a population-based cohort: The Framingham heart study. JAMA **271:** 840–844.
3. Sabatini, T., G.B. Frisoni, P. Barbisoni, *et al.* 2000. Atrila fibrillation and cognitive function: final results. Circulation **84:** 527–539.
4. Rozzini, R., T. Sabatini & M. Trabucchi. 1999. Chronic atrial fibrillation and low cognitive function. Stroke **30:** 190–191.
5. Kilander, L., B. Andren, H. Nyman, *et al.* 1998. Atrial fibrillation is an independent determinant of low cognitive function: A cross-sectional study in elderly men. Stroke **29:** 1816–1820.
6. Ott, A., M.M. Breteler, M.C. de Bruyne, *et al.* 1997. Atrial fibrillation and dementia in a population-based study. The Rotterdam study. Stroke **28:** 316–321.
7. Mäntylä, R., H.J. Aronen, O. Salonen, *et al.* 1999. Magnetic resonance imaging white matter hyperintensities and mechanism of ischemic stroke. Stroke **30:** 2053–2058.
8. Del Sette, M., M. Eliaziw, J.Y. Streifler, *et al.* 2000. Internal borderzone infarction: a marker for severe stenosis in patients with symptomatic internal carotid artery disease. For the North American Symptomatic Carotid Endarterectomy (NASCET) Group. Stroke **31:** 631–636.
9. Dastur, D.K. 1985. Cerebral blood flow and metabolism in normal human aging, pathological aging, and senile dementia. J. Cereb. Blood Flow Metab. **5:** 1–9.
10. Meyer, J.S., Y. Terayama & S. Takashima. 1993. Cerebral circulation in the elderly. Cerebrovasc. Brain Metabol. Rev. **5:** 122–146.
11. Tomlinson, B.E., G. Blessed & M. Roth. 1970. Observations on the brains of demented old people. J. Neurol. Sci. **11**(3): 205–242.
12. Ladurner, G., L.D. Iliff & H. Lechner. 1982. Clinical factors associated with dementia in ischemic stroke. J. Neurol. Neurosurg. Psychiat. **45:** 97–101.

13. TATEMICHI, T.K., D.W. DESMOND, M. PAIK, *et al.* 1993. Clinical determinants of dementia related to stroke. Ann. Neurol. **33:** 568–575.
14. LOEB, C., C. GANDOLFO & G. BINO. 1988. Intellectual impairment and cerebral lesions in multiple cerebral infarcts. A clinical-computed tomography study. Stroke **19**(5): 560–565.
15. CHARLETTA, D., P.B. GORELICK, T.J. DOLLEAR, *et al.* 1995. CT and MRI findings among African-Americans with Alzheimer's disease, vascular dementia, and stroke without dementia. Neurology **45**(8): 1456–1461.
16. GORELICK, P.B., J.A. BRODY, D.C. COHEN, *et al.* 1993. Risk factors for dementia associated with multiple cerebral infarcts: a case-control analysis in predominantly African-American hospital-based patients. Arch. Neurol. **50:** 714–720.
17. DESMOND, D.W., T.K. TATEMICHI, M. PAIK & Y. STERN. 1993. Risk factors for cerebrovascular disease as correlates of cognitive function in a stroke-free cohort. Arch. Neurol. **50:** 162–166.
18. ELIAS, M.F., A. BEISER, P.A. WOLF, *et al.* 2000. The preclinical phase of Alzheimer disease: a 22-year prospective study of the Framingham cohort. Arch. Neurol. **57**(6): 808–813.
19. LIS, C.G. & M. GAVIRIA. 1997. Vascular dementia, hypertension, and the brain. Neurol. Res. **19:** 471–480.
20. LIN, R.T., C.L. LAI, C.T TAI, *et al.* 1998. Cranial computed tomography in ischemic stroke patients with and without dementia-a prospective study. Kao Hsiung I Hsueh Ko Hsueh Tsa Chih **14:** 203–211.
21. SNOWDEN, D.A., L.H. GREINER, J.A. MORTIMER, *et al.* 1997. Brain infarction and the clinical expression of Alzheimer disease: the nun study. JAMA **277:** 813–817.
22. HOFMAN, A., A. OTT, M.M.B. BRETELER, *et al.* 1997. Atherosclerosis, Apoplipoprotein E, and prevalence of dementia and Alzheimer's disease in the Rotterdam study. Lancet **349:** 151–154.
23. HEYMAN, A., G.G. FILLENBAUM, K.A. WELSH-BOHMER, *et al.* 1998. Cerebral Infarcts in patients with autopsy-proven Alzheimer's Disease: CERAD, part XVIII. Neurology **51:** 159–162.
24. ESIRI, M.M. 2000 Which vascular lesions are of importance in vascular dementia? Ann. N.Y. Acad. Sci. **903:** 239–243.
25. OKAZAKI, H., T.J. REAGAN & R.J. CAMPBELL. 1979. Clinicopathologic studies of primary cerebral amyloid angiopathy. Mayo Clin. Proc. **54:** 22–31.
26. KULLER, L.H., L. SHEMANSKI, T. MANOLIO, *et al.* 1998 Relationship between ApoE, MRI findings, and cognitive function in the cardiovascular health study. Stroke **29:** 388–398.
27. HENOÑ, H., F. PASQUIER, I. DURIEU, *et al.* 1998. Medial temporal lobe atrophy in stroke patients: Relation to pre-existing dementia. J. Neurol. Neurosurg. Psychiatry. **65:** 641–647.
28. FOUNDAS, A.L., C.M. LEONARD, S.M. MAHONEY, *et al.* 1997 Atrophy of the hippocampus, parietal cortex, and insula in Alzheimer's disease: a volumetric magnetic resonance imaging study. Neuropsych. Neuropsychol. Behav. Neurol. **10:** 81–89.
29. JUOTTONEN, K., M.P. LAAKSO, K. PARTANEN & H. SOININEN. 1999 Comparative MR analysis of the entorrhinal cortex and hippocampus in diagnosing Alzheimer disease. AJNR. **20:** 139–144.
30. BARBER, R., A. GHOLKAR, P. SCHELTENS, *et al.* 1999. Apolipoprotein E epsilon4 allele, temporal lobe atrophy, and white matter lesions in late-life dementias. Arch. Neurol. **56**(8): 961–965.
31. STARKMAN, M.N. & D.E. SCHTEINGART. 1981 Neuropsychiatric manifestations of patients with Cushing's syndrome. Arch. Intern. Med. **141:** 215–219.
32. FORGET, H., A. LACROIX, M. SOMMA & H. COHEN. 2000. Cognitive decline in patients with Cushing's syndrome. J. Int. Neuropsychol. Soc. **6:** 20–29.
33. HELLEDAY, J., A. BARTFAI, E.M. RITZEN & M. FORSMAN. 1994. General intelligence and cognitive profile in women with congenital adrenal hyperplasia. Psychoneuroendocrinol. **19:** 343–356.
34. NASS, R. & S. BAKER. 1991. Androgen effects on cognition: congenital adrenal hyperplasia. Psychoneuroendocrinol. **16:** 189–201.

35. BEHL, C., F. LEZOUALC'H, T. TRAPP, et al. 1997. Glucocorticoids enhance oxidative stress-induced cell death in hippocampal neurons in vitro. Endocrinology **138:** 101–106.
36. BEHL, C. 1998. Effects of glucocorticoids on oxidative stress-induced hippocampal cell death: implications for the pathogenesis of Alzheimer's disease. Exp. Gerontol. **33:** 689–696.
37. HOWARD, S.A., A.Y. NAKAYAMA, S.M. BROOKE & R.M. SAPOLSKY. 1999 Glucocorticoid modulation of gp120-induced effects on calcium-dependent degenerative events in primary hippocampal and cortical cultures. Exp. Neurol. **158:** 164–170.
38. RABER, J. 1998 Detrimental effects of chronic hypothalamic–pituitary–adrenal axis activation, from obesity to memory deficits. Mol. Neurobiol. **18:** 1–22.
39. HAYASHI, Y., N. SEKIYAMA, S. NAKANISHI, et al. 1994. Anaylsis of agonist and antagonist activities of pheylglycine derivatives for different cloned metabotropic glutamate receptor subtypes. J. Neurosci. **14:** 3370–3377.
40. SCHOEPP, D.D., D.E. JANE & J.A. MONN. 1999. Pharmacological agents acting at subtypes of metabotropic glutamate receptors. Neuropharmacology **38:** 1431–1476.
41. CHOI, D.W. 1988 Calcium-mediated neurotoxicity: relationship to specific channel types and role in ischemic damage. Trends Neurosci. **11:** 465–469.
42. OLNEY, J.W., O.L. HO & V. RHEE. 1971. Cytotoxic effects of acidic and sulphur-containing amino acids on the infant mouse central nervous system. Exp. Brain Res. **14:** 61–76.
43. PEREZ-PINZON, M.A., C.M. MAIER, E.J. YOON, et al. 1995. Correlation of CGS 19755 neuroprotection against in vitro excitotoxicity and focal cerebral ischemia. J. Cereb. Blood Flow Metab. **15:** 865–876.
44. SAPOLSKY, R.M. 1996. Stress, glucocorticoids, and damage to the nervous system: the current state of confusion. Stress **1:** 1–19.
45. STEIN-BEHRENS, B.A., W.J. LIN & R.M. SAPOLSKY. 1994. Physiological elevations of glucocorticoids potentiate glutamte accumulation in the hippocampus. J. Neurochem. **63:** 596–602.
46. ARMANINI, M.P., C. HUTCHINS, B.A. STEIN & R.M. SAPOLSKY. 1990. Glucocorticoid endangerment of hippocampal neurons is NMDA-receptor dependent. Brain Res. **532:** 7–12.
47. LUPIEN, S.J., N.P. NAIR, S. BRIERE, et al. 1999. Increased cortisol levels and impaired cognition in human aging; implication for depression and dementia in later life. Rev. Neurosci. **10:** 117–139.
48. BLISS, T.V. & G.L. COLLINGRIDGE. 1993. A synaptic model of memory: long-term potentiation in the hippocampus. Nature **361:** 31–39.
49. SUEN, P.C., K. WU, J.L. XU, et al. 1998. NMDA receptor subunits in the postsynaptic density of rat brain: expression and phosphorylation by endogenous protein kinases. Brain Res. Mol. Brain Res. **59:** 215–228.
50. LIN, S.Y., K. WU, G.W. LEN, et al. 1999. Brain-derived neurotrophic factor enhances association of protein tyrosine phosphatase PTP1D with the NMDA receptor subunit NR2B in the cortical postsynaptic density. Brain Res. Mol. Brain Res. **70:** 18–25.
51. LYFORD, G.L., K. YAMAGATA, W.E. KAUFMANN, et al. 1995. Arc, a growth factor and activity-regulated gene, encodes a novel cytoskeleton-associated protein that is enriched in neuronal dendrites. Neuron **14:** 433–445.
52. D'ANGELO, E., P. ROSSI, S. ARMANO & V. TAGLIETTI. 1999. Evidence for NMDA and mGlu receptor-dependent long-term potentiation of mossy fiber-granule cell transmission in rat cerebellum. J. Neurophysiol. **81:** 277–287.
53. AIZENMANN, C.D. & D.J. LINDEN. 2000. Rapid, synaptically driven increases in the intrinsic excitability of cerebellar deep nuclear neurons. Nat. Neurosci. **3:** 109–111.
54. ARMANO, S., P. ROSSI, V. TAGLIETTI & E. D'ANGELO. 2000. Long-term potentiation of intrinsic excitability at the mossy fiber-granule cell synapse of rat cerebellum. J. Neurosci. **20:** 5208–5216.
55. MARINI, A.M. & A. NOVELLI. 1991. The glutamate uptake blocker DL-threo-3-hydroxyaspartate reduces NMDA receptor activation by glutamate in cultured neurons. Eur. J. Pharm. **194:** 131–132.

56. LEVI-MONTALCINI, R. 1987 The nerve growth factor 35 years later. Science **237:** 1154–1162.
57. HOHN, A., J. LEIBROCK, K. BAILEY & Y-A BARDE. 1990. Identification and characterization of a novel member of the nerve growth factor/brain-derived neurotrophic factor family. Nature **344:** 339–341.
58. MAISONPIERRE, P.C., L. BELLUSCIO, S. SQUINTO, et al. 1990. NT-3, BDNF, and NGF in the developing rat nervous system: parellel as well as reciprocal patterns of expression. Neuron **5:** 501–509.
59. JONES, K.R. & L.F. REICHARDT. 1990. Molecular cloning of a human gene that is a member of the nerve growth factor family. Proc. Natl. Acad. Sci. U.S.A. **87:** 8060–8064.
60. ROSENTHAL, A., D.V. GOEDDEL, T. NGUYEN, et al. 1990. Primary structure and biological activity of a novel human neurotrophic factor. Neuron **4:** 767–773.
61. HALLBOOK, F., C.F. IBANEZ & H. PERSSON. 1991 Evolutionary studies of the nerve growth factor family reveal a novel member abundantly expressed in Xenopus ovary. Neuron **6:** 845–858.
62. KAPLAN, D.R. & R.M. STEPHENS. 1994. Neurotrophin signal tranduction by Trk receptor. J. Neurobiol. **24:** 1404–1417.
63. RODRIGUEZ-TEBAR, A., G. DECHANT & Y.-A. BARDE. 1990. Binding of brain-derived neurotrophic factor to the nerve growth factor receptor. Neuron **4:** 487–492.
64. LAMBALLE, F., R. KLEIN & M. BARBACID. 1990. TrkC, a new member of the trk family of tyrosine protein kinases, is a receptor for neurotrophin-3. Cell **66:** 967–979.
65. KLEIN, R., S. JING, V. NANDURI, et al. 1991. The trk proto-oncogene encodes a receptor for nerve growth factor. Cell **65:** 189–197.
66. ACHESON, A., J.C. CONOVER, J.P. FANDL, et al. 1993. Brain-derived neurotrophic factor is a survival factor for cultured rat cerebellar granule cell neurons and protects them against glutamate-induced neurotoxicity. Eur. J. Neurosci. **5:** 1455–1464.
67. MARINI, A.M., S.J. RABIN, R.H. LIPSKY & I. MOCCHETTI. 1998. Activity-dependent release of brain-derived neurotrophic factor underlies the neuroprotective effect of N-methyl-D-aspartate. J. Biol. Chem. **273:** 29394–29399.
68. MARINI, A.M. & S.M. PAUL. 1992. N-methyl-D-aspartate receptor-mediated neuroprotection in cerebellar granule cells requires new RNA and protein synthesis. Proc. Natl. Acad. Sci. U.S.A. **89:** 6555–6559.
69. WIEGAND, F., W LIAO, C. BUSCH, et al. 1999. Respiratory chain inhibition induces tolerance to focal cerebral ischemia. J. Cereb. Blood Flow Metab. **19:** 1229–1237.
70. NATHANS, D., L.F. LAU, B. CHRISTY, et al. 1988. Genomic response to growth factors. Cold Spring Harbor Symposium on Quantitative Biology, **53**(2): 893–900.
71. BAEUERLE, P.A. & T. HENKEL. 1994. Function and activation of NF-κB in the immune system. Annu. Rev. Immunol. **12:** 141–179.
72. GUERRINI, L., F. BLASI & S. DENIS-DONINI. 1995. Syanptic activation of NF-κB by glutamate in cerebellar granule neurons *in vitro*. Proc. Natl. Acad. Sci. U.S.A. **92:** 9077–9081.
73. KALTSCHMIDT, C., B. KALTSCHMIDT & A.P. BAEUERLE. 1995. Stimulation of ionotropic glutamate receptors activates transcription factor NF-κB in primary neurons. Proc. Natl. Acad. Sci. U.S.A. **92:** 9618–9622.
74. GRILLI, M., M. PIZZI, M. MERNO & P. SPANO. 1996. Neuroprotection by aspirin and sodium salicylate through blockade of NF-κB activation. Science **274:** 1383–1385.
75. POST, A., F. HOLSBOER & C. BEHL. 1998. Induction of NF-κB activity during haloperidol-induced oxidative toxicity in clonal hippocampal cells: suppression of NF-κB and neuroprotection by antioxidants. J. Neurosci. **18:** 8236–8246.
76. TAGLIALATELA, G., R. ROBINSON & J.R. PEREZ-POLO. 1997. Inhibition of nuclear factor kappa B (NfkappaB) activity induces nerve growth factor-resistant apoptosis in PC12 cells. J. Neurosci. Res. **47:** 155–162.
77. BEG, A.A. & D. BALTIMORE. 1996. An essential role for NF-κB in preventing TNF-κ-induced cell death. Science **274:** 782–784.
78. KEMPERMANN, G., H.G. KUHN & F.H. GAGE. 1997. More hippocampal neurons in adult mice living in an enriched environment. Nature **386:** 493–495.

79. VAN PRAAG, H., G. KEMPERMANN & F.H. GAGE. 1999. Running increases cell proliferation and neurogenesis in the adult mouse dentate gyrus. Nat. Neurosci. **2:** 266–270.
80. KEMPERMANN, G., H.G. KUHN & F.H. GAGE. 1998. Experience-induced neurogenesis in the senescent dentate gyrus. J. Neurosci. **18:** 3206–3212.
81. KEMPERMANN, G. & F.H. GAGE. 1999. Experience-dependent regulation of adult hippocampal neurogenesis: effects of long-term stimulation and stimulus withdrawal. Hippocampus **9:** 321–332.
82. JOHANSSON, B.B. 2000. Brain plasticity and stroke rehabilitation. The Willis lecture. Stroke **31:** 223–230.
83. NELLES, G., G. SPIEKRAMANN, M. JUEPTNER, *et al.* 1999. Evolution of functional reorganization in hemiplegic stroke: a serial positron emission tomographic activation study. Ann. Neurol. **46:** 901–909.
84. NUDO, R.J. & K.M. FRIEL. 1999. Cortical plasticity after stroke: implications for rehabilitation. Rev. Neurol. (Paris) **155:** 713–717.
85. JORGENSEN, H.S., H. NAKAYAMA, H.O. RAASCHOU & T.S. OLSEN. 1999. Stroke. Neurologic and functional recovery the Copenhagen Stroke Study. Phys. Med. Rehabil. Clin. N. Am. **10:** 887–906.
86. CUADRADO, M.L., J.A. EGIDO, J.L. GONZALEZ-GUTIERREZ & E. VARELA-DE-SEIJAS. 1999. Bihemispheric contribution to motor recovery after stroke: a longitudinal study with transcranial Doppler ultrasonography. Cerebrovasc. Dis. **9:** 337–344.
87. RICHARDS, C.L., F. MALOUIN & C. DEAN. 1999. Gait in stroke: assessment and rehabilitation. Clin. Geriatr. Med. **15:** 833–855.
88. CRAMER, S.C. & E.P. BASTINGS. 2000. Mapping clinically relevant plasticity after stroke. Neuropharmacology **39:** 842–851.
89. CRAMER, S.C. 1999. Stroke recovery. Lessons from functional MR imaging and other methods of human brain mapping. Phys. Med. Rehabil. Clin. N. Am. **10:** 875–886.
90. ROSSINI, P.M., F PAURI, V PIZZELLA, *et al.* 1999. An integrative neuroimaging approach to restorative neurology. Electroenceph. Clin. Neurophysiol. Suppl. **49:** 19–24.
91. BEAULIEU, C., A. DE CRESPIGNY, D.C. TONG, *et al.* 1999. Longitudinal magnetic resonance imaging study of perfusion and diffusion in stroke: evolution of lesion volume and correlation with clinical outcome. Ann. Neurol. **46:** 568–578.
92. WOOLLEY, C.S., N.G. WEILAND, B.S. MCEWEN & P.A. SCHWARTZKROIN. 1997. Estradiol increases the sensitivity of hippocampal CA1 pyramidal cells to NMDA receptor-mediated synaptic input: correlation with dendritic spine density. J. Neurosci. **17:** 1848–1859.
93. MCEWEN, B.S., P. TANAPAT & N.G. WEILAND. 1999. Inhibition of dendritic spine induction on hippocampal CA1 pyramidal neurons by a nonsteroidal estrogen antagonist in female rats. Endocrinology **140:** 1044–1047.
94. YAMAKURA, T. & K. SHIMOJI. 1999. Subunit- and site-specific pharmacology of the NMDA receptor channel. Prog. Neurobiol. **59:** 279–98.

Part VII: NMDA Receptor Modification and Neuroprotection

QUESTIONS FOR DR. MARINI

From Dr. De-Maw Chuang

NFKB activation can be either proapoptotic or antiapoptotic, largely depending on the gene products that are affected. In an animal model of Huntington's disease, in which quinolinic acid is injected into the striatum to induce neurodegeneration, we found that NFKB activation is proapotptic and is involved in the induction of p53, c-myc, and cyclin D1. Therefore, I am wondering whether you have examined the effect of glutamate in cerebellar cells with respect to NFKB activation and the genes that are induced.

ANSWER: It seems that the converging signals the neurons receive from extracellular receptor activation or other modulators will ultimately depend upon whether the neuron will survive or die. For example, it has been shown that transient activation of NF-κB is not associated with cell death whereas more prolonged activation of NF-κB signals the neuron to die. It has already been shown that glutamate activates NF-κB in developing cultured cerebellar granule cells.

From Dr. Youdim

Could you please explain the neuroprotective action of NMDA in light of the toxic effects of glutamate. Is this concentration dependent? We have observed a very similar phenomenon with dopamine and apomorphine. At low concentration they are neuroprotective, but at high concentration they induce apoptosis. Since glutamate toxicity has an oxidative stress component you should examine whether NMDA has antioxidant action at low concentrations similar to that we observed with dopamine and apomonphine.

ANSWER: Yes, the neuroprotective effect of NMDA is concentration dependent. Subtoxic concentrations (30–100 μM) are protective whereas concentrations below 30 μM are not protective and concentrations above 100 μM are toxic. We agree that NMDA may have an antioxidant action.

From Dr. Abbracchio

What kind of cell death paradigms did you use to demonstrate NMDA-mediated protection in your model? Do your cells also express glutamate metabotropic receptors (which have been typically linked to neuroprotection) and, if so, do you think that these receptors may contribute to modulation of BDNF release or synthesis? A final comment to both Dr. Marini and Dr. Youdim's results. Besides glutamate, dopamine, and apomorphine, adenosine is another such agent, which can both mediate neuroprotection and cell death simply depending on concentration.

PART VII: QUESTIONS AND ANSWERS

ANSWER: We have shown that NMDA protects against the dopaminergic neurotoxin, 1-methyl-4-phenylpyridinium (MPP$^+$) and glutamate. MPP$^+$ kills granule cell neurons intracellularly, whereas glutamate kills these neurons through activation of NMDA receptors. Cultured cerebellar granule cells do express metabotropic glutamate receptors. It is quite possible that metabotropic glutamate receptors contribute to BDNF release and/or synthesis.

QUESTIONS FOR DR. BACHURIN

From Dr. Youdim

Your benzylamine derivative has low K_i and ED$_{50}$ for the glutamate receptor. These compounds are secondary and tertiary amines and they could be a substrate for MAO B, since benzylamine is a relatively good substrate for these enzyme. Did you check whether they are metabolized by MAO and would a MAO inhibitor affect their affinity?

ANSWER: Yes, we tested these compounds as MAO-B substrate and inhibitors. They were not metabolized by MAO-B and most of them show quite low inhibitor towards this enzyme.

The potency of glutamate receptor antagonism is related to its effect of producing psychotic episode and up regulation of dopamine neurotransmission. Did you examine these phenomena with your compounds?

ANSWER: No, we did not study these phenomena.

From Dr. Obrenovitch

Can you comment on the possibility that enhanced cognition NT1505 may be linked to positive modulation of AMPA-R function (i.e., like AMPA-kines)?

ANSWER: Yes, I also believe that the cognition-enhancing properties of NT1505 and its analogs related to their properties as positive modulators of AMPA receptors. We received the direct confirmation of their AMPAkines-like activities in electrophysiological experiments by comparing the ability of NT-1505 and cyclothiazide to effect glutamate- and kainate-induced currents. Both these compounds (NT1505 and cyclothiazide) demonstrated very close properties in stimulating glutamate- and kainate-induced currents in Purkinje cells, but the stimulating ability of NT1505 was more pronounced.

From Dr. Slikker

At higher doses, does your new agent produce "sedation" as does MK-801 at high doses?

ANSWER: The efficient dose of NT1505 in toxin-induced animal model of dementia was about 1–2 mg/kg. A slight obvious sedation-like effect of NT1505 was observed only in doses higher than 20 mg/kg.

From Dr. Manev

Your idea about mechanism of calcium ion penetration through NMDA channels seems very interesting; do you have any direct confirmation of this mechanism?

ANSWER: No, we have not yet confirmed it in the direct experiments. Our hypothesis was based only on results of docking analysis.

The Failure of Neuronal Protective Agents Versus the Success of Thrombolysis in the Treatment of Ischemic Stroke

The Predictive Value of Animal Models

SARAN JONAS,[a] VENKATESH AIYAGARI,[a] DORICE VIEIRA,[b] AND MIGUEL FIGUEROA

[a]*Department of Neurology, New York University School of Medicine, New York, U.S.A.*

[b]*Medical Library, New York University School of Medicine, New York, U.S.A.*

ABSTRACT: Agents claimed to be neuroprotective in animal stroke models have all failed in human trials. Thrombolysis has been reported as beneficial in animal and human stroke. We explore the reasons for this disparity, using a review of published results of agents tested both in animal stroke models and in human stroke trials. In animals the effect of neuroprotective agents and of thrombolytic agents on infarct size is time-dependent: early initiation of treatment works best; and benefit is progressively—and eventually totally—lost with increasing delay of time of first treatment. The animal data also show that, overall, the beneficial effects of the neuroprotective agents are weaker, and are totally lost sooner, than those of thrombolytics. The human data show that the failed trials of the neuroprotective agents had entry windows that went far beyond the windows of (any) success seen in tests of these agents in animals. By contrast, human thrombolysis trials uniformly restricted time of entry to windows in which these agents have shown beneficial effect in animals. In clinical stroke trials, neuroprotective agents failed to produce benefit because their effects at best are too weak, and they were used at times predictable from the animal models as too late. Thrombolytic therapy, which has a stronger effect than neuroprotective agents in animal models, was used clinically during the early window of optimal effectiveness, and produced beneficial results. "Too little/too late" is the recipe for failure in the treatment of ischemic stroke.

KEYWORDS: Neuroprotection; Animal stroke models; Thrombolysis; Ischemic stroke.

INTRODUCTION

In 1994 Grotta[1] asked: "Why have all neuronal protective drugs worked in animals but none so far in stroke patients?" In his discussion of the problems of testing such drugs in animal models, Grotta stated: "...while a positive effect may be found, the significance of the result is difficult to assess if there is no comparison with other

Address for correspondence: Saran Jonas, M.D., Department of Neurology, NYU School of Medicine, Bellevue Hospital Center, room 7W-11, 462 First Avenue, New York, NY 10016, U.S.A. Voice: 212-263-6347 or 212-263-7591; fax: 212-263-8228.
saran.jonas@med.nyu.edu

drugs or therapy, particularly since there is no 'gold standard' or accepted therapy which has a known clinical effect."

We submit that there is now a "gold standard" therapy with a known clinical effect in stroke: thrombolysis. Recognizing the effect of thrombolysis in humans, we herein define from the literature the effect of thrombolysis on infarct size in animal models of ischemic stroke. We then use this effect as the "animal model gold standard," against which we evaluate the animal results of nine putative neuronal protective drugs that have also been studied in human stroke trials.

THROMBOLYSIS IN HUMANS

Our analysis depends on the acceptance that thrombolysis is effective in the acute treatment of human ischemic stroke. As we show in this paper, Wardlaw et al.,[2] in their landmark 1997 meta analysis, demonstrated that thrombolysis lessened the likelihood of dependency or death in the 0–6 hour window and in the 0–3 hour window of treatment.

The 0–6 Hour Window

Wardlaw et al. studied the results of 12 trials of intravenous (iv) treatment with the thrombolytic agents tPA, urokinase, and streptokinase. In these 12 trials, 1297

TABLE 1. Animal studies of intravenous (or intra-arterial) administration of thrombolytic agents

First Author	Agent ( ), intra-arterial	First Treatment minutes after occlusion	Mean Infarct Size with Treatment, Relative to Control Size
Andersen, M. et al.[9]	t-PA	45	0.65
Carter, L. et al.[4]	t-PA	60	0.21
Carter, L.	t-PA	120	0.51
Del Zoppo, G. et al.[3]	(Urokinase)	180	0.22
Lekieffre, D. et al.[10]	t-PA	60	0.45
Overgaard, K. et al.[11]	t-PA	120	0.24
Sakurama, T. et al.[12]	t-PA	5	0.18
Sakurama, T.	t-PA	180	0.40
Sakurama, T.	t-PA	360	0.60
Sakurama, T.	Urokinase	5	0.25
Sereghy, T. et al.[13]	t-PA	120	0.20
Zhang, R. et al.[14]	(TNK t-PA)	120	0.60
Zhang, R.	(TNK t-PA)	240	0.82
Zhang, R.	t-PA	120	0.67
Zhang, R.	t-PA	240	1.10

thrombolysis patients and 1,270 control patients entered treatment in the 0–6 hour window. By the criterion of death or dependency at end of trial followup, the odds ratio favoring thrombolytic treatment was 0.75, with 95% CI (confidence interval) 0.63–0.88.

The 0–3 Hour Window

In their meta analysis, Wardlaw *et al.* gave the results for 1,007 patients (505 randomized to iv thrombolysis; 502 controls) who had begun treatment within three hours of stroke onset. By the criterion of death or dependency at end of followup, the odds ratio favoring thrombolytic treatment was 0.55, with 95% CI 0.42–0.71.

TABLE 2a. BW619C89, CGS 19755, clomethiazole

First Author	Agent	Time (in minutes) of First Treatment After Induction of Occlusion; Pre: pretreatment	Mean Infarct Size with Treatment, Relative to Control Size
Graham, S.H. *et al.*[16]	BW619C89	Pre	0.57
Graham, S.H.	BW619C89	15	0.60
Graham, S.H.	BW619C89	30	0.73
Graham, S.H.	BW619C89	45	0.79
Kawaguchi, K. *et al.*[17]	BW619C89	Pre	0.36
Kawaguchi, K.	BW619C89	30	0.52
Kawaguchi, K.	BW619C89	60	0.69
Kawaguchi, K.	BW619C89	125	0.98
Leach, M.J. *et al.*[18]	BW619C89	5	0.39
Smith, S.E. *et al.*[19]	BW619C89	Pre	0.46
Bullock, R. *et al.*[20]	CGS 19755	Pre	0.30
Bullock, R.	CGS 19755	60	0.78
Davis, M. *et al.*[21]	CGS 19755	Pre	0.70
Okada, M. *et al.*[22]	CGS 19755	Pre	0.22
Park, C.K. *et al.*[23]	CGS 19755	Pre	0.71
Simon, R.P. *et al.*[24]	CGS 19755	Pre	0.36
Simon, R.P.	CGS 19755	5	0.50
Simon, R.P.	CGS 19755	60	1.00
Snape, M.F. *et al.*[25]	Clomethiazole	5	0.69
Sydserff, S.G. *et al.*[26]	Clomethiazole	Pre	0.36
Sydserff, S.G.	Clomethiazole	70	0.50

THROMBOLYSIS IN ANIMALS

TABLE 1 summarizes the effects on infarct volume in animal studies in which thrombolytic agents (tPA, TNK tPA, urokinase) were first given intravenously or intra-arterially at various times after a cerebral artery was occluded by a thrombus. For 10 of 15 results, the time of initiation was 120 to 180 minutes after occlusion. Treatment was begun at 5 to 60 minutes in five instances. Mean infarct volume in treated animals relative to control animals was taken from the authors' statement or was calculated from the authors' numerical or graphic data. In the study of Del Zoppo et al.,[3] volume determination was from *in vivo* CT scanning (baboons). In the other studies (in rats, except for the rabbit study of Carter et al.[4]), volume was determined from brain slices after sacrifice.

It was noted that in four instances the investigators tested the same agent at several different times of initiation of treatment. Uniformly, the results show progressive loss of benefit with increasing delay.

RESULTS OF HUMAN CLINICAL TRIALS OF PUTATIVE NEUROPROTECTIVES

De Keyser et al.[5] reviewed the major completed (1999) phase III clinical trials of neuroprotective drugs in acute ischemic stroke. Their Table 1 lists the outcomes of

TABLE 2b. Citicoline, eliprodil, GM1 ganglioside, lubeluzole

First Author	Agent	Time (in minutes) of First Treatment After Induction of Occlusion; Pre: pretreatment	Mean Infarct Size with Treatment, Relative to Control Size
Andersen, M. et al.[9]	Citicoline	30	0.30
Aronowski, J. et al.[27]	Citicoline	15	0.98
Onal, M.Z. et al.[28]	Citicoline	60	0.93
Schabitz, W.R. et al.[29]	Citicoline	120	0.51
Gotti, B. et al.[30]	Eliprodil	5	0.64
Lekieffre, D. et al.[10]	Eliprodil	10	0.51
Simon, R.P. et al.[31]	GM1 ganglioside	Pre	0.62
Simon, R.P.	GM1 ganglioside	5	0.62
Simon, R.P.	GM1 ganglioside	15	0.82
Simon, R.P.	GM1 ganglioside	30	0.84
Aronowski, J.[32]	Lubeluzole	15	0.47
Culmsee, C. et al.[33]	Lubeluzole	1	0.79
Culmsee, C.	Lubeluzole	1	0.64
Culmsee, C.	Lubeluzole	180	0.77
De Ryck, M. et al.[34]	Lubeluzole	5	0.76

24 trials each involving at least 250 patients; 14 different agents were tested. The closing times of the entry windows were as follows: one trial: 4 hours; one trial: 5 hours; eight trials: 6 hours; four trials: 8 to 12 hours, three trials: 24 hours; seven trials: 48 hours. According to the analyses of De Keyser et al., no trial showed benefit.

RESULTS, IN ANIMALS, OF THE PUTATIVE NEUROPROTECTIVES TESTED IN HUMAN STROKE TRIALS

For nine of these fourteen agents (tested in clinical trials with 5–48 hour entry windows) we were able to find published reports of animal studies. TABLES 2 a, b, and c summarize these animal results: fifty in all. The format of these tables is that of TABLE 1 (The time "pre" represents administration of the agent before occlusion was made: 20 results. In 21 instances treatment of the animals was begun at 5 to 30 minutes. In the remaining nine instances, treatment was begun at 45 to 180 minutes). Where investigators tested more than one regimen of a given drug at a given time of first administration, we took the best result.

In seven instances the investigators tested the same agent at several different times of initiation of treatment. As has been noted for thrombolytics, the results uniformly show progressive loss of benefit with increasing delay.

TABLE 2c. Nimodipine, U74006F

First Author	Agent	Time (in minutes) of First Treatment After Induction of Occlusion; Pre: pretreatment	Mean Infarct Size with Treatment, Relative to Control Size
Backhauss, C. et al.[35]	Nimodipine	Pre	0.87
Barnett, G. et al.[36]	Nimodipine	Pre	0.84
Bielenberg, G. et al.[37]	Nimodipine	Pre	0.61
Cramer, W. et al.[38]	Nimodipine	Pre	0.84
Herz, R. et al.[39]	Nimodipine	Pre	0.75
Marinov, M. et al.[40]	Nimodipine	Pre	0.69
Prehn, J.H. et al.[41]	Nimodipine	Pre	0.87
Rickels, E. et al.[42]	Nimodipine	Pre	0.75
Sakaki, T. et al.[43]	Nimodipine	Pre	0.47
Sauter, A. et al.[44]	Nimodipine	15	0.83
Snape, M. et al.[25]	Nimodipine	5	1.11
Beck, T. et al.[46]	U74006F	Pre	0.69
Hellstrom, H. et al.[46]	U74006F	10	1.21
Park, C. et al.[47]	U74006F	15	0.66

COMPARISON BETWEEN THROMBOLYTIC AND NEUROPROTECTIVE DRUG RESULTS IN ANIMALS

The data from TABLES 1 and 2 are plotted in FIGURES 1–6 (Software: SPSS 10.0 for Windows: SPSS Inc. 233 South Wacker Drive, Chicago IL 60606-6307). We intended to collapse relative infarct sizes of 1.00 or greater into 1.00 so that all points would lie between the line $y = 0$ (total protection from infarction) and the line $y = 100$ (total failure of protective maneuver) on the plot. However, we found that the SPSS program drew the upper line at 110 for entries $y = 0.96$–1.00. The regression lines plotted into these domains. We considered that these plots implied that increasing the delay before treatment eventually led not only to no benefit, but to actual damage. To avoid this implication, we have collapsed all y values greater than 0.96 into 0.95. Similarly, pretreatment and treatment initiated at one minute led to a plot with a regression line sloping through negative time, which we consider meaningless. Accordingly, we collapsed all values $x =$ Pre and $x = 1$ minute into $x = 5$ minutes.

FIGURE 1 (derived from TABLE 1 data) plots relative infarct size against time of initiation of thrombolytic treatment. The mean regression line, with 95% confidence intervals, summarizes the results. At zero time the mean regression line for relative infarct size lies at 30% of control; at six hours (360 minutes) it lies at 73% of control.

FIGURE 2 superimposes the animal results of the nine drugs on the thrombolysis regression line with its 95% CI; the thrombolysis data points have been removed for clarity. Fifteen of the 50 drug results lie on or below the thrombolysis upper 95% CI curve. The other 35 results lie in the "no-fly" zone: the unfavorable domain above the 95% CI envelope for thrombolysis outcomes.

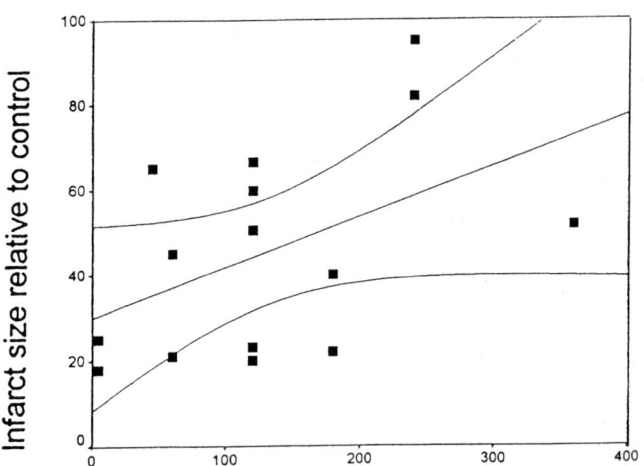

FIGURE 1. Relative infarct size in animals as a function of time of first administration of a thrombolytic agent, with regression line and its 95% confidence intervals.

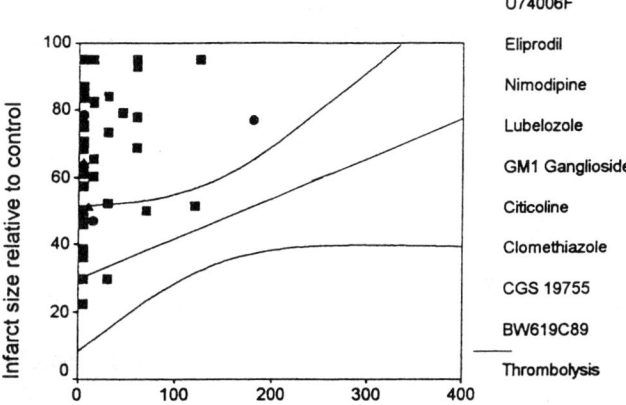

FIGURE 2. Relative infarct size in animals as a function of time of first administration of a variety of neuroprotective agents, compared with 95% confidence intervals of the effects of thrombolysis.

DISCUSSION

The data make clear that the clinical drug trial failures reflected the "too little/too late" qualities of the regimens. Thrombolytic treatment, shown to reduce infarct size in animals when given in the six-hour window, was studied within this window in

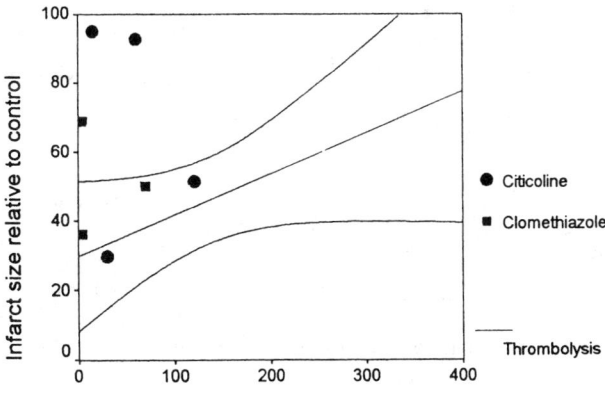

FIGURE 3. The potential value of citicoline and of clomethiazole in early stroke treatment: relative infarct size in animals as a function of time of first treatment, compared with 95% confidence intervals of the effects of thrombolysis.

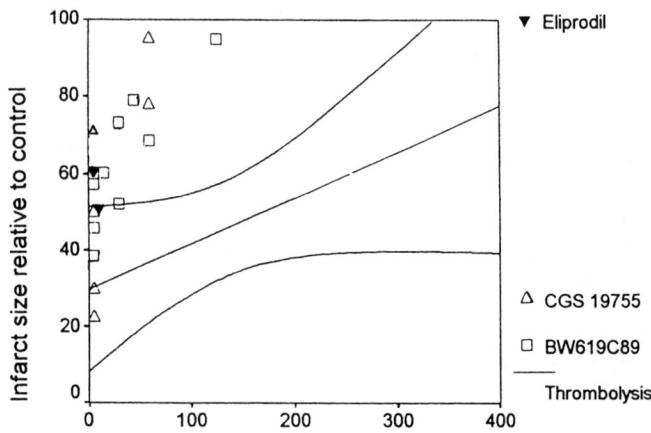

FIGURE 4. The potential value of eliprodil, of CGS19755 and of BW619C89 as prophylactic neuroprotectives: relative infarct size in animals as a function of time of first treatment, compared with 95% confidence intervals of the effects of thrombolysis.

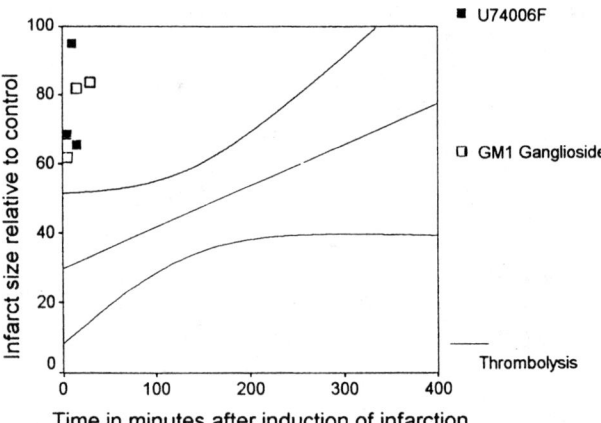

FIGURE 5. Graphic demonstration of the weakness of U74006F and of GM1 ganglioside when compared with thrombolysis (relative infarct size in animals as a function of time of first treatment, compared with 95% confidence intervals of the effects of thrombolysis), implying that these agents have no potential value as prophylactic neuroprotectives or as poststroke treatment.

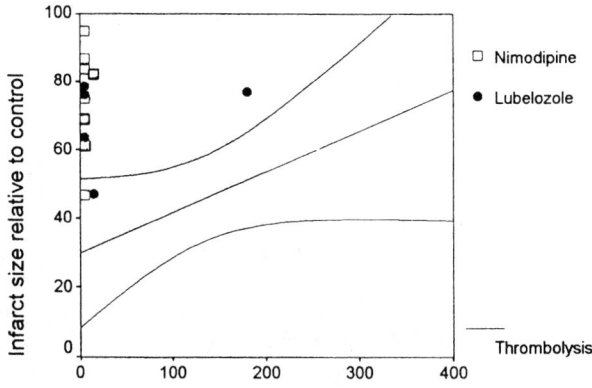

FIGURE 6. Graphic demonstration of the weakness of nimodipine and of libeluzole when compared with thrombolysis (relative infarct size in animals as a function of time of first treatment, compared with 95% confidence intervals of the effects of thrombolysis), implying that these agents are unlikely to have value as prophylactic neuroprotectives or as poststroke treatment.

clinical trials, demonstrating a beneficial outcome. In contrast, the published results of animal testing of the nine drugs do not extend beyond three hours (and with generally weak effect compared with thrombolysis), yet the clinical trial entry windows for these drugs range from five hours to as long as 48 hours. We contend that there is no experimental basis for proceeding to wide window clinical trial with any of these nine agents.

In our original publication in 1997,[6] reviewing the clinical and animal results of thrombolyic therapy and of four other treatments, including two neuroprotective agents, we stated: "In contrast to thrombolytic therapy, the animal study regression lines for the failed treatments with GM1, hemodilution, glucocorticoids, and nimodipine show much poorer zero time y values ... and all reach total failure ($y = 1.0$) between zero minutes and 2.9 hours. Thus, from the animal work, there is no reason to infer that these treatments can produce a neuroprotective effect, except when given very early. Therefore, the failure to see clinical benefit in stroke trials in which many patients entered long after closure of the windows for success in animals ... could be viewed as predicted by the animal results." Our current analysis adds the results of the seven drugs whose trial data have been published since our original communication; these additional results strongly support our original contention.

What we cannot judge is the ability of our model to predict success of a neuronal protective agent in human stroke. We will not be able to do this until we have the results of clinical trials with positive results. Clearly such success will require an agent that avoids the "too little/too late" problem. It would be highly desirable that the pharmacologists solve the "too little" aspect of the problem by coming up with more potent neuroprotective drugs. However, an approach requiring no pharmacologic breakthrough is available to us now: dealing with the "too late" issue through the use of clinical trial windows with closure times that are consonant with the windows for successful animal results with the given agent. The possibilities among the nine failed drugs are reviewed below.

FIGURE 3 raises the possibility that citicoline and clomethiazole might be useful stroke treatments if first given within 120 minutes. The animal results available are not consistent, however, and require extensive replication. If replication attempts push these agents decisively out of the "no fly" zone and into the domain under the upper 95% CI thrombolysis curve, then their clinical testing against tPA (in the appropriately defined window) would be reasonable.

The results shown in FIGURES 3 and 4 raise the possibility that BW619C89, CGS 19755, clomethiazole, and eliprodil might have value as prophylactic neuroprotectants (see our previous communications outlining this approach[7,8]). Again, the inconsistencies of the available animal results would demand extensive replication attempts before further clinical trial of any of these agents is considered.

The results shown in FIGURE 5 suggest that U74006F and GM 1 ganglioside have no potential for success: strength of effect is "too little", and even pretreatment would be "too late". The same is probably true for lubeluzole and nimodipine (see FIG. 6).

The approach to retesting these drugs (and to the screening of other drugs) requires rethinking. In the cited animal studies of neuroprotective agents, flow into the middle cerebral artery was typically blocked by mechanical devices: intraluminal suture, clamping. The thrombolytic agents were, of course, tested by clot insertion. Uniform direct comparison of putative neuronal protective agents with thrombolytics requires the clot model for the former as well as the latter.

ACKNOWLEDGMENTS

The work was supported by the generosity of Mr. & Mrs. John Furth, Mrs. Lita Greenwald and the late Mr. Stephen Greenwald, and Mr. & Mrs. Kenneth Hackel.

S.J. gives thanks to Richard Hanson, M.D., Department of Neurology, New York University School of Medicine, for bringing the paper of De Keyser *et al.* to his attention, and to Diego Grant, MS, Bellevue Hospital Center, NYC, for invaluable help with SPSS.

REFERENCES

1. GROTTA, J. 1994. Cerebrovasc. Dis. **4:** 115–120.
2. WARDLAW, J.M., C.P. WARLOW & C. COUNSELL. 1997. Lancet **350:** 607–614.
3. DEL ZOPPO, G.J., B.R. COPELAND, T.A. WALTZ, *et al.* 1986. Stroke **17:** 638–643.
4. CARTER, L.P., A.N. GUTHKELCH, J. OROZCO, *et al.* 1992. Stroke **23:** 883–888.
5. DE KEYSER, J., G. SULTER & P.G. LUITEN. 1999. Trend. Neurosci. **22:** 535–540.
6. JONAS, S., A.Q. TRAN, E. EISENBERG, *et al.* 1997. Ann. N.Y. Acad. Sci. **825:** 281–287.
7. FISHER, M., S. JONAS, R.L. SACCO, *et al.* 1994. Stroke **25:** 1075–1080.
8. GRIECO, G., M. D'HOLLOSY, A.T. CULLIFORD, *et al.* 1996. Stroke **27:** 858–874.
9. ANDERSEN, M., K. OVERGAARD, P. MEDEN, *et al.* 1999. Stroke **30:** 1464–1471.
10. LEKIEFFRE, D., J. BENAVIDES, B. SCATTON, *et al.* 1997. Brain Res. **776:** 88–95.
11. OVERGAARD, K., T. SEREGHY, H. PEDERSEN, *et al.* 1993. Neurol. Res. **15:** 344–349.
12. SAKURAMA, T., R. KITAMURA & M. KANEKO. 1994. Stroke **25:** 451–456.

13. SEREGHY, T., K. OVERGAARD & G. BOYSEN. 1993. Stroke **24:** 1702–1708.
14. ZHANG, R.L., Z.G. ZHANG & M. CHOPP. 1999. Neurology **52:** 273–279.
15. ZHANG, R.L., Z.G. ZHANG, L. ZHANG, et al. 2000. Stroke **31:** 341.
16. GRAHAM, S.H., J. CHEN, J. LAN, et al. 1994. J. Pharmacol. Exp. Therapeut. **269:** 854–859.
17. KAWAGUCHI, K. & S.H. GRAHAM. 1997. Brain Res. **749:** 131–134.
18. LEACH, M.J., J.H. SWAN, D. EISENTHAL, et al. 1993. Stroke **24:** 1063–1067.
19. SMITH, S.E., H. HODGES, P. SOWINSKI, et al. 1997. Neuroscience **77:** 1123–1135.
20. BULLOCK, R., J. MCCULLOCH, D.I. GRAHAM, et al. 1990. Stroke **21:** III32–36.
21. DAVIS, M., A.D. MENDELOW, R.H. PERRY, et al. 1994. Acta Neurochir. Suppl. **60:** 282–284.
22. OKADA, M., H. UEDA, M. KOMETANI, et al. 1997. Arzneimittel-Forsch. **47:** 703–705.
23. PARK, C.K., J. MCCULLOCH, J.K. KANG, et al. 1994. Acta Neurochir. **127:** 220–226.
24. SIMON, R. & K. SHIRAISHI. 1990. Ann. Neurol. **27:** 606–611.
25. SNAPE, M.F., H.A. BALDWIN, A.J. CROSS, et al. 1993. Neuroscience **53:** 837–844.
26. SYDSERFF, S.G., A.J. CROSS, K.J. WEST, et al. 1995. Brit. J. Pharmacol. **114:** 1631–1635.
27. ARONOWSKI, J., R. STRONG & J.C. GROTTA. 1996. Neurol. Res. **18:** 570–574.
28. ONAL, M.Z., F. LI, T. TATLISUMAK, et al. 1997. Stroke **28:** 1060–1065.
29. SCHABITZ, W.R., J. WEBER, K. TAKANO, et al. 1996. J. Neurol. Sci. **138:** 21–25.
30. GOTTI, B., D. DUVERGER, J. BERTIN, et al. 1988. J. Pharmacol. Exp. Therapeut. **247:** 1211–1221.
31. SIMON, R.P., J. CHEN & S.H. GRAHAM. 1993. J. Pharmacol. Exp. Therapeut. **265:** 24–29.
32. ARONOWSKI, J., R. STRONG & J.C. GROTTA. 1996. Stroke **27:** 1571–1576 and 1576–1577.
33. CULMSEE, C., V. JUNKER, P. WOLZ, et al. 1998. Eur. J. Pharmacol. **342:** 193–201.
34. DE RYCK, M., R. KEERSMAEKERS, H. DUYTSCHAEVER, et al. 1996. J. Pharmacol. Exp. Therapeut. **279:** 748–758.
35. BACKHAUSS, C., C. KARKOUTLY, M. WELSCH, et al. 1992. J. Pharmacol. Meth. **27:** 27–32.
36. BARNETT, G.H., B. BOSE, J.R. LITTLE, et al. 1986. Stroke **17:** 884–890.
37. BIELENBERG, G.W. & T. BECK. 1991. Brain Res. **552:** 338–342.
38. CRAMER, W.C. & G.P. TOOROP. 1998. Gen. Pharmacol. **30:** 195–200.
39. HERZ, R.C., D.J. DE WILDT & D.H. VERSTEEG. 1996. Eur. J. Pharmacol. **306:** 113–121.
40. MARINOV, M., H. WASSMANN & S. NATSCHEV. 1991. Neurol. Res. **13:** 77–83.
41. PREHN, J.H., M. WELSCH, C. BACKHAUSS, et al. 1993. Brain Res. **630:** 10–20.
42. RICKELS, E., M.R. GAAB, H. HEISSLER, et al. 1993. Zent. Neurochir. **54:** 3–12.
43. SAKAKI, T., S. TSUNODA & T. MORIMOTO. 1991. Neurosurgery **28:** 267–272.
44. SAUTER, A. & M. RUDIN. 1986. Stroke **17:** 1228–1234.
45. BECK, T. & G.W. BIELENBERG. 1991. Brain Res. **560:** 159–162.
46. HELLSTROM, H.O., A. WANHAINEN, J. VALTYSSON, et al. 1994. Acta Neurochir. **129:** 188–192.
47. PARK, C. K. & E.D. HALL. 1994. Brain Res. **645:** 157–163.

Low Molecular Weight Heparin and the Treatment of Ischemic Stroke

Animal Results, the Reasons for Failure in Human Stroke Trials, Mechanisms of Action, and the Possibilities for Future Use in Stroke

SARAN JONAS AND DAVID QUARTERMAIN

Department of Neurology New York University School of Medicine, New York, New York, U.S.A.

> KEYWORDS: Heparin; Low molecular weight heparin; Ischemic stroke; Animal results; Human stroke trials.

We have reviewed elsewhere the failure of nine putative pharmacologic neuroprotective agents (agents that interfere with the ischemic cascade) to improve the outcome in randomized clinical trials, in the treatment of ischemic stroke. We have contrasted these failures with the success of thrombolysis. We have shown that the failure of the nine agents was clearly predictable from their effects on cerebral infarct volume in animal stroke model studies. Thrombolysis, which reduces infarct volume in animals as late as six hours after induction of infarction, was studied in human trials with entry windows of up to six hours; meta-analysis shows success in this time window. In contrast, only five of the nine agents were studied in animals at 60 minutes or later, and none was studied past three hours. Only two of the five agents studied between one and three hours gave results comparable to those of thrombolysis; all nine agents can be projected to fail totally in animals by four hours. In view of the weakness and short duration of beneficial effect in animals, it seems obvious that the nine agents should have failed in human trials with entry windows ranging between four and 48 hours: the treatments were "too little/too late."

The low molecular weight heparin (LMWH), nadroparin, was reported to have been beneficial in a human stroke trial with a 48-hour entry window. However, a repeat trial failed to replicate this result. A 12-hour entry window trial of the LMWH danaproid also failed to show benefit. The doses of LMWH used in these failed clinical trials were 1–2 mg/kg/day.

Of three animal studies of LMWH 3–4.5 mg/kg/day, the two in which treatment was begun at 1–3 hours gave results inferior to those of thrombolysis. Only one—first treatment at five hours—gave results comparable to thrombolysis. In contrast, five animal studies of 9–30 mg/kg/day (including studies published by us) in which treatment was begun at one to six hours give an envelope of results congruent with the envelope of thrombolysis results in animals.

Address for correspondence: Department of Neurology, NYU School of Medicine, Bellevue Hospital Center, Rm. 7W-11, 462 First Avenue, New York, NY 10016, U.S.A. Voice: 212-562-2183; fax: 212-263-8228.

saran.jonas@med.nyu.edu

The above LMWH results are compatible with the following interpretations. The clinical LMWH trials, like the pharmacologic neuroprotectant trials, failed because the strength of the effect was too weak for the lateness of initiation of therapy. However, it appears that higher (by an order of magnitude) doses of LMWH, given in the thrombolysis window, are effective in animals. We suggest that they might also be effective under the same circumstances in humans. The proper approach here should be the testing of LMWH directly against tPA (the thrombolytic agent approved for use in human stroke). A key issue here is safety: the major complication of thrombolytic treatment in humans has been cerebral hemorrhage. We and others have found hemorrhage to be an increasing problem with increasing doses of LMWH in rats. Before high dose LMWH trials in humans could be considered, the complication rate of LMWH relative to tPA in animals must be defined.

With respect to a mechanism for the beneficial effect of LMWH in cerebral infarction in animals, we previously reviewed evidence[1] that the extracellular application of the LMWH enoxaparin (but not of unfractionated heparin) inhibits release of Ca^{2+} from intracellular sites in a glutamate model simulating the effects of ischemia on the neuron. This would make LMWH a pharmaclogic neuroprotectant.

The recent work of Zhang *et al.* from the laboratory of Michael Chopp (J. Neurosci. 1999. **19:** 10898–10907) has given support for the view that anticoagulant activity might have a beneficial effect in the first few hours of ischemia. Zhang *et al.* showed progressive obstruction of the cerebral microvascular circulation during the first four hours after embolic middle cerebral artery occlusion in the rat, with progressive microvascular plasma perfusion deficit. It is reasonable to postulate that this phenomenon, which clearly could result in conversion of viable penumbra into dead tissue, could be countered by an anticoagulant.

It is to be noted that the only agents successful in human stroke trials—thrombolytics, the LMWH nadroparin in the first clinical trial, and the defibrinating agent ancrod (a successful clinical trial with a three-hour entry window was reported in JAMA 2000. **283:** 2395–2403; the literature contains no animal data on the effect of ancrod on infarct size)—all have the potential for prevention/reversal of fibrin propagation in the cerebral microvasculature. Unfractionated heparin (UH) also shares this potential. UH has failed in clinical stroke trials, but—as was the case for the nine pharmacologic neuroprotectants as well as the LMWHs—the major human UH trial (Duke *et al.*[2]) had a very wide entry window; up to 48 hours.

The experience with thrombolytics and with ancrod indicates that a stroke must be treated early, and that the treatment should include an agent that prevents/lyses clots. The animal and clinical data give no evidence that a pharmacologic neuroprotectant used alone would have much value in a realistic time frame for treatment of human stroke. Would a combination of an anticlot agent and a neuroprotectant give a better result than an anticlot agent alone? If so, then LMWH, the only agent that combines these two qualities, has a unique potential advantage.

It should be pointed out that even in its action as a pharmacologic neuroprotectant, the LMWH enoxaparin is unique: although many agents block Ca^{2+} entry into the cell, only for enoxaparin is there evidence for inhibition of release of Ca^{2+} from intracellular sites. Since there is experimental evidence that perhaps 50% of the increase in cytosol Ca^{2+} concentration that occurs in ischemia is from intracellular sources, the failure of monotherapy with the calcium channel blocker nimodipine (no substantial benefit in animals, even with treatment within the first 15 minutes; no benefit in many human trials) is easy to understand. The combination of enoxaparin with nimodipine, however, could potentially provide more effective inhibition of cytosol Ca^{2+} release than either agent alone, as well as an anticlot action. The latter would bring the general benefits of protecting/increasing penumbral perfusion, with the added benefit of increasing the exposure of the penumbra to the calcium manipulating agents.

REFERENCES

1. JONAS, S., et al. 1997. Ann. N.Y. Acad. Sci. **825:** 389–393.
2. DUKE, et al. 1986. Ann. Intern. Med. **105:** 825–828.

Non-Pharmacologic (Physiologic) Neuroprotection in the Treatment of Brain Ischemia

ROLAND N. AUER

Departments of Pathology and Clinical Neurosciences, University of Calgary, Calgary, Alberta, Canada

ABSTRACT: Clinical trials for ischemic stroke have been characterized by a disappointing series of negative results, using a panoply of pharmacologic agents. This paper emphasizes five physiologic measures that can be taken to mitigate ischemic brain damage. These are (1) hypothermia, (2) insulin, (3) arterial hyperoxemia, (4) blood pressure control and (5) magnesium. Hypothermia is protective in both focal and global ischemia, even postischemically protecting against selective neuronal necrosis and infarction. The total equation for protection includes the (i) postischemic delay, (ii) depth, and (iii) duration of hypothermia. Insulin operates by lowering glucose levels to the normal range in focal ischemia. It is possible that very low glucose levels are detrimental in focal ischemia with paradoxical augmentation of the infarct size, and that spreading depression plays a role in this. Controlled arterial hyperoxemia seems effective experimentally in reducing infarct size, operating mechanistically by either a direct effect of oxygen, or vasoconstriction causing shunting of blood into the infarct, or both. Blood pressure is a critical determinant of infarct size, and raising blood pressure improves collateral blood flow and reduces stroke size. To be used clinically, however, hemorrhage must be ruled out. The most dramatic clinical effects of blood pressure are seen in aneurysm patients with vasospasm, where minor increases in blood pressure reverse temporary hemiparesis by reducing ischemia. Magnesium is likely the safest NMDA antagonist, with a long history of safe administration to pregnant women with eclampsia. There is potential interaction with insulin, in that magnesium causes hyperglycemia, which requires insulin to counteract it. Magnesium and insulin together have been shown effective in experimental brain ischemia. In the absence of safe and effective pharmacologic neuroprotection agents, clinical trials should be designed and launched to test these physiologic measures, singly and in combination, to reduce brain damage after ischemia.

KEYWORDS: Physiologic neuroprotection; Treatment of brain ischemia

INTRODUCTION

The decline in stroke mortality has slowed.[1] The incidence of stroke has even increased focally in some locations.[2] Brain ischemia (lack of blood flow), thus,

Address for correspondence: Dr. Roland N. Auer, Departments of Pathology & Clinical Neurosciences, University of Calgary, 3330 Hospital Drive N.W., Calgary, Alberta, Canada T2N 4N1. Voice: 403-220-6887; fax: 403-220-7054.
 rauer@ucalgary.ca

remains common, and comes in two general forms—global and focal. Global ischemia is epitomized by cardiac arrest in the human, which can show widespread necrosis in the neocortex, hippocampus, basal ganglia, and cerebellum after long term survival.[3] However, focal ischemia in the form of transient occlusion of the middle cerebral artery (MCA) is the more common form of human brain ischemia, affecting roughly 0.5 million people in the U.S.A. per year.[4]

The infarcts resulting from brain ischemia can be anywhere from lethal to silent.[5,6] The cost of stroke is roughly $200 annually per citizen in a Western country.[7] Thus, neuroprotection, if available, could reduce societal costs and personal illness considerably.

With this scenario for the prevalence of stroke, it is especially disappointing that clinical drug trials heretofore have all been negative, in spite of promising experimental data obtained in animals. Most clinical trials constitute pharmacologic intervention at a mechanism believed to be operant in generating tissue damage. Although the opinion has been expressed that there are problems in assessing efficacy in animal models,[8] it is likely that animal models are actually much more sensitive than human trials of efficacy. This, and other factors, has led to positive animal data, and negative human trials.[9,10]

It seems lesion size is easily altered in animals, by mechanisms that are somehow less important in the realm of human stroke. Thus, interfering with one mechanism pharmacologically may lead to a significant reduction of experimental infarct size, but the human trial at considerable expense, may subsequently be negative. It is, thus, germane to examine possible reasons for dissociation between animal data and human data in neuroprotection.

Ischemic Stroke: Effective Treatment in Animals, but Not Humans

With the exception of tissue plasminogen activator (*t*-PA), there is presently no clinically practiced treatment for ischemic stroke. Specifically, there is no method of tissue neuroprotection while opening the blood vessel with *t*-PA. Animal data have demonstrated a wealth of positive and promising results. However, inbred animals are inherently less variable than humans to begin with. On reading the literature one is struck by a tendency that studies in gyrencephalic animals such as dogs[11,12] or primates[13] tend to be less often positive than similar studies in small animals.

When a lesion is superimposed upon the naïve brain, the variability increases more in humans than in inbred animals. Even among animals, brain reparative and compensatory mechanisms in response to lesions of identical size, and in response to treatment, may not be identical from animal to animal.[14] Thus, there is increased variability as one progresses from the naïve prelesioned brain, to the initial generating lesion, to histology, and in turn to behavior (see FIGURE 1). Histology in animals is, therefore, an eminently sensitive method of detecting neuroprotection,[15] possibly contributing to some of the dissociation of animal results with human results. Humans are highly outbred, as opposed to inbred animals, and have intrinsically higher variability than animals.

Often, positive animal data have been obtained from global or forebrain ischemia models in rodents, and assumed to apply in human stroke. This is a formula for disappointment. The small, lissencephalic rodent brain has rheological and metabolic properties vastly different from the comparatively enormous gyrencephalic human

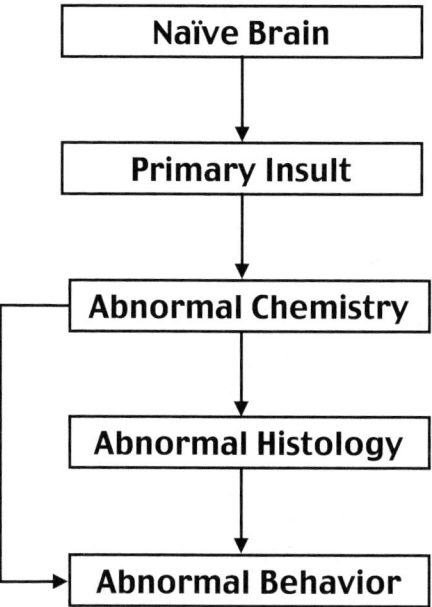

FIGURE 1. A primary insult, such as ischemia, is superimposed upon a naïve brain. This in turn results in almost immediate chemical changes. Some of these, over time, are critical in killing cells. The resulting abnormal histology, if widespread enough, will alter behavior. Variability increases as one goes down through the stages, and is greater in humans than animals, at all stages.

brain. Furthermore, the two-vessel occlusion (2-VO) in the rat,[16] 4-VO,[17] and the gerbil 2-VO model[18] all essentially produce forebrain, or global ischemia that mimics human cardiac arrest. Focal ischemia more closely mimics human ischemic stroke, but focal ischemia models have only been developed more recently.[19,20] Transferring findings from global ischemia results in rodents to focal ischemia in humans seems not prudent, in view of the vast differences.

Another potential factor giving rise to positive results in animals but not humans is that monotherapy affecting one mechanism may have a large effect seen superimposed upon an animal model, but a smaller effect in humans, with more, or differently weighted pathophysiologic factors involved. For example, calcium influx may be important in a rodent model, but relatively less important in a human where blood flow factors are more dominant in determining outcome. Thus, animal data may be based on a mechanism that has relatively increased in importance relative to the same mechanism in the human clinical situation.

Physiology, as opposed to pharmacology, may lead the way out of this treatment conundrum posed by pharmacologic monotherapy. In fact, manipulation of several physiologic parameters has been shown to be effective in cerebral ischemia: hypothermia, insulin, hyperoxemia, controlled hypertension, and magnesium. These will be discussed in turn.

Hypothermia

Hypothermia has been known to be effective for a long time. Infants abandoned in flowing streams in ancient times, were often found to be viable and could be rescued from cold water. More recently, hypothermia has been used to protect the brain during cardiac surgery.[21,22] A rediscovery of the hypothermic effect in experimental animals in the 1980s[23] led to a resurgence in interest to study hypothermia from an experimental viewpoint. Hypothermia is clearly effective not only in global, but also in focal ischemia.[24–26]

It is clear that the equation for neuroprotection is complicated,[23,27–29] but the following seems clear: first, postischemic hypothermia is effective, and the earlier it is instituted, the better the neuroprotection. Second, the depth of hypothermia is important, lower temperatures affording more neuroprotection. Third, the duration of hypothermia is critical, since cells may die if hypothermia is aborted too soon. Other variables include the rate of cooling and rate of rewarming, likely more important for toxicity than efficacy. Lastly interactions of hypothermia with other therapies (such as t-PA) exist. Each of these factors will now be discussed.

Mechanisms of hypothermic neuroprotection are not clear. However, hypothermia attenuates ischemia induced release of excitotoxic glutamate[30,31] and its coagonist, glycine.[32] Hypothermia diverts a greater proportion of glucose from the Embden-Meyerhof glycolytic pathway into the potentially neuroprotective pentose phosphate pathway.[33] Hypothermia does not seem to act by improving blood flow[34,35] nor by improving brain tissue pH.[36,37]

It should be noted that the effects of fever in ischemia are opposite those of neuroprotection, and fever is to be strictly avoided in human stroke. The curve relating

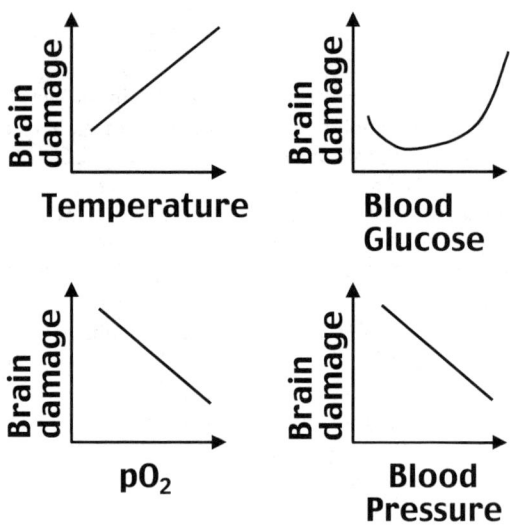

FIGURE 2. Crude dose–response curves for neuroprotective effects due to hypothermia, insulin, hyperoxemia, and blood pressure. Dose–response relationships for Mg^{2+} are unknown.

temperature to brain damage is likely progressively monotonic over low, normal, and high ranges (see FIGURE 2). We focus here on hypothermia.

(1) Delay of Hypothermia

Clearly, hypothermia cannot usually be instituted at the onset of the stroke. Symptoms must first occur, then transport to a hospital facility, and a decision must be made to institute cooling (in a trial, or eventually as routine clinical practice). This implies that hypothermic treatment must be intraischemic or postischemic (not preischemic) in most clinical situations. The operating room is a salient exception, where preischemic or intraischemic[24] hypothermia may be applied.

Data obtained from animals clearly indicate that postischemic hypothermia is effective. Unfortunately, most of the data have been obtained from global ischemia models (see above), produced by bilateral carotid occlusion. This mimics cardiac arrest more closely than focal ischemic stroke in the human. Nevertheless, studies of focal ischemia in animals indicate that hypothermia can be neuroprotective even if delayed by one hour[38] or by 30 minutes.[39] Other experiments show no effect with delayed hypothermia after focal ischemia.[40]

(2) Depth of Hypothermia

Early use of deep hypothermia in cardiac surgery indicated that the window of opportunity for surgical intervention could be greatly lengthened if the temperature was lowered. This allows operation on the heart, which could not be otherwise performed. Neurosurgery later picked up hypothermia clinically and developed the use of even more profound hypothermia, in some cases to 16°C.[41]

However, it is clear that neuroprotection occurs with even mild hypothermia.[23] Cooling to 34°C already offers profound benefit, as demonstrated in experimental animals. Human shivering begins as the temperature is reduced from 35° to 34°C, requiring the institution of measures to control shivering and its consequences. However, protection is obtained even at 35°C, a body core temperature that has been measured in boys who go swimming. It is clear that a clinical trial of even mild hypothermia may show a protective effect. Ventricular fibrillation in hypothermia has well been known,[42] and temperature gradients in the subendocradial zone may be critical. These gradients of temperature in the cardiac conducting system may themselves be influenced by rate of cooling/rewarming, rather than by the absolute temperature.

(3) Duration of Hypothermia

If hypothermia is prematurely aborted, cell death may subsequently occur.[29] However, cell death does not take as protracted and as prolonged a course in focal ischemia as in global ischemia. The well described phenomenon of delayed neuronal death, first noticed by Ito,[43] and later well described by Kirino in a series of seminal papers using gerbils[44] and rats,[45,46] all used global ischemia. Focal ischemia shows a much more rapid evolution of damage. Infarction at normothermia is essentially complete by 24 h. Ischemic edema, curiously, has a more protracted time course, peaking at roughly 48 h after the stroke. It is noteworthy that hypothermia has a beneficial effect on brain edema and intracranial pressure, in addition to tissue neuroprotection.

Focal ischemia, like global ischemia, is sensitive to the duration of postischemic hypothermia: neuroprotection was seen in one study with prolonged but not with brief (1 h) hypothermia.[47] In view of the animal data, it seems prudent to carry the duration of hypothermia to 48 h or 72 h, in human clinical trials.

(4) Interactions of Hypothermia

Tissue plasminogen activator clearly reduces injury in ischemia caused by blood clots both experimentally[48–51] and also in humans.[52]

Hypothermia not only interacts with the patients directly, but also interacts with other potential treatments. For example, the rate of clot dissolution with *t*-PA is slowed by hypothermia, but the beneficial effect of tissue neuroprotection due to the cooling seems to outweigh any delay in clot dissolution by *t*-PA, producing a net positive benefit.[53]

Other interactions of hypothermia can be envisaged. Most enzymes have their optimum activity in the body at normothermia. However, enzymes are variably slowed by hypothermia in their ability to catalyze body reactions, given by their Q_{10}. Hypothermia likely acts mostly via enzymes with larger Q_{10}. Since many drugs are eliminated enzymatically in the body, interactions of hypothermia with drug breakdown, conjugation, or elimination can easily be envisaged.

Insulin

Like hypothermia, insulin has been more extensively studied in global than in focal ischemia. It was first shown to be effective in the two-vessel occlusion model of Smith, whether insulin was given pre- or postischemically.[54] Subsequently, insulin was shown to produce a behavioral benefit as well as a histologic benefit.[55] The running audiogenic seizures seen in rodents,[56] which occur after global ischemia, are exacerbated by insulin-induced hypoglycemia in the postischemic rat.[54] Diazepam and low dose insulin ameliorate these seizures, as well as the accompanying brain damage due to the ischemia.[57]

Coadministration of glucose and insulin shows that hypoglycemia is not necessary for the neuroprotective effect of insulin in global ischemia,[58] suggesting a direct effect of insulin. The latter has been confirmed by direct intracerebral injection of insulin.[59]

However, the situation is different in focal ischemia, the more common form of ischemia in humans. Studies with insulin in focal ischemia have shown that it is also effective in reducing ischemic brain damage in the form of pan-necrosis (infarction).[60] Since glucose levels have an effect within the normal range in focal ischemia (as opposed to global ischemia), the question arises as to optimum glucose levels in transient focal ischemia when insulin is used. Results in cats suggests that normoglycemia is optimum,[61] implying a "U" shaped dose-response curve (FIG. 2).

Together, the above considerations from animal data suggest that a clinical neuroprotective trial of insulin should be undertaken in focal ischemic stroke.

Hyperoxemia

Animal results suggest that hyperbaric oxygen is either effective in preventing cerebral infarction or is ineffective and/or toxic.[62–69] Oxygen toxicity consists of tremor, myoclonus, convulsion, and pulmonary toxicity. Oxygen poisoning has been well documented in man.[71]

Oxygen is toxic due to free radical formation and this is often been deemed to account for an entity termed reperfusion injury.[72] Although the theoretical concept of oxygen-derived, free radical–induced reperfusion injury is attractive mechanistically, it seems that oxygen is beneficial during and after ischemia.[73] It is important to emphasize that this obtains even if hyperoxemia is instituted in reperfusion period. Clinical study of hyperbaric oxygenation in the treatment of stroke[74] reveals that hyperbaric oxygen is difficult to administer, and poorly tolerated. Hyperbaric oxygen can be demonstrated to be toxic in animals,[75,76] with alterations in mitochondria and axons.[77] Although central nervous system lesions require hyperbaric pressures of oxygen in adult animals, hyperoxia, even at normobaric pressure, can produce necrotizing lesions in the neonate.[79] Although hyperbaric oxygenation constricts the cerebral vasculature, these potentially detrimental effects seem outweighed by the benefit of oxygen, and vasoconstriction elsewhere may even pump blood into a "pressure passive" ischemic circulation (one that has lost autoregulatory capability). These data, plus recent experimental indications that infarction due to focal transient ischemia can be mitigated by high blood oxygen levels,[73] suggest that arterial hyperoxygenation should be explored as a systematic therapy for brain ischemia. It seems that oxygen influences ischemic brain damage roughly in a linear[73] fashion.

Blood Pressure During Ischemia

Early researchers in cerebral ischemia noticed that hypotension was critical in the recovery period after global ischemia.[81] Observations in the clinic also suggest that blood pressure during stroke progression is a major determinant of outcome, hypotension being detrimental.[82] Neurosurgeons have known for years that patients with subarachnoid hemorrhage and vasospasm-induced hemiparesis do better when the blood pressure is raised, even by 20 mm Hg.[83] Animal data shows that a 20-mm Hg difference in blood pressure can make an enormous difference in the size of the resulting infarct.

Concern over raising blood pressure with pharmacologic agents is based today on possibly increasing the rate of hemorrhage with coadministration of *t*-PA. Diapedetic hemorrhage seems a minor concern, but major lobar brain hemorrhage is obviously a major concern over iatrogenically raising blood pressure during ischemic stroke. Nevertheless, the data seem to indicate that blood pressure is a major determinant of stroke size, obviously by improved perfusion with higher arterial pressures.

Magnesium

Since the experiments in retinal preparations subjected to simulated ischemia by Ames *et al.* almost two decades ago, it was noticed that magnesium enhances recovery.[85] Magnesium has a long history of safe human use, having been successfully used for decades in the treatment of preeclampsia and eclampsia of pregnancy. Its safety in pregnant women contrasts with the poor safety profile of pharmacologic non-competitive NMDA antagonists such as MK-801. Psychotogenic and autonomic side effects have limited the use of MK-801 clinically. Magnesium, in contrast, is safe, and causes only mild hyperglycemia. This prompted one research group to study magnesium in MCA artery occlusion, finding reduction in infarct size even if administered one hour after the occlusion.[86,87] Insulin was used to compensate for the magnesium-induced hyperglycemia,[86,87] and insulin was found to be effective on its own or in combination.[88] Part of the benefit of magnesium may be its antiepileptic activity. Low magnesium enhances epileptiform activity[89] and epileptic activity is detrimental in cerebral ischemia. Magnesium levels are reduced in ischemic brain tissue, and are reduced further with energy failure and acidosis.[90] Clinical trials of magnesium in acute cerebral ischemia have begun.[91,92] In view of the safety profile of magnesium, it appears that the major question to the answer in these trials is efficacy.

EPILOGUE

In the absence of pharmacologic neuroprotective agents, we must do with what we have. Physiologic measures outlined in this article offer a way ahead, and are immediately available. Safety concerns are much better defined in these five physiologic manipulations than they are in new drugs. Hence, a clinical effort should be made to test these therapies individually, and then in combination. Drugs can be added at any time, if found to be effective. In view of the cost burden of stroke, inactivity in clinical trials of physiologic measures should be replaced by a flurry of activity to define the value, safety, and efficacy of manipulating physiologic parameters in the treatment of acute ischemic stroke.

REFERENCES

1. COOPER, R., *et al.* 1990. Slowdown in the decline of stroke mortality in the United States, 1978–1986. Stroke **21:** 1274–1279.
2. JØRGENSEN, H.S., *et al.* 1992. Marked increase of stroke incidence in men between 1972 and 1990 in Frederiksberg, Denmark. Stroke **23:** 1701–1704.
3. COLE, G. & V.A. COWIE. 1987. Long survival after cardiac arrest: case report and neuropathological findings. Clin. Neuropathol. **6:** 104–109.
4. GARRAWAY, W.M. *et al.* 1979. The declining incidence of stroke. N. Engl. J. Med. **300:** 449–452.
5. JØRGENSEN, H.S. *et al.* 1994. Silent infarction in acute stroke patients. Prevalence, localization, risk factors, and clinical significance: the Copenhagen Stroke Study. Stroke **25:** 97–104.
6. NORRIS, J.W. & C.Z. ZHU. 1992. Silent stroke and carotid stenosis. Stroke **23:** 483–485.
7. TERENT, A., *et al.* 1994. Costs of stroke in Sweden. A national perspective. Stroke **25:** 2363–2369.
8. CORBETT, D. & S. NURSE. 1998. The problem of assessing effective neuroprotection in experimental cerebral ischemia. Prog. Neurobiol. **54:** 531–548.
9. GROTTA, J. 1995. Why do all drugs work in animals but none in stroke patients? 2. Neuroprotective therapy. J. Intern. Med. **237:** 89–94.
10. DEL ZOPPO, G.J. 1995. Why do all drugs work in animals but none in stroke patients? 1. Drugs promoting cerebral blood flow. J. Intern. Med. **237:** 79–88.
11. D'ALECY, L., *et al.* 1986. Dextrose containing intravenous fluid impairs outcome and increases death after eight minutes of cardiac arrest and resuscitation in dogs. Surgery **100:** 505–511.

12. RUIZ, E., et al. 1986. Cerebral resuscitation after cardiac arrest using hetastarch hemodilution, hyperbaric oxygenation and magnesium ion. Resuscitation **14:** 213–223.
13. AUER, R.N., et al. 1996. MK-801 in primate focal ischemia. Mol. Chem. Neuropathol. **29:** 193–210.
14. ROD, M.R., I.Q. WHISHAW & R.N. AUER. 1990. The relationship of structural ischemic brain damage to neurobehavioural deficit: the effect of postischemic MK-801. Can. J. Psychol. **44:** 196–209.
15. DE COURTEN-MYERS, G.M., et al. 1998. Stroke assessment: morphometric infarct size versus neurologic deficit. J. Neurosci. Meth. **83:** 151–157.
16. SMITH, M.-L., R.N. AUER & B.K. SIESJÖ. 1984. The density and distribution of ischemic brain injury in the rat after 2–10 minutes of forebrain ischemia. Acta Neuropathol. (Berl.) **64:** 319–332.
17. PULSINELLI, W.A. & J.B. BRIERLEY. 1979. A new model of bilateral hemispheric ischemia in the unanesthetized rat. Stroke **10:** 267–272.
18. LEVINE, S. & H. PAYAN. 1966. Effects of ischemia and other procedures on the brain and retina of the gerbil (*Meriones unguiculatus*). Exp. Neurol. **16:** 255–262.
19. TAMURA, A., et al. 1981. Focal cerebral ischemia in the rat: 1. Description of technique and early neuropathological consequences following middle cerebral artery occlusion. J. Cereb. Blood Flow Metab. **1:** 53–60.
20. LONGA, E.Z., et al. 1989. Reversible middle cerebral artery occlusion without craniectomy in rats. Stroke **20:** 84–91.
21. BIGELOW, W.C. 1959. Methods for inducing hypothermia and rewarming. Ann. N.Y. Acad. Sci. **80:** 522–532.
22. BIGELOW, W.G., J.C. CALLAGHAN & J.A. HOPPS. 1950. General hypothermia for experimental intracardiac surgery. Ann. Surg. **132:** 531–537.
23. BUSTO, R., et al. 1987. Small differences in intraischemic brain temperature critically determine the extent of ischemic neuronal injury. J. Cereb. Blood Flow Metab. **7:** 729–738.
24. GOTO, Y., et al. 1993. Effects of intraischemic hypothermia on cerebral damage in a model of reversible focal ischemia. Neurosurgery **32:** 980–989.
25. JIANG, Q., et al. 1994. The effect of hypothermia on transient focal ischemia in rat brain evaluated by diffusion- and perfusion-weighted NMR imaging. J. Cereb. Blood Flow Metab. **14:** 732–741.
26. KARIBE, H., et al. 1994. Mild intraischemic hypothermia reduces postischemic hyperperfusion, delayed postischemic hypoperfusion, blood-brain barrier disruption, brain edema, and neuronal damage volume after temporary focal cerebral ischemia in rats. J. Cereb. Blood Flow Metab. **14:** 620–627.
27. BUSTO, R., et al. 1989. Postischemic moderate hypothermia inhibits CA1 hippocampal ischemic neuronal injury. Neurosci. Lett. **101:** 299–304.
28. COLBOURNE, F. & D. CORBETT. 1994. Delayed and prolonged post-ischemic hypothermia is neuroprotective in the gerbil. Brain Res. **654:** 265–272.
29. COLBOURNE, F. & D. CORBETT. 1995. Delayed postischemic hypothermia: a six month survival study using behavioral and histologic assessments of neuroprotection. J. Neurosci. **15:** 7250–7260.
30. ILLIEVICH, U.M., et al. 1994. Effects of hypothermic metabolic suppression on hippocampal glutamate concentrations after transient global cerebral ischemia. Anesth. Analg. **78:** 905–911.
31. WINFREE, C.J., et al. 1996. Mild hypothermia reduces penumbral glutamate levels in the rat permanent focal cerebral ischemia model. Neurosurgery **39:** 1216–1222.
32. BAKER, A.J., et al. 1991. Hypothermia prevents ischemia-induced increases in hippocampal glycine concentrations in rabbits. Stroke **22:** 666–673.
33. KAIBARA, T., et al. 1999. Hypothermia: depression of tricarboxylic acid cycle flux and evidence for pentose phosphate shunt upregulation. J. Neurosurg. **90:** 339–347.
34. LO, E.H. & G.K. STEINBERG. 1992. Effects of hypothermia on evoked potentials, magnetic resonance imaging, and blood flow in focal ischemia in rabbits. Stroke **23:** 889–893.

35. MORIKAWA, E., et al. 1992. The significance of brain temperature in focal cerebral ischemia: histopathological consequences of middle cerebral artery occlusion in the rat. J. Cereb. Blood Flow Metab. **12:** 380–389.
36. BACHER, A., J.Y. KWON & M.H. ZORNOW. 1998. Effects of temperature on cerebral tissue oxygen tension, carbon dioxide tension, and pH during transient global ischemia in rabbits. Anesthesiology **88:** 403–409.
37. KOZLOWSKI, P., et al. 1997. Effect of temperature in focal ischemia of rat brain studied by ^{31}P and ^{1}H spectroscopic imaging. Magn. Reson. Med. **37:** 346–354.
38. BAKER, C.J., S.T. ONESTI & R.A. SOLOMON. 1992. Reduction by delayed hypothermia of cerebral infarction following middle cerebral artery occlusion in the rat: a time-course study. J. Neurosurg. **77:** 438–444.
39. KARIBE, H., et al. 1994. Delayed induction of mild hypothermia to reduce infarct volume after temporary middle cerebral artery occlusion in rats. J. Neurosurg. **80:** 112–119.
40. MOYER, D.J., F.A.WELSH & E.L. ZAGER. 1992. Spontaneous cerebral hypothermia diminishes focal infarction in rat brain. Stroke **23:** 1812–1816.
41. SYLVESTER, E.J. 1993. The healing blade: a tale of neurosurgery. *In* The Healing Blade. M. Abraham, Ed. Simon & Schuster, New York.
42. COVINO, B.G. & H.E. D'AMATO. 1962. Mechanism of ventricular fibrillation in hypothermia. Circ. Res. **10:** 148–155.
43. ITO, U., et al. 1975. Experimental cerebral ischemia in Mongolian gerbils. I. Light microscopic observations. Acta Neuropathol. (Berl.) **32:** 209–223.
44. KIRINO, T. 1982. Delayed neuronal death in the gerbil hippocampus following ischemia. Brain Res. **239:** 57–69.
45. KIRINO, T. & K. SANO. 1984. Selective vulnerability in the gerbil hippocampus following transient ischemia. Acta Neuropathol. (Berl.) **62:** 201–208.
46. KIRINO, T., et al. 1986. Treatable ischemic neuronal damage in the gerbil hippocampus. No To Shinkei **38:** 1157–1163.
47. YANAMOTO, H., et al. 1996. Mild postischemic hypothermia limits cerebral injury following transient focal ischemia in rat neocortex. Brain Res. **718:** 207–211.
48. BEDNAR, M.M., et al. 1990. Tissue plasminogen activator reduces brain injury in a rabbit model of thromboembolic stroke. Stroke **21:** 1705–1709.
49. ZIVIN, J.A., et al. 1985. Tissue plasminogen activator reduces neurological damage after cerebral embolism. Science **230:** 1289–1292.
50. BENES, V., et al. 1990. Effect of intra-arterial tissue plasminogen activator and urokinase on autologous arterial emboli in the cerebral circulation of rabbits. Stroke **21:** 1594–1599.
51. LYDEN, P.D., et al. 1989. Tissue plasminogen activator-mediated thrombolysis of cerebral emboli and its effect on hemorrhagic infarction in rabbits. Neurology **39:** 703–708.
52. MARLER, J., et al. 1995. Tissue plasminogen activator for acute ischemic stroke—The National Institute of Neurological Disorders and Stroke rt-PA Stroke Study Group. N. Engl. J. Med. **333:** 1581–1587.
53. MEDEN, P., et al. 1994. The influence of body temperature on infarct volume and thrombolytic therapy in a rat embolic stroke model. Brain Res. **647:** 131–138.
54. VOLL, C.L. & R.N. AUER. 1988. The effect of post-ischemic blood glucose levels on the density and distribution of ischemic brain damage in the rat. Ann. Neurol. **24:** 638–646.
55. VOLL, C.L., I.Q. WHISHAW & R.N.AUER. 1989. Postischemic insulin reduces spatial learning deficit following transient forebrain ischemia in the rat. Stroke **20:** 646–651.
56. MCCOWN, T.J., et al. 1984. Electrically elicited seizures from the inferior colliculus: a potential site for the genesis of epilepsy? Exp. Neurol. **86:** 527–542.
57. VOLL, C.L. & R.N. AUER. 1991. Postischemic seizures and necrotizing ischemic brain damage: Neuroprotective effect of postischemic diazepam and insulin. Neurology **41:** 423–428.
58. VOLL, C.L. & R.N. AUER. 1991. Insulin attenuates ischemic brain damage independent of its hypoglycemic effect. J. Cereb. Blood Flow Metab. **11:** 1006–1014.

59. ZHU, C.Z. & R.N. AUER. 1994. Intraventricular administration of insulin and IGF-1 in transient forebrain ischemia. J. Cereb. Blood Flow Metab. **14:** 237–242.
60. HAMILTON, M.G., B.I. TRANMER & R.N. AUER. 1995. Insulin reduction of cerebral infarction due to transient focal ischemia. J. Neurosurg. **82:** 262–268.
61. DE COURTEN-MYERS, G.M., et al. 1994. Normoglycemia (not hypoglycemia) optimizes outcome from middle cerebral artery occlusion. J. Cereb. Blood Flow Metab. **14:** 227–236.
62. BURT, J.T., J.P. KAPP & R.R. SMITH. 1987. Hyperbaric oxygen and cerebral infarction in the gerbil. Surg. Neurol. **28:** 265–268.
63. CORKILL, G., et al. 1985. Videodensitometric estimation of the protective effect of hyperbaric oxygen in the ischemic gerbil brain. Surg. Neurol. **24:** 206–210.
64. IWATSUKI, N., et al. 1994. Hyperbaric oxygen combined with nicardipine administration accelerates neurologic recovery after cerebral ischemia in a canine model. Crit. Care Med. **22:** 858–863.
65. KOL, S., et al. 1993. Hyperbaric oxygenation for arterial air embolism during cardiopulmonary bypass. Ann. Thorac. Surg. **55:** 401–403.
66. KRAKOVSKY, M., et al. 1998. Effect of hyperbaric oxygen therapy on survival after global cerebral ischemia in rats. Surg. Neurol. **49:** 412–416.
67. SHIOKAWA, O., et al. 1986. Hyperbaric oxygen therapy in experimentally induced acute cerebral ischemia. Undersea Biomed. Res. **13:** 337–344.
68. TAKAHASHI, M., et al. 1992. Hyperbaric oxygen therapy accelerates neurologic recovery after 15-minute complete global cerebral ischemia in dogs. Crit. Care Med. **20:** 1588–1594.
69. WEINSTEIN, P.R., G.G. ANDERSON & D.A. TELLES. 1987. Results of hyperbaric oxygen therapy during temporary middle cerebral artery occlusion in unanesthetized cats. Neurosurgery **20:** 518–524.
70. DONALD, K.W. 1947. Oxygen poisoning in man. Part I. Br. Med. J. **1:** 667–672.
71. DONALD, K.W. 1947. Oxygen poisoning in man. Part II. Br. Med. J. **1:** 712–717.
72. ARONOWSKI, J., R. STRONG & J.C. GROTTA. 1997. Reperfusion injury: demonstration of brain damage produced by reperfusion after transient focal ischemia in rats. J. Cereb. Blood Flow Metab. **17:** 1048–1056.
73. MIYAMOTO, O. & R.N. AUER. 2000. Hypoxia, hyperoxia, ischemia and brain necrosis. Neurology **54:** 362–371.
74. ANDERSON, D.C., et al. 1991. A pilot study of hyperbaric oxygen in the treatment of human stroke. Stroke **22:** 1137–1142.
75. BALENTINE, J.D. 1968. Clinicopathological features of limb paralysis induced in rats by hyperbaric oxygen exposure. J. Neuropathol. Exp. Neurol. **27:** 163.
76. BALENTINE, J.D. & B.B. GUTSCHE. 1966. Central nervous system lesions in rats exposed to oxygen at high pressure. Am. J. Pathol. **48:** 107–127.
77. BALENTINE, J.D. 1974. Ultrastructural pathology of hyperbaric oxygenation in the central nervous system, observations in the anterior horn gray matter. Lab. Invest. **31:** 580–592.
78. BALENTINE, J.D. 1968. Pathogenesis of central nervous system lesions induced by exposure to hyperbaric oxygen. Am. J. Pathol. **53:** 1097–1109.
79. AHDAB-BARMADA, M., et al. 1986. Hyperoxia produces neuronal necrosis in the rat. J. Neuropathol. Exp. Neurol. **45:** 233–246.
80. BERGO, G.W. & I. TYSSEBOTN. 1995. Effect of exposure to oxygen at 101 and 150 kPa on the cerebral circulation and oxygen supply in conscious rats. Eur. J. Appl. Physiol. **71:** 475–484.
81. CANTU, R.C., et al. 1969. Hypotension: a major factor limiting recovery from cerebral ischemia. J. Surg. Res. **9:** 525–529.
82. JØRGENSEN, H.S., et al. 1994. Effect of blood pressure and diabetes on stroke in progression. Lancet **344:** 156–159.
83. KASSELL, N.F., et al. 1982. Treatment of ischemic deficits from vasospasm with intravascular volume expansion and induced arterial hypertension. Neurosurgery **11:** 337–343.
84. ZHU, C.Z. & R.N. AUER. 1995. Graded hypotension and MCA occlusion duration: effect in transient focal ischemia. J. Cereb. Blood Flow Metab. **15:** 980–988.

85. AMES, III, A. & F.B. NESBETT. 1983. Pathophysiology of ischemic cell death: I. Time of onset of irreversible damage; importance of the different components of the ischemic insult. Stroke **14:** 219–226.
86. IZUMI, Y., *et al.* 1991. Reduction of infarct volume by magnesium after middle cerebral artery occlusion in rats. J. Cereb. Blood Flow Metab. **11:** 1025–1030.
87. IZUMI, Y., *et al.* 1992. Insulin protects brain tissue against focal ischemia in rats. Neurosci. Lett. **144:** 121–123.
88. SCHMID-ELSAESSER, R., *et al.* 1999. Combination drug therapy and mild hypothermia: a promising treatment strategy for reversible, focal cerebral ischemia. Stroke **30:** 1891–1899.
89. AVOLI, M., *et al.* 1991. Epileptiform activity induced by low extracellular magnesium in the human cortex maintained *in vivo*. Ann. Neurol. **30:** 589–596.
90. HELPERN, J.A., *et al.* 1993. Acute elevation and recovery of intracellular [Mg^{2+}] following human focal cerebral ischemia. Neurology **43:** 1577–1581.
91. MUIR, K.W. & K.R. LEES. 1995. A randomized, double-blind, placebo-controlled pilot trial of intravenous magnesium sulfate in acute stroke. Stroke **26:** 1183–1188.
92. MUIR, J.K., *et al.* 1999. Postinjury magnesium treatment attenuates traumatic brain injury-induced cortical induction of p53 mRNA in rats. Exp. Neurol. **159:** 584–593.

T_2-Weighted MRI Correlates with Long-Term Histopathology, Neurology Scores, and Skilled Motor Behavior in a Rat Stroke Model

GENE C. PALMER,[a] JAMES PEELING,[b] DALE CORBETT,[c] MARC R. DEL BIGIO,[b] AND THOMAS J. HUDZIK[a]

[a]*AstraZeneca R&D Boston, 3 Biotech, One Innovation Dr. Worcester, Massachusetts, U.S.A.*

[b]*Departments of Pharmacology/Therapy and Pathology, University of Manitoba, Winnipeg, Manitoba, Canada R3E OW3*

[c]*Basic Medical Science, Faculty of Medicine, Memorial University, St. John's, Newfoundland, Canada A1B 3V6*

ABSTRACT: The intraluminal suture model of transient middle cerebral artery occlusion (MCAO) in the Sprague Dawley strain of rats characteristically results in an inconsistently sized brain lesion. The purpose of the investigation reported here was to determine whether there were strong point-to-point correlations between the degree of cortical lesion size, as assessed *in vivo* using T_2-weighted magnetic resonance imaging (MRI) and corresponding cortical lesion size using routine histopathological techniques. Moreover, we aimed to investigate if cortical lesion size as determined by these two modalities correlates with neurological and/or skilled motor deficits observed in individual animals. Baseline behavioral scores were obtained on the animals prior to receiving 60 min of transient MCAO. Following MCAO, animals were tested for 1–21 days for neurological deficits. T_2-weighted MRIs of the cortex were taken at two and seven days post-MCAO. At 30 and 60 days the rats were retested for forelimb dexterity in the staircase test. Subsequently, the cortex was examined for histopathological damage. Indeed, there were highly significant correlations between lesion size determined by MRI and histopathology. The degree of cortical damage observed in the T_2-weighted MRI, as well as the size of the histopathological lesions were, in turn, highly correlated with the degrees of deficiencies observed in the composite neurological assessments and with the deficits involving skilled use of the contralateral forepaw (damaged side).

KEYWORDS: Stroke; Transient focal ischemia; MRI; Histopathology; Behavior.

INTRODUCTION

A major requirement for evaluation of promising compounds to treat stroke in both patients and animal models is the development of a methodology to correlate and monitor efficacy on an acute and long-term basis. Under rigidly controlled laboratory conditions such a means is possible. However, in the emergency room the

Address for correspondence: Gene C. Palmer, Ph.D., AstraZeneca R&D Boston, 3 Biotech, One Innovation Drive, Worcester, MA 01605, U.S.A. Voice: 508-890-8142.
eugene.palmer@astrazeneca.com

physician is faced with an unknown situation(s), such as, the timing of the occurrence, the exact location of the occluded artery(s), the completeness of the occlusion, state of consciousness, the condition, age, and health of the patient, and other underlying diseases. Previously, we[1,2] and others[3,4] have successfully employed the monofilament suture model of middle cerebral artery occlusion (MCAO) in the Wistar rat in order to evaluate promising compounds for advancement to clinic trials. However, when we attempted to develop a similar model in the Sprague Dawley strain of rat, we noticed that despite the use of rigid controls for the experimental conditions, the size of the ensuing cortical infarcts as assessed with T_2-weighted magnetic resonance imaging (MRI) and ultimately histopathology varied considerably among animals. Could this model then be used to mimic the situation seen by the physician in the emergency room and would the model be applicable to long-term follow up? The investigation reported here employed 60 min of transient MCAO in the Sprague Dawley rat with the overall goal to determine whether there were point-to-point correlations with respect to both short term (neurological evaluations and cortical T_2-weighted MRI) and long-term (forelimb dexterity and cortical histopathology) sequelæ.

METHODS

Appropriate references are made to the details of methods that have been well described by others and us elsewhere.

Animals

Approval of the protocol was obtained from the three animal care committees overseeing the study (Astra Arcus USA, Memorial University, and the University of Manitoba). As mandated by the Canadian Council on Animal Care, strict adherence of their policy was observed for the treatment of animals. Male Sprague Dawley rats were obtained from Charles River Laboratories, Montreal, Quebec.

Protocol Outline

Rats were received at Memorial University and acclimatized to the animal care facility for seven days.

1. Staircase Test. On day eight the food-deprivation schedule (animals maintained at 85% of their feeding weight) was started and the animals were tested daily, for eight days at two trials per day, in the staircase task to establish baseline scores for forelimb dexterity which involved retrieval and eating of Noyes food pellets. The data obtained were based on the average number of pellets consumed over the last three test sessions.[3,5,6]

2. The rats were next shipped to the University of Manitoba for surgery, T_2-weighted MRI, and postsurgical neurological evaluations.

3. The duration of transient MCAO was 60 min (intraluminal suture method) with crude motor assessments (measurement of circling behavior and contralateral forelimb deficits when suspended by the tail) taken at 60 min postreflow. During surgery and recovery care was taken to prevent any central or peripheral hypothermia. Motor

impairment was rated on a scale of 0 (no impairment) to 4 (severe circling or forelimb deficits).[1,2,6-8]

4. T_2-weighted MR images were taken at two and seven days post-MCAO.[1]

5. Assessments for neurological impairment were evaluated prior to surgery and at 1, 2, 4, 7, 14, and 21 days post-MCAO; testing paradigms included: (1) ipsilateral circling behavior; (2) forelimb flexion; and (3) beam walking ability (see Ref. 1 and legend to FIGURE 4).

6. At about 23 days post-MCAO rats were returned to Memorial University for postsurgical evaluation of forelimb dexterity. Thus, at about 30 and about 60 days post-MCAO, animals were food deprived and retested on the staircase test (five days at two trials per day).

7. The brains were perfusion-fixed and sent to the University of Manitoba for histopathological analyses.

8. All aspects of the experiment were coded and run blind.

Histopathology

Rats were overdosed with pentobarbital and perfusion fixed via the left cardiac ventricle with 250 ml of 4% ice cold paraformaldehyde in 0.1 mol/L phosphate buffer. After removal the brain was fixed for one week in the same mixture maintained at 4°C. The cerebrum was cut into 6–7 coronal slices (about 1.5 mm thick) followed by paraffin embedding. Histological sections (6 μM) were prepared and stained with hematoxylin and esosin (H&E). Two contiguous areas, namely areas of cortical infarction (loss of all tissue elements) and areas of intact tissue with selective neuronal loss were mapped separately onto standardized drawings of brain coronal sections. Histopathological analyses were performed at high magnification by experienced observers blind as to the findings from the T_2-weighted MRI and behavioral data. The area of damage in the cortical regions was determined by computerized planimetry. The final values were represented as damage area relative to the intact brain area (no attempt was made to extrapolate total volumes of tissue loss). When comparing with the MR images we assumed that the MR changes would occur either in areas that went on to infarct, or the total area that exhibited damage.

Statistics

Data were plotted using regression line analysis and point-to-point correlations of the data were made using the Pearson product moment correlation test (SigmaStat, Jandel Scientific, San Rafael, California).

RESULTS

Observations from the Individual Experiments

Animals

A total of, 77 rats were prepared for surgery. Twenty-one died, eleven within two hours of surgery, and ten at later times, to six days. Rats dying within two hours were assumed to have died from surgical complications and were removed from the study. At initial arrival, the mean weight ± SD of the animals was 264 ± 13.6 grams. At the time of MCAO surgery, the mean weight had increased to 323 ± 15.5 grams. At the times of testing for forelimb dexterity the mean weights were: 30 days post-MCAO, 435 ± 35; and 60 days post-MCAO, 498 ± 49 for a total mean weight gain of 234 ± 48 grams.

Crude Motor Behavior

Testing for crude motor behavior at one hour postreperfusion has been described to be a useful technique in the Wistar rat to eliminate incompletely lesioned animals from a study.[1,2,8] A close inspection of the present data found no correlations between ultimate lesion severity (cortical T_2-weighted MRI or cortical histopathology) and the degree of crude motor deficits exhibited by the animals immediately following the MCAO. Most scores ranged from a value of 2 to 3.

Neurological Assessment of Motor Behavior from 1–21 Days Post-MCAO

Prior to MCAO the control neurological test scores ranged from 0.00 to 0.75. The mean ± SD composite baseline score was 0.18 ± 0.19. Following MCAO the highest (most severely impaired) possible composite neurological behavioral score would be 10. In the present study the highest individual score attained by any animal was 8, occurring on post-MCAO Days 1 and 2. The range in test scores was, however, considerable because on these testing days, scores of 0.0 were observed in some rats signifying a complete lack of neurological deficit. Over the ensuing testing periods to 21 days post-MCAO the deficit scores gradually improved. For example, at 21 days post-MCAO the scores ranged from 0.0 to 4.0. The greatest deficits occurred at Days 1 and 2 post-MCAO (mean deficits were 3.3 ± 2.4 and 3.2 ± 2.4, respectively). These composite neurological test scores again gradually improved to a mean deficit score of 1.6 ± 1.2 at Day 21 post-MCAO. The trend for improvement was significant. The p value (Student's two-tailed t test), comparing the difference between Day 1 post-MCAO and Day 21 post-MCAO, was less than 0.001, $n = 45$.

T_2-Weighted MRI

FIGURE 1 is an example of a T_2-weighted MRI for a severely damaged rat brain taken at two days post-MCAO. The interanimal range in cortical lesion volumes (expressed as cm^3) was considerable, from rats exhibiting no damage to those with severe damage (ranges, 0 to 0.35 on Day 2; and 0 to 0.22 on Day 7). The mean ± SD for the cortical lesion on Day 2 was 0.10 ± 0.12, a value twice as large as the mean lesion volume of 0.05 ± 0.07 on Day 7.

Staircase Test

Contralateral—Left Forepaw. The left forepaw is contralateral to the lesion (occlusion of the right common carotid artery) and represents the damaged side. During the baseline testing session an animal retrieved and consumed a mean ± SD of 16.2 ± 2.1 Noyes pellets. When testing resumed at approximately 30 and 60 days following MCAO, the overall mean number of Noyes pellets retrieved and consumed was significantly reduced: 30 days post-MCAO, 8.2 ± 7.0; and 60 days, 8.6 ± 6.8 ($p < 0.0001$, Student's two-tailed t test). The range of Noyes pellets retrieved and consumed varied considerably from zero in the most severely lesioned animals, to 20 in the animals with mild lesions.

Ipsilateral—Right Forepaw. The right forepaw is ipsilateral to the lesion and is therefore not expected to sustain the degree of deficits observed for the left forepaw. During pretesting the mean ± SD number of Noyes pellets (16.4 ± 1.9) reached for and consumed by the right forepaw on the staircase test was similar to scores for the left forepaw. During retesting after MCAO the mean number of pellets reached for and consumed was slightly reduced: 30 days, 15.5 ± 4.4; and 60 days, 15.6 ± 4.5. The range of pellets eaten and consumed was from 0 to 21. However, the 0 score only occurred in one animal, the actual range being 8 to 21.

Histopathology

Sixty days after MCAO two types of histopathological changes were observed in the cerebral cortex of the rat brains. Regions of infarction exhibited tissue loss with small cavity formation or atrophy of tissue with only some blood vessels and glial cells remaining. Regions of selective neuronal loss in the cortex appeared hypercellular, because the larger neurons were lost and there was an increase in the number of small astroglial cells. This condition existed alone or surrounding an infarcted area. In the latter circumstance there was usually a band of involved tissue not more than 1 mm in width. It was clear that there were three cohorts of rats, those with hypothalmic damage, those with hypothalamic + striatal damage, and those with

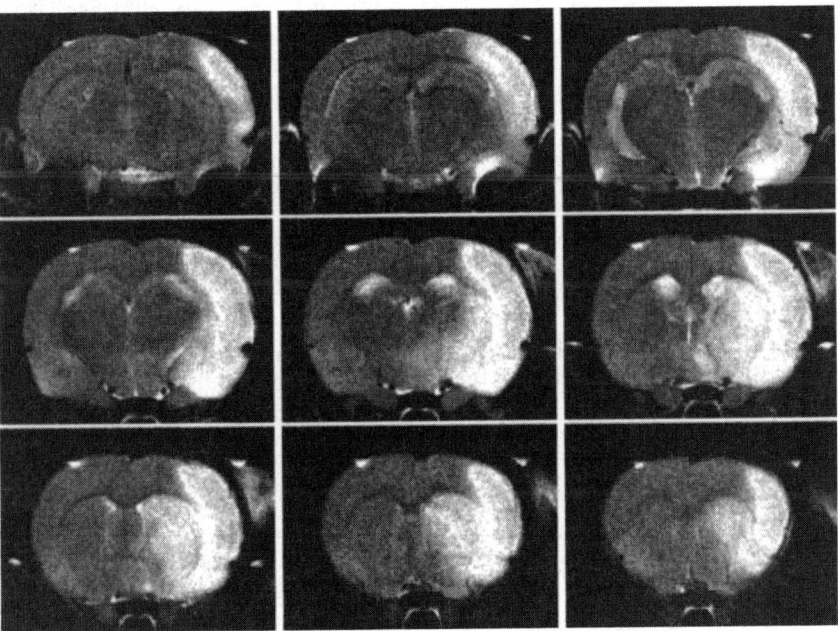

FIGURE 1. Representative set of T_2-weighted MR images spanning the brain of a severely lesioned Sprague Dawley rat taken at two days after 60 min transient MCAO. The ischemic lesion corresponds to the hyperintense region on each image.

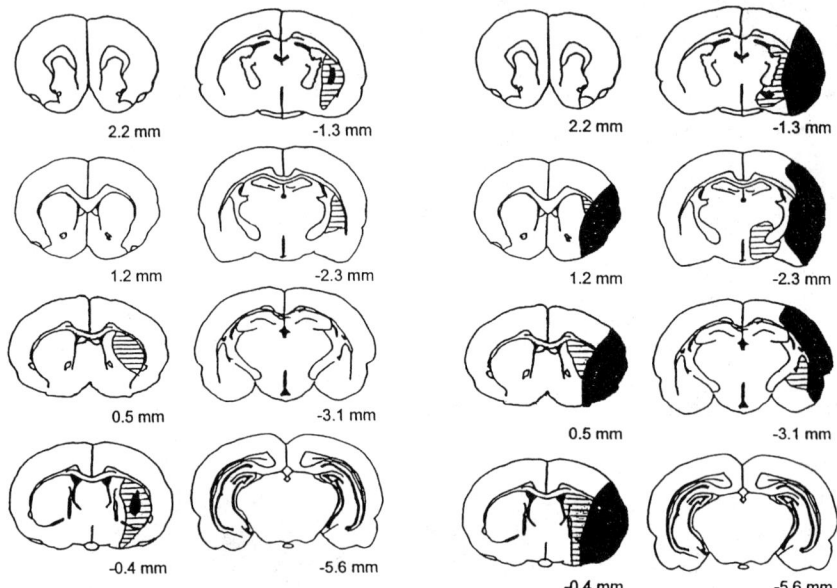

FIGURE 2. Representative examples of the maps of histopathologically defined brain injury 60 days after 60 minutes of transient MCAO in Sprague Dawley rats. Lesions of moderate size were confined to the striatum and hypothalamic region. Larger lesions extended laterally into the cerebral cortex including the amygdala and piriform cortex. The *cross hatched* regions mark areas with selective neuronal (and perhaps some glial) cell loss. The *solid regions* mark areas of necrosis with loss of all tissue elements, cavitation, macrophages, and atrophy. The *left panel* is representative of an animal with a slight to moderate sized lesion. The *right panel* is representative of a severely lesioned animal.

hypothalamic + striatal + cortical damage (see FIGURE 2). Areas of infarction and total damage (infarction + selective neuronal loss) were determined in the cortex. Relative areas of cortical infarct ranged from 0.0% to 56.12% and relative areas of total cortical injury ranged from 0.0% to 64.65%.

Correlations (Pearson Product Moment Correlation Test)

Histopathology

Correlations between the individual data points from the cortical histopathology comprising the infarct volume plus the region of selective neuronal loss were made with the following experimental parameters. The values are the respective correlation coefficients with associated p values and the number of animals used in that particular determination.

1. Animal Weight Gain To Histopathology. The ensuing relationship between weight gain and lesion size was significant; that is, the bigger the lesion the smaller the weight gain: correlation coefficient, -0.383; $p = 0.0134$; $n = 41$.

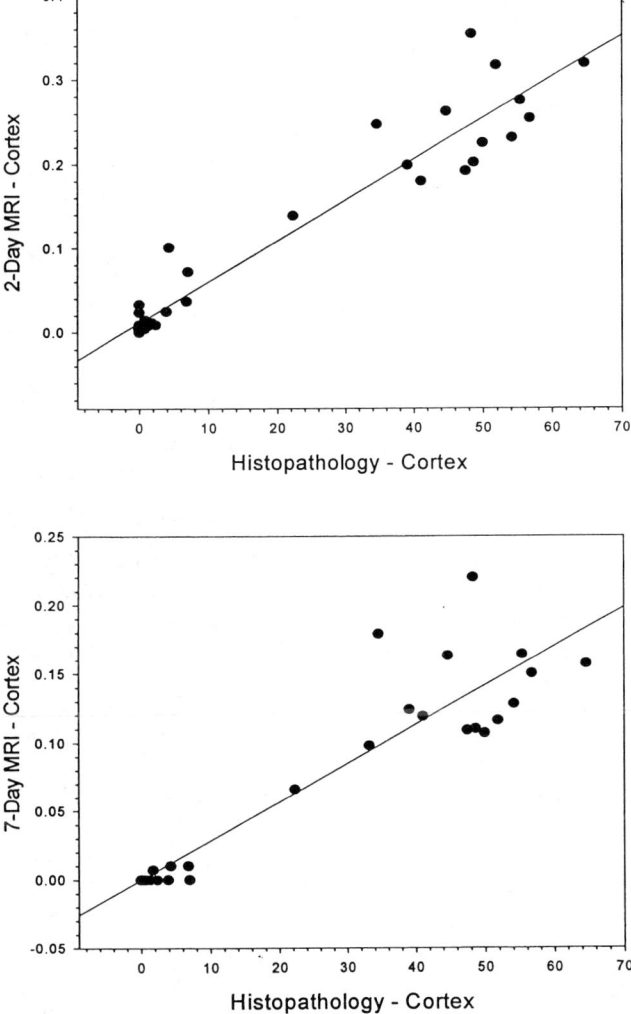

FIGURE 3. Pearson product moment correlations comparing total area of histopathological injury (volume of infarct plus region of selective neuronal loss) in the cortex with the corresponding T_2-weighted MRI cortical lesion volumes (cm^3) taken at two days (0.96; $p = 1.3 \times 10^{-20}$, $n = 36$) and seven days post-MCAO (0.94; $p = 9.8 \times 10^{-20}$, $n = 41$).

2. *Cortical T_2-Weighted MRI at Days 2 and 7 Post-MCAO.* The degree of cortical histopathology was highly correlated with the respective cortical lesion volumes seen on T_2-weighted MRI taken at two and seven days post-MCAO. The respective correlation coefficients, were 0.96 ($p = 1.3 \times 10^{-20}$, $n = 36$) for the Day 2 MRI and 0.94 ($p = 9.8 \times 10^{-20}$, $n = 41$) for the Day 7 MRI (see FIGURE 3).

3. *Neurological Behavior.* Individual neurological scores were obtained from testing on days 1, 2, 4, 7, 14, and 21 post-MCAO. These scores were highly correlated with cortical histopathology ($n = 45$).

Day 1:	0.75, $p = 2.4 \times 10^{-8}$	Day 7:	0.76, $p = 1.4 \times 10^{-8}$
Day 2:	0.72, $p = 1.8 \times 10^{-7}$	Day 14:	0.61, $p = 2.7 \times 10^{-5}$
Day 4:	0.72, $p = 1.5 \times 10^{-7}$	Day 21:	0.62, $p = 8.3 \times 10^{-5}$

4. *Forelimb Dexterity (Staircase Test).* The scores from the left forepaw representing the damaged side were highly correlated to cortical histopathology. On the other hand, the degree of correlation of cortical damage to the scores for the right forepaw was not as impressive ($n = 45$),

Left forepaw, 30 day score:	-0.71, $p = 1.7 \times 10^{-7}$	Right forepaw, 30 day score:	-0.42, $p = 0.007$
Left forepaw, 60 day score:	-0.74, $p = 4.4 \times 10^{-8}$	Right forepaw, 60 day score:	-0.29, $p = 0.07$

T_2-Weighted MRI

Correlation of the individual data points from the cortical infarct volumes obtained by T_2-weighted MRI at two and seven days post-MCAO were made with the following experimental parameters. The values listed below are the respective correlation coefficients with associated p values and the number of animals used in that particular determination.

1. *Neurological Behavior.* Individual neurological scores were obtained from test days 1, 2, 4, 7, 14, and 21 post-MCAO. These scores were highly correlated with T_2-weighted MRI. FIGURE 4 represents a regression line analysis of the average of the individual composite scores from testing on days 1–21.

Day 2, T_2-weighted MRI ($n = 36$)

Day 1:	0.76, $p = 6.4 \times 10^{-8}$	Day 7:	0.77, $p = 3.0 \times 10^{-8}$
Day 2:	0.75, $p = 4.0 \times 10^{-7}$	Day 14:	0.63, $p = 5.0 \times 10^{-7}$
Day 4:	0.74, $p = 3.0 \times 10^{-7}$	Day 21:	0.61, $p = 3.0 \times 10^{-4}$

Day 7, T_2-weighted MRI ($n = 41$)

Day 1:	0.75, $p = 2.0 \times 10^{-8}$	Day 7:	0.76, $p = 1.0 \times 10^{-8}$
Day 2:	0.72, $p = 1.0 \times 10^{-7}$	Day 14:	0.62, $p = 2.0 \times 10^{-5}$
Day 4:	0.73, $p = 1.0 \times 10^{-7}$	Day 21:	0.62, $p = 9.0 \times 10^{-5}$

2. *Forelimb Dexterity (Staircase Test).* The 30 and 60 day testing scores from the left forepaw representing the damaged side were highly correlated with the volume of cortical infarct assessed by T_2-weighted MRI at two and seven days post-MCAO (see FIGURE 5). On the other hand, the degree of correlation of cortical infarct determined by T_2-weighted MRI to the scores for the right forepaw was not as impressive.

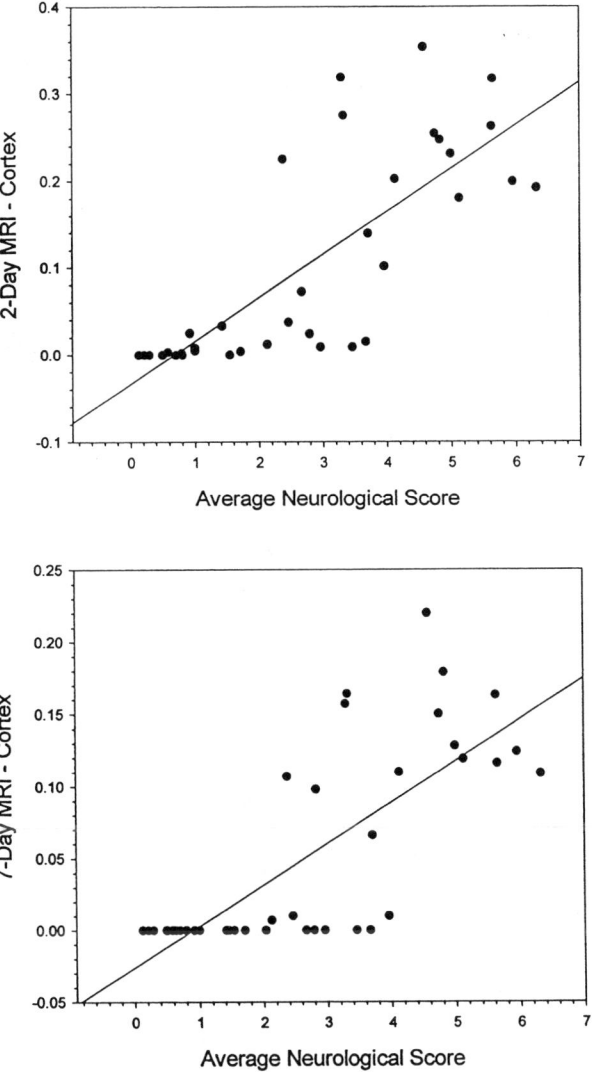

FIGURE 4. Pearson product moment correlations comparing the cortical T_2-weighted MRI lesion volumes (cm^3) taken at two days and seven days post-MCAO with the average of the individual composite neurological scores at 1, 2, 4, 7, 14, and 21 days post-MCAO. The average correlations were: two days, 0.71, $p = 4.2 \times 10^{-7}$, $n = 36$; and seven days, 0.72, $p = 2.6 \times 10^{-7}$, $n = 41$. The Neurological testing battery consisted of the following three tests with a possible overall neurological deficit score of 10: (1) Ipsilateral circling—graded from 0 (no circling) to 4 (continuous circling). (2) Forelimb flexion—measures the flexion of the paralytic forelimb when a rat is held by the tail above a plain surface (graded from 0 for a rat extending both forelimbs towards the surface to 2 for a rat showing severe flexion encompassing the entire forelimb). (3) Beam walking—graded from 0 for an animal that readily traversed a 2.4 cm wide beam, to 4 for a rat unable to accomplish the task.

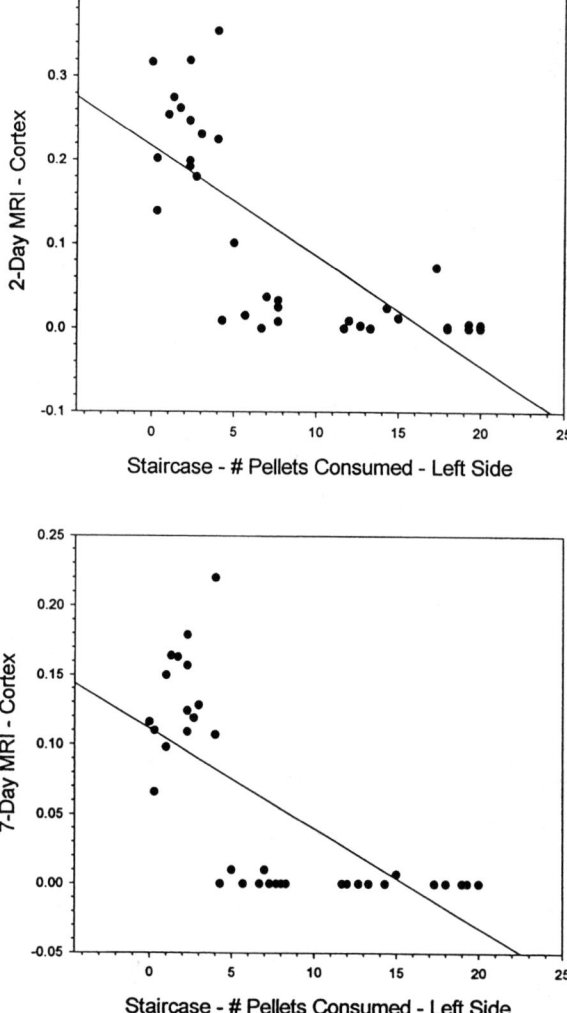

FIGURE 5. Pearson product moment correlations comparing the cortical lesion volumes (cm^3), obtained with T_2-weighted MRI at two and seven days post-MCAO, with deficits in contralateral forepaw dexterity during retesting on the staircase apparatus at 60 days post-MCAO. The correlation coefficients and associated p values are: (1) T_2 at two days versus 60 day staircase: -0.76; $p = 8.6 \times 10^{-8}$; $n = 36$; (2) T_2 at seven days versus 60 day staircase: -0.71; $p = 2.3 \times 10^{-7}$; $n = 41$.

Day 2, T$_2$-weighted MRI ($n = 36$),

Left forepaw, 30 day score:	$-0.73, p = 4.0 \times 10^{-7}$	Right forepaw, 30 day score:	$-0.42, p = 0.01$
Left forepaw, 60 day score:	$-0.76, p = 8.6 \times 10^{-8}$	Right forepaw, 60 day score:	$-0.30, p = 0.07$

Day 7, T$_2$-weighted MRI ($n = 41$),

Left forepaw, 30 day score:	$-0.70, p = 2.8 \times 10^{-7}$	Right forepaw, 30 day score:	$-0.41, p = 0.007$
Left forepaw, 60 day score:	$-0.71, p = 2.3 \times 10^{-7}$	Right forepaw, 60 day score:	$-0.31, p = 0.049$

3. Day 2 versus Day 7 MRI. Despite the reduction in the cortical infarct volume (due to changes in brain swelling) as determined by T$_2$-weighted MRI at seven days, the point-to-point correlations between two and seven days were highly significant: correlation coefficient, $0.97, p = 1.4 \times 10^{-21}, n = 36$.

DISCUSSION

Model

Notable observations from the present investigation involving transient 60 min MCAO in the Sprague Dawley rat model were the high degrees of correlation between the degree of cortical infarct volumes, assessed by T$_2$-weighted MRI at two and seven days post-MCAO, and the degree of total cortical histopathology, observed after 60 days post-MCAO. These two test parameters were, in turn, highly correlated with the degree of deficits observed in neurological scores from 1–21 days post-MCAO, as well as with the degree of deficiencies in contralateral forepaw dexterity occurring at 30 and 60 days post-MCAO. The degree of neurological damage following induction of transient MCAO with the intraluminal suture technique in rats is strain sensitive.[2] Wistar rats have in general been the strain of choice because of good survivability rates and consistent lesion volumes observed to two hours of MCAO.[1–4,6,8] The Lister Hooded strain likewise gives consistent sized lesions, but survival rates diminish after 60 min duration of MCAO.[2] Because of inconsistencies in lesion size and subsequent behavioral measures, Sprague Dawley rats are, therefore, not desirable for evaluation of therapeutic agents under rigidly controlled laboratory conditions. Nevertheless, the model afforded us the opportunity to look at point-to-point correlations of data taken from the individual animals under differing experimental conditions to two months in time.

Magnetic Resonance Imaging

Whereas in most experimental models of stroke, histopathology has been the mode of choice for evaluation of therapeutic efficacy of experimental drugs,[2–4,9–11] evaluation of such treatment regimens would be enhanced with the use of MRI. Both diffusion weighted and T$_2$-weighted MRI are useful noninvasive tools to determine the extent of *in vivo* brain pathology following stroke. Diffusion weighted MRI is especially sensitive to early ischemic changes in the brain whereas T_2-weighted images obtained at 24–48 h postischemia accurately measure mature infarcted tissue.[8,12–15] The extent and time course of both MRI indices depends upon the severity and regional location of the lesion.[8,12] In testing for efficacy of drugs in animal stroke models involving transient focal ischemia, rarely have histopathological outcomes been compared to MRI.[1,11,14–16] The major reason for this is that most drug investigations are acute in nature—less than 48 h[3,4,9,11,14–16]—and infrequently extend to one week[4,5,10,15] or longer.[1,2,17–19] Moreover, logistical problems may limit the use of MRI after one week.[1] Recently J.B.W. Marshall (personal communication) addressed this problem in experiments with the marmoset model of permanent MCAO. In this case T_2-weighted MRI was conducted at 3, 10, and 20 weeks and matched with histopathology.

Behavior

In general there has been a lack of agreement as to which long-term behavioral measures adequately represent the sequelæ of stroke, nor have such paradigms been used to evaluate many potential therapeutic agents. In most cases, neurological testing was administered within one week or less following ischemia.[3,4,10] As the present results and other investigations[1,3,4] indicate, neurological assessments improve with time. Strain differences may account for variability in outcome as well.[1] The goal of therapy is an extended quality-of-life issue. Thus, the aim of investigators should be to determine valid behavioral indices applicable to long-term follow up in patients. Measurement of complex behavioral function over time requires considerable effort because of the requirement for establishing baseline scores for comparison of outcomes during subsequent retesting. Another factor hampering such investigations is animal survivability.[20] Such tests that have reported success in the past have been memory tasks with gerbils and rats,[18,20] complex operant procedures and methamphetamine-induced rotation in rats,[6] forelimb dexterity in rats or marmosets (this paper and Refs. 1, 6, and 19), and spatial deficits in marmosets.[19] Except for the marmoset (J.W.B. Marshall, personal communication) and one equivocal study in the spontaneously hypertensive rat involving improvement in the ipsilateral forepaw dexterity,[1] these investigations lack confirmation of lesion volume *in vivo* using MRI. In a clinical study, the degree of infarct volume (T_2-weighted MRI) measured in humans within 72 h of diagnosis of occlusion of the middle cerebral artery was predictive of long term prognosis. The Scandinavian Stroke Scale was, however, not predictive of outcome.[21] Another investigation evaluated patients experiencing their first stroke. At 30 days the extent of the T_2-weighted lesion intersecting with the corticospinal tract was related to the magnitude of motor deficits.[22]

CONCLUSION

The transient MCAO model is considered by many investigators to be among the most valid representation of the pathology of stroke seen in humans.[23] The present model, in many instances, is reflective of the situation faced by the emergency room clinician, in which patients present with a wide variety of symptoms owing to the severity and location of the occlusion(s). The utilization of histological and T_2-weighted MRI techniques provides corroboration of pathology to functional, behavioral, and neurological modalities. Use of this long-term approach for drug evaluation should impart a rationale to the important question posed by investigators as to whether a drug actually prevents or merely delays subsequent neural or behavioral damage.[17,18]

REFERENCES

1. CREGAN, E.F., J. PEELING, et al. 1997. [(S)-Alpha-phenyl-2-pyridine-ethanamine dihydrochloride], a low affinity uncompetitive N-methyl-D-aspartic acid antagonist, is effective in rodent models of global and focal ischemia. J. Pharmacol. Exp. Ther. **283:** 1412–1424.
2. BIALOBOK, P., E.F. CREGAN, et al. 1999. Efficacy of AR-R 15896AR in the rat monofilament model of transient middle cerebral artery occlusion. J. Stroke Cerebrovasc. Dis. **8:** 388–397.
3. MUELLER, A.L., L.D. ARTMAN, et al. 1999. NPS 1506, a novel NMDA receptor antagonist and neuroprotectant. Ann. N.Y. Acad. Sci. **890:** 450–457.
4. BELAYEV, L., R. BUSTO, et al. 1995. HU-211, a novel noncompetitive N-methyl-D-aspartate antagonist, improves neurological deficit and reduces infarct volume after reversible focal cerebral ischemia in the rat. Stroke **26:** 2313–2320.
5. GRABOWSKI M., P. BRUNDIN & B.B. JOHANSSON. 1993. Paw-reaching, sensorimotor and rotational behavior after brain infarction in rats. Stroke **24:** 889–895.
6. HUDZIK, T.J., A. BORRELLI, et al. 2000. Long-term functional end points following middle cerebral artery occlusion in the rat. Pharmacol. Biochem. Behav. **65:** 553–562.
7. LONGA, E.Z., P.R. WEINSTEIN, et al. 1989. Reversible middle cerebral artery occlusion without craniotomy in rats. Stroke **20:** 84–91.
8. JIANG, Q., M. CHOPP, et al. 1997. The temporal evolution of MRI tissue signatures after transient middle cerebral artery occlusion in rat. J. Neurol. Sci. **145:** 15–23.
9. ARONOWSKI, J., R. STRONG & J.C. GROTTA. 1996. Combined neuroprotection and reperfusion therapy for stroke. Effect of lubeluzole and diaspirin cross-linked hemoglobin in experimental focal ischemia. Stroke **27:** 1571–1577.
10. KATSUTA, K., H. NAKANISHI, et al. 1995. The neuroprotective effect of the novel noncompetitive NMDA antagonist, FR115427 in focal cerebral ischemia in rats. J. Cereb. Blood Flow Metab. **15:** 345–348.
11. PEREZ-PINZON, M.A., C.M. MAIER, et al. 1995. Correlation of CGS 19755 neuroprotection against in vitro excitotoxicity and focal cerebral ischemia. J. Cereb. Blood Flow Metab. **15:** 865–876.
12. KNIGHT, R.A., M.O. DERESKI, et al. 1994. Magnetic resonance imaging assessment of evolving focal cerebral ischemia: Comparison with histopathology in rats. Stroke **25:** 1252–1262.
13. WARACH, S., M. BOSKA & K.M.A. WELCH. 1997. Pitfalls and potential of clinical diffusion-weighted MR imaging in acute stroke. Stroke **28:** 481–482.
14. LAURENT, D., M. EIS, et al. 1996. Reduction of excitotoxicity-induced brain damage by the competitive NMDA antagonist CGP 40116: A longitudinal study using diffusion-weighted imaging. Neurosci. Lett. **213:** 209–212.
15. MINEMATSU, K., M. FISHER, et al. 1993. Effects of a novel NMDA antagonist on experimental stroke rapidly and quantitatively assessed by diffusion-weighted MRI. Neurology **43:** 397–403.
16. SUTHERLAND, G.R., J.T. PERRON, et al. 2000. AR-R 15896AR reduces cerebral infarct volume following focal ischemia in cat. Neurosurgery **46:** 710–720.
17. COLBOURNE, F., H. LI & A.M. BUCHAN. 1999. Continuing postischemic neuronal death in CA1. Influence of ischemia duration and cytoprotective doses of NBQX and SNX-111 in rats. Stroke **30:** 662–668.

18. COLBOURNE, F. & D. CORBETT. 1995. Delayed postischemic hypothermia: a six month survival study using behavioral and histological assessments of neuroprotection. J. Neurosci. **15:** 7250–7260.
19. MARSHALL, J.W.B., A.J. CROSS & R.M. RIDLEY. 1999. Functional benefit from clomethiazole treatment after focal cerebral ischemia in a nonhuman primate species. Exp. Neurol. **156:** 121–129.
20. ORDY, J.M., B. VOLPE, et al. 1992. Pharmacological effects of remacemide and MK-801 on memory and hippocampal CA1 damage in the rat four-vessel occlusion (4-VO) model of global ischemia. In The Role of Neurotransmitters in Brain Injury. M. Globus & W.D. Dietrich, Eds.: 83–92. Plenum Press, New York.
21. SAUNDERS, D.E., A.G. CLIFTON & M.M. BROWN. 1995. Measurement of infarct size using MRI predicts prognosis in middle cerebral artery infarction. Stroke **26:** 2272–2276.
22. PINEIRO, R., S.T. PENDLEBURY, et al. 2000. Relating MRI changes to motor deficit after ischemic stroke by segmentation of functional motor pathways. Stroke **31:** 672–679.
23. SIESJO, B.K. 1992. Pathophysiology and treatment of focal cerebral ischemia. Part II: mechanism of damage and treatment. J. Neurosurg. **77:** 337–354.

On the Relationship Between Plasma Concentrations of Drugs and Outcome of Stroke Studies in Laboratory Animals

STEPHEN H. CURRY

Department of Pharmacology and Physiology, University of Rochester, New York, U.S.A.

ABSTRACT: In assessing plasma concentrations of drugs in relation to neuroprotective effect, emphasis should be placed on measured or calculated concentrations during the window of opportunity for effect, rather than at the end of the experiment. Unbound (plasma free) concentrations should be especially considered as should brain penetration to the stroked area. Problem-solving exercises should include *post hoc* assessment of dosing residues and proof of exposure. The shape of the graph of response versus concentration in plasma is very steep, giving the impression of an all-or-none effect. Although higher doses lead to greater effects, attempts to statistically correlate plasma level and infarct size are likely to be unsuccessful. There is strong evidence that the pharmacokinetic properties of drugs are affected by the physiological consequences of ischemia.

KEYWORDS: Plasma concentration; Stroke studies; Laboratory animals; Pharmacokinetics.

INTRODUCTION

Data from experimental stroke models in laboratory animals often show confusing discrepancies. Drugs may be active in one or more, but not all, models. There may be differences between investigators, unexpected positives and negatives, and lack of measurable dose dependency of response. It is possible that some of these differences relate to pharmacokinetic factors in drug response. However, research with stroke models still infrequently involves conduct of experiments in which blood samples are collected for drug analysis.

There are at least three principal objectives to sampling: (1) proof of exposure; (2) adequate experimental design—timing of pharmacodynamic assessment in relation to drug bioavailability, and halflife; and (3) facilitation of comparison across species, especially from animals to humans. In this review, the role of pharmacokinetic factors in response to the experimental drug AR-R15896 is evaluated as a case study, using published literature.[1–7] Details of the pharmacology and clinical potential of AR-R15896 are to be found in earlier publications in this series,[8–9] as well as in a recent clinical paper.[10]

Address for correspondence: Stephen H. Curry, P.O. Box 24733 Rochester, NY 14624, U.S.A. Voice: 716-259-7088.
  stephenhcurry@earthlink.net

It is an article of faith in pharmacology that drug responses, and drug concentrations in plasma, both relate to dose. Hence, responses must relate to concentrations. However, there are many variations on this theme, particularly when considering drugs in experimental models in which responses are rarely "graded" in the classical pharmacological sense and in which collection of sufficient blood samples with pharmacokinetic analysis in mind is especially difficult. In a perfect world, we would expose our test species to a square wave of drug concentration (see FIGURE 1). At a chosen time, dosing would commence and immediately induce a previously chosen target concentration, which would be maintained for a predetermined period, after which the concentration would fall instantly to zero. Responses would be measured repeatedly during this time, and related to target concentration, duration of exposure, and total exposure (concentration × time). Data of this type could then be compared across species. Additionally, the brain-to-plasma (B/P) ratio would be the same across species, and unaffected by ischemia.

In reality, this square wave exposure is unattainable. Even if the dosing is by intravenous injection, metabolic and tissue distribution considerations dictate that the graph of concentration will show a growth and decay pattern. Intravenous infusions may induce a plateau. If the dosing is extravascular, the growth and decay patterns do not incorporate a plateau, although some semblance of a plateau might be achievable by combining a loading dose with a maintenance dose.

There are other potential problems:
- The target concentration, especially in humans, may not be known.
- The kinetics of the drug, especially if they are affected by cardiac output and

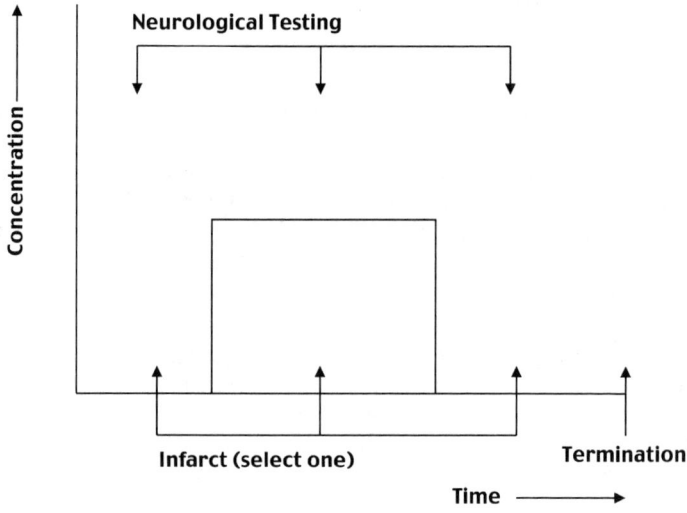

FIGURE 1. Representation of a desirable, but unattainable square wave exposure to a drug in an experiment involving induction of ischemia, neurological testing, and termination for histology. Induction of ischemia (infarct) can be before, during, or after drug treatment. Ischemia can be permanent or transient.

hepatic perfusion, may not be known in the stroked animals and may not be predictable from unstroked animals.

- Sampling difficulties are common. There may be physical limitations on access to a vein in multicathetered animals, the timing of samples within a complex protocol may be difficult, the volume collectable may be limited by the blood volume of a small animal, and there may be legal limitations, especially in the case of primates.

- Brain uptake, which is heavily dependent on perfusion, may be severely affected by the stroke.

CAT: MCAO

In their investigation in cats, Sutherland et. al.[1] first conducted an experiment in three, non-stroked cats to determine the approximate halflife and apparent volume of distribution of the drug, AR-R15896. Each cat was sampled at three times. The data from this experiment were used to calculate the dosing regimen for an experiment in which cats received a single intravenous infusion of the drug over 15 minutes from the 30–45 minute interval within a 90 minute ischemic period. The dose was designed to give a maximum concentration (C_{max}) at the end of the infusion of 1,500 ng·ml^{-1} (see FIGURE 2). In the event, the mean measured concentration was a significant, 17 percent below prediction, at 1,240 ng·ml^{-1}, although the halflife was similar in both stroked and non-stroked animals. The model was one of transient ischemia, and there were highly significant effects seen by MRI and histopathology measures,

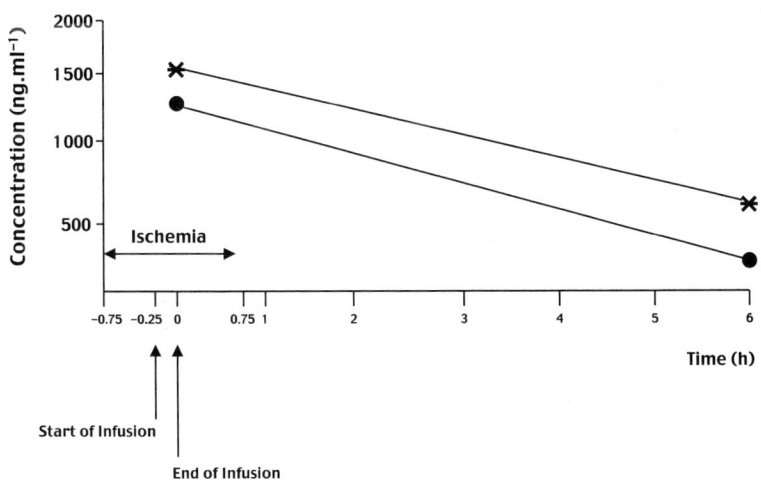

FIGURE 2. Design and pharmacokinetic outcome of an MCAO experiment in cats. (Design, *asterisks*; outcome, *solid circles*.)

with a reduction of infarct size compared with drug-free controls of approximately 75 percent. The concentration of 1,500 ng·ml^{-1} has become established as a target for further stroke studies in both laboratory animals and patients.[2] This is believed to be the only published study in which the plasma levels have been *measured* at an early time, at the end of the infusion, during an ischemia experiment with a controlled regimen. The levels represent C_{max} at the time of maximum concentration (T_{max}). Protein binding of the drug was also assessed to be 50 to 60 percent. Work with cats is, of course, severely limited by number of animals available.

RAT: MCAO/SHR

Much of the selection of dosing regimens with AR-R15896 for rats can be traced back to pharmacokinetic studies with 100 mg·kg^{-1} subcutaneous (s.c.) doses in non-stroked rats.[3] Dose proportionality in pharmacokinetic area under the curve (AUC), dose independence of T_{max}, and input and output rate constants were assumed. In the MCAO/SHR experiments, rats received 10 mg·kg^{-1} doses of AR-R15896 at 0.5, 4.5, and 12 h after induction of MCAO. Pharmacokinetic modeling (see FIGURE 3 for a representation of the model, without the third dose, and with the outcome) showed that the concentration reached 1,500 ng·ml^{-1} at 2.1 h, rising later to a maximum of approximately 3,000 ng·ml^{-1} at the five hour point. The model was validated by means of single sample measurements in seven stroked rats at 6.5 h (FIG. 3). The

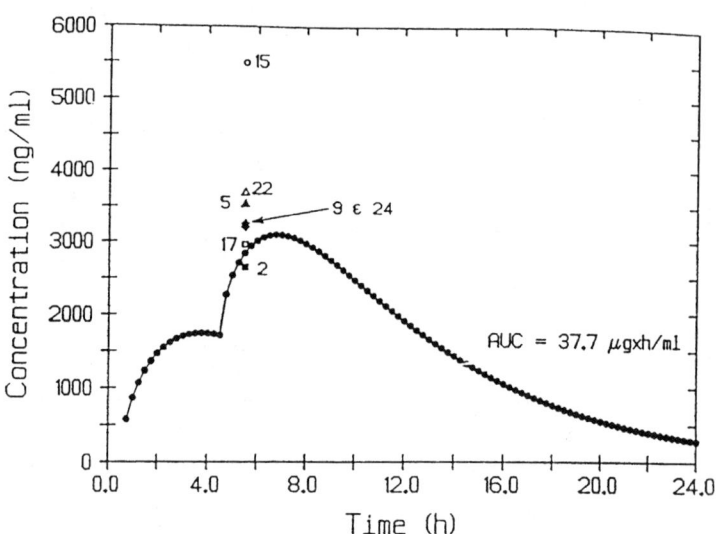

FIGURE 3. Design and pharmacokinetic outcome of an MCAO/SHR experiment in rats. The graph shows the modeled data (*solid circles*) and individual measured data points in seven rats with unique numbers. The AUC is the pharmacokinetic area under the curve. (Reproduced, with permission, from Ref. 2.)

effects of the drug on infarct size, MRI parameters, and neurological scores were statistically significant.

RAT: MALONATE MODEL

In this experiment, subcutaneous doses of 9 mg·kg^{-1} were based on the pharmacokinetic data used in modeling the MCAO/SHR dosing regimen. The plasma concentrations designed to occur were consistent with a concentration of 1,500 ng·ml^{-1} at approximately two hours after a single dose. In the experiment, some of the rats received doses at the time of the malonate injection and 210 minutes later. Others received doses 30 minutes before or five minutes after malonate injection, and again 210 minutes after the malonate injection. The actual C_{max} was calculated to have been 1,778 ng·ml^{-1} at four hours after the first dose and 3,186 ng·ml^{-1} four hours later. In the protocol in which the first dose was given 30 minutes before the malonate dose, the effect was highly significant.

RAT: INTRACEREBRAL HEMORRHAGE MODEL

This experiment used subcutaneous loading doses and minipumps, based on the s.c. pharmacokinetics discussed earlier. The plan was to induce and maintain concentrations in plasma of approximately 2,000 ng·ml^{-1} for one week. The actual measured levels were 2,490 ± 632 (S.E.M.) ng·ml^{-1} at four hours and 1,860 ± 270 (S.E.M.) ng·ml^{-1} at seven days, in accord with the model. The experiment showed improvement in quality of life indices, but not neurological outcome.

RAT: MONOFILAMENT MODEL

In this experiment, AR-R15896 was administered at three exposure levels (low, medium, and high) over seven days, involving intravenous loading doses (over 30 minutes) followed by subcutaneous minipump maintenance doses (from two hours onward). The loading doses were calculated from known intravenous kinetics in non-stroked animals, with the intention of inducing target concentrations of 400, 1,500, and 3,000 ng·ml^{-1} as early as possible, and maintaining them for the entire investigation period. Ischemia lasted for two hours, starting five minutes before the start of the loading dose. The outcome was mean concentrations (Cp_{ss}) of 682, 1,885, and 2,682 ng·ml^{-1} measured at seven days after the low, medium, and high exposure levels, respectively, in reasonable agreement with the models.

For the purpose of this review, the clearance of the drug in these studies was calculated using the actual measured concentrations as Cp_{ss} values in the equation:

$$Cp_{ss} = \text{Input Rate/Clearance}$$

and the concentrations were recalculated at times up to and including seven days using standard pharmacokinetic equations. This model showed that the concentrations rose to an initial peak at the end of the 30 minute infusion, fell to a trough just

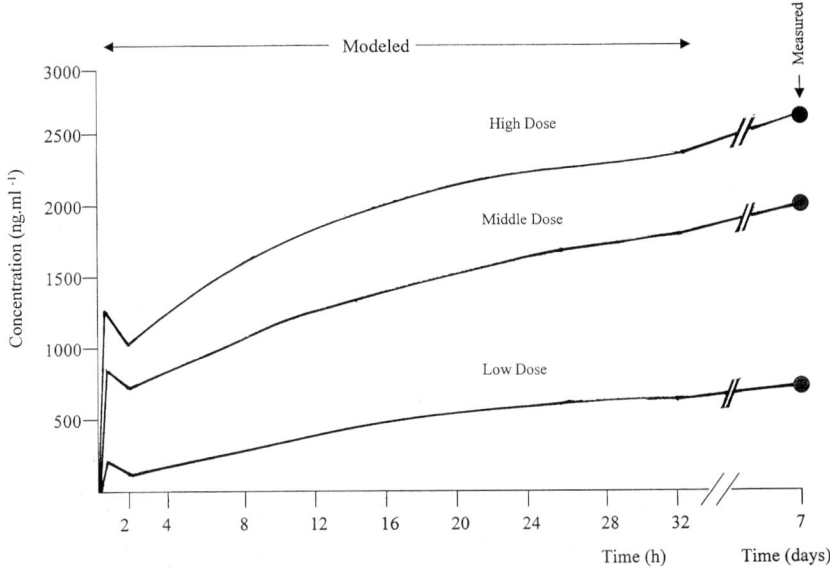

FIGURE 4. Post-hoc pharmacokinetic modeling of a rat monofilament experiment at three exposure levels involving loading and maintenance doses. Modeled data are indicated by *solid lines*. Measurements were at seven days.

before starting the subcutaneous minipump dose, and then rose to the steady state concentrations measured over the ensuing seven days.

The modeled data is shown in FIGURE 4. The high dose regimen induced approximately 1,250 ng/ml at the end of the infusion, and 1,500 ng·ml^{-1} at about seven hours; the C_{max} was 2,682 ng·ml^{-1} at seven days. The middle dose regimen achieved 1,500 ng·ml^{-1} at approximately 18 hours and 1,885 ng·ml^{-1} at seven days. The lower doses never achieved 1,500 ng·ml^{-1}, reaching 682 ng·ml^{-1} at seven days.

Except for a small effect of the lower dose on infarct size in subcortical areas, only the highest dose produced significant effects. This supports the view that late achievement of plasma concentrations that are adequate when achieved earlier is of little, if any, value. This study illustrates, in stark reality, the need to consider both *thresholds* for *effect* and *when* they are reached in designing stroke model experiments adequately. It does *not* tell us that concentrations in excess of 2,682 ng·ml^{-1} are needed for useful effect.

MARMOSETS

In these studies, the investigators were faced with a need to obtain as much data as possible from a species with limited numerical availability, and in which only a moderate amount of blood could be taken, for legal reasons. The experiment involved induction of permanent ischemia, treatment with the drug for 48 hours, extensive neurological testing, and termination for histology after ten weeks. A three-step pharmacokinetic approach was used:

1. A study of the kinetics of the compound in four non-stroked marmosets, each given the same dose of AR-R15896, but sampled in an incomplete block design at different times to give a composite pharmacokinetic picture in the four marmosets as a group.

2. Pharmacokinetic analysis of this data and calculation of a dosing regimen designed to induce and maintain 1,500 ng·ml^{-1} for 48 hours; this dose was given and samples were collected for assay at 48 hours.

3. Administration of a further dose of AR-R15896 to three stroked animals in the 24 hours before termination, after all neurological testing was complete, with collection of blood samples at three times, at early times after the dose, for comparison with predictions.

The objective of the experiment was to determine, if possible, whether AR-R15896 was active in a primate model of permanent ischemia at 1,500 ng·ml^{-1} in plasma. Since only 15–20 marmosets were available for both pharmacokinetic and stroke studies, a positive effect could have been taken as a positive, but a negative effect would have indicated no effect or an inadequate experimental design. It was, therefore, imperative to optimize the experimental design first. In the event, the drug was effective against hemineglect and reduced infarct size, but showed no significant effect on motor function at the single dose level studied.

The plasma levels in the initial three animals gave a terminal phase halflife at approximately 5.5 hours. Using this data, it was determined that a loading dose of 4.5 mg·kg^{-1} followed by a maintenance dose of 1.1 mg·kg^{-1}·h^{-1} would give a mean concentration of 1,500 ng·ml^{-1}. In the event, the 48-hour concentrations were in the

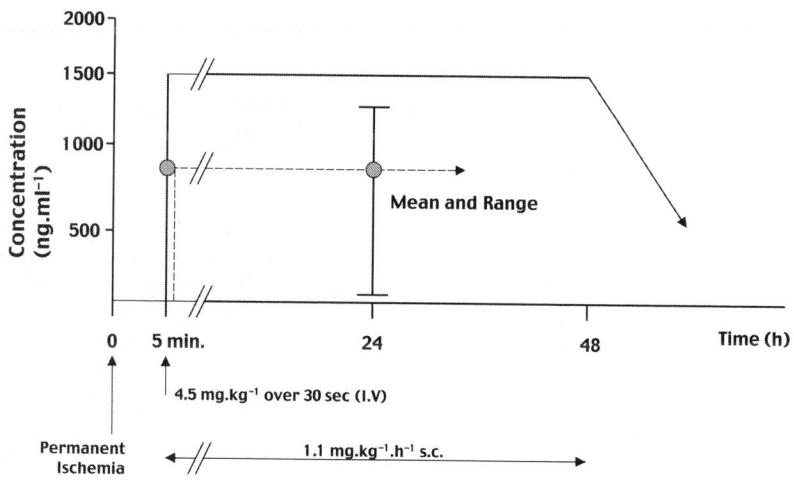

FIGURE 5. Design and pharmacokinetic outcome of a marmoset pMCAO experiment. The *solid line* is the modeled data. The outcome is indicated by *solid circles*. —, Target; --, achieved.

range of 400–1,200 ng·ml^{-1}, 30 percent below the prediction. The concentrations at the early times were in the same range (see FIGURE 5).

The loading doses were intravenous, whereas the maintenance doses were by means of minipumps. Dosing solutions were assayed *post hoc,* as a check on pharmaceutical compounding and minipump delivery characteristics, an essential control experiment in the event that outcome fails to meet expectations. Concentrations in these materials were correct. However, since it was not possible to evaluate the volume delivered from the minipumps, it was not possible to determine whether the 30 percent low concentrations resulted from faster metabolism in the stroked animals, or slower release from the minipumps. The latter is more likely. It was concluded that the marmosets experienced an effective concentration at both early and late times after induction of the infarct.

BRAIN-TO-PLASMA RATIO

The various studies quoted already involved no analysis of brain uptake. However, it has been shown in rats and cats that the brain-to-plasma ratio is constant and in excess of three, with the brain-to-plasma free ratio approximately twice this. "Free" indicates "unbound" concentration in plasma, or concentration in plasma water. This extrapolates to brain concentrations of approximately 4–10 μg·g^{-1}, or approximately 22.5–50 micromolar. It is not known at present the extent to which this represents bound drug and/or drug in the ISF bathing the NMDA receptors. Theoretical considerations suggest that ISF concentrations are the same as, or somewhat higher than those in plasma water, which exceed the EC$_{50}$ in functional assays.

It has been shown that AR-R15896 penetrates the ischemic side of the brain to the same extent as the non-ischemic side, albeit more slowly, in that full penetration was recorded at 60 minutes post dose, but there was somewhat less than full penetration at five minutes.

IS THERE A GRADED RELATIONSHIP BETWEEN PLASMA CONCENTRATION OF DRUG AND INFARCT SIZE?

The fact that control animals (drug concentration zero) show larger infarcts than do treated animals (drug concentration greater than zero) is a plasma concentration–related effect. However, each animal provides a single unit of information to the data set, so in any individual the response is more "quantal" than "graded". The question as to whether a population can provide graded data has been examined by studying the correlation between neurological score and/or infarct size and plasma concentration in marmosets and cats. In marmosets, in spite of the high significance of the effect, albeit in a small population of animals, no correlation was significant.

In the cats, linear, logarithmic, and exponential relationships were examined, using plasma concentrations and concentrations in plasma water. The correlation between infarct size (volume) and concentration in plasma water (plasma free) over the concentration range 300–800 ng·ml^{-1}, was the strongest, with a statistically significant correlation coefficient of -0.706 ($p < 0.02$), fitting the equation:

$$y = ae^{b \cdot x}$$

in which y is the infarct size, a is the intercept on the y-axis (response at zero concentration), b is the regression constant, and x is concentration. However, the calculated value of a was approximately twice the measured infarct size in controls. A probit analysis was therefore conducted, relating the measured infarct size (as percent of mean control, such that no response is 100 and no infarct is zero), and concentration in plasma water. The graph of probits versus concentration was a steep straight line, indicating that there are large increases in effect towards a maximum once the threshold for effect is reached. FIGURE 6 shows plots of the logarithmic relationship (equation above), and the relationship based on percent of controls. This figure indicates that there is a threshold concentration in plasma water of approximately 150 ng·ml^{-1}. Above that threshold, the response rises steeply to exceed 90 percent remission above 400 ng·ml^{-1}. There is, therefore, evidence of a graded effect, but the chances of finding a response in the middle third of the relationship are low—thus, it will commonly appear that the response is of a quantal, or "all-or-nothing" type, even in a population of animals. This may explain why no correlation *per se* between effect and concentration was observed in the marmosets, and supports the practice of defining a threshold concentration above which the concentration needs to rise to give a high probability of success. In the case of AR-R15896, a plasma free threshold concentration of 500 ng·ml^{-1} in cats should give more than 90% reduction in infarct size approximately 50% of the time. This translates into approximately 1,000 ng·ml^{-1} total plasma concentration. If this were to be the threshold for effect in rats also, then, in the monofilament model, the threshold was firmly established by the thirty minute point at the highest dose, was reached at

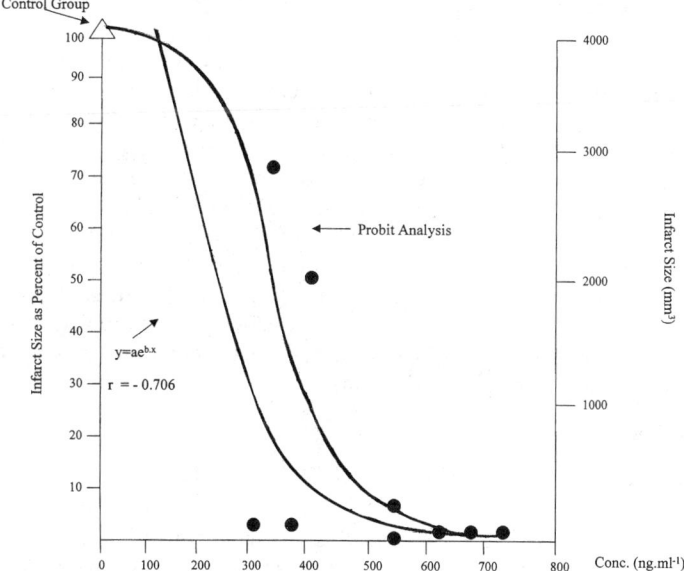

FIGURE 6. Infarct size against unbound drug concentration in cats, expressed as both volume and percent of control.

approximately seven hours during the middle dose regimen, and was never reached with the low dose.

PROBLEM SOLVING: THE GERBIL

In a *post hoc* analysis, it was shown that lack of activity of AR-R15896 in the gerbil model almost certainly resulted from a very short halflife (below 0.5 h).

THRESHOLD FOR EFFECT AND WINDOW OF OPPORTUNITY

Previously in this paper, emphasis was placed on the need to consider threshold concentrations for effect and when they are reached, together. The expression "window of opportunity" captures the concept that there is finite time after the induction of an infarct during which drug treatment must be initiated if it is to be successful. From the cat data, we can conclude that there is a steeply graded effect, and that a plasma free concentration of approximately 500 ng·ml^{-1}, translating to a total plasma concentration of approximately 1,000 ng·ml^{-1}, is the threshold, but there was no window of opportunity study in the cat. In the marmoset, a mean total concentration of 827.4 ng·ml^{-1} was found to be effective with no demonstration of a graded effect or window of opportunity study. The only published data on the window of opportunity for this compound was obtained by Buchan (see Ref. 2). There was a significant effect when drug treatment was delayed one hour after the start of ischemia, but not when two hours elapsed, in rats. Extrapolating the threshold from the cat to the rat, the rat monofilament model data, in part because three dose levels were used, suggests that the window of opportunity could be as much as 6–7 hours. It seems that the coordinates (threshold concentration × window of opportunity) for a high chance of a robust effect of this compound, which may depend on the persistence of raised glutamate in the body following trauma and which are probably different between species, are approximately 1,000 ng·ml^{-1} × y hours, with y (the window of opportunity) highly species dependent, possibly over the range of 1–12 hours. The critical, unanswered question is: What is the window of opportunity in humans? There lies the dilemma of drug development. It is just not possible to organize therapeutic trials in patients with multiple treatment groups of sufficient magnitude to examine different dose levels, let alone different potential windows of opportunity, and the best possible prediction is needed from grossly inadequate animal studies and only rudimentary knowledge of the mode of action.

CONCLUDING COMMENTS

The studies discussed in this review illustrate the fact that pharmacokinetic predictions from unstroked to stroked animals, from animal species to animal species, and from animals to humans are extremely difficult when stroke models are considered. There is strong evidence that the stroke procedure perturbs normal drug disposition, including minipump delivery, especially, perhaps, in models of permanent ischemia. However, it is not clear which of the pharmacokinetic processes are perturbed, and this is clearly an area awaiting intensive research, with implications for expectations regarding pharmacokinetics in patients. A clear correlation between drug concentration and experimental outcome should not be expected, and even setting some minimum concentration level will often be fraught with difficulty. Careful choice of dosing regimen is essential and prestroke and poststroke pharmacokinetic testing in order to relate outcome to measured concentrations is preferable to assumption of adherence to a model. No project will be complete without consideration of penetration by the drug into ischemic brain and binding to plasma protein. No problem-solving exercise will be successful without evaluation of dosing solutions and proof of exposure.

ACKNOWLEDGEMENTS

I wish to thank my former colleagues at AstraZeneca for their major contributions to, and publication of the work on which this review is based, and Dr. Garnette Sutherland and his colleagues for the opportunity to reanalyze the cat data. AR-R 15896 has also been referred to as AR-L 15896 and may have been given the designation AR-C 15896 in some literature. Additionally, it sometimes appears with an additional AR as a suffix, indicating hydrochloride salt.

REFERENCES

1. SUTHERLAND, G.R., J.T. PERRON, P. KOZLOWSKI & D.J. MCCARTHY. 2000. AR-R15896 AR reduces cerebral infarction volumes after focal ischemia in cats. Neurosurgery **46:** 710–719.
2. CREGAN, E.F., J. PEELING, D. CORBETT, et al. 1997. [(S)-alpha-phenyl-2-pyridine-ethenamine dihydrochloride], a low affinity uncompetitive N-methyl-D-aspartic acid antagonist, is effective in rodent models of global and focal ischemia. J. Pharm. Exp. Therap. **283:** 1412–1424.
3. LEES, K.R., A.G. DYKES, A. SHARMA, et al. 1999. AR-R15896 AR in patients with acute stroke—a dose finding study. Eighth European Stroke Conference.
4. GREEN, J.G., R.H.P. PORTER & J.T. GREENAMYRE. 1996. AR-L15896, a novel N-methyl-D-aspartate receptor ion channel antagonist: neuroprotection against mitochondrial metabolic toxicity and regional pharmacology. Exp. Neurol. **137:** 66–72.
5. PALMER, G.C., R.J. MURRAY, C.L. CRAMER, et al. 1999. [S]-AR-R15896AR—a novel anticonvulsant: acute safety, pharmacokinetic and pharmacodynamic properties. J. Pharm. Exp. Therap. **288:** 121–132.
6. BIALOBOK, P., E.F. CREGAN, S. SYDSERFF, et al. 1999. Efficacy of AR-R15896 AR in the rat monofilament model of transient middle cerebral artery occlusion. J. Stroke Cerebrovas. Dis. **8:** 388–397.
7. MARSHALL, J.W.B., E.J. JONES, K.J. DUFFIN, et al. 2001. AR-R 15896 AR, a low affinity, use dependent, NMDA antagonist, is neuroprotective in a primate model of stroke. J. Stroke Cerebrovas. Dis. **9:** 303–312.
8. PALMER, G.C., E.F. CREGAN, P. BIALOBOK, et al. 1999. The low-affinity, use-dependent NMDA receptor antagonist AR-R15896 AR: an update of progress in stroke. Ann. N.Y. Acad. Sci. **890:** 406–420.
9. PALMER, G.C., J.A. MILLER, E.F. CREGAN, et al. 1997. Low-affinity NMDA receptor antagonists: the neuroprotective potential of ARL15896 AR. Ann. N.Y. Acad. Sci. **825:** 220–231.

10. LEES, K.R., A.G. DYKER, A. SHARMA, *et al.* 2001. Tolerability of the low-affinity, use-dependent NMDA antagonist AR-R15896AR in stroke patients. A dose-ranging study. Stroke **32:** 466–472.

Combination Therapy Stroke Trial
rt-PA ± Lubeluzole

JAMES GROTTA FOR THE TRIAL INVESTIGATORS

Department of Neurology, The University of Texas-Houston Medical School, Houston, Texas, U.S.A.

KEYWORDS: Therapy; Stroke trial; rt-PA; Lubeluzole; Thrombolysis; Neuroprotection.

BACKGROUND

A neuroprotective drug may be safe and effective if given very early and in combination with rt-PA to acute stroke patients. No clinical trial has yet tested this hypothesis.

OBJECTIVE

To assess the feasibility, safety, and efficacy of simultaneously combining the neuroprotective drug lubeluzole with rt-PA.

METHOD

Patients who qualified for and received rt-PA within three hours of symptom onset were randomly allocated 1:1 to lubeluzole (7.5 mg iv over one hour, then continuous five-day infusion of 10 mg/d) or placebo. Infusion of study medication was started before the end of the one hour rt-PA infusion. Inclusion criteria were the same as FDA-approved guidelines for rt PA, plus NIHSS greater than 5, and absence of serious ventricular arrhythmia, AV block, or QT > 450 msec. EKG was continuously monitored until 48 h posttreatment. The primary outcomes were adverse events, especially hemorrhage and severe arrhythmia, and functionality as determined by the Barthel Index partitioned into greater than 70, 0–70, and dead.

RESULTS

Eighty-nine patients were randomized at 34 centers over eight months. The study was terminated by the sponsor before the planned enrollment of 200 patients when

Address for correspondence: Dr. James Grotta, Department of Neurology, The University of Texas-Houston Medical School, 6431 Fannin Street, MSB 7.004, Houston, TX 77030, U.S.A. Voice: 713-500-7100; fax: 713-500-7019.

a concurrent phase 3 trial of lubeluzole versus placebo, given up to eight hours post stroke, was negative. In our study, mean NIHSS was 14.5, and mean time from symptom onset to rt-PA was 2.5 h and to randomization to lubeluzole or placebo was 3.2 h. Mortality was 26%, ICH occurred in 10%, and serious adverse events in 51%. There were no differences between the two treatment groups in any of these variables, outcomes or in Barthel Index or other measures of functionality.

CONCLUSION

Combining neuroprotective drugs such as lubeluzole simultaneously with rt-PA is feasible and safe. The efficacy of this strategy should be evaluated in an adequately powered clinical trial.

Part VIII: Ischemic Stroke: Current Therapy

QUESTIONS FOR THE PANEL

From Dr. Slikker

Can MRI (both T2 and flow [perfusion or diffusion] MRI) be used to monitor the extent of ischemia or other brain damage in animal models in real time? Can MRI be used in the clinic to monitor the extent of stroke damage?

ANSWER: (Palmer) Yes, both T2-weighted MRI and diffusion weighted imaging have been used to monitor stroke damage over time. Dr. Garnette Sutherland used both parameters to assess damage in stroked cats treated with the low affinity use dependent NMDA antagonist, AR-R 15896. The inherent problem with laboratory animals is the use of anesthesia while they are in the magnet, but they can be evaluated at intervals over a period of several days. Several clinical investigators have indicated the potential use of diffusion weighted MRI to select patients for clinical studies. Diffusion weighted MRI gives the earliest information on the degree of damage, whereas T2-weighted images indicate mature lesions. Thus, a patient presenting with a severe T2-weighted image might not be included in a potential drug study whose goal is acute treatment to prevent further damage.

From Dr. Rosin

Dr. Auer, many patients present with a raised BP at the onset of an ischemic stroke. Do they necessarily do better than those with normal BP at that time? If the BP is raised (reactive to brain stem ischemic or other mechanisms) is it acceptable to lower it to within normal limits by therapy?

ANSWER: (Auer) Not in my view. Recovery after stroke has been linked to BP, and lower BP, even by 20 mm Hg, favors a poorer stroke-in-progression type of course (H.S. Jørgensen, *et al.* Effect of blood pressure and diabetes on stroke in progression. Lancet **344**: 156–159, 1994). Our animal data also shows that 20 mm Hg makes a different in infarct size (C.Z. Zhu and R.N. Auer. Graded hypotension and MCA occlusion duration: effect in transient focal ischemia. J. Cereb. Blood Flow Metab. **15**: 980–988, 1995). Neurosurgeons already know about the importance of blood pressure in ischemia, and have (successfully) treated their vasospastic ischemia with induced hypertension (N.F. Kassell. Treatment of ischemic deficits from vasospasm with intravascular volume expansion and induced arterial hypertension. Neurosurgery **11**: 337–343, 1982) but perhaps neurologists are too timid to do this in stroke! If hypertension is not to be carried out, I would certainly not induce hypotension in an ischemic stroke patient. We must remember that the lifelong tissue deficit is determined at the time of the stroke. Nature is using the Cushing reflex for a very good reason.

(Palmer) [answering with a question]: Would there not be a problem with the so-called "steal effect" where the viable tissue with its intact vasomotor control of its vasculature receives most of the blood supply leaving the ischemic tissue in worse condition?

(Grotta) We know that perfusion pressure as reflected in mean arterial blood pressure (MAP) will determine perfusion of ischemic regions so that lowering blood pressure below a certain level will aggravate ischemia. Conversely, CBF studies in man have shown that raising MAP will improve perfusion to penumbral regions. Anecdotally this is associated with better outcome, particularly in those patients who are fluctuating with internal watershed (large subcortical) strokes, but this has never been proven by a prospective randomized analysis. There is also no evidence that patients presenting with elevated MAPs do better. In the NINDS rt-PA trial, in patients with SBP > 180 or DBP > 110, MAP was carefully lowered to 130 mm HG before thrombolytic treatment. This was not associated with clinical deterioration and such patients still benefited from thrombolytic therapy without increased risk of hemorrhage.

Dr. Jonas: (1) In consequence of the gradual deterioration in stroke-in-evolution often over hours—12 or more—is this not a situation in which neuroprotection is likely to be efficacious, more than in an acute onset of neurobological deficit? (2) Since ancrod, a defibrinating agent, was successful in a trial of human stroke in combination with tPA it may produce a positive additive effect. However, would not the combination of a defibrinating drug with a thrombolytic drug significantly enhance the danger of hemorrhage? This may necessitate lowering the dose of each, thereby decreasing the therapeutic efficiency.

ANSWER: (Jonas) No, the clinical trials had wide windows of entry and therefore must have included evolving strokes; no benefit has been seen. It has been recorded in a mixed-up form; we should not include it.

(Grotta) I agree that deterioration in the first 12 hours may be due to mechanisms responsive to so-called "neuroprotective" measures. Some such deterioration may be due to hemorrhage, edema, or complications such as hypotension, heart failure, and pneumonia, but in most cases the cause is undetermined and may involve cytotoxic processes occurring in the penumbra. Such an unexplained deteriorating clinical course may identify those patients with substantial penumbral tissue at risk, who are exactly the patients who may benefit most from cytoprotective therapy.

(Grotta) Ancrod was not given with rt-PA in the trial showing its efficacy. I agree, giving it with rt-PA may be risky, just as giving streptokinase with aspirin or heparin has resulted in increased hemorrhage. However, it is possible that some form of antithrombotic therapy (ancrod, GP 2b3a antagonists, thrombin inhibitors, etc.) might be combined with low doses of rt-PA to improve results.

From Dr. Weinstock

With reference to Dr. Trembly's suggestion that drugs like Viagra may be potentially useful for increasing blood supply to penumbral regions after stroke, I would like to propose that the release of nitric oxide and vasodilatation could be confined to the cerebral cortex and hippocampus by the administration of cholinesterase inhibitors, thereby avoiding the "steal effect". The blood vessels in these brain regions are innervated by cholinergic neurons arising in the nucleus basalis of Meynert and in the septum. Electrical stimulation of these nuclei in rats and monkeys increases blood flow by releasing acetylcholine, which activates M_3 receptors to release NO. Cholinesterase inhibitors have been shown to increase blood flow to these brain regions selectively, but not to others or in the periphery and they do not lower systemic blood pressure.

ANSWER: (Auer) Sounds good in theory. Someone must try it in practice, though. I would start in experimental animals on this one.

From Dr. Manev

Recent experiments in rats have confirmed the hypothesis that circadian rhythm may influence the outcome of stroke (lesion size). Would it be possible to analyze the available clinical data to look for possible difference in the outcome of stroke in relation to the time of onset?

ANSWER: (Auer) Sleep stroke has been looked for (A.A. Glynn. Vascular diseases of the nervous system—A series of 315 cases. Br. Med. J. **1**: 1216–1219, 1956). Stroke onset during sleep may relate to increased platelet aggregation in some patients (C.L. Voll, N. Chetty, and P. Atkinson. Platelet aggregability in sleep-related stroke. Can. J. Neurol. Sci. **16**: 71–77, 1989).

(Grotta) This is a good idea and to my knowledge has not been done, though I may not be aware of all the literature on this topic.

From Dr. von Lubitz

The study of Dr. Grotta *et al.* is most alluring both in terms of good medicine and a very civilized approach: a glass of brandy and a cup of coffee either prior or after stroke seem to minimize its consequences. The questions remains whether the FDA will approve something as simple and civilized as this combination.

The issue of vasodilation in treatment of stroke has been addressed in the days when adenosine was less than popular. Using P-sulphephenytheophyhine (8-δ1P), a peripheral A_1 receptor antagonist, prior to ischemia, and N^6-cyclo-peutyladlenosine (CPA, a selective agonist of A_1 receptor) we and other investigators have shown a clear improvement of the outcome in several measures. The result may be the consequence of the well-documented neuroprotection and vasodilation, both of which result from adenosine A_1 receptor stimulation.

ANSWER: (Auer) Just do not lower the blood pressure. See above.

(Grotta) I believe the FDA would agree to carefully carried out safety studies of this combination. The more difficult question would be labeling and controlling its use if it were eventually shown to be effective. We are intrigued by the effect of caffeine on adenosine, but remember that caffeine alone, without ethanol, was not particularly effective, and we also found no elevation of CBF, at least as measured by laser doppler. However, we agree that we need more thorough testing of the effect of ethanol and caffeine on cerebral hemodynamics.

From Dr. Slikker

With the multicenter clinical trail of neonatal head cooling in progress and the favorable outcome of the small clinical study of body cooling for stroke victims, it would appear that hypothermia is a high visibility approach for neuroprotection and has a long use history in cardiovascular surgery. As demonstrated in cell culture experiments in a poster presentation at this meeting (William Slikker, III, *et al.*, Free Radical Biology and Medicine, 2001). Hypothermia enhances Bcl-2 expression and protects against oxidative stress–induced cell death in CHO cells, and hypothermia (even mild 35 or 32°C) will significantly increase Bcl-2 (anti-apoptoic) and gluthuthine peroxidase (antioxidative) levels. What will be the role of hypothermia in the future of neuroprotection? What is the possible mechanism of hypothermia in neuroprotection?

ANSWER: (Auer) Hypothermia may act to reduce acidosis, glutamate release, or shunt more glucose through the pentose phosphate pathway, among many mechanisms. Each of these would be helpful to tissue survival. Protons kill tissue in sufficient concentration (M. Nedergaard, *et al.* Acid-induced death in neurons and glia. J. Neurosci. **11**: 2489–2497, 1991; R.P. Kraig, *et al.* Hydrogen ions kill brain at concentrations reached in ischemia. J. Cereb. Blood Flow Metab. **7**: 379–386, 1987). Glutamate is the archetypal excitotoxin (J.W. Olney. Brain lesions, obesity, and other disturbances in mice treated with monosodium glutamate. Science **164**: 719–721, 1969). The metabolic products of the pentose shunt are likely to be neuroprotective (T. Kaibara, *et al.* Hypothermia: depression of tricarboxylic acid cycle flux and evidence for pentose phosphate shunt upregulation. J. Neurosurg. **90**: 339–347, 1999).

(Grotta) I think that hypothermia will eventually be at least part of an effective neuroprotective strategy. It is probably the most consistently effective strategy across various models not only of stroke, but cardiac arrest, head trauma, and intraoperative protection. However, current means to accomplish hypothermia are cumbersome and take too long. I am optimistic that this strategy will get a pragmatic and economic boost as catheter companies and other device manufacturers get interested in engineering new approaches to efficiently and rapidly achieve hypothermia.

From Dr. Reiter

During the presentations, I heard no consideration of gender or age in efficacy of drug treatment for cerbrovascular accident. Usually, animal studies for stroke models are conducted on young male animals. Stroke in humans is more frequently an advanced age condition and occurs in males and females. Could you comment on the influence of gender and age on the efficacy of treatment outcome and how tests in young male animals are helpful in identifying potential treatments in humans?

ANSWER: (Auer) Age makes a slight difference in animals, mainly in focal ischemia, where it worsens outcome (G.R. Sutherland, G.A. Dix, and R.N. Auer. Effect of age in rodent models of focal and forebrain ischemia. Stroke **27**: 1663–1668, 1996). Sex (which one, not how much) is important in stroke in view of the neuroprotective effect of estrogens (R. Rusa, *et al.* 17β-estradiol reduces stroke injury in estrogen-deficient female animals. Stroke **30**: 1665–1670, 1999). The basic biology of stroke is likely to be the same, but graded along age gradients, and with a step difference between the sexes.

(Palmer) In our studies with the low-affinity, use-dependent NMDA antagonist, ARR15896, we did one study with older, heavier Lister hooded rats (certainly not aged rats though). A single dose of the drug was administered 70 min after the onset of one hour of transient middle cerebral artery occlusion. Cortical and subcortical neuroprotection was assessed 23 hours later and significant neuroprotection was achieved. I agree that both sexes, as well as aged animals should be evaluated.

(Grotta) Age is an important clinical determinant of outcome in general, with older patients having worse outcome. However, at least in the NINDS rt-PA trial, neither age nor gender predicted lack of response or increased risk from therapy. Therefore, I expect that benefits might be less, but therapy will still be of some value in older male brains.

From Dr. Bowyer

Does the fact that anesthetics are administered in animal models of stroke affect the outcome of neuroprotective agents with respect to either their pharmacokinetics or inhibition of infarct size?

ANSWER: (Auer) Anesthetics have been looked at as neuroprotectants (T.R. Ridenour, *et al.* Comparative effects of propofol and halothane on outcome from temporary middle cerebral artery occlusion in the rat. Anesthesiology **76:** 807–812, 1992), since they depress cerebral metabolism. It may work in the recovery period, too (T. Kuroiwa, P. Bonnekoh, and K.-A. Hossmann. Therapeutic window of halothane anesthesia for reversal of delayed neuronal injury in gerbils: relationship to postischemic motor hyperactivity. Brain Res. **563:** 33–38, 1991), a point relevant to clinical hypothermia trials pondering the necessity for anesthesia during hypothermia. The anesthesia itself may be of benefit at that time.

(Palmer) The use of anesthetics can always be reckoned to be a confounding problem. In earlier work with the four-vessel occlusion technique of transient forebrain ischemia, the animal was operated on 24 h before the sutures were tightened and we did see neuroprotection with two of our compounds, remacemide and ARR15896. Moreover Drs. Buchan and Pulsinelli previously found essentially the same results with compounds they tested. Within the past few years animal welfare regulations have rightly mandated the animals be knocked out with halothane while the sutures are tightened.

(Curry) Adequate systematic studies relating such factors as age, gender and anesthetics, and for that matter biochemical effects such as acidosis, glucose or insulin changes, on responses to neuroprotective drugs in small animal models, appear not to have been carried out. Indeed, some anesthetics may themselves be neuroprotective in these models. There is, of course, a huge experimental design problem when multiple factors are studied in experiments in which each animal gives basically one unit of data. The differences in responses seen in different models and laboratories, and the amazingly poor extrapolation from animals to man, may relate, at least in part, to such factors, which may in turn relate to pharmacokinetic differences.

COMMENT

From Dr. von Lubitz

Splendid point. If treatment is restricted to a very narrow space of implementations and if it is effective in a subset of a subset of the affected population, the treatment is essentially worthless in the centered of its clinical applicability.

The neuroprotective effects of adenosine A_1 receptor agonists have been often related to their profound metabolism depressant impact and the resulting hypothermia. Interestingly, the NIH has finally realized the role of hypothermia in resuscitation, and hypothermia research related to stroke, cardiac dysfunction, surgery, and so forth, may now have increased chances of receiving support.

QUESTIONS FOR DR. JONAS

From Dr. Slikker

Dr. Grotta's data presented in the poster session indicated that tolerance to (or frank reversal of) beneficial effect from alcohol plus caffeine to decrease infarct size developed. Would tolerance develop to the prophylactic treatment with heparin that you suggest?

ANSWER: Clinically there is no loss of response to continuous iv unfractionated heparin or subcutaneous low molecular weight heparin administration over several weeks in humans. Also, there is no loss of response if a patient is given heparin on another occasion months or years later. There are no long-term data on oral heparin.

Thrombolysis, according to your Figure 1 (infarct size relative to control vs. time), has a positive effect to reduce infarct size (0.8 or better) even when treatment is not introduced until 5–6 h after infarct indication. What is the underlying physiology behind this positive effect?

ANSWER: I think benefit comes from the opening up or the preservation of the penumbral circulation.

From Dr. Paule

I think I recall from your presentation at this meeting four years ago that the compound remacemide was perhaps the only "neuroprotective agent" that showed any efficacy in human trials—yet it was not even mentioned in your presentation at this year's meeting. Can you address this issue?

ANSWER: A very good point. I did not include this in the review because the positive clinical trial was of cognitive protection during cardiopulmonary bypass, not stroke treatment. I believe that the cognitive decline in bypass is from multiple tiny emboli, and that a beneficial effect suggests potential value in stroke. However, it is a long way from one positive result in bypass to acceptance of efficacy in stroke.

QUESTION FOR DR. GROTTA

From Dr Banik

Ray Bartus in his stroke model studies showed a significant reduction in infarct size following treatment with calpain inhibitor AK-295. I wonder what kind of calpain inhibitor you used in your studies and could you comment on this?

ANSWER: I believe the compound was MDL28170 from Merrill Dow.

PART VIII: QUESTIONS AND ANSWERS

QUESTIONS FOR DR. PALMER

From Dr. Manev

Is there any information on the possible explanation for greater variability in the stroke outcome in Sprague Dawley versus Wister rats? Identifying the reason for this may point to a potential target for stroke therapy.

ANSWER: Someone, I am unable to remember who, said there was more extensive collateral circulation in the Sprague Dawley rats.

From Dr. Slikker

In the food-reinforced forelimb dexterity (staircase test), are the results modified by food intake/weight gain?

ANSWER: In order for the test to work, the rats must be maintained at 80–85% of their free-feeding weight and after each day's testing period they are given only 10 grams of rat food. Slight hunger is a strong motivator for the animals to reach for and retrieve food pellets.

Rat Model of Autism Spectrum Disorders

Genetic Background Effects on Borna Disease Virus-Induced Developmental Brain Damage

M.V. PLETNIKOV,[a,b] M.L. JONES,[a,b] S.A. RUBIN,[b] T.H. MORAN,[a] AND K.M. CARBONE[a,b,c]

[a]*Department of Psychiatry, Johns Hopkins University School of Medicine, Baltimore, Maryland U.S.A.*

[b]*LPRVD/DVP/ OVRR/CBER/FDA, Bethesda, Maryland, U.S.A.*

[c]*Department of Medicine, Johns Hopkins University School of Medicine, Baltimore, Maryland U.S.A.*

KEYWORDS: Rat model; Autism spectrum disorder; Genetic background; Borna disease; Brain damage.

During the past ten years we have been characterizing the first virus-induced Autism Spectrum Disorders (ASD) model, based on Lewis rats (Lew) infected with Borna disease virus (BDV) at birth.[1] Infection of newborn rats with this persistent, RNA virus induces characteristic neuroanatomical, neurochemical, neuroimmune, and behavioral deficits without an associated cellular inflammatory response in the brain. These deficits show strong correlation with abnormalities detected in children with ASD, including neuronal cell dropout in the cerebellum and hippocampus, abnormalities in serotonin neurotransmitter systems and social (abnormal play), emotional (chronic anxiety), and cognitive (decreased spatial learning and memory) behaviors.[2–8]

Genes or genetic background influence the development of ASD or the expression of ASD symptoms in the majority of children with these disease syndromes.[9–12] However, only in a subset of patients has ASD been shown to be the result of a purely genetic disease (e.g., Rett's syndrome, Fragile X, and phenylketonuria). Therefore, it is likely that most cases of ASD are due to a complex interaction between environmental insults and the host's genetic background. Thus, we sought to use our virus-induced model of ASD to explore the effects of genetic background on expression of disease.

We studied the effects of genetic background on developmental brain injury by comparing neuroanatomical, neurochemical and behavioral disturbances in inbred Lew, inbred Fisher 344 (Fi), and outbred Sprague-Dawley (SD) rats neonatally infected with BDV. BDV-induced cerebellar hypoplasia and hippocampal dentate gyrus degeneration appeared to be more severe in Fi and was less severe in SD rats, when compared to Lew rats. In all strains, BDV produced locomotor hyperactivity, stereotypies, and disturbed emotionality when tested at postnatal day (PND) 180 in an open field test. At PND 30, infected Fi344 rats demonstrated attenuated non-play and play social activity, whereas infected Lew rats exhibited abnormally increased non-play interaction and deficient play activity. Neonatal BDV infection differentially altered regional brain monoamine concentrations in three strains of six-month old rats. In hippocampus and cerebellum, 5-HT content was increased in Fi and Lew but not SD-infected rats. In frontal cortex, levels of 5-HT were increased only in infected Lew and remained unaffected in infected Fi and SD rats. BDV increased norepinephrine (NE) content in cerebellum and did not alter NE levels in the cortex of any strain. In the hippocampus, NE was increased in infected Fi rats only. Dopamine content in the caudate-putamen and nucleus accumbens was not changed by neonatal BDV infection in any of the strains. Thus, neonatal BDV infection induces selective neuroanatomical, neurochemical and behavior abnormalities in rats with different genetic backgrounds. These data indicate that neonatal BDV infection of the rat brain can be a valuable animal model for studying the pathogenic mechanisms of developmental brain and behavior damage.

REFERENCES

1. CARBONE, K.M., *et al.* 1991. Borna disease: association with a maturation defect in the cellular immune response. J. Virol. **65:** 6154–6164.
2. BAUTISTA, J.R., *et al.* 1995. Developmental injury to the cerebellum following perinatal Borna disease virus infection. Brain Res. Der. Brain Res. **90:** 45–53.
3. BAUDSTA, J.R., *et al.* 1994. Early and persistent abnormalities in rats with neonatally acquired Borna disease virus infection. Brain Res. Bull. **34:** 31–40.
4. PLETNIKOV, M., *et al.* 2000. Effects of neonatal rat BDV infection on the postnatal development of brain monoaminergic systems. Dev. Br. Res. **119:** 179–185.
5. PLETNIKOV, M., *et al.* 1999. Persistent neonatal Borna disease virus (BDV) infection of the brain muses chronic emotional abnormalities in adult rats. Physiol. Behav. **66:** 823–831.
6. PLETNIKOV, M., *et al.* 1999. Developmental brain injury associated with abnormal play behavior in neonatally Borna disease virus (BDV) infected Lewis rats: a model of autism. Behav. Brain Res. **100:** 30–45.
7. RUBIN, S.A., *et al.* 1999. Borna disease virus-induced hippocampal dentate gyms damage is associated with spatial learning and memory deficits. Brain Res. Bull. **48:** 23–30.
8. PLATA-SALAMAN, *et al.* 1999. Persistent Borna disease virus infection of neonatal rats causes brain regional changes of taRNAs for cytokines, cytokine receptor components and neuropeptides. Br. Res. Bull. **49:** 441–51.
9. LAMB, J.A., *et al.* Autism: recent molecular genetic advances. Hum. Mol. Genet. **9:** 861–868.
10. RUTTER, M. 2000. Genetic studies of autism: from the 1970s into the millennium. J. Abnorm. Child Psychol. **28:** 3–14.
11. FILLPEK, P.A., *et al.* 1999. The screening and diagnosis of autistic spectrum disorders. J. Autism Der. Disord. **29:** 439–484.
12. HENDREN, R.L., *et al.* Review of neuroimaging studies of child and adolescent psychiatric disorders from the past 10 years. J. Am. Acad. Child Adol. Psych. 2000. **39:** 815–828.

Antiapoptotic Properties of Rasagiline, N-Propargylamine-1(R)-aminoindan, and Its Optical (S)-Isomer, TV1022

WAKAKO MARUYAMA,[a] MOUSSA B.H. YOUDIM,[b] AND MAKOTO NAOI[c]

[a]*Department of Basic Gerontology, National Institute for Longevity Sciences, Obu, Aichi, Japan*

[b]*Technion-Israel Institute of Technology, Faculty of Medicine, Eve Topf and NPF Centers of Excellence for Neurodegenerative Diseases Research, Haifa, Israel*

[c]*Department of Brain Sciences, Institute of Applied Biochemistry, Mitake, Gifu, Japan*

ABSTRACT: Rasagiline and structurally related propargylamines protected dopaminergic SH-SY5Y cells from apoptosis induced by 6-OHDA and peroxynitrite-generating SIN-1. It was suggested that the intracellular mechanism of the neuroprotection is related to the stabilization of mitochondrial membrane potential, as indicated by use of a fluorescent indicator, JC-1. The opening of the permeability transition pore (PTP) was prevented by rasagiline, even in isolated mitochondria. The activation of apoptotic cascade by the oxidative stress and neurotoxins, such as activation of caspase 3 and DNA fragmentation, was also inhibited by pretreatment with rasagiline. These propargylamines may prevent or rescue declining neurons induced by mitochondrial apoptotic cascade and may be applicable as "neuroprotective agents" in aging and age-related neurodegenerative disorders, such as Parkinson's and Alzheimer's diseases.

KEYWORDS: Apoptosis; Parkinson's disease; Rasagiline; Neuroprotection; Monoamine oxidase B; Monoamine oxidase B inhibitor; 6-Hydroxydopamine; Peroxynitrite; Mitochondrial membrane potential.

INTRODUCTION

At the cellular level, Parkinson's disease (PD) is characterized by the selective death of melanin-containing dopamine neurons in the specified brain region, the substantia nigra pars compacta. Observations on the pathology of parkinsonian brains suggest that oxidative stress and apoptosis may play a role in depletion of dopamine neurons in PD.[1–3] However, by no means has it been established that apoptosis is a major mode of neuronal death in PD.[4] Nevertheless, apoptosis was demonstrated in the brain of various neurodegenetaive diseases,[5] such as Alzheimer's[6] and Huntington's disease.[7]

Address for correspondence: M. Naoi, Department of Brain Sciences, Institute of Applied Biochemistry, Yagi Memorial Park, Mitake, Gifu 505-0116, Japan. Voice: + 81 574 67 5500; fax: + 81 574 67 5310.
mnaoi@quartz.ocn.ne.jp

Apoptosis is a form of programmed neuronal cell death characterized by morphological features; cytoplasmic shrinking, chromatin condensation due to DNA fragmentation, loss of mitochondrial membrane potential ($\Delta\Psi$m), changes in plasma membrane composition, and formation of apoptotic bodies.[8] In apoptosis and programmed cell death, cells themselves participate and the well-conserved machinery mediates the death process from the induction, propagation, and execution. Various kinds of stimuli have been proposed by *in vivo* and *in vitro* experiments to induce apoptosis in dopaminergic neurons, dopamine,[9] dopaminergic neurotoxins, 1-methyl-4-phenyl-1,2,3,6-tetrahydropyridine (MPTP),[10] 1-methyl-4-phenylpyridinium ion (MPP$^+$),[11] and 6-hydroxydopamine (6-OHDA).[12] Apoptosis was induced in human dopaminergic SH-SY5Y cells by an endogenous neurotoxin, *N*-methyl(*R*) salsolinol [*N*M(*R*)Sal],[13] a most possible candidate of endogenous neurotoxins involved in the pathogenesis of PD,[14] and reactive oxygen (ROS) and nitrogen species (RNS), such as peroxynitrite (ONOO$^-$) and nitric oxide.[15] Apoptotic cell death was confirmed by activation of caspase 3, an executor of apoptosis, formation of nuclear DNA ladder, a biochemical hallmark of apoptosis, and also morphological changes.[16]

(−)Deprenyl (selegiline, *N*, α-dimethyl-*N*-2-propynylbenzeneethanolamine) was found to prevent apoptotic DNA damage induced by *N*M(*R*)Sal[13] and RNS.[15] Selegiline is a potent and selective inhibitor of type B monoamine oxidase [MAO-B],[17] and it was proposed that it delays the progression of PD.[18] Neuroprotection by selegiline was further suggested by a fact that it prevented development of parkinsonism through inhibition of MPTP oxidation into MPP$^+$ by MAO-B.[19] The DATATOP studies were carried out based on the idea that selegiline and an antioxidant, α-tochophenol, delay the progression of the disease.[20] However, the results of DATATOP study did not support neuroprotection and prevention of the disease progression, and the clinical responses were considered as being symptomatic. In addition, selegiline is metabolized into bioactive amphetamine and methamphetamine, giving rise to adverse effects in patients treated with selegiline. On the other hand, *in vivo* and *in vitro* experiments of animal or cell models of PD indicate that selegiline suppresses cell death in dopaminergic neurons. We examined the structure–activity relationship among aminoindan-derived propargylamines to determine the chemical structure required for the neuroprotection. Among the compounds so far studied, the most potent neuroprotective agent was the anti-Parkinson drug, rasagiline, (*N*-propargyl-1(*R*)-aminoindan, Teva Pharmaceutical Co. Netanya, Israel; DATATOP, 2000, in press), a selective irreversible MAO-B inhibitor.[21–23] Furthermore, rasagiline unlike selegiline, is not a vasoactive amine and not metabolized into amphetamine like compounds. Its metabolite is aminoindan.

However, the detailed mechanism underlying neuroprotection by propargylamine derivatives remains to be elucidated. This study describes the intracellular mechanism of rasagiline action to protect SH-SY5Ycells from apoptosis induced by ROS generating 6-OHDA and peroxynitrite generating SIN-1 (*N*-morpholinosydnonimine, Dojindo, Kumamoto, Japan). SH-SY5Y cells, where only type A, but not type B MAO, is expressed, were used as a model of dopamine neurons, to exclude the effect of MAO-B inhibition from the neuroprotective function. The antiapoptotic function of propargylamines was studied in relation to the apoptotic cascade, opening of permeability transition pore (PTP), activation of caspases, and fragmentation of nucleosomal DNA. The results so far suggest that mitochondria have a central role in determining death and survival of cells against apoptosis induced by oxidative stress and neurotoxins, and the protection afforded by propargylamines.

RESULTS

Prevention of Apoptotic Cell Death

SH-SY5Y cells were incubated with 2.5–40 μM of 6-OHDA, or 100 μM–1 mM of SIN-1, and cell death was quantitatively analyzed by fluoromicroscopy after staining with propidium iodide (PI) or Hoechst 33342. After an 18-h treatment with 100 μM SIN-1, almost all the cells were stained with PI (data, not shown), which does not permeate into intact cells. By staining with Hoechst 33342, condensed and fragmented nuclei were observed that were typical for apoptosis (see FIGURE 1). Phase contrast microscopy demonstrated the morphological change of the cells with retarded dendrites and shrunk cell bodies. 6-OHDA and SIN-1 induced apoptotic cell death at low concentrations, but they also induced necrotic cell death at higher concentrations.

Pretreatment with rasagiline protected cells from apoptosis, as shown in FIGURE 1. The number of cells with apoptotic nuclei was almost the same as in the control and the morphology of the cells did not change, as shown by phase contrast microscopy. The protection of the cells by rasagiline required incubation at least 20 min before the treatment with 6-OHDA or SIN-1, as reported elsewhere in apoptosis induced by NM(R)Sal.[13] Incubation, simultaneously or after the treatment with the apoptosis inducers, did not prevent cell death.

The activity–structure relationship of propargylamine analogues was studied by semiquantitative analysis with a "Comet" assay.[24] The cells were treated for 20 min with various concentrations of propargylamines and then incubated with 250 μM SIN-1 for 30 min. The cells were embedded in agarose, lysed an alkaline, and subjected to electrophoresis. The fragmented DNA electrophoresed faster than the intact nuclei, forming a "comet tail" on staining with ethidium bromide. The comet tail length (length of migrated DNA from the tailing edge of nucleus) was measured as an indicator of DNA damage. As is summarized in TABLE 1, rasagiline (N-propargyl-1(R)(+)-aminoindan) is the most potent in preventing DNA damage, and prevented DNA damage at 10 to 1 nM, whereas (–)deprenyl was effective only at concentrations higher than 1 μM. The (S)-enantiomer of N-propargyl-1-aminoindan, TV1022, was also effective in protecting the cells, but less potent than the (R)-enantiomer, rasagiline: it prevented apoptosis at 1 μM–100 nM. Other derivatives of propargylamines with an aralkyl (selegiline derivatives) or an alkyl group (N-aliphatic derivatives) also protected the cells. The (S)-enantiomer of deprenyl and N-hexyl-N-methylpropargylamine also prevented cell death, but they were less potent than the (R)-enantiomers. It should be noted that these (S)-enantiomers are very poor inhibitors of MAO-B, suggesting that inhibition of MAO-B does not contribute to the antiapoptotic activity or potency. The desmethyl derivative of (–)deprenyl, rasagiline, and N-heptyl-N-propargylamine protected the cells as potently as the N-methylated derivatives, indicating a methyl group at the N position is not required for the neuroprotection. On the other hand, pargyline (N-methyl-N-2-propargylbenzylamine) and clorgyline [N-methyl-N-propargyl-3(2, 4-dichlorophenoxy)propylamine] did not prevent the apoptotic cell death. These results suggest that a propargyl group is essentially required in addition

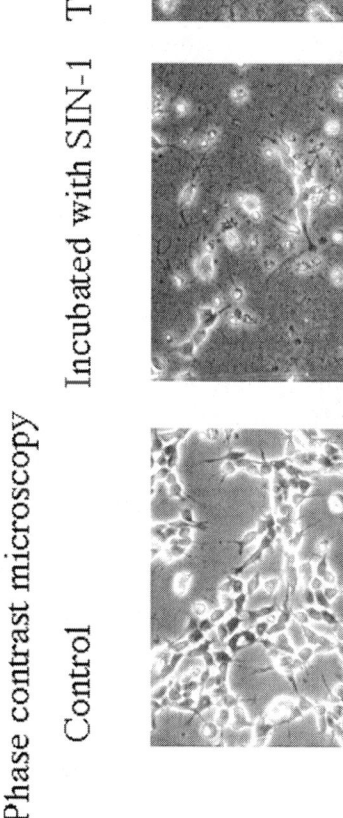
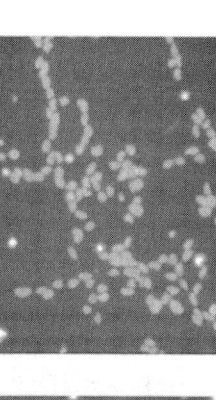
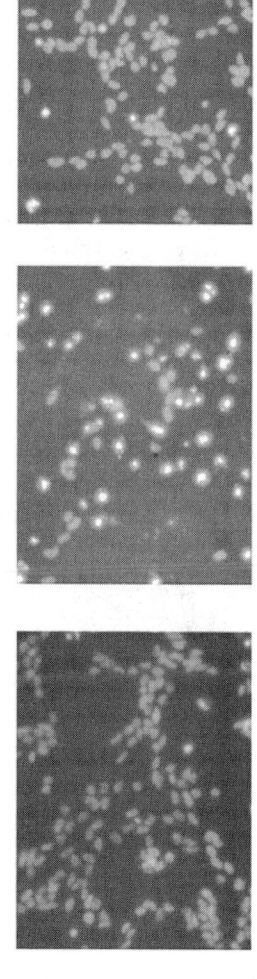

FIGURE 1. Apoptotic cell death induced by a peroxynitrite generating agent, SIN-1. SH-SY5Y cells were incubated with 100 μM SIN-1 for 18 hours with or without preincubation with 1 μM rasagiline for 20 min. By staining with Hoechst 33342, most of the cells treated with SIN-1 showed condensed and fragmented nuclei, whereas preincubation with rasagiline completely suppressed the apoptotic damage of nuclei.

to a hydrophobic group, such as a cyclic benzene ring, aralkyl, or alkyl group, with an adequate separation.

Effects of Rasagiline on Apoptotic Cascade

Mitochondrial permeability transition pore (PTP) is considered to control the initiation step of apoptotic death process.[25] It is a polyprotein complex formed in the contact site between the inner and the outer mitochondrial membranes, and functions as a Ca^{2+}-, voltage-, pH-, and redox gated channel. As an indicator of changes in PTP, mitochondrial membrane potential, $\Delta\Psi m$, was studied in the cells treated with neurotoxins and SIN-1 by use of a fluorescent probe, 5,5',6,6'-tetrachloro-1,1'3,3'-tetraethylbenzimidazolyl-carbocyanine iodide (JC-1, Molecular Probes, Eugene, OR, USA). JC-1 as monomer shows green fluorescence and inside the intact mitochondrial membrane it aggregates (J-aggregates) to show red fluorescence. SH-SY5Y cells cultured on poly-L-lysine-coated six-well flasks were treated with JC-1, and then 6-OHDA or SIN-1 was added. The red fluorescence of J-aggregates detected at 580 nm with excitation at 520 nm, and green fluorescence of monomeric JC-1 at 515 nm with excitation at 460 nm were observed. The relative intensity of red to green fluorescence, representing the ratio of J-aggregates to monomeric JC-1, was determined as an indicator of $\Delta\Psi m$ changes.

The relative fluorescence intensity of red/green represents J-aggregates in intact mitochondria and it declined in a time- and dose-dependent way in cells treated with 6-OHDA or SIN-1. Pretreatment of the cells with rasagiline or other propargylamines completely prevented collapse in $\Delta\Psi m$ by SIN-1 or 6-OHDA, as is shown in FIGURE 2.

TABLE 1. Antiapoptotic activity of propargylamines

Compound	Concentration Required for the Effect
(−)Deprenyl (selegiline)	1 μM–100 nM
(+)Deprenyl	10 μM
(−)Desmethyldeprenyl	10–1 nM
(R)(+)-N-Propargyl-1-aminoindan (Rasagiline)	10–1 nM
(S)(−)-N-Propargyl-1-aminoindan (TV 1022)	1 μM–100 nM
Aminoindan	not active
(R)-N-(2-Heptyl)-N-methylpropargylamine	1 μM–100 nM
(S)-N-(2-Heptyl)-N-methylpropargylamine	10 μM
(R)-N-(2-Heptyl)-propargylamine	1 μM–100 nM
(R)-3-(2-Heptylamine)-N-methylpropionic acid	not active
Clorgyline	not active
Pargyline	not active

NOTE: SH-SY5Y cells were incubated with various concentrations of propargylamines for 20 min, then treated with 100 μM SIN-1 for 30 min. Apoptosis was quantitatively determined by a "Comet" assay and the concentration required for the complete protection of the cells from apoptosis was obtained.

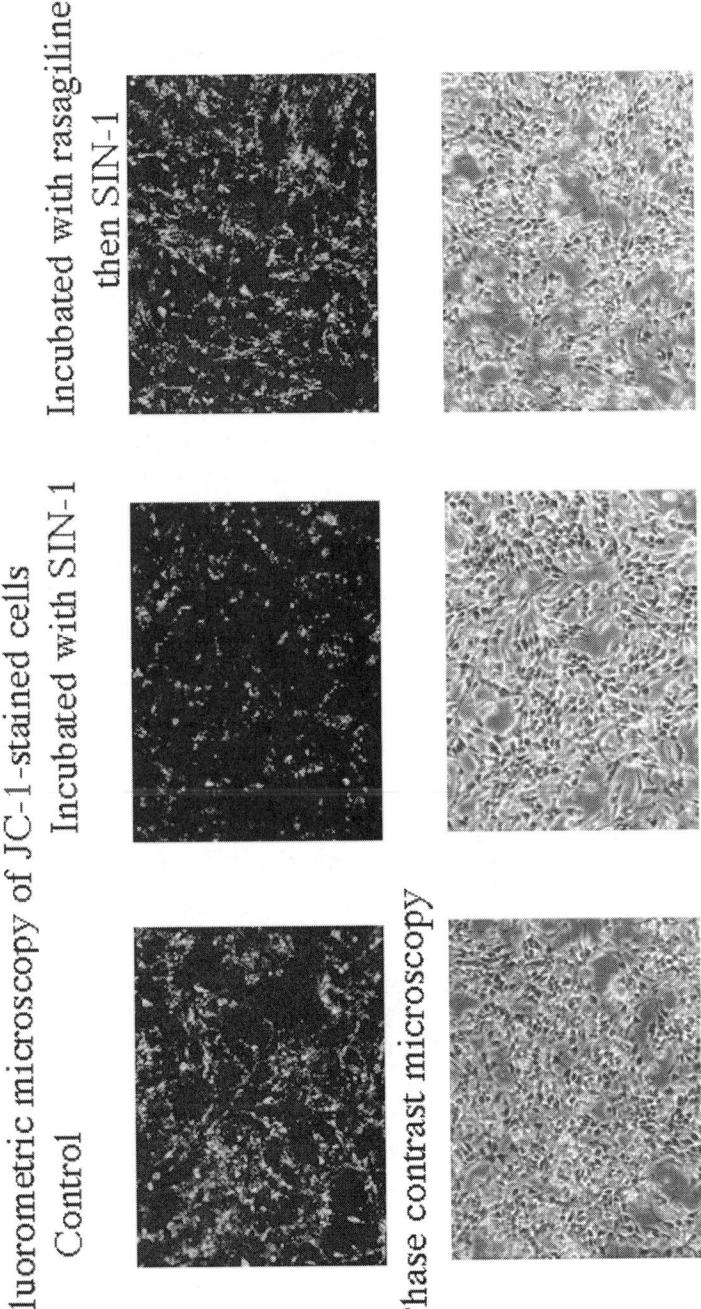

FIGURE 2. Effect of SIN-1 and rasagiline-preincubation on $\Delta\Psi$m. The cells were stained with a fluorescence indicator, JC-1, and the change in red fluorescence of J-aggregates were followed by fluorescence microscopy. After 30 min treatment with SIN-1, the red fluorescence of J-aggregates declined markedly, but preincubation of the cells with 1 μM rasagiline prevented collapse in $\Delta\Psi$m. Phase contrast microscopy showed almost intact features of the cells, even after treatment with SIN-1.

Recently the effect of apoptosis inducers on PTP was examined using isolated mitochondria (Akao et al., in preparation). Opening of PTP was visualized by release of trapped Rhodamine 123 from mitochondria, and the apoptosis inducers opened PTP in a same way as with the whole cells. Pretreatment of the cells with rasagiline delayed the opening of PTP in a dose-dependent manner. These results suggest that the observed loss of $\Delta\Psi m$ was induced by mitochondria themselves and did not require other cellular components.

Activated caspases are considered to be the executors of apoptosis, which was confirmed in the cells incubated with NM(R)Sal,[16] or SIN-1.[26] Caspases are a family of cysteine proteases and are considered to be the heart of biochemical pathway in the tightly regulated process of apoptosis.[27] Caspases are activated by stepwise proteolysis of the proenzymes. Caspase 3 is considered to function as the final executor in activating DNAase and cleaving DNA. The activation of caspase 3 was examined in the cells incubated with the toxins or SIN-1. The activity of caspase-3, measured with a substrate, acetyl Asp-Glu-Val-Asp-4-methyl-cumaryl-7-amide (Ac-DEVD-MCA), increased in a time-dependent way by treatment with SIN-1, reached a maximal level six hours after the treatment, and declined thereafter.[26] Activation of caspase 3 was also confirmed in cells treated with NM(R)Sal by Western blot analysis.[16] The active form of 17-kDa caspase 3 was detected in cells incubated with NM(R)Sal. The amount of the active form increased to a maximum after 24 hours of incubation, and then declined. In control cells, only the proenzyme with the molecular weight of 32 kDa was detected. The involvement of caspase 3 in apoptosis was further confirmed by use of an inhibitor of caspase 3, acetyl Asp-Glu-Val-Asp aldehyde (Ac-DEVD-CHO, Peptide Institute, Osaka, Japan). The inhibitor blocked DNA ladder formation in a dose-dependent way, whereas untreated cells showed typical DNA fragmentation after exposure to NM(R)Sal.

Pretreatment with rasagiline suppressed the activation of caspase 3 in cells treated for 72 hours with the apoptosis inducers SIN-1, 6-OHDA, and NM(R)Sal. In cells pretreated with rasagiline, the activity of caspase 3 was similar, 10.22 ± 1.78 pmol/min/mg protein to control values, 10.74 ± 2.50, after 72 hours of incubation. By contrast, in NM(R)Sal-treated cells the value increased by almost 1.6-fold, to 16.74 ± 3.64 pmol/min/mg protein.

DISCUSSION

Rasagiline and its optical isomer, TV1022, protected the cells from apoptosis induced by 6-OHDA and SIN-1, which suggests that they may suppress a common apoptotic pathway in cells. Previous observations have shown that rasagiline and selegiline analogues protect the various types of neurons against different insults; dopamine neurons from MPP^{+}[28] and 6-OHDA,[29] noradrenergic neurons from DSP-4,[30] and cholinergic neurons from AF64A.[31] The exact mechanism underlying the antiapoptotic actions of rasagiline and its analogues remains an enigma, even though numerous hypotheses have been proposed. We found that rasagiline prevented decline in $\Delta\Psi m$ and the following activation of apoptogenic cascade. These data provide clear evidence that the neuroprotection by rasagiline does not depend on inhibition of MAO-B, as described above. This is supported by the finding that TV1022, the optical (*S*)-isomer of rasagiline, which is neither an inhibitor of MAO-A nor of -B, has the same neuroprotective, antiapoptotic activity, and prevents a decrease in $\Delta\Psi m$, as rasagiline.[32] Direct scavenging of reactive oxygen species by rasagiline can be excluded, since *in vitro* experiments do not support this. The action of rasagiline and TV1022 may depend on increased synthesis of proteins, Bcl-2, or trophic factors.[32] This is supported by the findings that the antiapoptotic actions of rasagiline and TV1022 in undifferentiated PC12 cells were prevented by inhibitors of transcriptional (actinomycin) and translational (cycloheximide).[33] Western blot measurements have shown that both drugs prevent the decrease in Bcl-2 and Cu-Zn SOD1 during serum and NGF withdrawal in partially neuronally differentiated PC12 cells. Furthermore chronic treatment with rasagiline *in vivo* induces the increase of SOD and catalase in brains and other tissues of rats.[34] However, our data with isolated mitochondria suggest that mitochondria themselves determine the initiation of apoptosis, which is antagonized by rasagiline and other propargylamines. At present, we hypothesize that, as an immediate function, rasagiline and its analogues may directly activate antiapoptotic or suppress proapoptotic cascade machinery by interacting with protein molecules in or near to PTP that have a tertiary structure similar to the active site of MAO-B.

Further studies are in progress to determine the mechanism underlying the prevention of apoptosis by propargylamines, and also to develop new methods to protect or rescue declining neurons in PD and other neurodegenerative diseases.

ACKNOWLEDGMENT

This work was supported by a grant-in-aid for Scientific Research on Priority Area from the Ministry of Education, Science and Education, Japan, (M.N.), and a Grant for Longevity Sciences from the Ministry of Health and Welfare, Japan. We wish to thank Teva Pharmaceutical Co., Netanya, Israel for the supply of rasagiline and TV 1022. Support of National Parkinson Foundation (Miami) and Eve Top Neurodegenerative Diseases Center and Golding Parkinson Funds (Technion) are gratefully acknowledged.

REFERENCES

1. MOCHIZUKI, H., G. GOTO, H. MORI & Y. MIZUNO. 1996. Histochemical detection of apoptosis in Parkinson's disease. J. Neurol Sci. **137**: 120–123.
2. ANGLADE, P., S. VYAS & F. JAVOY-AGID, *et al.* 1997. Apoptosis and autophagy in nigral neurons of patients with Parkinson's disease. Histol. Histopatho. **12**: 25–31.
3. GERLACH, M., D. BEN-SHACHER, P. RIEDERER & M.B.H. YOUDIM. 1994. Altered brain metabolism of iron as a cause of neurodegenerative diseases? J. Neurochem. **63**: 793–807.
4. BURKE, R.E. & N.G. KHOLODILOX. 1998. Programmed cell death: Does it play a role in Parkinson's disease? Ann. Neurol. **44**(Suppl. 1): S126–133.
5. PETTMANN, B. & C.E. HENDERSON. 1998. Neuronal cell death. Neuron **20**: 633–647.
6. COTMAN, C.W. & A.J. ANDERSON. 1995. A potential role for apoptosis in neurodegeneration and Alzheimer's disease. Mol. Neurobiol. **10**: 19–45.
7. PORTERA-CAILLIAU, C., J.C. HEDREEN, D.L. PRICE & V.E. KOLIATSOS. 1995. Evidence for apoptotic cell death in Huntington's disease and excitotoxic animal models. J. Neurosci. **15**: 3775–3787.

8. THOMPSON, C.B. 1995. Apoptosis is the pathogenesis and treatment of disease. Science **267:** 1456–1462.
9. ZIV, H., E. MELAMED, N. NARDI, et al. 1994. Dopamine induces apoptosis-like cell death in cultured chick sympathetic neurons—a possible novel pathogenic mechanism in Parkinson's disease. Neurosci. Lett. **170:** 176–140.
10. TATTON, N.A. & S.J. KISH. 1997. In situ detection of apoptotic nuclei in the substantia nigra compacta of 1-methyl-4-phenyl-1,2,3,6-tetrahydropyridine-treated mice using terminal deoxynucleotidyl transferease labeling and acridine orange staining. Neurosci. **77:** 1037–1048.
11. DIPASQUALE, B., A.M. MARINI & R.J. YOULE. 1991. Apoptosis and DNA degradation induced by 1-methyl-4-phenylpyridinium in neurons. Biochem. Biophys. Res. Commun. **181:** 1442–1448.
12. WALKINSHAW, G. & C.M. WATERS. 1994. Neurotoxin-induced cell death in neuronal PC12 cells is mediated by induction of apoptosis. Neurosci. **63:** 975–987.
13. MARUYAMA, W., M. NAOI, T. KASAMATSU, et al. 1997. An endogenous dopaminergic neurotoxin, N-methyl-(R)salsolinol, induces DNA damage in human dopaminergic neuroblastoma SH-SY5Y cells. J. Neurochem. **69:** 322–329.
14. NAOI, M. & W. MARUYAMA. 1999. N-Methyl(R)salsolinol, a dopamine neurotoxin, in Parkinson's disease. Adv. Neurol. **46:** 259–264.
15. MARUYAMA, W., T. TAKAHASHI & M. NAOI. 1998. (–)-Deprenyl protects human dopaminergic neuroblastoma SH-SY5Y cells from apoptosis induced by peroxynitrite and nitric oxide. J. Neurochem. **70:** 2510–2515.
16. AKAO, Y., Y. NAKAGAWA, W. MARUYAMA, et al. 1999. Apoptosis induced by an endogenous neurotoxin, N-methyl(R)salsolinol, is mediated by activation of caspase 3. Neurosci. Lett. **267:** 153–156.
17. KNOLL, J., Z. ECSERY, K. KELEMEN, et al. 1965. Phenylisopropyl-methylpropinylamine (E-250), a new spectrum psychic energizer. Arch. Int. Pharmacodyn. Ther. **155:** 154–164.
18. BIRKMAYER, W., J. KNOLL, P. RIEDERER, et al. 1985. Increased life expectancy resulting from addition of L-deprenyl to Madopar treatment in Parkinson's disease. J. Neural Transm. **64:** 113–127.
19. HEIKKILA, R.E., L. MANZINO, F.S. CABBAT & R.C. DUVOISIN. 1985, Protection against the dopaminergic neurotoxicity of 1-methyl-4-phenyl-1,2,3,6-tetrahydropyridine by monoamine oxidase inhibitors. Nature **311:** 467–469.
20. PARKINSON STUDY GROUP. 1989. DATATOP: a multicenter controlled clinical trial in early Parkinson's disease. Arch. Neurol. **46:** 1052–1060.
21. KALIR, A., A. SABBAGH & M.B.H. YOUDIM. 1978. Selective acetylenic "suicide" and reversible inhibitors of monoamine oxidase types A. and B. Br. J. Pharmacol. **73:** 55–64.
22. FINBERG, J.P.M., M. TENNE & M.B.H. YOUDIM. 1981. Tyramine antagonistic properties of AGN 1135-an irreversible inhibitor of monoamine oxidase type B. Br. J. Pharmacol. **73:** 65–74.
23. FINBERG, J.P.M., I. LAMENDORF, M. WEINSTOCK, et al. 1999. Pharmacology of rasagiline (N-propargyl-1R-aminoindan). Adv. Neurol. **80:** 495–499.
24. OSTLING, O. & K.J. HOHANSON. 1984. Microelectrophoretic study of radiation-induced DNA damage in individual mammalian cells. Biochem. Biophys. Res. Commun. **123:** 291–298.
25. KROEMER, G., P.X. PETIT, N. ZAMZAMI, et al. 1995. The biochemistry of apoptosis. FASEB. J. **9:** 1277–1287.
26. OH-HASHI, K., W. MARUYAMA, H. YI, et al. 1999. Mitogen-activated protein kinase pathway mediates peroxynitrite-induced apoptosis in human dopaminergic neuroblastoma SH-SY5Y cells. Biochem. Biophys. Res. Commun. **263:** 504–509.
27. ENARI, M., H. SAKAHIRA, O. YOKOYAMA, et al. 1998. A caspase-activated DNase that degrades DNA during apoptosis, and its inhibitor ICAD. Nature **391:** 43–50.
28. MYTILLINEOU, C. & G. COHEN. 1985. Deprenyl protects dopaminergic neurons from neurotoxic effects of 1-methyl-4-phenylpyridiniun ion. J. Neurochem. **45:** 1951–1953.

29. SALONEN, H., A. HAAPALINNA, E. HEINONEN, *et al.* 1996. Monoamine oxidase B inhibitor selegiline protects young and aged rat peripheral sympathetic neurons against 6-hydroxydopamine-induced neurotoxicity. Acta Neuropathol. **91:** 466–474.
30. FINNEGAN, K.T., J.J. SKRATT, I. IRWIN, *et al.* 1990. Protection against DSP-4 induced neurotoxicity by deprenyl is not related to its inhibition of MAO B. Eur. J. Pharmacol. **184:** 119–126.
31. BRONZETTI, E., L. FELICI, F. FERRANTE & B. VASECCHI. 1992. Effect of ethylcholine mustard aziridinium (AF64A) and on monoamine oxidase-B-inhibitor L-deprenyl on the morphology of rat hippocampus. Int. J. Tissue React. **XIV:** 175–182.
32. YOUDIM, M.B.H., J.S. WADIA & W.G. TATTON. 1999. Neuroprotective properties of the antiparkinson drug rasagiline and its optical S-isomer. Neurosci. Lett. **55:** S45.
33. TATTON, W.G., W.J.H. JU, D.P. HOLLAND, *et al.* 1994. (–)-Deprenyl reduced PC12 cell apoptosis by inducing new protein synthesis. J. Neurochem. **63:** 1572–1575.
34. CARRILLO, M.C., C. MINAMI, K. KIYANI, *et al.* 2000. Enhancing effect of rasagiline on superoxide dismutase (SOD) and catalase (CAT) activities in the dopaminergic system in the rat. Life Sci. **67:** 577–585.

Rescue of Dying Neurons by (R)-Deprenyl in the MPTP-Mouse Model of Parkinson's Disease Does Not Include Restoration of Neostriatal Dopamine

STEFANUS J. STEYN, KAY CASTAGNOLI, AND NEAL CASTAGNOLI, JR.

*Harvey W. Peters Center for the Study of Parkinson's Disease,
Department of Chemistry, Virginia Tech., Blacksburg, Virginia, U.S.A.*

ABSTRACT: Chronic (8- to 10-week) administration of the selective, potent, and irreversible monoamine oxidase B inhibitor (R)-deprenyl has been shown to increase the tyrosine hydroxylase immunoreactivity in the substantia nigra of mice that had been treated three days earlier with a neurotoxic dose of the parkinsonian-inducing agent 1-methyl-4-phenyl-1,2,3,6-tetrahydropyridine (MPTP). This reported rescuing of lesioned nigrostriatal cell bodies by (R)-deprenyl prompted us to investigate if this (R)-deprenyl treatment also could restore neostriatal dopamine levels that are depleted by MPTP. The results of these experiments show that long term (8 or 10 weeks) treatment with (R)-deprenyl beginning three days post MPTP administration did not result in restoration of depleted neostriatal dopamine levels in C57BL/6 mice. We conclude that, although (R)-deprenyl may rescue MPTP-injured nigrostriatal neurons, it does not lead to functional recovery of these neurons as measured by the restoration of neostriatal dopamine levels.

KEYWORDS: (R)-deprenyl; Neuroprotection; Dopamine; MPTP; C57BL/6 mouse; Regeneration; Spontaneous recovery; Parkinson's disease; Neuronal rescue; Monoamine oxidase.

INTRODUCTION

The two hallmark characteristics of idiopathic Parkinson's disease (PD) are a decrease in striatal dopamine levels[1] (functional deficit) and a decrease in the numbers of nigrostriatal neurons[2] (structural deficit). Both of these parkinsonian characteristics have been observed in humans[3,4] and in animals[5,6] exposed to the parkinsonian-inducing agent 1-methyl-4-phenyl-1,2,3,6-tetrahydropyridine (MPTP [1] see SCHEME 1). The neurotoxic effects of MPTP are dependent on the monoamine oxidase B[7,8] (MAO-B) catalyzed oxidation of this cyclic allylamine to the 2,3-dihydropyridinium intermediate MPDP$^+$ (2), which undergoes further oxidation to form the ultimate neurotoxin, the 1-methyl-4-phenylpyridinium species MPP$^+$ (3).[9,10] Pretreatment of animals with (R)-deprenyl, a selective, potent, and irreversible

---

Address for correspondence: Neal Castagnoli, Jr., Ph.D. Harvey W. Peters Center Department of Chemistry, Virginia Tech., Blacksburg, VA 24061-0212, U.S.A. Voice: 540-231-8200; fax: 540-231-8890.
ncastagnoli@chemserver.chem.vt.edu

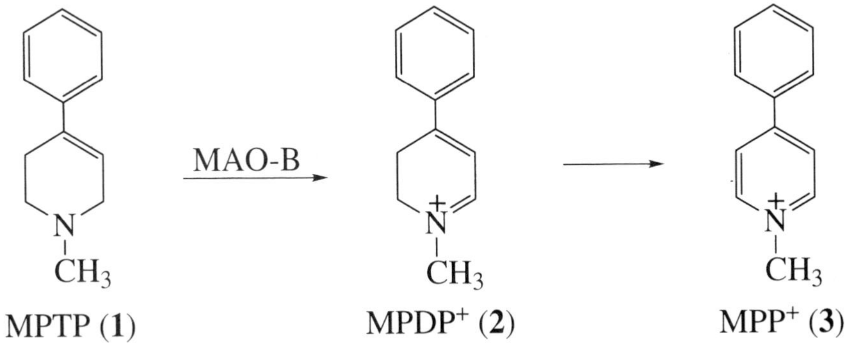

SCHEME 1. Structures of compounds discussed in the text.

inhibitor of MAO-B, prevents the MPTP-mediated neuronal injury by blocking MPP$^+$ formation.[9,10]

The nigrostriatal toxicity of MPP$^+$ is also dependent on its selective uptake by the dopamine transporter[11] (DAT), which is located mainly in the nerve terminals[12] of dopaminergic neurons.[13] Subsequently, MPP$^+$ accumulates in the mitochondria[14] where it inhibits complex I of the electron transport chain.[15] Striatal nerve terminal degeneration[16] is evident within 24 hours post MPTP exposure.[17] The degeneration induced by MPP$^+$ progresses in a retrograde fashion that results in loss of the cell bodies located in the substantia nigra.[18] The time required for maximum loss of cell bodies has been reported to be from 3[19] or 4[20] days to 20 days[21] post-MPTP exposure.

Current pharmacotherapies for the treatment of Parkinson's disease focus mainly on increasing endogenous dopamine concentrations[22] through the administration of L-DOPA[23] together with a peripheral decarboxylase inhibitor.[24–26] (R)-deprenyl also is of therapeutic value, particularly during the early stages of the disease.[27] Results from a large clinical study, the DATATOP study,[28,29] indicate that patients treated with (R)-deprenyl have a lower cumulative probability of reaching the study end point (defined as the need to initiate L-DOPA therapy) over the first 600 days of the study than do patients not treated with (R)-deprenyl. Additionally, some postmortem findings indicate a reduced loss of nigrostriatal neurons in (R)-deprenyl-treated patients suggesting that this compound may have neuroprotective properties.[30]

These neuroprotective properties of (R)-deprenyl were also observed in the MPTP-mouse model of Parkinson's disease. The initiation of chronic, low dose (R)-deprenyl (0.25 mg/kg/48 hours) treatment three days after administering a neurotoxic dose of MPTP, resulted in an increase in tyrosine hydroxylase immunoreactivity in the substantia nigra of eight-week-old C57BL/6 mice.[21,31] An important feature of this experiment is that the (R)-deprenyl treatment was initiated three days after MPTP treatment, long after the period during which MPTP is undergoing MAO-B-catalyzed metabolic bioactivation to the neurotoxic MPP$^+$ species.[32] This finding has led to the thesis that (R)-deprenyl treatment can rescue dying neurons.[21,31]

In addition to the well-characterized MAO-B inhibitory properties of (R)-deprenyl, the following observations have been reported. Several authors have stated that (R)-deprenyl displays trophic-like support of axotomized rodent facial motoneurons.[33–35] It has also been reported to reduce apoptosis in a variety of cells via mechanisms that are also independent of its MAO-B inhibition properties.[36–39]

These properties of (R)-deprenyl and the possibility that neurons may be capable of dynamic change throughout life[40] have prompted us to examine if (R)-deprenyl treatment in older (36 week) C57BL/6 mice also can lead to a restoration of neostriatal dopamine levels in MPTP-lesioned animals. Results from previous studies established that a single, 10 mg/kg dose of (R)-deprenyl in 32-week-old C57BL/6 mice is neuroprotective only when administered prior to three hours post MPTP treatment.[41] Expression of the putative neurotrophic properties of (R)-deprenyl, however, may require chronic drug administration. In the present study we have examined the effects of (R)-deprenyl on neostriatal dopamine levels using the protocols that are reported to rescue dying neurons in the MPTP mouse model.[21,31,34]

MATERIALS AND METHODS

Caution! MPTP is a known nigrostriatal neurotoxin and should be handled following established procedures.[42] Protocols for all animal experiments were reviewed and approved by the Virginia Tech Animal Care Committee and follow the *Guide for the Care and Use of Laboratory Animals*, 7th edition, 1996, National Research Council.

Effects of Chronic (R)-*Deprenyl Treatment (1 mg/kg/48 h) on Neostriatal Dopamine Levels in MPTP-Lesioned Mature Mice*

A group of 24 mature (36-week old) male C57BL/6 mice were injected intraperitoneally with MPTP·HCl (35 mg/kg). Three days later, 12 of these MPTP treated mice received (R)-deprenyl·HCl (1.0 mg/kg) intraperitoneally every second day for 10 weeks, yielding a total of 31 (R)-deprenyl·HCl administrations per mouse. Another 12 MPTP-treated mice received saline according to the same treatment schedule. These animals, together with four control mice were sacrificed 48 hours after the last (R)-deprenyl or saline administration. Their brains were removed rapidly, the neostriata dissected on ice, and neostriatal dopamine concentrations estimated using an HPLC assay with electrochemical detection as reported elsewhere.[43]

Comparison of the Effect of a Chronic Low Dose Regimen of (R)-*Deprenyl (0.25 mg/kg/48 hours) in Young and Mature MPTP-Lesioned Mice*

A group of 24 mature (32-week-old) and 24 young (6-week-old) male C57BL/6 mice were injected intraperitoneally with MPTP·HCl (40 mg/kg). A second group of 24 mice from each age group received a single saline injection. Three days later, eight MPTP-treated mice from each age group and eight saline-treated mice from each age group received (R)-deprenyl·HCl (0.25 mg/kg) intraperitoneally every second day for eight weeks, yielding a total of 27 (R)-deprenyl·HCl administrations per mouse. Seven days post MPTP treatment, eight MPTP-treated and eight saline-treated mice from each age group were sacrificed and their neostriatal dopamine levels estimated as described above. Eight weeks later (24 hours after the last [R]-deprenyl administration) the remaining mice (eight per group), which included mice treated with MPTP only, saline only, saline + (R)-deprenyl, and MPTP + (R)-deprenyl, were sacrificed and the neostriatal dopamine levels were measured.

RESULTS AND DISCUSSION

As discussed above, substantial experimental evidence argues that (R)-deprenyl can rescue MPTP-damaged neurons by a mechanism that is independent of its inhibition of the MAO-B-catalyzed bioactivation of MPTP to the neurotoxic pyridinium species MPP^+. The extent to which this effect of (R)-deprenyl may lead to a functional recovery of MPTP lesioned nigrostriatal neurons through restoration of neostriatal dopamine levels is the focus of this report. The possibility that (R)-deprenyl alone could alter brain dopamine levels has been raised by the report of Thiffault et al.,[44] who observed a 4.5-fold increase in neostriatal dopamine levels over baseline in six-week-old C57BL/6 mice following chronic treatment with (R)-deprenyl. Since an increase in dopamine levels that is independent of the recovery of injured neurons would complicate the interpretation of our experimental results, we have repeated these studies, first in mature (32-week-old) animals and subsequently in young (seven-week-old) animals. We found no evidence of altered neostriatal dopamine levels in any of the animals studied even though the MAO-B activity was depressed to only 20% of the control value.[45] Consequently, if neostriatal dopamine levels are restored by (R)-deprenyl treatment initiated three days post-MPTP, this effect is not a direct effect of (R)-deprenyl on unlesioned neurons.

These findings led the way to our investigations into the dopamine restorative properties of (R)-deprenyl in the MPTP mouse model. As shown in FIGURE 1, mature (36-week-old) male C57BL/6 mice treated with MPTP·HCl (35 mg/kg) followed three days later by saline injections every 48 hours for 10 weeks had a neostriatal dopamine concentration of 73.5 ± 7.5 pmol/mg wet weight. MPTP-treated mice that, starting three days later, also received (R)-deprenyl (1 mg/kg) every 48 hours had a neostriatal dopamine concentration of 78.6 ± 12.2 pmol/mg wet weight. These values, which are not statistically different, correspond to 58% and 62%, respectively, of neostriatal dopamine concentrations observed in control mice (FIG. 1). Consequently, in mature mice, chronic, low dose (R)-deprenyl treatment does not restore the MPTP-mediated depletion of neostriatal dopamine.

It is possible that the 36-week-old mice used in this study were not responsive to the trophic-like actions of (R)-deprenyl. This possibility would be consistent with the reduced neuronal regenerative properties observed in older mice.[46] Consequently, we extended our investigations to include young (six-week-old) mice. A cohort of 32-week-old mice were included for comparison. In these studies, we also used an (R)-deprenyl dose (0.25 mg/kg/48 hours) which was reported to have neuronal rescue properties[21,31] in young (eight-week-old) MPTP-treated C57BL/6 mice. Taking into account that (R)-deprenyl was reported to have neuronal trophic-like properties in rodents at dosages as low as 0.01 mg/kg every other day,[33] the 0.25

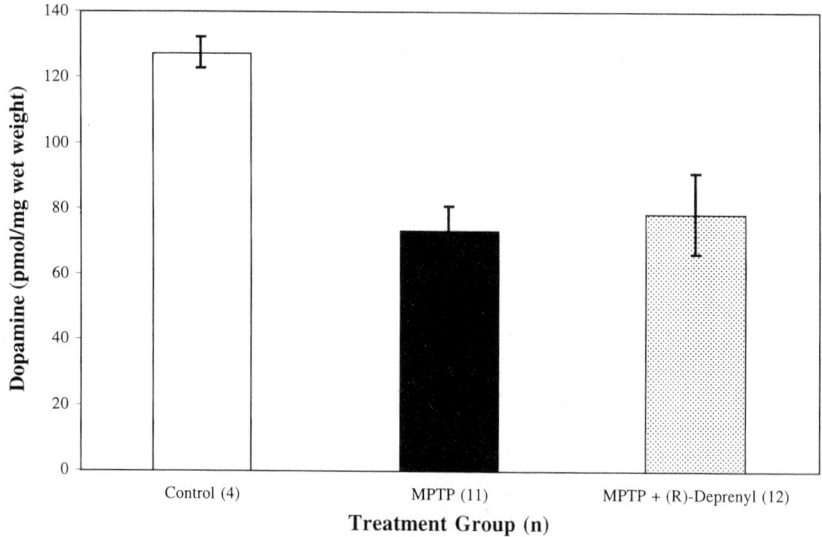

FIGURE 1. Effect of chronic (R)-deprenyl treatment (1 mg/kg every second day for 10 weeks) on neostriatal dopamine levels in MPTP-lesioned mice. The results are expressed as mean ± SD.

mg/kg/48 hour dose used in these studies was thought to be sufficient to display the reported trophic-like properties as well.

As part of the control data, we measured neostriatal dopamine levels in young and mature mice seven days after treatment with MPTP (40 mg/kg) and after treatment with saline only. We observed a difference in basal neostriatal dopamine levels between the 33-week-old mature (138.6 ± 7.5 pmol/mg wet weight) and seven-week-old young (109.2 ± 6.2 pmol/mg wet weight) mice (see FIGURE 2A). Seven weeks after these measurements, the basal neostriatal dopamine level of mature mice (now 40 weeks old) was 148.4 ± 6.3 pmol/mg wet weight whereas that of the young mice (now 15 weeks old) was 140.4 ± 10.4 pmol/mg wet weight (FIG. 2A). Thus, in the developing young mice the increase in dopamine levels leads to significantly different neostriatal dopamine levels ($p < 0.005$) over the time course of the experiment, whereas the mature mice do not display a significant alteration. This finding of increased neostriatal dopamine levels over time for young mice is in agreement with results from a previous report.[47] Consequently, the natural maturation process in young mice must be accounted for when evaluating either spontaneous increases in neostriatal dopamine or drug-mediated increases in neostriatal dopamine in MPTP lesioned animals.

Differences between mature and young mice were also evident when evaluating the effect the same MPTP dose had on the depletion of neostriatal dopamine seven days post MPTP administration (FIG. 2B). Treatment of mature mice with MPTP·HCl (40 mg/kg) led to a 70% reduction in neostriatal dopamine in comparison with aged-matched controls, whereas this treatment in young mice led to a reduction of only 35%. Mouse brain MAO-B levels increase with age[47] and these higher

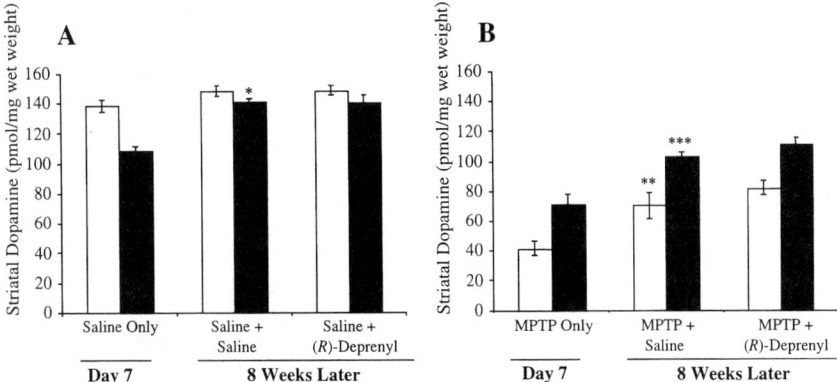

FIGURE 2. Effect of time, (R)-deprenyl- and MPTP-treatment on neostriatal dopamine levels in mature (□) versus young (■) mice. Results are expressed as mean ± SE. **A.** Comparison of basal neostriatal dopamine levels between mice at day 7 and eight weeks later following chronic treatment (27 administrations) with either saline or (R)-deprenyl. **B.** Comparison between neostriatal dopamine levels in mice seven days after MPTP-treatment (day 7) and after eight weeks of chronic treatment (27 administrations) with either saline or (R)-deprenyl; $p < 0.005$ from day 7, saline-only-treated controls. **Dopamine levels significantly different ($p < 0.05$) from day 7 MPTP only treated mature mice. ***Dopamine levels significantly different ($p < 0.005$) from day 7 MPTP-only-treated young mice.

MAO-B levels could have led to a greater lesion in mature mice due to the enhanced MAO-B-catalyzed bioactivation of MPTP.

The neurochemical changes induced by MPTP in mice do not appear to be long lasting. In young (6–12-week) C57BL/6 mice, the spontaneous recovery in neostriatal dopamine levels of up to 70% of control values is seen within three months after MPTP injection.[16,48,49] In older C57BL/6 mice, there is an age related decline in the potential for spontaneous neostriatal dopamine recovery after MPTP exposure. The recovery has been reported to vary from a 10% increase in neostriatal dopamine levels relative to control values during three months in 28 week old C57BL/6 mice[49] to very little recovery in 52 week old C57BL/6 mice during three months.[48]

If one were to compare the neostriatal dopamine levels of MPTP-treated young mice at eight weeks post treatment with their historic controls (day 7, FIG. 2B), it appears as though young mice were capable of nearly complete spontaneous recovery over eight weeks following MPTP treatment. However, by comparing the neostriatal dopamine of these MPTP-treated mice with age-matched controls, thereby taking the maturation process into account, the young mice had spontaneous dopamine recovery of only 8%. Mature mice had neostriatal dopamine recovery of 17%. This indicates the importance of including time-matched controls when evaluating spontaneous and/or drug-induced recovery following MPTP treatment. In the mouse, substantia nigra neurons are units capable of dynamic change throughout life.[40] However, it has been reported that older mice have reduced regenerative properties.[46] Nevertheless, we observed greater spontaneous recovery in mature compared to young mice (FIG. 2B). The mechanisms for this recovery are not clear. Arguably, spontaneous recovery following MPTP treatment in mice might involve regenerative sprouting of dopaminergic nerve terminals,[16] which could be due to the reported dynamic changes seen with aging.[40]

Irrespective of the maturation processes in young mice and the dynamic changes possible in older mice, chronic low dose (R)-deprenyl treatment at concentrations that were reported to be capable of rescuing neuronal cells, did not have an effect on neostriatal dopamine levels in either young or mature MPTP-treated mice (FIG. 2B). Consequently, neither chronic low doses of (R)-deprenyl treatment to C57BL/6 mice at concentrations that were reported to provide trophic support[34] to neurons, nor concentrations that were reported to be capable of rescuing substantia nigra cell bodies[21] (structural recovery), were able to induce functional recovery in MPTP-treated mice at the dosages and time periods of these experiments. The inability of mammalian central nervous system neurons to regenerate has been noted previously,[50] but the mechanisms responsible for this inability are not yet clear.

SUMMARY

In this study, we have found that chronic (R)-deprenyl treatment (1 mg/kg/48 hours for 10 weeks) initiated three days post-MPTP treatment (35 mg/kg) of mature mice at a dose that was reported to have trophic-like properties did not result in any increases in neostriatal dopamine (functional recovery) in comparison with mice treated similarly with saline. A chronic low dose (R)-deprenyl treatment protocol (0.25 mg/kg/48 hours) that was reported to rescue dying substantia nigra cell bodies following MPTP treatment in young mice also did not result in increases in neostriatal dopamine (functional recovery) in either young or mature mice in comparison with saline-treated controls. However, some spontaneous recovery in dopamine levels in both age groups, was observed. We also found that, as young mice mature, their neostriatal dopamine levels increase. Consequently, the natural maturation processes in young mice must be accounted for when evaluating either spontaneous or drug induced functional (and perhaps structural) recovery following MPTP treatment. We conclude that, although (R)-deprenyl may display trophic-like properties towards peripheral neurons, and also may rescue MPTP injured nigrostriatal neurons, (R)-deprenyl treatment does not lead to restoration of neostriatal dopamine (functional recovery).

ACKNOWLEDGMENTS

This work was supported by the Harvey W. Peters Center for the Study of Parkinson's Disease. Aspects of this work were presented at the 9th North American International Society for the Study of Xenobiotics Meeting, cosponsored by the American Chemical Society—Division of Chemical Toxicology, Nashville, TN, October, 1999.

REFERENCES

1. HORNYKIEWICZ, O. 1973. Parkinson's disease: from brain homogenate to treatment. Fed. Proc. **32:** 183–190.
2. ALVORD, E.C., JR, L.S. FORNO, J.A. KUSSKE, et al. 1974. The pathology of Parkinsonism: a comparison of degenerations in cerebral cortex and brainstem. Adv. Neurol. **5:** 175–193.
3. BALLARD, P.A., J.W. TETRUD & J.W. LANGSTON. 1985. Permanent human parkinsonism due to 1-methyl-4-phenyl-1,2,3,6-tetrahydropyridine (MPTP): seven cases. Neurology. **35:** 949–956.
4. LANGSTON, J.W., P. BALLARD, J.W. TETRUD & I. IRWIN. 1983. Chronic Parkinsonism in humans due to a product of meperidine-analog synthesis. Science **219:** 979–980.
5. HEIKKILA, R.E., A. HESS & R.C. DUVOISIN. 1984. Dopaminergic neurotoxicity of 1-methyl-4-phenyl-1,2,5,6-tetrahydropyridine in mice. Science **224:** 1451–1453.
6. ROYLAND, J.E. & J.W. LANGSTON. 1998. MPTP: A dopaminergic neurotoxin. In Highly Selective Neurotoxins: Basic and Clinical Applications. R.M. Kostrzewa, Ed.: 141–194. Humana Press, New Jersey.
7. CASTAGNOLI, N. JR., K. CHIBA & A.J. TREVOR. 1985. Potential bioactivation pathways for the neurotoxin 1-methyl-4-phenyl-1,2,3,6-tetrahydropyridine (MPTP). Life Sci. **36:** 225–230.
8. CHIBA, K., A. TREVOR & N. CASTAGNOLI, JR. 1984. Metabolism of the neurotoxic tertiary amine, MPTP, by brain monoamine oxidase. Biochem. Biophys. Res. Commun. **120:** 574–578.
9. HEIKKILA, R.E., L. MANZINO, F.S. CABBAT & R.C. DUVOISIN. 1984. Protection against the dopaminergic neurotoxicity of 1-methyl-4-phenyl-1,2,5,6-tetrahydropyridine by monoamine oxidase inhibitors. Nature **311:** 467–469.
10. MARKEY, S.P., J.N. JOHANNESSEN, C.C. CHIUEH, et al. 1984. Intraneuronal generation of a pyridinium metabolite may cause drug-induced parkinsonism. Nature **311:** 464–467.
11. JAVITCH, J.A., R.J. D'AMATO, S.M. STRITTMATTER & S.H. SNYDER. 1985. Parkinsonism-inducing neurotoxin, N-methyl-4-phenyl-1,2,3,6-tetrahydropyridine: uptake of the metabolite N-methyl-4-phenylpyridine by dopamine neurons explains selective toxicity. Proc. Natl. Acad. Sci. U.S.A. **82:** 2173–2177.
12. MILLER, G.W., J.K. STALEY, C.J. HEILMAN, et al. 1997. Immunochemical analysis of dopamine transporter protein in Parkinson's disease. Ann. Neurol. **41:** 530–539.
13. CILIAX, B.J., G.W. DRASH, J.K. STALEY, et al. 1999. Immunocytochemical localization of the dopamine transporter in human brain. J. Comp. Neurol. **409:** 38–56.
14. RAMSAY, R.R. & T.P. SINGER. 1986. Energy-dependent uptake of N-methyl-4-phenylpyridinium, the neurotoxic metabolite of 1-methyl-4-phenyl-1,2,3,6-tetrahydropyridine, by mitochondria. J. Biol. Chem. **261:** 7585–7587.
15. NICKLAS, W.J., I. VYAS & R.E. HEIKKILA. 1985. Inhibition of NADH-linked oxidation in brain mitochondria by 1-methyl-4-phenyl-pyridine, a metabolite of the neurotoxin, 1-methyl-4-phenyl-1,2,5,6-tetrahydropyridine. Life Sci. **36:** 2503–2508.
16. RICAURTE, G.A., J.W. LANGSTON, L.E. DELANNEY, et al. 1986. Fate of nigrostriatal neurons in young mature mice given 1-methyl-4-phenyl-1,2,3,6-tetrahydropyridine: a neurochemical and morphological reassessment. Brain Res. **376:** 117–124.
17. LINDER, J.C., H. KLEMFUSS, & P.M. GROVES. 1987. Acute ultrastructural and behavioral effects of 1-methyl-4-phenyl-1,2,3,6-tetrahydropyridine (MPTP) in mice. Neurosci. Lett. **82:** 221–226.
18. HERKENHAM, M., M.D. LITTLE, K. BANKIEWICZ, et al. 1991. Selective retention of MPP^+ within the monoaminergic systems of the primate brain following MPTP administration: an in vivo autoradiographic study. Neuroscience **40:** 133–158.
19. MITSUMOTO, Y., A. WATANABE, A. MORI & N. KOGA. 1998. Spontaneous regeneration of nigrostriatal dopaminergic neurons in MPTP-treated C57BL/6 mice. Biochem. Biophys. Res. Commun. **248:** 660–663.
20. JACKSON-LEWIS, V., M. JAKOWEC, R.E. BURKE & S. PRZEDBORSKI. 1995. Time course and morphology of dopaminergic neuronal death caused by the neurotoxin 1-methyl-4-phenyl-1,2,3,6-tetrahydropyridine. Neurodegeneration **4:** 257–269.

21. TATTON, W.G. & C.E. GREENWOOD. 1991 Rescue of dying neurons: a new action for deprenyl in MPTP Parkinsonism. J. Neurosci. Res. **30:** 666–672.
22. HORNYKIEWICZ, O. 1974. The mechanisms of action of L-dopa in Parkinson's. Life Sci. **15:** 1249–1259.
23. COTZIAS, G.C. 1968. L-Dopa for Parkinsonism. N. Engl. J. Med. **278:** 630.
24. PAPAVASILIOU, P.S., G.C. COTZIAS, S.E. DÜBY, et al. 1972. Levodopa in Parkinsonism: potentiation of central effects with a peripheral inhibitor. N. Engl. J. Med. **286:** 8–14.
25. MARSDEN, C.D., P.E. BARRY, J.D. PARKES & K.J. ZILKHA. 1973. Treatment of Parkinson's disease with levodopa combined with L-alpha-methyldopahydrazine, an inhibitor of extracerebral DOPA decarboxylase. J. Neurol. Neurosurg. Psychiatry **36:** 10–14.
26. PINDER, R.M., R.N. BROGDEN, P.R. SAWYER, et al. 1976. Levodopa and decarboxylase inhibitors: a review of clinical pharmacology and use in the treatment of parkinsonism. Drugs **11:** 329–377.
27. CSANDA, E., M. TÁRCZY, A. TAKÁTS, et al. 1983. L-Deprenyl in the treatment of Parkinson's disease. J. Neural. Transm. Suppl. **19:** 283–290.
28. THE PARKINSON STUDY GROUP. 1989. Effect of deprenyl on the progression of disability in early Parkinson's disease. N. Engl. J. Med. **321:** 1364–1371.
29. THE PARKINSON STUDY GROUP. 1993. Effects of tocopherol and deprenyl on the progression of disability in early Parkinson's disease. N. Engl. J. Med. **328:** 176–183.
30. RINNE, J.O., M. RÖYTTÄ, L. PALJÄRVI, et al. 1991. Selegiline (deprenyl) treatment and death of nigral neurons in Parkinson's disease. Neurology **41:** 859–861.
31. TATTON, W.G. 1993. Selegiline can mediate neuronal rescue rather than neuronal protection. Mov. Disord. **8**(Suppl. 1): S20–S30.
32. CASTAGNOLI, K., S. PALMER, A. ANDERSON, et al. 1997. The neuronal nitric oxide synthase inhibitor 7-nitroindazole also inhibits the monoamine oxidase-B-catalyzed oxidation of 1-methyl-4-phenyl-1,2,3,6-tetrahydropyridine. Chem. Res. Toxicol. **10:** 364–368.
33. ANSARI, K.S., P.H. YU, T.P. KRUCK & W.G. TATTON. 1993. Rescue of axotomized immature rat facial motoneurons by R(−)-deprenyl: stereospecificity and independence from monoamine oxidase inhibition. J. Neurosci. **13:** 4042–4053.
34. OH, C., B. MURRAY, N. BHATTACHARYA, et al. 1994. (−)-Deprenyl alters the survival of adult murine facial motoneurons after axotomy: increases in vulnerable C57BL strain but decreases in motor neuron degeneration mutants. J. Neurosci. Res. **38:** 64–74.
35. SALO, P.T. & W.G. TATTON. 1992. Deprenyl reduces the death of motoneurons caused by axotomy. J. Neurosci. Res. **31:** 394–400.
36. PATERSON, I.A. & W.G. TATTON. 1998. Antiapoptotic actions of monoamine oxidase B inhibitors. Adv. Pharmacol. **42:** 312–315.
37. PATERSON, I.A., A.J. BARBER, D.L. GELOWITZ & C. VOLL. 1997. (−)Deprenyl reduces delayed neuronal death of hippocampal pyramidal cells. Neurosci. Biobehav. Rev. **21:** 181–186.
38. TATTON, W.G. & R.M. CHALMERS-REDMAN. 1996. Modulation of gene expression rather than monoamine oxidase inhibition: (−)-deprenyl-related compounds in controlling neurodegeneration. Neurology **47**(Suppl. 3): S171–S183.
39. TATTON, W.G., W.Y. JU, D.P. HOLLAND, et al. 1994. (−)-Deprenyl reduces PC12 cell apoptosis by inducing new protein synthesis. J. Neurochem. **63:** 1572–1575.
40. GREENWOOD, C.E., W.G. TATTON, N.A. SENIUK & F.G. BIDDLE. 1991. Increased dopamine synthesis in aging substantia nigra neurons. Neurobiol. Aging **12:** 557–565.
41. CASTAGNOLI, K., S. PALMER & N. CASTAGNOLI, JR. 1999. Neuroprotection by (R)-deprenyl and 7-nitroindazole in the MPTP C57BL/6 mouse model of neurotoxicity. Neurobiol. **7:** 135–149.
42. PITTS, S.M., S.P. MARKEY, D.L. MURPHY & A. WEISZ. 1986. Recommended practices for the safe handling of MPTP. In MPTP: A Neurotoxin Producing a Parkinsonian Syndrome. S.P. Markey, N. Castagnoli, Jr., A.J. Trevor & I.J. Kopin, Eds.: 703–716. Academic Press, New York.

43. VAN DER SCHYF, C.J., K. CASTAGNOLI, S. PALMER, et al. 2000. Melatonin fails to protect against long-term MPTP-induced dopamine depletion in mouse striatum. Neurotox. Res. **1:** 261–269.
44. THIFFAULT, C., L. LAMARRE-THÉROUX, R. QUIRION & J. POIRIER. 1997. L-Deprenyl and MDL72974 do not improve the recovery of dopaminergic cells following systemic administration of MPTP in mouse. Mol. Brain Res. **44:** 238–244.
45. STEYN, S.J., K. CASTAGNOLI, S. STEYN & N. CASTAGNOLI, JR. 2001. Selective inhibition of MAO-B through chronic low dose (*R*)-deprenyl treatment in C57BL/6 mice has no effect on basal neostriatal dopamine levels. Exp. Neurol. **168:** 434–436.
46. DATE, I., S.Y. FELTEN, J.A. OLSCHOWKA & D.L. FELTEN. 1990. Limited recovery of striatal dopaminergic fibers by adrenal medullary grafts in MPTP-treated aging mice. Exp. Neurol. **107:** 197–207.
47. IRWIN, I., K.T. FINNEGAN, L.E. DELANNEY, et al. 1992. The relationships between aging, monoamine oxidase, striatal dopamine and the effects of MPTP in C57BL/6 mice: a critical reassessment. Brain Res. **572:** 224–231.
48. DATE, I., D.L. FELTEN & S.Y. FELTEN. 1990. Long-term effect of MPTP in the mouse brain in relation to aging: neurochemical and immunocytochemical analysis. Brain Res. **519:** 266–276.
49. NISHI, K., T. KONDO & H. NARABAYASHI. 1989. Difference in recovery patterns of striatal dopamine content, tyrosine hydroxylase activity and total biopterin content after 1-methyl-4-phenyl-1,2,3,6-tetrahydropyridine (MPTP) administration: a comparison of young and older mice. Brain Res. **489:** 157–162.
50. SCHWAB, M.E. & D. BARTHOLDI. 1996. Degeneration and regeneration of axons in the lesioned spinal cord. Physiol. Rev. **76:** 319–370.

Gene Therapy for Treatment of Cerebral Ischemia Using Defective Herpes Simplex Viral Vectors

MIDORI A. YENARI,[a,b,c] THEODORE C. DUMAS,[d,e] ROBERT M. SAPOLSKY,[b,d,e] AND GARY K. STEINBERG[a,b,c,d]

Departments of [a]Neurosurgery, [b]Neurology, and [c]Stanford Stroke Center, Stanford University Medical Center, Stanford, California, U.S.A.

[d]Program in Neurosciences, and [e]Department of Biological Sciences, Stanford University, Stanford, California, U.S.A.

ABSTRACT: Significant advances have been made over the past few years concerning the cellular and molecular events underlying neuron death. Recently, it is becoming increasingly clear that some of genes induced during cerebral ischemia may actually serve to rescue the cell from death. However, the injured cell may not be capable of expressing protein at high enough levels to be protective. One of the most exciting arenas of such interventions is the use of viral vectors to deliver potentially neuroprotective genes at high levels. Neurotropic herpes simplex viral (HSV) strains are an obvious choice for gene therapy to the brain, and we have used bipromoter vectors that are capable of transferring various genes to neurons. Using this system in experimental models of stroke, cardiac arrest, and excitotoxicity, we have found that it is possible to enhance neuron survival against such cerebral insults by overexpressing genes that target various facets of injury. These include energy restoration by the glucose transporter (GLUT-1), buffering calcium excess by calbindin, preventing protein malfolding or aggregation by stress proteins and inhibiting apoptotic death by BCL-2. We show that in some cases, gene therapy is also effective after the onset of injury, and also address whether successful gene therapy necessarily spares function. Although gene therapy is limited to the few hundred cells the vector is capable of transfecting, we consider the possibility of such gene therapy becoming relevant to clinical neurology in the future.

KEYWORDS: Gene therapy; Cerebral ischemia; Excitotoxin; Heat shock protein; Herpes virus.

INTRODUCTION

Recent scientific advances in the area of stroke and neurodegeneration have led to the discovery of specific cellular events that occur during necrosis and apoptosis. It is now possible to therapeutically target these events with the hope of rescuing brain cells from death. The past decade has seen a variety of pharmaceutical agents

---

Address for correspondence: Midori A. Yenari, M.D., Dept. of Neurosurgery, Stanford University, MSLS Building, 1201 Welch Rd. P304, Stanford, CA 94305-5487, U.S.A. Voice: 650-736-1482; fax: 650-736-1949.
yenari@Stanford.edu

designed to limit excitotoxicity, inflammation, and free radical generation. Another strategy employed by a few groups is that of gene therapy, or the transfer of genetic material into host cells with the intent of expressing the protein of interest. Several studies have shown that cerebral ischemia alters gene expression and that some of the induced genes may serve a protective or damaging role. Among the many genes that have been identified to participate in ischemia, those possessing neuroprotective properties may be excellent candidates for gene therapy. Of the various candidate genes, we have shown that neurons can be rescued from various insults through the overexpression of the glucose transporter (GLUT-1), the calcium-buffering protein calbindin D28K (CaBP), the antiapoptotic protein BCL-2, and the stress protein HSP70. We summarize our experience with gene transfer therapy to cerebral neurons using defective herpes simplex viral vectors in various experimental models of cerebral ischemic and neurotoxic injury.

THE CELLULAR CASCADE OF NEURON DEATH FOLLOWING INJURY

The cascade of events mediating necrotic and apoptotic elements of neuron death[1,2] suggests specific points to be targeted in therapy (see FIGURE 1). Following the deprivation of its metabolic substrates, synaptic concentrations of excitatory amino acids (EAAs) in the brain rise into the excitotoxic range. Excessive EAAs,

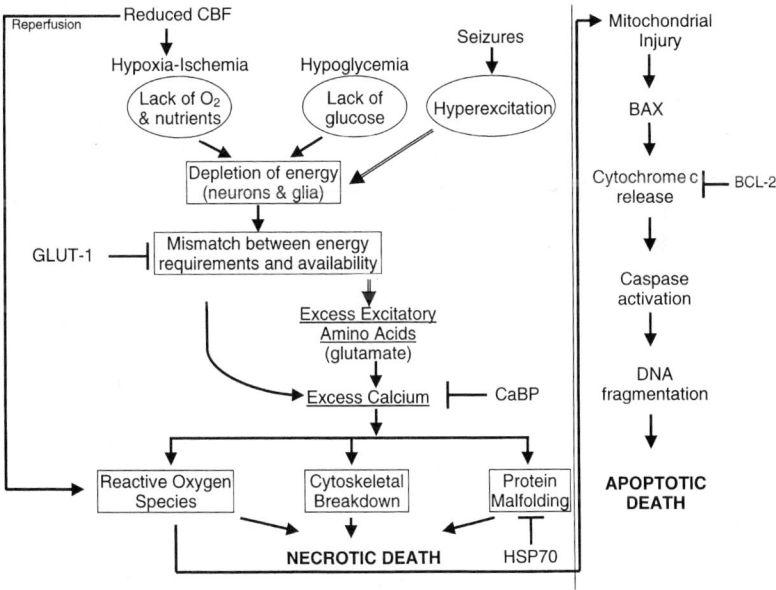

FIGURE 1. Schematic representation of the pathways mediating necrotic and apoptotic death following acute neurological insults (*arrows*), and sites of demonstrated therapeutic efficacy of gene transfer approaches (*T-squares*). Abbreviations: CBF, cerebral blood flow; GLUT-1, glucose transporter; CaBP, calbindin D28K; HSP70, heat shock protein 70.

namely glutamate, exert their excitotoxic effects through NMDA and non-NMDA receptors, and increase influx of calcium and sodium. Additional calcium accumulation results from internal release from the endoplasmic reticulum and mitochondria. This excess of free cytosolic calcium is central to the excitotoxic hypothesis, whereby calcium activates the generation of reactive oxygen species (ROS), cytoskeletal proteolysis, and the malfolding and aggregation of proteins, collectively leading to cell death. ROS are thought to be particularly damaging when reduced cerebral blood flow is followed by reperfusion.

Recently, it has been recognized that apoptotic, or programmed cell death occurs in a subset of neurons.[3] Whereas necrotic cell death typically involves non-specific DNA degradation, nuclear pyknosis, loss of membrane integrity, and mitochondrial swelling, apoptosis involves chromosome condensation, internucleosomal DNA fragmentation, membrane blebbing, and changes in surface antigens. Current evidence suggests that for ischemic and excitotoxic insults, ROS-induced damage to mitochondria results in activation, dimerization and membrane translocation of the proapoptotic protein BAX. This leads to cytochrome c release from mitochondria which, in turn, activates proteolytic enzymes called caspases. This eventually results in the pathway of events leading to the various changes that characterize apoptosis. Among the genes involved in apoptotic death, some are known to inhibit this process. One is example is the antiapoptotic gene, BCL-2, a membrane-bound protein now known to antagonize cytochrome c release.[4] Although this picture of neuron death is highly simplified, it does suggest potential targets for the genetic interventions.[2] For example, expression of the glucose transporter (GLUT-1) could enhance transport of the neuron's metabolic substrate during a period where the availability is critically low. Intracellular calcium can be buffered by overexpressing high levels of endogenous binding proteins. Protein malfolding and aggregation could be prevented by upregulating molecular chaperones such as the heat shock proteins. Finally, apoptotic death could be inhibited by increasing the amount of antiapoptotic proteins.

VIRAL VECTOR CONSTRUCTION AND CHARACTERIZATION

Viral vectors are one means by which genetic material can be transferred into host cells. Herpes simplex virus is a natural choice for gene transfer into the adult central nervous system since the virus is neurotropic.[5,6] Replication incompetent strains have been developed containing mutations within genes responsible for cytotoxicity.[7] The $d120$ strain contains a 4.1-kb deletion mutation in $\alpha 4$, an immediate early gene essential for early and late gene induction.[7] Without $\alpha 4$, the virus in incapable of performing replication, capsid synthesis, and assembly.[8] Our group uses an amplicon ("cloning amplifying") based system for vector generation. The construction of amplicon plasmids and vectors have been described in detail elsewhere.[9–14] A plasmid containing the gene(s) of interest and a replication-incompetent helper virus are cotransfected into a cell line (E5) that contains all the necessary components for virus replication and assembly. The transfected cell line produces progeny virion containing either the plasmid of interest or helper virus. After removing cellular debris, the vector (concatemers of the plasmid and helper virus) is said to be purified, and suitable for *in vivo* gene transfer study. After direct parenchymal injection, the vector is largely taken up by neurons, although ependymal cells will take up the viral vector after intraventricular injection. In fact, with direct injection into hippocampus and striatum, approximately 70% of transfected cells colocalize with neuronal markers.

The vectors used in our studies are referred to as "bipromoter". That is, two separate promoters drive expression of two different genes.[10] An advantage of this system is that both a reporter gene (e.g., the *E. coli lacZ* gene) and the gene of interest are coexpressed within targeted cells. Since many insults often induce the gene of interest within the cells themselves, transgene expression due to transfection and endogenous gene expression can be difficult to differentiate. With these vectors, the problem is circumvented because transfected cells also overexpress a reporter gene, in this case β-galactosidase (β-gal), the gene product of *lacZ*. Transgene expression begins within 4–6 h after injection and peaks at about 12–24 h.[14] Expression is typically limited to a few days, but the duration of expression may depend on the tissue halflife of the protein itself, rather than the duration of transcription and translation. For example, the tissue halflife of βgal is particularly long and it may persist as long as seven days, whereas HSP70 protein persists for only one day.[14,15] For cerebral ischemia and other acute nervous system insults, the issue of long term expression from these vectors may not be so critical, since the temporal therapeutic window for stroke treatment is finite, and for most interventions, relatively brief. Therefore, stroke treatments using gene therapy may require transgene expression for only a few hours to a few days. On the other hand, recent data has also shown that ischemic injury evolves over a few days to weeks depending on the model studied; therefore, the prolonged detection of the reporter protein, βgal may serve as an advantage in stroke models in that the "tagged" cells can be identified regardless of whether the transgene of interest is still expressing. A limitation of this vector system concerns the extent of transfection. With a transfection efficiency of approximately 10%, only a few hundred neurons can be expected to take up the plasmid, and about 80% of transfected cells are contained within a 0.5 mm radius of the injection site.[14,16] Although this amplicon based system is unlikely to be applied at the clinical level, we have nevertheless adapted this system for gene transfer studies using various *in vivo* models of necrotic and apoptotic injury.

THERAPEUTIC TARGETS FOR GENE THERAPY

We now review our experience using viral vectors to alter the various steps in cell death. Because viral vector injection can only transfect a few hundred cells at best, and the genes we have studied to date act intracellularly or within the membrane, we have adapted standard injury models to accommodate this limitation.

In the focal cerebral ischemia model, a 3-0 surgical suture with the tip rounded and enlarged by flaming is inserted into the common carotid artery and advanced to occlude the ostium of the middle cerebral artery (MCA). Reperfusion can be controlled simply by withdrawing the suture. Injury to the ipsilateral striatum is consistent even after brief periods of MCA occlusion, and injection and expression of the viral vector to this brain region is consistent and reproducible.[15,17,24] Because the transgene reaches peak expression 9–12 h after injection, the vector is stereotaxically injected into each striatum approximately 6–15 h prior to ischemia onset. The MCA is occluded for a relatively brief period of time (1 h) to ensure some ischemic striatal damage, but not so much as to hinder transcriptional and translational machinery of the vector and host cell. 48 h later, the brains are harvested, and coronal sections are stained with X-gal (a chromogenic substrate for β-galactosidase) and cresyl violet to identify vector targeted cells and to delineate the extent of ischemic injury, respectively. The number of X-gal-positive neurons are counted in each striatum and the total number of positive cells in the ischemic striatum are normalized to the number of positive cells in the contralateral striatum (see FIGURE 2). Because the number of cells transfected by the vectors is small (a few hundred), it is unlikely that overall infarct size will be altered. In fact, to ensure that any observed neuroprotection is not due to an imbalance in the overall damage between the treated and control groups, the infarct size is both measured and given a semiquantitative score.[11,15,17,24]

For the kainic acid and global cerebral ischemia models, a similar logic is applied. Kainic acid administration (either by parenteral administration or direct parenchymal injection) leads to destruction of the neurons in the hippocampal formation, especially CA3/4, whereas eight minutes of bilateral common carotid artery occlusion plus hypotension leads to selective death within hippocampal CA1. Vector is injected into the hippocampus followed by the application of the respective

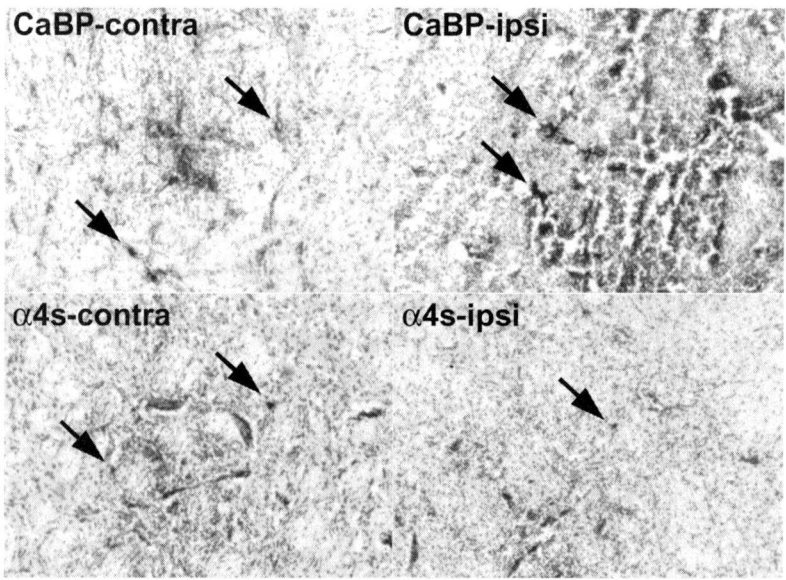

FIGURE 2. Calbindin (CaBP) overexpression protects striatal neurons from a stroke. Several surviving vector-targeted, X-gal-stained (*arrows*) neurons are observed within the nonischemic (**CaBP-contra**) and ischemic (**CaBP-ipsi**) striata of an animal injected with the CaBP vector. In the striata of an animal given control vector, X-gal-stained neurons are seen within targeted striatal neurons on the nonischemic side (**α4s-contra**), whereas fewer X-gal-stained cells are observed on the ischemic side (**α4s-ipsi**).

hippocampal insult approximately 12 h later. To determine the percentage of surviving transfected hippocampal neurons, the number of remaining X-gal-positive neurons is normalized to the number of transfected neurons in injected sham controls. To ensure that the extent of damage is equal between control and treated groups, total cell densities are computed within the different subsectors.[11,15]

Gene Therapy to Improve Cellular Energetics

One of the earliest attempts at neuronal gene therapy attempted to bolster neuronal energetics during insults via overexpression of the rat brain glucose transporter (GLUT-1). Using 2-deoxy glucose autoradiography, injection of this vector did lead to localized increased uptake of glucose.[16] In fact, such overexpression is broadly protective, reducing neuron loss in different injury models.[13,18] At the whole animal model level, overexpression of GLUT-1 following transient focal cerebral ischemia[17] resulted in 105% improvement in survival. Overexpression in the kainic acid model was also protective within dentate granule cells which received the GLUT-1 vector. GLUT-1 overexpression also protected against the electron transport uncoupler, 3-acetylpyridine (3-AP), that is preferentially toxic to the dentate gyrus *in vivo*.[19] A few studies have examined subsequent downstream events leading to neuronal sparing. During an insult, GLUT-1 overexpression blunts the decline in ATP concentrations and metabolism, decreases glutamate release and cytosolic calcium concentrations.[20,21]

Gene Therapy to Inhibit Intracellular Calcium Excess

Several pharmacological studies have also targeted the excess of cytosolic calcium. At the gene therapy level, possible strategies would include overexpression of a calcium binding protein, or antisense targeting of a calcium channel. To date, all relevant studies have overexpressed the calcium-binding protein calbindin D28K,[22-26] chosen because of the correlation in a number of studies of its presence with neuronal survival.[27-29] Such overexpression protects against a variety of insults. Vector-mediated CaBP overexpression has been previously shown to alter neuronal synaptic responses, consistent with a calcium-buffering function.[30] In our prior *in vitro* studies, the CB vector has also been demonstrated to reduce intracellular calcium responses.[23] Consistent with the notion that calcium overload is damaging, we then found that CaBP overexpression protected cultured hippocampal neurons from various ischemia-like insults such as hypoglycemia and excitotoxin exposure, including *N*-methyl-D-aspartate, kainate and glutamate; however, CaBP did not protect against cyanide toxicity, suggesting that the protective effects are not effective against mitochondrial toxins, or that CaBP cannot protect against such severe insults.[22,23] At the *in vivo* level, we did find hippocampal neuron protection against 3-AP,[26] a different mitochondrial toxin, suggesting that the lack of protection against cyanide *in vitro* was more likely due to the insult severity. More recently, we have applied this vector to the experimental stroke model and kainic acid, and found that CaBP overexpression also protected against these insults (see FIGURES 2 and 3).

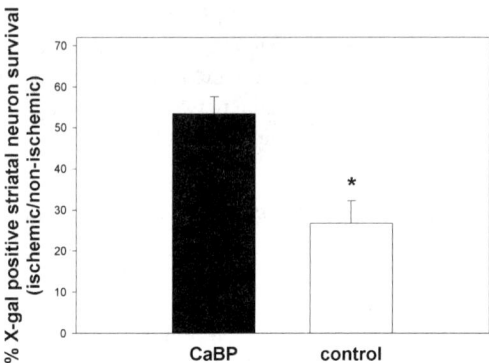

FIGURE 3. Calbindin overexpression in a model of experimental stroke. Fifteen hours prior to ischemia onset 19 rats were injected with calbindin (CaBP) or control vector. Animals were subjected to one hour of transient focal cerebral ischemia followed by 47 h of reperfusion. Brain sections were stained with X-gal to detect vector-transfected cells, then counter stained with cresyl violet to delineate regions of infarction. The bar graph shows the percentage of X-gal-stained neurons remaining in the ischemic striatum relative to the contralateral nonischemic striatum. (*$p < 0.001$).

Gene Transfer of Stress Proteins

Gene transfer therapy can also be used to address certain biological questions that cannot ordinarily be addressed by other means. Other adverse consequences of calcium excess following insults includes incorrect intracellular targeting of nascent proteins, their failure to function due to malfolding, and the potential for malfolded proteins to aggregate. It is in this realm that heat shock proteins, as molecular chaperones, are most likely to exert their protective actions. Thus, heat shock protein overexpression is also a potential strategy, complicated by the lack of consensus as to whether any of the expressed genes have a physiological role during these acute insults. The published reports have been concerned mainly with the inducible form of HSP70 (also referred to by HSP72), the most robustly induced of stress proteins. Some have argued that the induction of HSP70 served a protective role against ischemic insults because expression was most robust in the most resistant cell populations.[31] Induction of HSP70 was also correlated with the phenomenon of induced tolerance.[32] Others found that stress proteins were expressed in cells regardless of their fate, and argued that these proteins were merely an epiphenomenon that served no neuroprotective role.[33] In our hands, HSP70 overexpression failed to protect against an excitotoxic insult in primary neuronal cultures, but did protect against other insults in neuronal and astrocytic cultures including heat shock[9] and ischemia-like conditions,[34,35] as well as against ischemic and excitotoxic insults *in vivo*.[15] These observations have since been replicated using transgenic animal models, but with some variability.[36–38] Nevertheless, the majority of data support an active protective role for the heat shock proteins, although there may be strict limits to this protection.[39]

Gene Therapy to Alter Apoptotic Death

It is now established, that to some extent, apoptotic, or programmed, energy-dependent death also occurs during cerebral ischemic injury.[3] Among the various genes involved include those that inhibit apoptosis. Perhaps the most studied of these antiapoptotic genes is *bcl-2,* which encodes a 26-kD protein that is the mammalian analogue of the nematode gene, *ced-9.* A membrane-bound protein, BCL-2 inhibits apoptosis by blocking cytosolic translocation of mitochondrial cytochrome c,[4,40] and subsequent activation of downstream caspases. BCL-2 may also have different sites of action given that it protects against a variety of insults,[41] both necrotic[42–45] and apoptotic.[11,46–48] BCL-2 is induced by cerebral ischemia,[49–51] and we and others have shown that BCL-2 overexpression protects against cerebral ischemia either

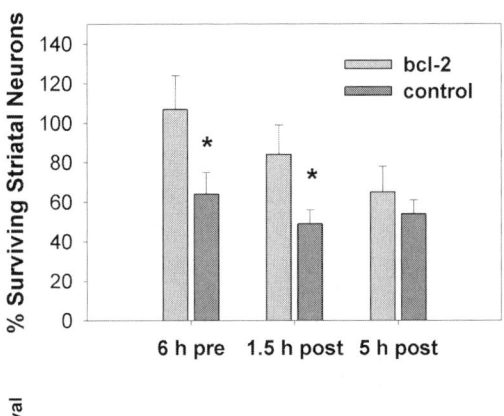

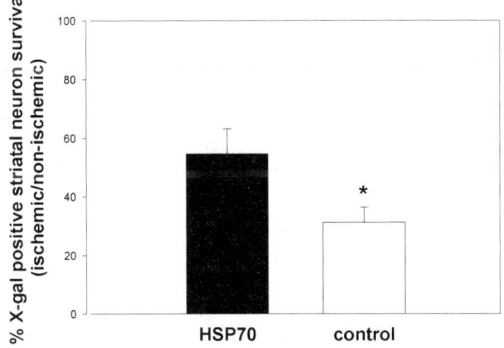

FIGURE 4. BCL-2 and HSP70 overexpression protects striatal neurons when vector is administered after ischemia onset. Rats were subjected to one hour of transient focal cerebral ischemia followed by 47 h of reperfusion. Vectors were injected into the striata at various times prior to or after MCA occlusion. Brain sections were stained with X-gal to detect vector transfected cells, then counter stained with cresyl violet to delineate regions of infarction. Bar graphs show the percentage of X-gal-stained neurons remaining in the ischemic striatum relative to the contralateral nonischemic striatum. BCL-2 overexpression improves neuron survival when delivered 1.5 h, but not five hours after ischemia onset (**A**). HSP70 improves neuron survival when applied 30 min after ischemia onset (**B**). (*$p < 0.05$).

by vector-mediated overexpression (see FIGURE 4A),[11,52,53] or in transgenic animal models.[54] BCL-2 does not appear to protect against all insults, however. Although it demonstrates marked protection by several groups against cerebral ischemia and trauma, it does not appear to be effective against severe necrotic insults such as exposure to the electron transport uncoupler 3-AP.[55]

POSTINSULT GENE TRANSFER AND NEURON FUNCTION

In order for gene therapy to have direct clinical relevance, the various vectors should also be examined for neuroprotection when given after the insult. Furthermore, it will be increasingly important to determine whether the rescued neuron also functions normally. To these ends, our groups have begun to address these issues in our models.

Postinsult Gene Transfer

Although we have shown that preinsult gene therapy can ameliorate neuronal death against cerebral ischemia, most clinicians would agree that pretreatment of stroke has limited utility. Therefore, it is important to investigate whether postinsult gene therapy is also feasible. Certainly, this approach may depend on the nature of the gene used. Apoptotic death may explain the observed delayed injury in ischemia; therefore, BCL-2 could be viewed as a rational choice. In fact, we first found that the BCL-2 vector applied following ischemia onset improved striatal neuron survival when vector was delayed 1.5 h, but not by 5 h (FIG. 4A).[52,56] Given that gene expression from these vectors begins 4–6 hours following injection[52] with peak protein expression 12–24 hours later,[9] this implies that neuroprotection from BCL-2 is possible when overexpression is applied as late as 6.5–13.5 hours following the onset of ischemia in the rodent. A similar observation was also noted by Jia et al.,[43] who found that BCL-2 protected cultured neurons when applied eight hours after glutamate exposure.

Surprisingly, genes that interfere early in the ischemic cascade protect even when applied postinsult. In the *in vivo* excitotoxin model, GLUT-1 vector protected the hippocampus even when administered 30 minutes after kainic acid administration, but not four hours.[20] Hippocampal neuron cultures exposed to hypoglycemia were protected when GLUT-1 was applied one hour later,[20] but the therapeutic window for CaBP in the same model was only 30 min.[23] No protection was observed for either vector when administered at four hours, however.

The mechanisms of protection for HSP70 are only beginning to emerge.[39,57] Although a widely held concept is that HSP70 protects in the early phases of injury by maintaining protein structure and preventing aggregation, other work has shown that HSP70 may also exert an antiapoptotic effect downstream of caspase-3.[58] Therefore, HSP70 may also provide neuroprotection when given postinsult as well. In fact, in a preliminary study, we found that HSP70 overexpression applied 30 min after MCA occlusion in our focal cerebral ischemia paradigm did lead to improved striatal neuron counts (FIG. 4B).[59] The mechanisms and the temporal therapeutic window for HSP70 neuroprotection are currently under intensive study.

Gene Transfer and Neuron Function

Another critical issue is whether a neuron, when saved from death because of the protective effects of some transgene, actually functions appropriately afterward. In the studies described to this point, however, "protection" consists of improved numbers of surviving cells, with no demonstrated preservation of neuronal function. Certainly, the hope of successful gene therapy is for naught if the rescued neuron does not work properly. Recently, our laboratories have embarked on detailed physiological testing[60] to determine whether gene therapy results in recovery of neurological function. In an initial study, we first established that injection of the vectors themselves did not alter synaptic transmission.[60] Using the model of KA toxicity, the GLUT-1 or the BCL-2 vector was used to overexpress these respective proteins within the hippocampus. We then delivered KA with active vector to the ipsilateral hippocampus, and control vector to the contralateral hippocampus. The extent of injury was measured as the percent of the entire CA3 subregion that was lesioned. We established conditions where overexpression of either GLUT-1 or BCL-2 was equally protective, decreasing excitotoxicity to equivalent extents in hippocampal cultures and the hippocampus. Excitatory postsynaptic potential (EPSP) to fiber potential (FP) ratios and posttetanic potentiation (PTP) were measured to determine differences in baseline synaptic strength and short term plasticity, respectively. PTP is one form of short-term plasticity found in the hippocampus, whereby increases in EPSP reflects an increase in transmitter release due to residual calcium in the presynaptic terminal following the high frequency stimulation. Electrophysiological function improved after GLUT-1 gene transfer, but not after BCL-2 (see FIGURE 5) suggesting that there are two routes of protection involving different intermediary steps. GLUT-1 overexpression constituted an early intervention (in that the time window of saving postinsult was one hour), maintained neuronal metabolism longer during the insult, and decreased ROS accumulation.[61] In contrast, BCL-2 overexpression constituted a late intervention (with a time window of saving postinsult of at least six hours) and had no sparing effect on metabolism or ROS accumulation.

At the behavioral level, spatial memory was tested using the Morris water maze in animals subjected to KA injury.[61] After determining conditions in which hippocampal neuron sparing was equivalent for GLUT-1 and BCL-2, vector and KA were delivered to the hippocampus, then rats were subjected to training three days later. Memory function was assessed at five days. All animals showed learning in the visual discrimination and spatial versions of the maze as a decrease in the latency to find the platform, indicating that the KA and vector transfection did not create visuospatial or motivational deficits. By five days (two days after training), animals receiving control vector and KA showed increased latencies to platform finding compared to sham-lesioned controls. GLUT-1-, but not BCL-2-injected, animals showed significant decreases in this latency. Therefore, GLUT-1, but not BCL-2 overexpression, improved memory function in KA-lesioned rats. Thus, a gene therapy intervention that spares neurons from insult-induced death does not necessarily translate into a sparing of function.

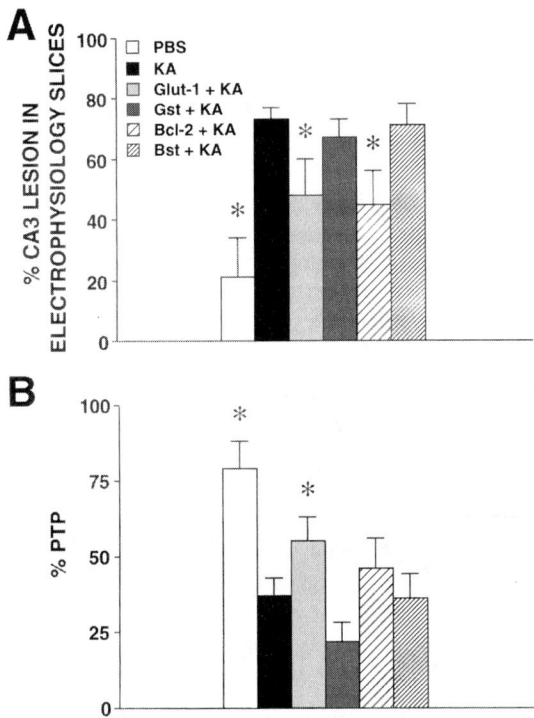

FIGURE 5. Electrophysiology is preserved after gene transfer with GLUT-1, but not BCL-2. **A.** Shows the mean CA3 lesion size in slices used for electrophysiological recording. The extent of injury in CA3 is expressed relative to the entire length of CA3. Recordings were made three days after delivery of kainic acid (KA) and vector to the dentate gyrus. **B.** Shows the mean percent PTP of the mossy fiber excitatory postsynaptic potential (EPSP) slope for each group. The stimulating electrode is placed in the dentate granule cell layer and recording electrode is in the stratum lucidum of CA3. Percent PTP for each slice was measured as the average of the first two responses following tetanization (equated to the mean change in the response within the first minute after induction). (*$p < 0.05$ vs. pooled controls; PBS, phosphate-buffered solution; KA, kainic acid; Glut-1, glucose transporter vector; Gst, control vector identical to the Glut-1, but contains a stop codon early in the sequence for Glut-1; Bst, control vector identical to the Bcl-2 vector, but contains a stop codon early in the sequence for Bcl-2; PTP, posttetanic potentiation.)

FUTURE DIRECTIONS

Gene Transfer in Other Models of Brain Injury

The main *in vivo* models studied in our laboratories include the focal cerebral ischemia model and kainic acid model. However, we have expanded some of our work to address questions such as whether gene therapy might protect against global insults, or whether gene therapy to the ischemic penumbra is possible. These questions are valid to the extent that global ischemic injury may be different from focal. Certainly, certain glutamate antagonists do not appear to be effective against global ischemia, although they are protective in focal.[62] Early clinical trials for stroke treatment have focused on targeting the ischemic penumbra, the brain region most amenable to salvage, particularly in the case of irreversible vessel occlusion.[63]

We are currently studying the potential for gene therapy in global ischemia models, a model that simulates cerebral injury due to cardiac arrest. By applying the model of bilateral carotid artery occlusion plus hypotension, we are capable of generating reproducible injury restricted to the hippocampal CA1 subsector. We applied a similar paradigm as the kainic acid model using the HSP70 vector. Our preliminary studies suggest that the number of HSP70-transfected CA1 neurons are higher than that of control vector–transfected (see FIGURE 6). Paralleling the kainic acid studies, future studies in this model may include the application of behavior testing to determine whether neuronal function is rescued with gene therapy in ischemia models.

Penumbral gene therapy makes logical sense given that a majority of pharmaceuticals are designed to target this region. In order to determine whether gene overexpression can rescue in the ischemic penumbra, we employed a model of distal MCA occlusion as described by Chen et al.[64] In this model, a reproducible cortical infarct is produced in the territory of the MCA. Since the borderzones are quite consistent between experiments, we applied vector to this region (and a comparable contralateral area in the nonischemic cortex) and used permanent distal MCA occlusion plus two hours of temporary bilateral carotid artery occlusion. From our preliminary experiments, we show that cortical penumbral vector targeting is possible (see FIGURE 7).

Gene Transfer to Alter Overall Effects of Injury

As previously indicated, the majority of our studies have focused on intracellular or membrane bound proteins whose actions are likely restricted to the local environment in which they are expressed. Given the limited number of cells the vector is capable of transfecting, overall lesion size cannot be altered by our current approach. However, future approaches may capitalize on the tendency for viral vectors to transfect the ependyma following intraventricular injection. By identifying genes whose protein product is a diffusible substance, ependymal generation and secretion of a potentially protective protein could act remotely. This has previously been shown by Betz et al.,[65] who used an adenoviral vector to overexpress the antiinflammatory cytokine, interleukin receptor antagonist (IL-1ra) within ependymal cells. Following permanent ligation of the MCA, they found that such gene therapy could attenuate overall infarct size.

Gene Transfer to Human Tissues

Certainly, a progressive goal in these studies would be to ultimately apply this technology to humans. Although there are certain safety issues that need to be addressed before such studies in clinical stroke can commence, some preliminary *ex vivo* work is promising. Using fresh human brain tissue from surgical specimens, stable metabolism could be maintained for at least 24 h by incubating in artificial CSF. Using our control vectors, we were able to transfect human brain specimens within 20 minutes of removal.[66] However, long-term studies and *in vivo* safety in humans

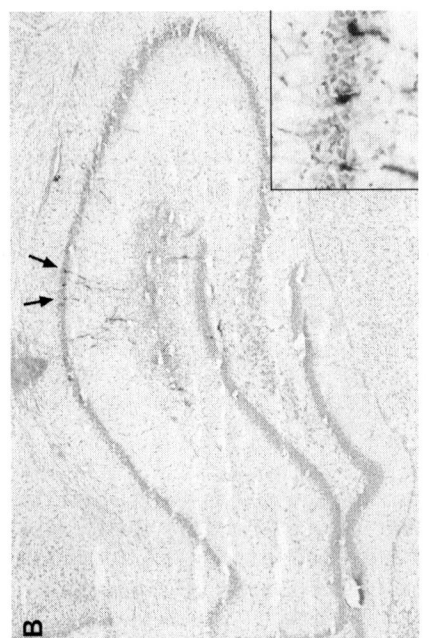
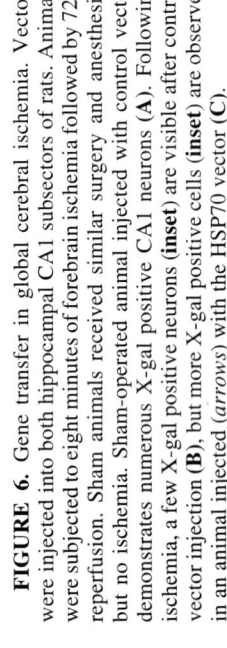
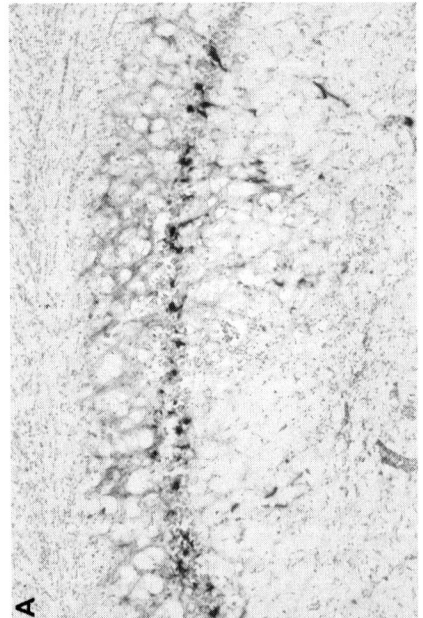
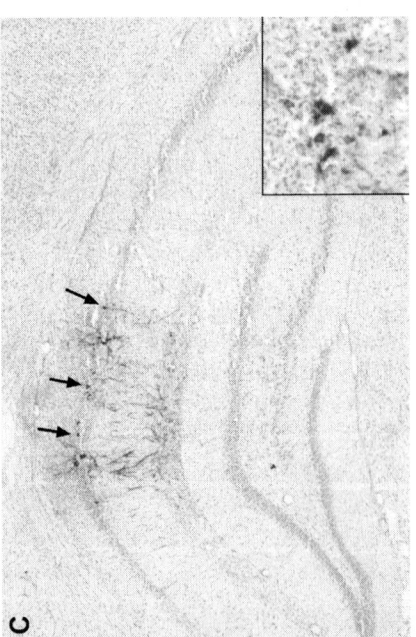

FIGURE 6. Gene transfer in global cerebral ischemia. Vectors were injected into both hippocampal CA1 subsectors of rats. Animals were subjected to eight minutes of forebrain ischemia followed by 72 h reperfusion. Sham animals received similar surgery and anesthesia, but no ischemia. Sham-operated animal injected with control vector demonstrates numerous X-gal positive CA1 neurons (**A**). Following ischemia, a few X-gal positive neurons (**inset**) are visible after control vector injection (**B**), but more X-gal positive cells (**inset**) are observed in an animal injected (*arrows*) with the HSP70 vector (**C**).

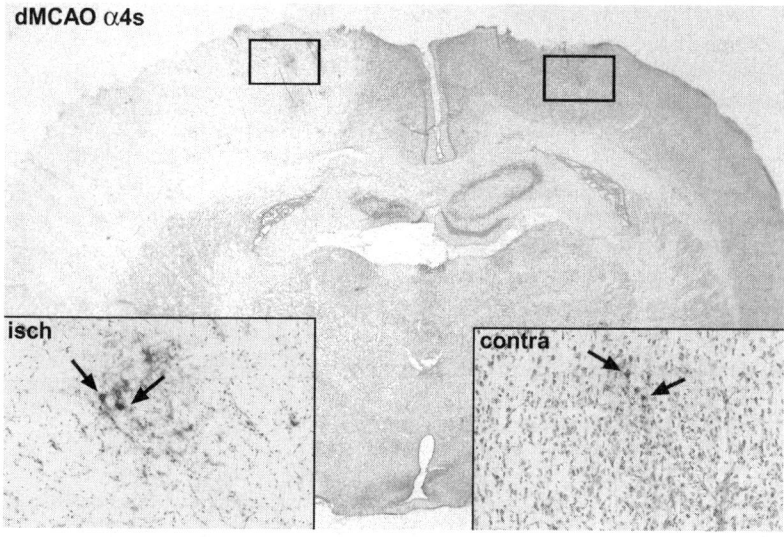

FIGURE 7. Penumbral gene transfer. Vectors were injected into penumbral zones of the cortex supplied by the middle cerebral artery (MCA). Twelve to fifteen hours later, animals were subjected to permanent distal MCA occlusion plus bilateral carotid artery occlusion for two hours. Brains were harvested 48 h later, and stained with X-gal and cresyl violet. Several transfected (*arrows*) neurons are visible within both the ischemic (**isch**) and nonischemic cortices (**contra**).

is not yet known for these vectors. Early clinical experience with adenoviral vectors for treatment of cystic fibrosis has shown that transgene expression is transient, and patients may develop an inflammatory response.[67] Whether defective HSV vectors can reactivate latent viral infection in humans is also not known. Retroviral vectors are being studied in patients with brain tumors, but the results of these studies have yet to be published. Unfortunately, the recent decision by the FDA to terminate clinical gene therapy trials due to a tragic outcome in one study has raised concerns about the safety of viral vectors.[68] The future may bring effective nonviral strategies for gene transfer such as liposomes or mechanical transfer.

CONCLUSIONS

We show that gene transfer therapy in experimental stroke is feasible, although it is currently limited by the extent to which these viral vectors can infect and the route of administration. We find that these techniques provide insights into the molecular biology and pathophysiology of cerebral ischemia that are complementary to studies employing transgenic animals. Moreover, the finding that some of the genes induced in response to cerebral injury are neuroprotective when expressed at high levels, offer new possibilities in the development of novel pharmaceuticals. We and others are beginning to show that gene transfer therapy for treatment of cerebral ischemia and variety of other diseases has potential, although a number of questions and issues remain before it can be implemented in humans.

ACKNOWLEDGMENTS

This work was supported by NIH NINDS Grants PO1 NS 37520 (G.K.S., R.M.S., M.A.Y.) and K08 NS01860 (M.A.Y.), American Heart Association Beginning Grant in Aid (M.A.Y.), the Adler Foundation (R.M.S.), Bernard and Ronni Lacroute (G.K.S.), the William Randolph Heart Foundation (G.K.S.), and John and Dodie Rosekrans (G.K.S.). We would also like to extend our thanks to the many members of the Sapolsky, Steinberg, and Yenari Laboratories at Stanford University who contributed significantly to the data discussed here.

REFERENCES

1. GREEN, D.R. & J.C. REED. 1998. Mitochondria and apoptosis. Science **281**: 1309–1312.
2. SAPOLSKY, R.M. & G.K. STEINBERG. 1999. Gene therapy using viral vectors for acute neurologic insults. Neurology **53**: 1922–1931.
3. BREDESEN, D.E. 1995. Neural apoptosis. Ann. Neurol. **38**: 839–851.
4. KLUCK, R.M., E. BOSSY-WETZEL, D.R. GREEN, *et al.* 1997. The release of cytochrome c from mitochondria: a primary site for Bcl-2 regulation of apoptosis. Science **275**: 1132–1136.
5. GLORIOSO, J.C., W.F. GOINS, C.A. MEANEY, *et al.* 1994. Gene transfer to brain using herpes simplex virus vectors. Ann. Neurol. **35**(Suppl.): S28–34.
6. GELLER, A.I., M.J. DURING & R.L. NEVE. 1991. Molecular analysis of neuronal physiology by gene transfer into neurons with herpes simplex virus vectors. Trends Neurosci. **14**: 428–432.
7. DELUCA, N.A., A.M. MCCARTHY & P.A. SCHAFFER. 1985. Isolation and characterization of deletion mutants of herpes simplex virus type 1 in the gene encoding immediate-early regulatory protein ICP4. J. Virol. **56**: 558–570.
8. FINK, D.J., N.A. DELUCA, W.F. GOINS, *et al.* 1996. Gene transfer to neurons using herpes simplex virus based vectors. Ann. Rev. Neurosci. **19**: 265–287.
9. FINK, S.L., L.K. CHANG, D.Y. HO, *et al.* 1997. Defective herpes simplex virus vectors expressing the rat brain stress-inducible heat shock protein 72 protect cultured neurons from severe heat shock. J. Neurochem. **68**: 961–969.
10. HO, D.Y. 1994. Amplicon-based herpes simplex virus vectors. Meth. Cell Biol. **43**: 191–219.
11. LAWRENCE, M.S., D.Y. HO, G.H. SUN, *et al.* 1996. Overexpression of Bcl-2 with herpes simplex virus vectors protects CNS neurons against neurological insults *in vitro* and *in vivo*. J. Neurosci. **16**: 486–496.
12. HO, D.Y., M.S. LAWRENCE, T.J. MEIER, *et al.* 1995. Using of herpes virus vectors for protection from necrotic neuron death. *In* Viral Vectors. M.G. Kaplitt & A.D. Loewy, Eds.: 133–155. Academic Press, New York.
13. HO, D.Y., T.C. SAYDAM, S.L. FINK, *et al.* 1995. Defective herpes simplex virus vectors expressing the rat brain glucose transporter protect cultured neurons from necrotic insults. J. Neurochem. **65**: 842–850.
14. YENARI, M.A., S.L. FINK, M.S. LAWRENCE, *et al.* 1998. Gene transfer therapy for cerebral ischemia. *In* Pharmacology of Cerebral Ischemia. J. Krieglstein, Ed.: 453–465. Medpharm Scientific Publishers, Stuttgart.
15. YENARI, M.A., S.L. FINK, G.H. SUN, *et al.* 1998. Gene therapy with HSP72 is neuroprotective in rat models of stroke and epilepsy. Ann. Neurol. **44**: 584–591.

16. Ho, D.Y., E.S. Mocarski & R.M. Sapolsky. 1993. Altering central nervous system physiology with a defective herpes simplex virus vector expressing the glucose transporter gene. Proc. Natl. Acad. Sci. U.S.A. **90:** 3655–3659.
17. Lawrence, M.S., G.H. Sun, D.M. Kunis, et al. 1996. Overexpression of the glucose transporter gene with a herpes simplex viral vector protects striatal neurons against stroke. J. Cereb. Blood Flow Metab. **16:** 181–185.
18. Fink, S.L., D.Y. Ho, J. McLaughlin, et al. 2000. An adenoviral vector expressing the glucose transporter protects cultured striatal neurons from 3-nitropropionic acid. Brain Res. **859:** 21–25.
19. Dash, R., M. Lawrence, D. Ho, et al. 1996. A herpes simplex virus vector overexpressing the glucose transporter gene protects the rat dentate gyrus from an antimetabolite toxin. Exp. Neurol. **137:** 43–48.
20. Lawrence, M.S., D.Y. Ho, R. Dash, et al. 1995. Herpes simplex virus vectors overexpressing the glucose transporter gene protect against seizure-induced neuron loss. Proc. Natl. Acad. Sci. U.S.A. **92:** 7247–7251.
21. Robert, J.J., V. Bouilleret, V. Ridoux, et al. 1997. Adenovirus-mediated transfer of a functional GAD gene into nerve cells: potential for the treatment of neurological diseases. Gene Ther. **4:** 1237–1245.
22. Meier, T.J., D.Y. Ho, T.S. Park, et al. 1998. Gene transfer of calbindin D28k cDNA via herpes simplex virus amplicon vector decreases cytoplasmic calcium ion response and enhances neuronal survival following glutamatergic challenge but not following cyanide. J. Neurochem. **71:** 1013–1023.
23. Meier, T.J., D.Y. Ho & R.M. Sapolsky. 1997. Increased expression of calbindin D28k via herpes simplex virus amplicon vector decreases calcium ion mobilization and enhances neuronal survival after hypoglycemic challenge. J. Neurochem. **69:** 1039–1047.
24. Yenari, M.A., M. Minami, G.H. Sun, et al. 2001. Calbindin D28K overexpression protects striatal neurons from transient focal cerebral ischemia. Stroke **32:** 1028–1035.
25. Kindy, M., J. Yu, R. Miller, et al. 1996. Adenoviral vectors in ischemic injury. In Pharmacology of Cerebral Ischemia. J. Krieglstein, Eds. Medpharm Scientific Publishers, Stuttgart.
26. Phillips, R.G., T.J. Meier, L.C. Giuli, et al. 1999. Calbindin D28K gene transfer via herpes simplex virus amplicon vector decreases hippocampal damage *in vivo* following neurotoxic insults. J. Neurochem. **73:** 1200–1205.
27. Goodman, J.H., C.G. Wasterlain, W.F. Massarweh, et al. 1993. Calbindin-D28k immunoreactivity and selective vulnerability to ischemia in the dentate gyrus of the developing rat. Brain Res. **606:** 309–314.
28. Mattson, M.P., B. Rychlik, C. Chu, et al. 1991. Evidence for calcium-reducing and excito-protective roles for the calcium-binding protein calbindin-D28k in cultured hippocampal neurons. Neuron **6:** 41–51.
29. Mattson, M.P., B. Cheng, S.A. Baldwin, et al. 1995. Brain injury and tumor necrosis factors induce calbindin D-28k in astrocytes: evidence for a cytoprotective response. J. Neurosci. Res. **42:** 357–370.
30. Chard, P.S., J. Jordan, C.J. Marcuccilli, et al. 1995. Regulation of excitatory transmission at hippocampal synapses by calbindin D28k. Proc. Natl. Acad. Sci. U.S.A. **92:** 5144–5148.
31. Nowak, T.S., Jr. & M. Jacewicz. 1994. The heat shock/stress response in focal cerebral ischemia. Brain Pathol. **4:** 67–76.
32. Chen, J., S.H. Graham, R.L. Zhu, et al. 1996. Stress proteins and tolerance to focal cerebral ischemia. J. Cereb. Blood Flow Metab. **16:** 566–577.
33. Chopp, M., Y. Li, M.O. Dereski, et al. 1991. Neuronal injury and expression of 72-kDa heat-shock protein after forebrain ischemia in the rat. Acta Neuropathol. (Berl.) **83:** 66–71.
34. Xu, L. & R.G. Giffard. 1997. HSP70 protects murine astrocytes from glucose deprivation injury. Neurosci. Lett. **224:** 9–12.
35. Papadopoulos, M.C., X.Y. Sun, J. Cao, et al. 1996. Over-expression of HSP-70 protects against combined oxygen-glucose deprivation. Neuroreport **7:** 429–432.

36. PLUMIER, J.C., A.M. KRUEGER, R.W. CURRIE, et al. 1997. Transgenic mice expressing the human inducible Hsp70 have hippocampal neurons resistant to ischemic injury. Cell Stress Chaperones **2:** 162–167.
37. RAJDEV, S., K. HARA, Y. KOKUBO, et al. 2000. Mice overexpressing rat heat shock protein 70 are protected against cerebral infarction. Ann. Neurol. **47:** 782–791.
38. YENARI, M.A., J.E. LEE, M. EMOND, et al. 1999. Transgenic mice which overexpress HSP70 are protected against some but not all central nervous system insults. Stroke **30:** 233.
39. YENARI, M.A., R.G. GIFFARD, R.M. SAPOLSKY, et al. 1999. The neuroprotective potential of heat shock protein 70 (HSP70). Mol. Med. Today **5:** 525–531.
40. YANG, J., X. LIU, K. BHALLA, et al. 1997. Prevention of apoptosis by Bcl-2: release of cytochrome c from mitochondria blocked. Science **275:** 1129–1132.
41. ZHONG, L.T., T. SARAFIAN, D.J. KANE, et al. 1993b. bcl-2 inhibits death of central neural cells induced by multiple agents. Proc. Natl. Acad. Sci. U.S.A. **90:** 4533–4537.
42. BEHL, C., L.D. HOVEY, S. KRAJEWSKI, et al. 1993. BCL-2 prevents killing of neuronal cells by glutamate but not by amyloid beta protein. Biochem. Biophys. Res. Commun. **197:** 949–956.
43. JIA, W.W., Y. WANG, D. QIANG, et al. 1996. A bcl-2 expressing viral vector protects cortical neurons from excitotoxicity even when administered several hours after the toxic insult. Brain Res. Mol. Brain Res. **42:** 350–353.
44. KANE, D.J., T. ORD, R. ANTON, et al. 1995. Expression of bcl-2 inhibits necrotic neural cell death. J. Neurosci. Res. **40:** 269–275.
45. ZHONG, L.T., D.J. KANE & D.E. BREDESEN. 1993. BCL-2 blocks glutamate toxicity in neural cell lines. Brain Res. Mol. Brain Res. **19:** 353–355.
46. MAH, S.P., L.T. ZHONG, Y. LIU, et al. 1993. The protooncogene bcl-2 inhibits apoptosis in PC12 cells. J. Neurochem. **60:** 1183–1186.
47. KANE, D.J., T.A. SARAFIAN, R. ANTON, et al. 1993. Bcl-2 inhibition of neural death: decreased generation of reactive oxygen species. Science **262:** 1274–1277.
48. ALLSOPP, T.E., S. WYATT, H.F. PATERSON, et al. 1993. The proto-oncogene bcl-2 can selectively rescue neurotrophic factor-dependent neurons from apoptosis. Cell **73:** 295–307.
49. CHEN, J., S.H. GRAHAM, M. NAKAYAMA, et al. 1997. Apoptosis repressor genes Bcl-2 and Bcl-x-long are expressed in the rat brain following global ischemia. J. Cereb. Blood Flow Metab. **17:** 2–10.
50. GILLARDON, F., C. LENZ, K.F. WASCHKE, et al. 1996. Altered expression of Bcl-2, Bcl-X, Bax, and c-Fos colocalizes with DNA fragmentation and ischemic cell damage following middle cerebral artery occlusion in rats. Brain Res. Mol. Brain Res. **40:** 254–260.
51. MATSUSHITA, K., T. MATSUYAMA, K. KITAGAWA, et al. 1998. Alterations of Bcl-2 family proteins precede cytoskeletal proteolysis in the penumbra, but not in infarct centres following focal cerebral ischemia in mice. Neuroscience **83:** 439–448.
52. LAWRENCE, M.S., G.H. SUN, D.Y. HO, et al. 1997. Herpes simplex viral vectors expressing Bcl-2 are neuroprotective when delivered following a stroke. J. Cereb. Blood Flow Metab. **17:** 740–744.
53. LINNIK, M.D., P. ZAHOS, M.D. GESCHWIND, et al. 1995. Expression of bcl-2 from a defective herpes simplex virus-1 vector limits neuronal death in focal cerebral ischemia. Stroke **26:** 1670–1674.
54. MARTINOU, J.C., M. DUBOIS-DAUPHIN, J.K. STAPLE, et al. 1994. Overexpression of BCL-2 in transgenic mice protects neurons from naturally occurring cell death and experimental ischemia. Neuron **13:** 1017–1030.
55. PHILLIPS, R.G., M.S. LAWRENCE, D.Y. HO, et al. 2000. Limitations in the neuroprotective potential of gene therapy with Bcl-2. Brain Res. **859:** 202–206.
56. YENARI, M.A., M.S. LAWRENCE, G.H. SUN, et al. 1996. Herpes simplex viral vectors expressing Bcl-2 are neuroprotective against focal cerebral ischemia. *In* Pharmacology of Cerebral Ischemia. J. Krieglstein, Ed.: 537–543. Wissenschaftliche Verlagsgesellschaft mbH, Stuttgart.
57. SHARP, F.R., S.M. MASSA & R.A. SWANSON. 1999. Heat-shock protein protection. Trends Neurosci. **22:** 97–99.

58. JAATTELA, M., D. WISSING, K. KOKHOLM, et al. 1998. Hsp70 exerts its anti-apoptotic function downstream of caspase-3-like proteases. EMBO J. **17:** 6124–6134.
59. RINGER, T., S.L. FINK, D.Y. HO, et al. 2000. HSP72 overexpression protects striatal neurons when delivered after experimental stroke. Soc. Neurosci. Abst. **26:** 15.
60. DUMAS, T., J. MCLAUGHLIN, D. HO, et al. 1999. Delivery of herpes simplex virus amplicon-based vectors to the dentate gyrus does not alter hippocampal synaptic transmission *in vivo*. Gene Ther. **6:** 1679–1684.
61. MCLAUGHLIN, J., B. ROOZENDAAL, T. DUMAS, et al. 2000. Sparing of neuronal function post-seizure with gene therapy. Proc. Natl. Acad. Sci. U.S.A. **97:** 12804–12809.
62. YENARI, M.A. & G.K. STEINBERG. 1998. Pharmacological advances in cerebrovascular protection. *In* Current Techniques in Neurosurgery. I.A. Awad, Ed.: 98–116. Current Science, Philadelphia.
63. DE KEYSER, J., G. SULTER & P.G. LUITEN. 1999. Clinical trials with neuroprotective drugs in acute ischaemic stroke: are we doing the right thing? Trends Neurosci. **22:** 535–540.
64. CHEN, S.T., C.Y. HSU, E.L. HOGAN, et al. 1986. A model of focal ischemic stroke in the rat: reproducible extensive cortical infarction. Stroke **17:** 738–743.
65. BETZ, A.L., G.Y. YANG & B.L. DAVIDSON. 1995. Attenuation of stroke size in rats using an adenoviral vector to induce overexpression of interleukin-1 receptor antagonist in brain. J. Cereb. Blood Flow Metab. **15:** 547–551.
66. BOTTINO, C.J., S.A. HOWARD, G. STEINBERG, et al. 2000. Model for examining neuroprotection in human brain tissue. Soc. Neurosci. Abs. **26:** 266.
67. CRYSTAL, R.G. 1995. Transfer of genes to humans: early lessons and obstacles to success. Science **270:** 404–410.
68. BALTER, M. 2000. Gene therapy on trial. Science **288:** 951–957.

Hypothermia-Induced Ischemic Tolerance

Characteristics and Candidate Mechanisms

KEVIN S. LEE, MATTHEW J. ANZIVINO, MASATOSHI YUNOKI, AND DANIEL DECKER

Departments of Neuroscience and Neurological Surgery, University of Virginia, Charlottesville, Virginia, U.S.A.

KEYWORDS: Hypothermia; Ischemic tolerance; Characteristics.

Delayed tolerance to ischemic injury has been described in response to a variety of preconditioning stimuli. However, most of these preconditioning stimuli are inherently dangerous and/or possess limited safety margins. Recently, we have shown that hypothermia can serve as a safe and effective preconditioning stimulus for inducing ischemic tolerance. Elsewhere we have reviewed current findings from our laboratory concerning fundamental characteristics of hypothermia-induced tolerance. The timing, depth and duration of hypothermia that is capable of establishing tolerance has been described, as have ongoing studies aimed at elucidating cellular and molecular mechanisms of delayed tolerance. Microarray analyses are being used to identify consensus alterations in gene expression induced by disparate preconditioning stimuli. The consensus list of candidate genes is being used to evaluate beneficial and deleterious changes in gene products that occur during the period of tolerance. Identification of known and novel genes contributing to tolerance will facilitate the future development of gene-based therapies for treating ischemic neuronal injury.

Effect of L-Carnitine Pretreatment on 3-Nitropropionic Acid–Induced Inhibition of Rat Brain Succinate Dehydrogenase Activity

ZBIGNIEW K. BINIENDA, NATALYA V. SADOVOVA, ROBERT L. ROUNTREE, ANDREW C. SCALLET, AND SYED F. ALI

Division of Neurotoxicology, National Center for Toxicological Research/FDA, Jefferson, Arkansas, U.S.A.

ABSTRACT: L-Carnitine (LC) plays an important regulatory role in the mitochondrial transport of long chain free fatty acids (FFA). 3-Nitropropionic acid (3-NPA) is known to induce cellular energy deficit and oxidative stress–related neurotoxicity via an irreversible inhibition of mitochondrial succinate dehydrogenase (SDH). In the present study, activity of SDH was measured in order to evaluate neuroprotective effects of LC against the 3-NPA-induced neurotoxicity. Male, CD Sprague-Dawley rats, three months old, were injected with either 50 or 100 mg/kg of LC, i.p., 30 min prior to 3-NPA (30 mg/kg, s.c.) or with 3-NPA alone. The activity of brain SDH was quantified spectrophotometrically in caudate nucleus (CN), frontal cortex (FC), and hippocampus (HIP) 60 min after the 3-NPA injection. The SDH activity in the animals treated with 3-NPA alone was 38% (CN), 50% (FC), and 36% (HIP) that of saline controls. Pretreatment with LC prior to 3-NPA injection attenuated decreases of SDH activity by approximately 15 and 29% (LC low and high dose, respectively). Despite the attenuation of SDH inhibition, the activity of SDH in these regions remained significantly lower in treated than in control rats ($p < 0.05$). It appears that the protective effect of LC against 3-NPA-induced oxidative stress cannot be explained by the direct action of LC to interfere with the SDH inhibition but are rather achieved by LC actions downstream of the SDH inhibition.

KEYWORDS: L-Carnitine; Neuroprotection; Rat brain; Succinate dehydrogenase activity.

INTRODUCTION

The L-lysine derivative, L-carnitine (3-hydroxy-4-trimethylaminobutyric acid, LC), plays an important regulatory role in the mitochondrial transport of long chain free fatty acids (FFA). The metabolism of FFA via β-oxidation is a key source of energy in peripheral tissues. Despite low level of β-oxidation in brain, LC is actively transported through the blood–brain barrier and accumulates in neural cells.[1,2] A major modulatory role for LC and its acetylated form, acetyl-L-carnitine (ALC), in neural function may be played through the carnitine-mediated transfer of acetyl

Address for correspondence: Dr. Zbigniew K. Binienda, Division of Neurotoxicology, HFT-132, FDA/NCTR, Jefferson, AR 72079-9502, U.S.A. Voice: 870-543-7920; fax: 870-543-7745.
   zbinienda@nctr.fda.gov

groups for acetylcholine synthesis, as well as its influence on signal transduction pathways and gene expression.[1,3,4]

The ALC administration following ischemia-hypoxia has been reported to lower lactate and induce a faster recovery of cerebral ATP.[5] In the situation of cellular energy deficit, LC and ALC modulate cerebral metabolism of energy sources other than glucose, ketone bodies and FFA.[6] It has also been shown that pretreatment with L-carnitine prevents mitochondrial damage induced in the rat choroid plexus by medium chain fatty acids.[7] Chronic ALC administration increased specific enzyme activities like cytochrome oxidase in brain cerebral cortex.[8]

The *in vitro* study has shown that LC attenuated cell damage, induced by the mitochondrial succinate dehydrogenase (SDH) inhibitor, 3-nitropropionic acid (3-NPA), as assessed by lactate dehydrogenase (LDH) release in cultured rat cortical cells.[9] We showed elsewhere a protective effect of LC in adult rats exposed to 3-NPA, assessed by the activity of antioxidant enzymes.[10] In the study reported here, activity of mitochondrial SDH was measured in order to evaluate neuroprotective effects of LC against 3-NPA-induced neurotoxicity.

MATERIAL AND METHODS

Animal Preparation

Adult, male, CD Sprague-Dawley rats were obtained from the FDA/National Center for Toxicological Research breeding colony. They were housed individually with *ad libitum* access to food and water and kept under controlled environmental conditions (temperature 22°C, relative humidity 45–55%, 12-h light/dark cycle with lights out at 18:00 h). Rats were separated into control ($n = 3$), 3-NPA ($n = 5$), L-carnitine low dose ($n = 5$), and L-carnitine high dose ($n = 5$) groups. They were injected with 3-NPA alone (30 mg/kg s.c.) or either a low (50 mg/kg) or high (100 mg/kg) dose of L-carnitine administered i.p. 30 min prior to 3-NPA. Control rats received vehicle (0.1 M potassium phosphate buffer). All animals were sacrificed by decapitation 60 min after 3-NPA administration. Brains were removed immediately, dissected into CN, FC, and HIP on dry ice, and stored at −80°C until the enzyme analysis.[11]

Preparation of Brain Mitochondria

The preparation of rat brain mitochondria was carried out at 4°C, as described by Ringler and Singer.[12] The brain tissue was homogenized for 12–15 sec with a Sonifier cell disrupter model W-350 in five volumes (based on original starting material) of ice cold 0.25 M sucrose, previously adjusted to pH 7.6 with 1 M K_2HPO_4. The resulting homogenate was centrifuged at $1,600\,g$ for 12 min in a Hermle model Z 360K centrifuge. The supernatant was decanted and centrifuged at $20,000\,g$ for 12 min in an Eppendorf model 5810R centrifuge. The supernatant from this centrifugation was discarded, and the residue was washed once with the homogenizing medium and again centrifuged at $20,000\,g$ for 20 min. The mitochondrial suspension was subjected to two cycles of freezing and thawing to abolish the permeability barrier of mammalian mitochondria to phenasine methosulphate (PMS). The protein content was determined by the Bradford method.[13] The enzyme activity was expressed as milligram of formazan formed per minute per milligram protein.

Enzyme Assay

The SDH activity was measured spectrophotometrically using PMS-INT (2-p-iodophenyl-3p nitrophenyl-5-phenyltetrazolium) as the electron acceptor system, as described elsewhere.[14,15] The reaction mixture was prepared by adding the stock solutions of phosphate buffer (0.40 ml), succinate (0.05 ml), KCN (0.05 ml), tetrazolium salt (0.25 ml), and gelatin (0.1 ml) into 10-ml glass tubes. After shaking, the tubes were placed in a constant temperature water bath at 37°C; then PMS (0.05 ml) and mitochondrial suspension (0.05 ml) were added. The incubations were carried out for 15 min. The reaction was stopped by transferring the tubes into ice. The color density was read within the following 10–30 min in a Beckman model DU 640 spectrophotometer at A 510 nm, using a 1-cm optical path cuvette. Since some color developed in the absence of succinate, the absorbance in experiment was corrected by subtracting the values for blanks, which consisted of all components of the incubation mixture except succinate. All determinations were made in duplicate. A calibration curve was obtained as described elsewhere.[14]

Statistical Analysis

Data were analyzed using one way analysis of variance (ANOVA) followed by the Dunnett correction for multiple comparisons to detect differences between control and treated rats with significance indicated when $p < 0.05$.

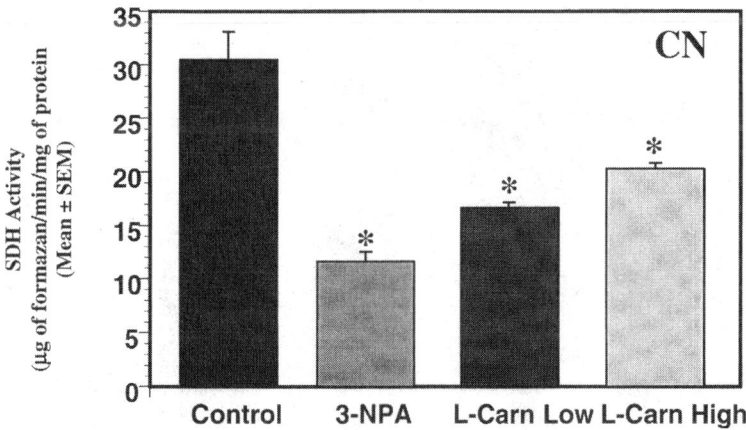

FIGURE 1. Effect of pretreatment with L-carnitine on the 3-NPA inhibition of succinate dehydrogenase activity in the caudate nucleus (CN) of adult male rats. Control, vehicle; 3-NPA, 3-nitropropionic acid (30 mg/kg s.c.); L-Carn Low, low dose L-carnitine (50 mg/kg i.p.); L-Carn High, high dose L-carnitine (100 mg/kg i.p.); injected 30 min prior to 3-NPA administration. Mean ± SEM *$p < 0.05$ significantly different from control.

RESULTS

The SDH activity evaluated by spectrophotometry in the CN of control rats was 30.4 µg/min/mg. The activity decreased to 38% of control level in rats treated with 3-NPA alone. The inhibition was attenuated by pretreatment with LC at both low and high dose to 55% and 66%, respectively (see FIGURE 1). Compared with control, the SDH activity in FC of rats exposed to 3-NPA alone was inhibited to 50% (see FIGURE 2). Pretreatment with LC at low and high doses prior to 3-NPA attenuated the decreases to 62% and 79%, respectively. In the HIP, 3-NPA treatment led to inhibition of SDH activity to 36% of control level that was attenuated by pretreatment with LC, at both low and high dose, to 50% and 65%, respectively (see FIGURE 3). Despite the attenuation, activity of SDH in these regions remained significantly lower than control ($p < 0.05$).

DISCUSSION AND CONCLUSIONS

The primary mechanism of 3-NPA toxic action involves the irreversible inhibition of succinate dehydrogenase (SDH) present in the Krebs cycle and complex II of the mitochondrial electron transport chain.[16] Bioenergetic defects are critical in a neurotoxic cascade of events that trigger the activation of NMDA receptors leading to intracellular calcium overload, generation of radical oxygen species, neuronal damage, and death. Inhibition of SDH activity interferes with the electron cascade, and interrupts oxidative phosphorylation, thus leading to mitochondrial dysfunction and cellular energy deficit. Likewise, 3-NPA induced inhibition of SDH resulted in a decrease in ATP production,[17] and oxidative stress has been shown to play a role in 3-NPA-induced neurotoxicity.[18,19]

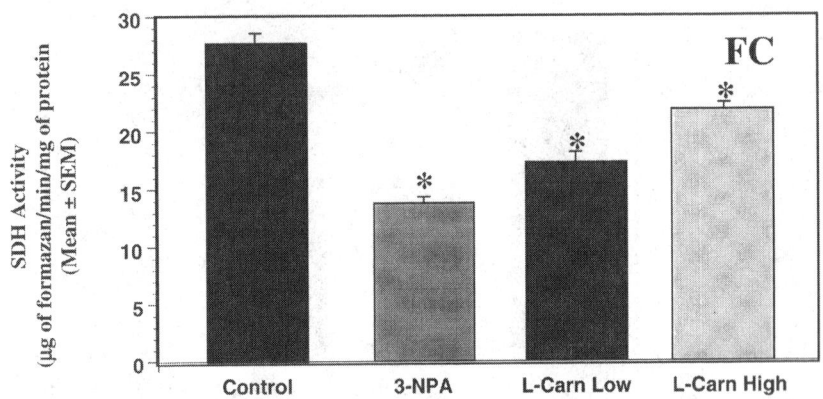

FIGURE 2. Effect of pretreatment with L-carnitine on the 3-NPA inhibition of succinate dehydrogenase activity in the hippocampus (HIP) of adult male rats. Control, vehicle; 3-NPA, 3-nitropropionic acid (30 mg/kg s.c.); L-Carn Low, low dose L-carnitine (50 mg/kg i.p.); L-Carn High, high dose L-carnitine (100 mg/kg i.p.); injected 30 min prior to 3-NPA administration. Mean ± SEM *$p < 0.05$ significantly different from control.

It appears that the protective mechanism of carnitines has a multifactorial character. Promising neuroprotective effects of carnitines were shown during metabolic inhibition induced by ammonia intoxication[20] or ischemia-hypoxia.[5] The authors suggest that neuroprotective effects were achieved through mitochondrial energy enhancement by using alternative energy sources, such as ketone bodies. As the concentration of carnitines decrease in ischemic brain, therapeutic administration of LC and ALC might enhance oxidative metabolism preventing lactate accumulation. This effect may not be sufficient to overcome the severe insult. Studies have failed to show positive effect of LC on infarct size volume in focal cerebral ischemia.[21] On the other hand, behavioral improvement observed in rats exposed to cyanide was not associated with increased brain metabolism. In this model, increased use of FFA prevented accumulation of fatty acyl-Co-As and, therefore, the preservation of membrane phospholipids was proposed to underlie the observed neuroprotection.[22]

We have found that pretreatment with L-carnitine prior to 3-NPA exposure prevented the oxidative stress–associated increase in the activity of free radical–scavenging enzymes observed 60–120 min after the administration of 3-NPA alone.[10] Data gathered in our study, suggest that intraperitoneal L-carnitine pretreatment prior to 3-NPA injection may prevent, to some extent, the SDH inhibition induced by 3-NPA. On the other hand, the attenuation of 3-NPA inhibition did not prevent the activity of brain SDH from falling significantly below the control level. The study has shown that chronic partial inhibition of brain SDH by 3-NPA to 50–60% of control values was sufficient to induce striatal degeneration.[23]

The LC was protective to the rat cortical cells against 3-NPA neurotoxicity[9] and as shown here, it did not significantly prevent SDH inhibition. Therefore, it appears that direct or indirect interference of LC with 3-NPA does not play a major role in the mechanism underlying protective effect of LC against 3-NPA toxicity. In this

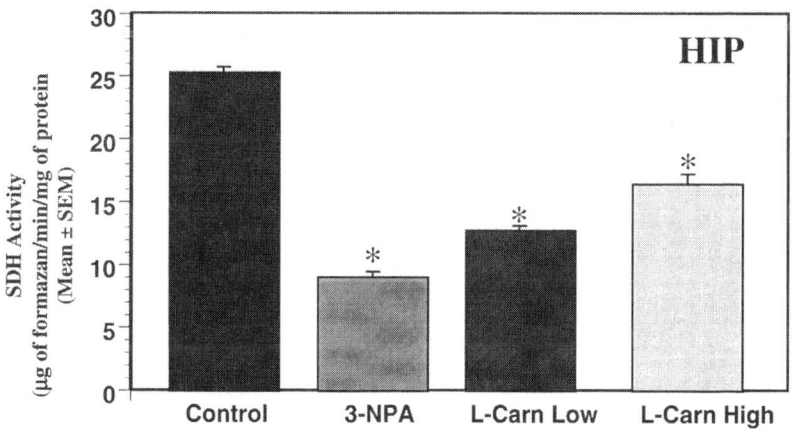

FIGURE 3. Effect of pretreatment with L-carnitine on the 3-NPA inhibition of succinate dehydrogenase activity in the frontal cortex (FC) of adult male rats. Control, vehicle; 3-NPA, 3-nitropropionic acid (30 mg/kg s.c.); L-Carn Low, low dose L-carnitine (50 mg/kg i.p.); L-Carn High, high dose L-carnitine (100 mg/kg i.p.); injected 30 min prior to 3-NPA administration. Mean ± SEM *$p < 0.05$ significantly different from control.

case, neuroprotection is, rather, achieved due to LC actions downstream of the SDH inhibition that result in enhancement of mitochondrial energy metabolism and prevention of oxidative stress. Other factors may play a role in the complex mechanism of neuroprotection mediated by carnitines. Possible wide therapeutic use of carnitines merits further research on those compounds.

REFERENCES

1. SHUG, A.L., M.J. SCHMIDT, G.T. GOLDEN & R.T. FARIELLO. 1982. The distribution and role of carnitine in the mammalian brain. Life Sci. **31:** 2869–2874.
2. MROCZKOWSKA, J.E., H.-J. GALLA, M.J. NALECZ & K.A. NALECZ. 1997. Evidence for an asymmetrical uptake of L-carnitine in the blood–brain barrier in vitro. Biochem. Biophys. Res. Comm. **241:** 127–131.
3. WAWRZENCZYK, A., K.A. NALECZ & M.J. NALECZ. 1994. Synergistic effect of choline and carnitine on acetylcholine synthesis in neuroblastoma NB-2a cells. Biochem. Biophys. Res. Commun. **202:** 354–359.
4. NALECZ, K.A. & M.J. NALECZ. 1996. Carnitine—a known compound, a novel function in neural cells. Acta Neurobiol. Exp. **56:** 597–609.
5. AURELI, T., A. MICCHELI, M.E. DI COCCO, et al. 1994. Effect of acetyl-L-carnitine on recovery of brain phosphorus metabolites and lactic acid level during reperfusion after cerebral ischemia in the rat. Study by 31P- and 1H-NMR spectroscopy. Brain Res. **643:** 92–99.
6. AURELI, T., M.E. DI COCCO, C. PUCCETTI, et al. 1998. Acetyl-L-carnitine modulates glucose metabolism and stimulates glycogen synthesis in rat brain. Brain Res. **796:** 75–81.
7. KIM, C.S., C.R. ROE & W.W. AMBROSE. 1990. L-Carnitine prevents mitochondrial damage induced by octanoic acid in the rat choroid plexus. Brain Res. **536:** 335–338.
8. GORINI, A., A. D'ANGELO & R.F. VILLA. 1998. Action of L-acetylcarnitine on different cerebral mitochondrial populations from cerebral cortex. Neurochem. Res. **23:** 1485–1491.
9. VIRMANI, M.A., R. BISELLI, A. SPADONI, et al. 1995. Protective actions of L-carnitine and acetyl-L-carnitine on the neurotoxicity evoked by mitochondrial uncoupling or inhibitors. Pharmacol. Res. **32:** 383–389.
10. BINIENDA, Z., J.R. JOHNSON, A.A. TYLER-HASHEMI, et al. 1999. Protective effect of L-carnitine in the neurotoxicity induced by the mitochondrial inhibitor 3-nitropropionic acid (3-NPA). Ann. N.Y. Acad. Sci. **890:** 173–178.
11. GLOWINSKI, J. & L.L. IVERSON. 1966. Regional studies of catecholamines in the rat brain. I. The disposition of ^{3}H-norepinephrine, ^{3}H-dopamine, and ^{3}H-dopa in various regions of the brain. J. Neurochem. **13:** 655–669.
12. RINGLER, R.L. & T.P. SINGER. 1962. Alpha-L-Glycerophosphate dehydrogenase from pig brain. In Methods in Enzymology, Vol.5. R.W. Eastbrook & M.E. Pullman, Eds.: 432–439. Academic Press, New York.
13. BRADFORD, M.M. 1976. A rapid and sensitive method for the quantitation of microgram quantities of protein utilizing the principle of protein-dye binding. Analyt. Biochem. **72:** 248–254.
14. NACHLAS, M.M., S.I. MARGULIES & A.M. SELIGMAN. 1960. A colorimetric method for the estimation of succinic dehydrogenase activity. J. Biol. Chem. **235:** 499–503.
15. PENNINGTON, R.J. 1961. Mitochondrial succinate-tetrazolium reductase and adenosine triphosphatase. Biochem. J. **80:** 649–655.
16. ALSTON, T.A., MELA, L. & H.J. BRIGHT. 1977. 3-nitropropionic acid, the toxic substance of Indigofera, is a suicide inactivator of succinate dehydrogenase. Proc. Natl. Acad. Sci. U.S.A. **74:** 3767–3771.
17. LUDOLPH, A.C., M. SEELIG, A. LUDOLPH, et al. 1992. 3-Nitropropionic acid decreases cellular energy levels and causes neuronal degeneration in cortical explants. Neurodegeneration **1:** 155–161.

18. BEAL, M.F., E. BROUILLET, B.G. JENKINS, et al. 1993. Neurochemical and histologic characterization of striatal excitotoxic lesions produced by the mitochondrial toxin 3-nitropropionic acid. J. Neurosci. **13:** 4181–4192.
19. BINIENDA, Z., C. SIMMONS, S. HUSSAIN, et al. 1998. Effect of acute exposure to 3-nitropropionic acid on activities of endogenous antioxidants in the rat brain. Neurosci. Lett. **251:** 173–176.
20. BELLEI, M., D. BATTELLI, D.M. GUARRIERO, et al. 1989. Changes in mitochondrial activity caused by ammonium salts and the protective effect of carnitine. Biochem. Biophys. Res. Comm. **158:** 181–188.
21. SLIVKA, A., D. SILBERSWEIG & W. PULSINELLI. 1990. Carnitine treatment for stroke in rats. Stroke **21:** 808–811.
22. BLOKLAND, A., J. BOTHMER, W. HONIG & J. JOLLES. 1993. Behavioral and biochemical effects of acute central metabolic inhibition: effects of acetyl-l-carnitine. Eur. J. Pharmacol. **235:** 275–281.
23. BROUILLET, E., M.-C. GUYOT, V. MITTOUX, et al. 1998. Partial inhibition of brain succinate dehydrogenase by 3-nitropropionic acid is sufficient to initiate striatal degeneration in rat. J. Neurochem. **70:** 794–805.

Methamphetamine-Induced Dopaminergic Neurotoxicity: Role of Peroxynitrite and Neuroprotective Role of Antioxidants and Peroxynitrite Decomposition Catalysts

SYED Z. IMAM,[a] JAMAL EL-YAZAL,[a] GLENN D. NEWPORT,[a]
YOSSEF ITZHAK,[b] JEAN L. CADET,[c] WILLIAM SLIKKER, JR.,[a]
AND SYED F. ALI[a]

[a]*Neurochemistry Laboratory, Division of Neurotoxicology,
National Center for Toxicological Research/FDA, Jefferson, Arkansas, U.S.A.*

[b]*Department of Psychiatry and Behavioral Sciences,
University of Miami School of Medicine, Miami, Florida, U.S.A.*

[c]*Molecular Neuropsychiatry Section, National Institute on Drug Abuse,
Intramural Research Program, Baltimore, Maryland, U.S.A.*

ABSTRACT: Oxidative stress, reactive oxygen (ROS), and nitrogen (RNS) species have been known to be involved in a multitude of neurodegenerative disorders such as Parkinson's disease (PD), Alzheimer's disease (AD), and amyotrophic lateral sclerosis (ALS). Both ROS and RNS have very short half-lives, thereby making their identification very difficult as a specific cause of neurodegeneration. Recently, we have developed a high performance liquid chromatography/electrochemical detection (HPLC/EC) method to identify 3-nitrotyrosine (3-NT), an *in vitro* and *in vivo* biomarker of peroxynitrite production, in cell cultures and brain to evaluate if an agent-driven neurotoxicity is produced by the generation of peroxynitrite. We show that a single or multiple injections of methamphetamine (METH) produced a significant increase in the formation of 3-NT in the striatum. This formation of 3-NT correlated with the striatal dopamine depletion caused by METH administration. We also show that PC12 cells treated with METH has significantly increased formation of 3-NT and dopamine depletion. Furthermore, we report that pretreatment with antioxidants such as selenium and melatonin can completely protect against the formation of 3-NT and depletion of striatal dopamine. We also report that pretreatment with peroxynitrite decomposition catalysts such as 5, 10,15,20-tetrakis(*N*-methyl-4′-pyridyl)porphyrinato iron III (FeTMPyP) and 5, 10, 15, 20-tetrakis (2,4,6-trimethyl-3,5-sulfonatophenyl) porphinato iron III (FETPPS) significantly protect against METH-induced 3-NT formation and striatal dopamine depletion. We used two different approaches, pharmacological manipulation and transgenic animal models, in order to further investigate the role of peroxynitrite. We show that a selective neuronal nitric oxide synthase (nNOS) inhibitor, 7-nitroindazole (7-NI), significantly protect against the formation of 3-NT as well as striatal dopamine depletion. Similar

Address for correspondence: Syed F. Ali, Ph.D. Head, Neurochemistry Laboratory Division of Neurotoxicology, HFT-132 National Center for Toxicological Research/FDA 3900 NCTR Rd., Jefferson, AR 72079, U.S.A. Voice: 870-543-7123; fax: 870-543-7745.
sali@nctr.fda.gov

results were observed with nNOS knockout and copper zinc superoxide dismutase (CuZnSOD)–overexpressed transgenic mice models. Finally, using the protein data bank crystal structure of tyrosine hydroxylase, we postulate the possible nitration of specific tyrosine moiety in the enzyme that can be responsible for dopaminergic neurotoxicity. Together, these data clearly support the hypothesis that the reactive nitrogen species, peroxynitrite, plays a major role in METH-induced dopaminergic neurotoxicity and that selective antioxidants and peroxynitrite decomposition catalysts can protect against METH-induced neurotoxicity. These antioxidants and decomposition catalysts may have therapeutic potential in the treatment of psychostimulant addictions.

KEYWORDS: Methamphetamine; Dopaminergic neurotoxicity; Peroxynitrite; Antioxidant; Decomposition catalysts.

INTRODUCTION

The central nervous system (CNS) is highly susceptible to damage by a variety of biological agents. This problem is enhanced by the fact that neurons and neuron-derived cells, with few exceptions, do not renew themselves, so a gradual reduction in these essential elements throughout life is unavoidable. During the past few years, it has become evident that neurodegeneration in the CNS share common mechanisms of pathogenecity. Among these mechanisms, free radical–induced oxidative stress, mitochondrial dysfunction, and excitotoxic processes are pivotal in cell death. The role of the free radical nitric oxide (NO) in CNS morphogenesis and developmental synaptic plasticity is well documented.[1] In the nervous system, three highly homologous isoforms of nitric oxide synthase (NOS) catalyze the formation of NO from L-arginine and molecular oxygen.[2] These isoforms are endothelial (eNOS), inducible (iNOS), and neuronal (nNOS). Whereas the calcium-dependent nNOS isoform is constitutively expressed in brain cells,[3,4] the iNOS isoform is only expressed after pathological stimuli, including ischemia, viral, and bacterial infections or trauma.[5–7] Therefore, our success in determining the pathogenesis of dopaminergic damage or neurodegeneration, as well as in developing neuroprotective therapies that target the NO pathway is contingent on our elucidating which of the NOS isoenzymes contribute to the production of the NO involved in dopaminergic neuron degeneration.

NO is a simple inorganic radical that exhibits diverse physiological functions, including the regulation of neurotransmitter and vascular tone.[8] It is closely linked to the pathogenesis of many inflammatory and degenerative disorders. Many of the physiological actions of NO are mediated by cyclic guanosine monophosphate (cGMP)–dependent pathways by activating guanylate cyclase.[8,9] Secondary oxidants derived from NO and leading to S-nitrosylation and nitration of endogenous biomolecules can also be involved in various physiological and pathophysiological phenomena through cGMP-independent pathways.[10–13] When NO reacts with superoxide it yields peroxynitrite (ONOO$^-$), a highly reactive oxidant species.[14] ONOO$^-$ crosses the lipid membranes at a rate significantly higher than the rates of its known decomposition pathways,[15] indicating that this oxidant, unlike reactive radicals such as $O_2^{-\cdot}$ or ·HO, can travel distances of cellular dimensions. The nitration of tyrosyl residues has been shown to be a stable biochemical marker of OONO$^-$ production both *in vitro* and *in vivo*.[16] The biological significance of tyrosine nitration is a subject of great interest. In addition, tyrosine nitration causes inactivation of several enzymes,[17,18] influences tyrosine phosphorylation–mediated signal transduction,[19] and induces apoptosis through impairment of microtubule formation.[20]

Methamphetamine (METH) is a well known drug of abuse and neurotoxin that may cause long-lasting changes in the CNS dopaminergic pathway.[21] It causes neurotoxicity in rodents and non-human primates by producing long-term depletion of dopamine (DA) and its metabolites,[22,23] decreasing the number of high affinity DA uptake sites[24] and decreasing the activity of tyrosine hydroxylase (TH) in striatum.[25] DA, TH, and the DA transporter are reduced in the postmortem striatum of chronic METH users.[26] Although the cellular and molecular mechanisms involved in METH-induced toxicity are not completely understood, a role for oxygen (ROS) and nitrogen-based (RNS) radicals is well supported by several studies in the literature.[27] The present review focuses on the oxidative mechanisms supporting the combined role of ROS and RNS in METH-induced dopaminergic damage and we further provide insight into the role of various antioxidants as a neuroprotective agent.

ROS-BASED OXIDATIVE MECHANISMS IN METH-INDUCED NEUROTOXIC DAMAGE

Oxidative mechanisms have now been implicated in a number of pathological states that affect both the central and the peripheral nervous systems. Oxidative stress has also been implicated in the pathology of various neurodegenerative disorders. The brain is thought to be particularly vulnerable to such damage because of its high rate of oxygen consumption, and the presence of high levels of polyunsaturated fatty acids as substrates for lipid peroxidation. Cadet and Brannock,[27] in 1998, discussed the possibility that some neuronal cells might be especially sensitive to the effects of increased oxygen-based radicals derived during states that caused increased DA metabolism. The oxygen-based radical theory suggests that formation of toxic radicals from DA might be the main determinant of neurotoxicity of METH.[27] These ideas are supported by results obtained from copper/zinc-superoxide dismutase-transgenic mice, which are protected against the toxic effects of METH. Both acute and chronic administration of METH caused a marked decrease in DA levels mainly in the non-Tg mice, and these results do support the view that oxidative stress is an important mechanistic event in the development of these changes.[28] Furthermore, METH has also been reported to increase hydroxyl radicals in the brain in a time-dependent manner,[29] produce oxidative stress[30] and produce free radicals.[31] Recently, Subramaniam *et al.*[32] reported that toxic doses of METH alter the free radical–scavenging enzymes in non-transgenic and transgenic superoxide dismutase (SOD) mice. The effects of the drug on the reduction of antioxidant enzymes activities, such as the activities of SODs, glutathione peroxidase, and catalase, and the increase in lipid peroxidation observed in non-transgenic mice provide further substantiation for a role of oxygen-based radicals in the deleterious effects of METH.[32] In a recent study with PC12 cells, it was demonstrated that overexpression of glutathione peroxidase in PC12 cells resulted in protection against METH toxicity.[33] Furthermore, repeated doses of METH have been found to decrease cortical and striatal glutathione peroxidase activity in mice.[32] As evidenced from the previous studies, METH might be damaging the DAergic system via two major pathways, one by inducing hyperthermia and a second via the generation of oxygen-mediated radicals and oxidative stress. Both of these pathways may be interacting with each other, or one is influencing the other to result in the damage.

RNS-BASED OXIDATIVE MECHANISM IN METH-INDUCED NEUROTOXIC DAMAGE

Recently, in addition to oxygen-based radicals, a role for excitatory amino acids (EAA) transmission in METH-induced neurotoxicity was suggested by two major findings. First, blockade of the N-methyl-D-aspartate (NMDA) type of glutamate receptors by dizocilpine (MK-801) attenuates METH-induced neurotoxicity.[22] Second, repeated administration of METH causes an increase in glutamate release in the striatum.[34] The fact that EAA receptors modulate amphetamine-induced behavioral sensitization further supports the role of glutamatergic neurotransmission in the effects of these psychostimulants.[35] NMDA receptor activation is thought to be associated with the stimulation of neuronal nitric oxide synthase (nNOS) isoform. Interaction of glutamate, for instance, with the NMDA receptor complex opens channels that admit calcium ions into the cell; binding of calcium ions to calmodulin activates nNOS, which produces nitric oxide (NO), and subsequently elicits the accumulation of cyclic guanosine nucleotide monophosphate (cGMP).[36]

Therefore, it can be concluded that the METH-induced DA neurotoxicity is mediated through free radicals, either by superoxide radicals or nitric oxides. However, which radical mediates these neurotoxic effects is still unclear.

ROLE OF PEROXYNITRITE IN METH-INDUCED NEUROTOXIC DAMAGE

The generation of both ROS and RNS is clearly evident during METH-induced neurotoxic damage. Therefore, the possibility of the interaction of these two free radical species during METH intoxication and the formation of $OONO^-$ can be one of the mechanisms by which METH might induce dopaminergic damage. To gain insight into this mechanism, we explored the possibility of the generation of $OONO^-$ in animals treated with METH. Recently, we reported that the METH treatment that caused dopaminergic neurotoxicity also produced 3-nitrotyrosine (3-NT), an *in vivo* biomarker for $OONO^-$ production, in the mouse striatum.[37] We also reported that METH induces dopaminergic damage both *in vitro* and *in vivo* by generating $OONO$.[38] Therefore, based on these concurrent studies, there are ample evidence for the role of $OONO^-$ in METH-induced dopaminergic neurotoxicity (see FIGURE 1, A–D).

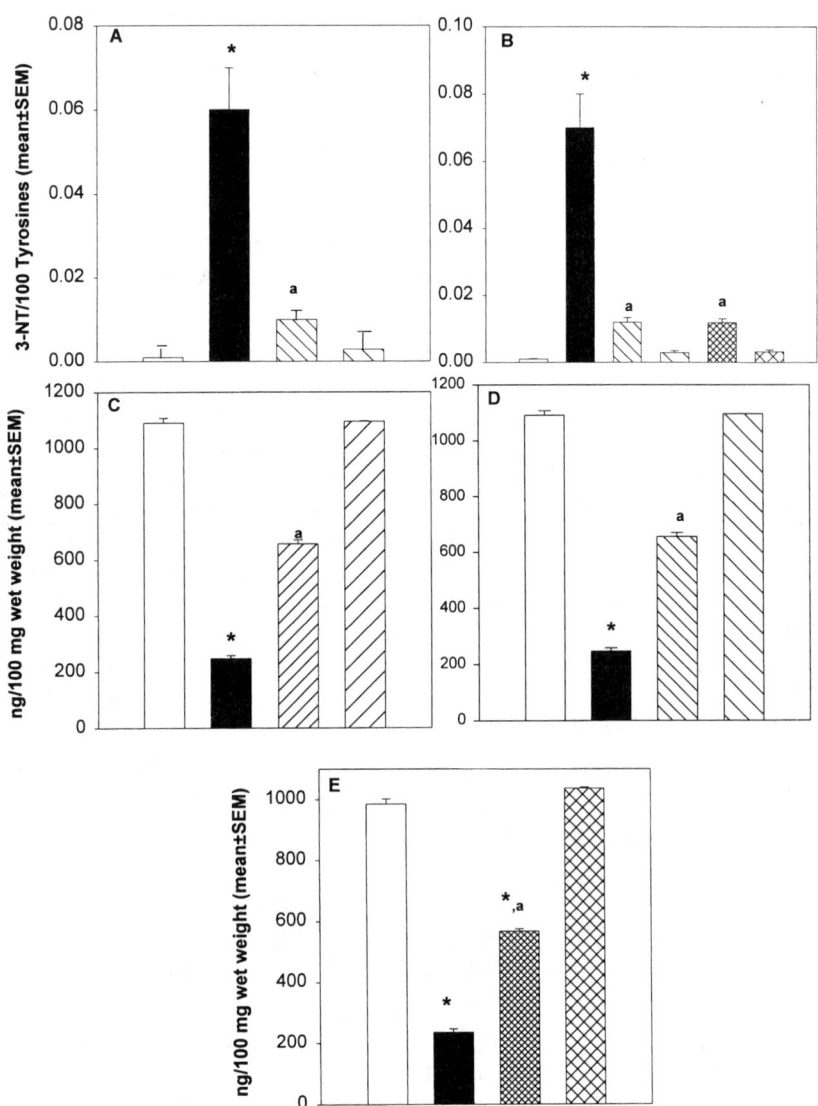

FIGURE 1. Protection against METH-induced 3-NT formation and dopamine depletion in mouse striatum by peroxynitrite decomposition catalysts FeTMPyP (**A** and **C**), FeTPPS (**B** and **D**), and by nNOS inhibitor 7-NI (**B** and **E**). Each value is mean ± S.E.M. derived from eight animals/group; *$p < 0.05$, significantly different from control group; $^a p < 0.05$, significantly different from METH group; ▭, control; ■, METH; ▨, METH+FeTMPyP; ▨, FeTMPyP; ▨, METH+FeTPPS; ▨, FeTPPS; ▨, METH+7-NI; ▨, 7-NI. (Adapted and modified from Refs. 37 and 49.)

NEUROPROTECTIVE ROLE OF VARIOUS ANTIOXIDANTS IN METH-INDUCED PEROXYNITRITE GENERATION AND DOPAMINERGIC NEUROTOXICITY

Peroxynitrite has been reported to inactivate glutathione peroxidase (GPx) by oxidizing the selenocysteine of the GPx.[39] Interaction of METH with GPx has also been documented both *in vitro* and *in vivo*. In a recent study with PC12 cells, it was demonstrated that overexpression of GPx in PC12 cells resulted in protection against METH cytotoxicity.[33] Additionally, repeated doses of METH were found to decrease cortical and striatal GPx activity in mice.[32]

Selenium (Se) is an essential dietary component for mammals, including humans. One of its well understood functions is that it is present in the active center of GPx, an antioxidant enzyme that scavenges various peroxides and protects membrane

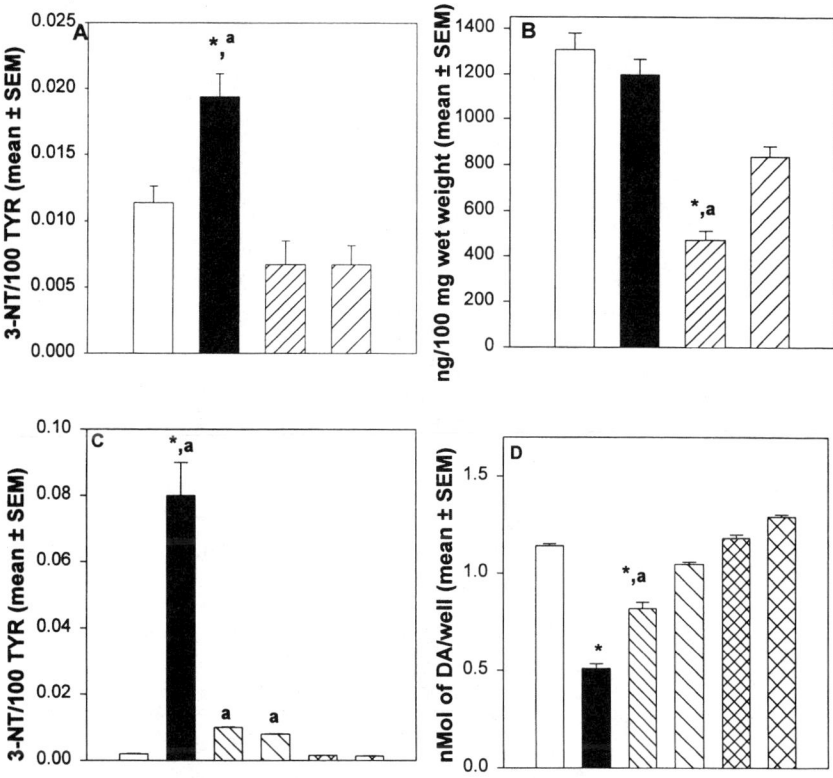

FIGURE 2. Effects of selenium on METH induced 3-NT formation and dopaminergic neurotoxicity in mouse striatum (**A** and **B**) and PC12 cells (**C** and **D**). In *in vivo* studies, each value is mean ± S.E.M. derived from eight animals/group. In *in vitro* studies, each value is mean ± S.E.M. derived from three different sets of experiments. *$p < 0.05$, significantly different from control group; [a]$p < 0.05$, significantly different from METH group; ▭, control; ▬, METH; ▨, Se; ▱, Se+METH; ▧, METH+Se 10 μM; ▨, METH+Se 20 μM; ▩, Se 10 μM; ▨, Se 20 μM. (Adapted and modified from Ref. 38.)

lipids and macromolecules from oxidative damage.[40] Therapy with Se and other antioxidants result in positive clinical responses in various neurodegenerative diseases associated with increased oxidative damage.[41] GPx and some Se-containing compounds have also been shown to protect against $OONO^-$-mediated oxidations and nitrations. Various selenoproteins like ebselen, selenomethionine, and selenocysteine have been designated as peroxynitrite reductases in model oxidation and nitration reaction mediated by $OONO^-$.[42] The role of Se as an antioxidant agent, as well as its presence in the active center of the enzyme glutathione peroxidase, prompted us to investigate the effect of Se on METH-induced dopaminergic neurotoxicity that might result from oxidative stress. METH-induced dopaminergic toxicity and the generation of peroxynitrite were significantly protected both *in vitro* and *in vivo* by Se supplementation. Se supplementation significantly protected against METH-induced dopaminergic toxicity and the generation of $OONO^-$ in PC12 cell cultures, as well as in mouse striatum (see FIGURE 2).

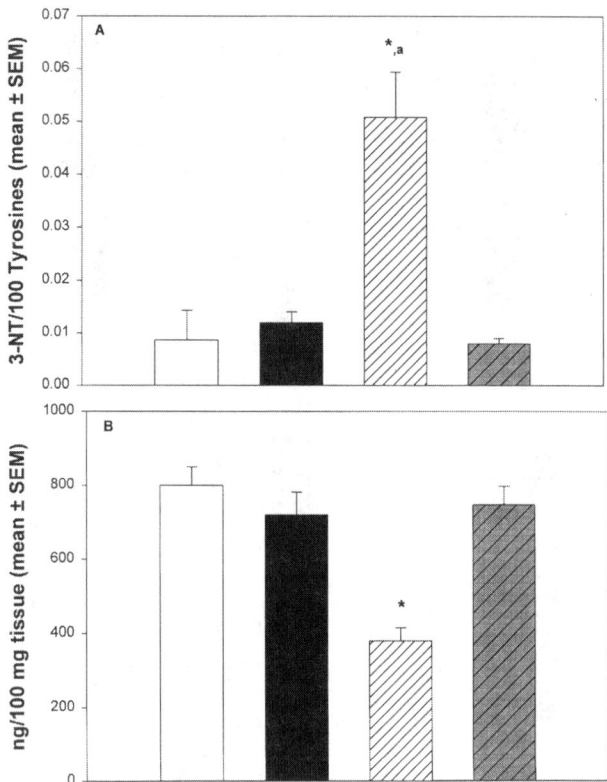

FIGURE 3. Effects of melatonin on METH-induced 3-NT formation (**A**) and dopaminergic neurotoxicity (**B**) in mouse striatum. Each value is mean ± S.E.M. derived from eight animals/group; *$p < 0.05$, significantly different from control group; $^a p < 0.05$, significantly different from METH group; ▭, control; ■, melatonin; ▨, METH; ▨, melatonin+METH.

Melatonin is a natural hormone produced by the pineal gland and thought to regulate the circadian rhythm. Recent studies suggest that melatonin acts also as free radical scavenger and antioxidant.[43] It is believed that melatonin donates an electron and directly detoxifies free radicals, such as the highly toxic hydroxyl radical.[44] *In vivo* and *in vitro* studies showed that melatonin protects cells, tissue, and organs against oxidative damage caused by various free radical generating agents.[45] As an antioxidant, this hormone is effective in protecting nuclear DNA, membrane lipids, and possibly cytosolic proteins from oxidative damage.[46] In our studies, we have shown that melatonin significantly protected the depletion of dopamine and its metabolites in mice after the repeated administration of METH.[47] Furthermore, in a follow up study, pretreatment with melatonin provided significant protection against METH-induced generation of OONO$^-$ and 3-NT formation in mouse striatum (see FIGURE 3).

NEURONAL NITRIC OXIDE SYNTHASE INHIBITORS AND PEROXYNITRITE DECOMPOSITION CATALYSTS AS NEUROPROTECTIVE AGENTS IN METH-INDUCED DOPAMINERGIC DAMAGE

Blocking the formation of NO or effective scavenging or decomposition of peroxynitrite may potentially provide a therapeutic intervention in METH-induced dopaminergic neurotoxicity. Use of peroxynitrite scavengers like uric acid has been reported to provide protection against experimental allergic encephalomyelitis.[48] Thus, the ability to either block the formation of NO, or to intercept and decompose peroxynitrite, may represent a novel and critical point of therapeutic intervention in pathophysiological cases associated with the overproduction of superoxide and nitric oxide.

Recently, it was reported that a selective nNOS inhibitor, 7-nitroindazole (7-NI) protects against the METH-induced dopaminergic neurotoxicity,[24] which implicates the role of nitric oxide radicals in METH neurotoxicity. We also reported that 7-NI protected against METH-induced OONO$^-$ generation in mouse striatum[49] (FIG. 1, B and E).

Stern and coworkers[50] discovered that various iron porphyrins catalyze the efficient decomposition of OONO$^-$ to nitrate under physiological conditions.[51] These iron porphyrins have profound activity in biological models of OONO$^-$-related disease states, and they have been investigated as therapeutic agents for diseases in which OONO$^-$ has been implicated.[50] Some of these iron porphyrins are 5,10, 15,20-tetrakis(*N*-methyl-4'-pyridyl)porphyrinato iron III (FeTMPyP) and 10,15,20-tetrakis (2,4,6-trimethyl-3,5-sulfonatophenyl) porphinato iron III (FeTPPS). In our studies, both FeTMPyP and FeTPPS significantly protected against METH-induced dopaminergic damage as well as OONO$^-$ formation in the striatum of animals treated with the repeated dose of METH[37,49] (FIG. 1, A–D).

GENETIC MANIPULATION OF METH-INDUCED DOPAMINERGIC DAMAGE

It is evident from the previous studies that OONO$^-$ is involved in the neurotoxic damage caused by METH administration. Since the formation of OONO$^-$ requires both superoxide and nitric oxide, the manipulation of either one of these free radical species appears to provide significant protective effects. We and others have reported that METH-induced dopaminergic neurotoxicity is attenuated in mice overexpressing CuZnSOD or human manganese superoxide dismutase (hMnSOD).[28,52] We have also shown that mice overexpressing CuZnSOD are protected against METH-induced OONO$^-$ production.[53]

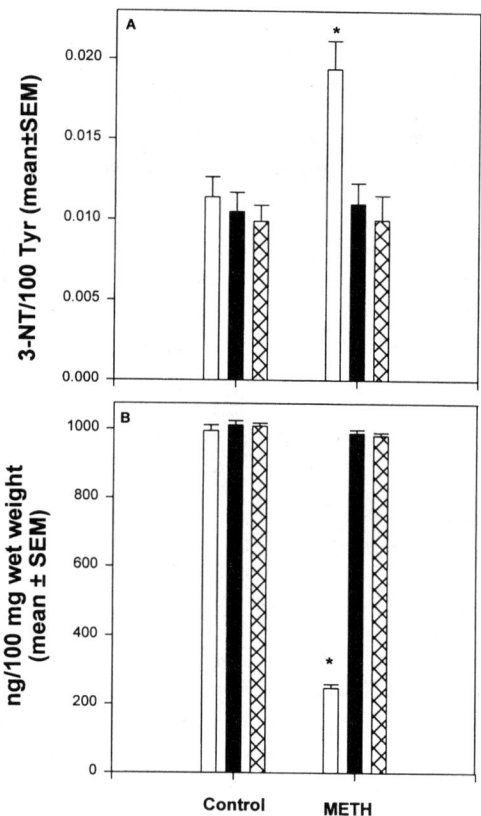

FIGURE 4. 3-Nitrotyrosine levels (**A**) and dopamine concentration (**B**) in the striatum of adult male C57BL/J6, nNOS knockout, and CuZnSOD overexpressed transgenic mice after METH administration. Each value is mean ± S.E.M. derived from eight animals/group; *$p < 0.05$, significantly different from control group; ☐, control; ■, nNOS knockout; ▨, CuZnSOD overexpressed transgenic. (Adapted and modified from Ref. 53.)

Similarly, we have shown that animals lacking nNOS gene are significantly protected against METH induced dopmainergic neurotoxicity.[59] Furthermore, we suggested that nNOS knockout mice are also protected against METH-induced peroxynitrite generation (see FIGURE 4).

PROPOSED MECHANISM FOR SELECTIVE TYROSINE NITRATION IN THE ENZYME TYROSINE HYDROXYLASE BY PEROXYNITRITE IN METH-INDUCED DOPAMINERGIC DAMAGE

Recently, Ara et al.[54] reported that nitration of tyrosine hydroxylase (TH) caused the inactivation of tyrosine hydroxylase in PC12 cell cultures exposed to peroxynitrite and MPP$^+$, as well as mice treated with MPTP. Several other studies also suggested that the nigrostriatal damage in the MPTP model of Parkinson's disease may be mediated by the generation of peroxynitrite, a reactive species formed by the nearly diffusion limited reaction of nitric oxide with superoxide.[55,56] These studies led to the hypothesis that peroxynitrite may be involved in the dopaminergic neurotoxicity induced by METH. Our studies have confirmed the role of OONO$^-$ in METH-induced dopaminergic neurotoxicity.

To provide further insight into our hypothesis of peroxynitrite-mediated inactivation of TH, we carried out a Protein Data Bank (PDB) survey for the crystal structure of the enzyme TH reported in the literature. We found two crystal structures of the enzyme coded 1TOH and 2TOH, as published in 1997[57] and 1998,[58] respectively. The PDB crystal structure of the enzyme 1TOH (see FIGURE 5A) revealed that it has several tyrosine moieties. Some of these tyrosine moieties are clustered around the active center of the enzyme (FIG. 5B). Tyrosine 371, 314, and 289 are within 10Å bond length of the active center of the enzyme (FIG. 5C). The oxygen atom of Tyr 371 has a bond length of 5.02Å from the iron atom of the active center. Nitration of the tyrosine 371 results in the reduction of bond length between the iron atom of the active center and the oxygen atom of the tyrosine residue 371. This reduction of the bond length creates a steric hindrance at the active center of the enzyme that contributes either to the reversible or irreversible inactivation of the enzyme. This might be the case after the generation of peroxynitrite in the striatum of animals treated with METH. At present, this explanation stays at a hypothetical stage. Additional studies are underway in our laboratory to explore this hypothesis of the specific nitration of these tyrosine moieties of the enzyme tyrosine hydroxylase.

SUMMARY AND CONCLUSION

The *in vitro* and *in vivo* studies reported here clearly support the hypothesis that peroxynitrite plays an important role in METH-induced dopaminergic neurotoxicity and selective antioxidants, selective nNOS inhibitors and peroxynitrite decomposition catalysts can protect against METH-induced neurotoxicity. 3-Nitrotyrosine can be use a useful biomarker for the peroxynitrite generation and should be investigated in other neurodegenerative disorders.

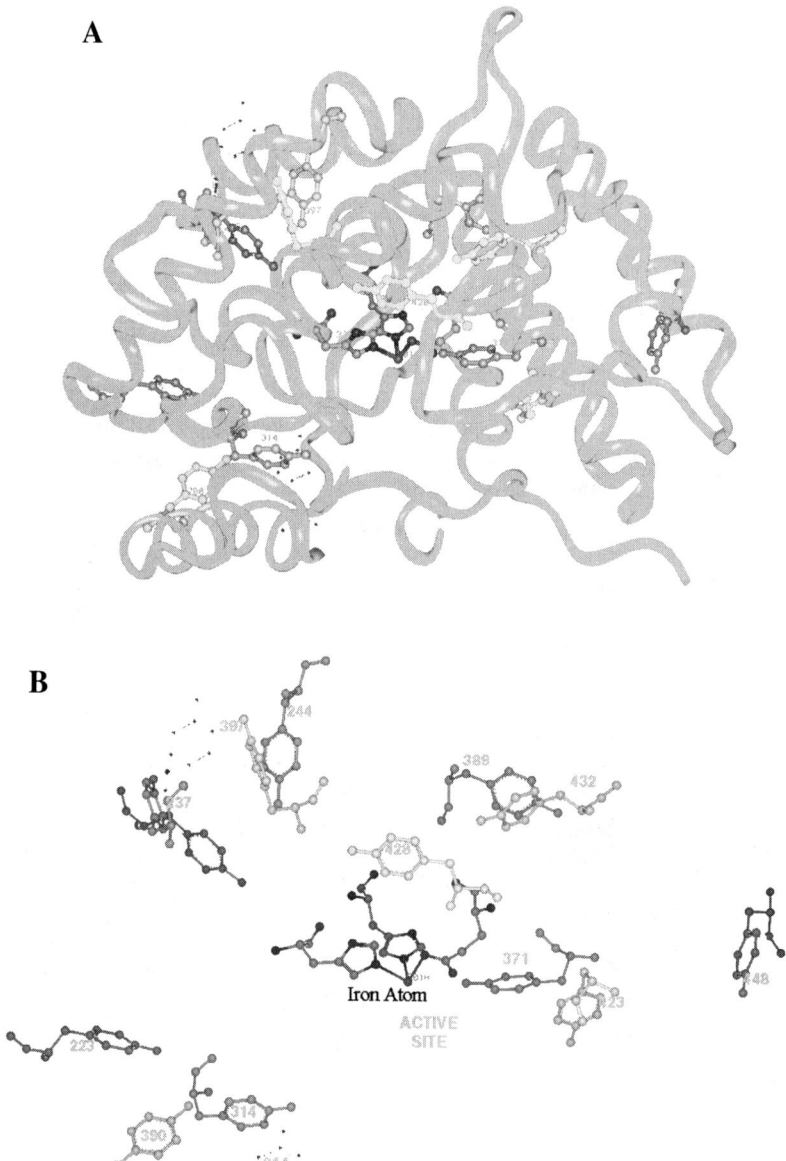

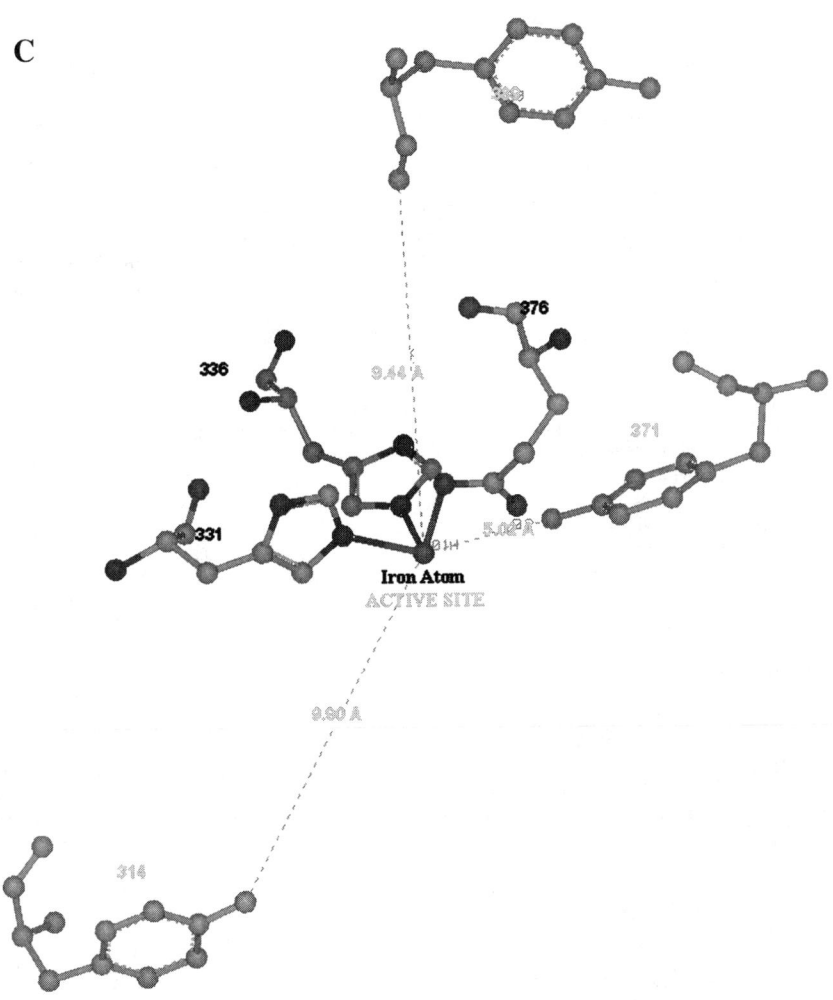

FIGURE 5. Proposed hypothesis of the selective nitration of specific tyrosine moieties in the enzyme tyrosine hydroxylase based on the crystal structure of the enzyme (**A**) obtained from protein data bank. The active site and the tyrosine moieties in the enzyme are represented (**B**). The bond length of the closest tyrosine moieties to the active center are represented (**C**). See the text for explanation.

REFERENCES

1. CRAMER, K.S., C.A. LEAMEY & M. SUR. 1998. Nitric oxide as a signalling molecule in visual system development. Prog. Brain Res. **118:** 101–114.
2. DAWSON, T.M. & V.L. DAWSON. 1996. Nitric oxide synthase: role as a neurotransmitter in the brain and endocrine system. Annu. Rev. Med. **47:** 219–227.
3. BREDT, D.S., C.E. GLATT, P.L. HUANG, et al. 1991. Nitric oxide synthase protein and mRNA are discretely localized in neuronal populations of the mammalian CNS together with NADPH diaphorase. Neuron **7:** 615–624.
4. HUANG, P.L., T.M. DAWSON, S.D. BREDT, et al. 1993. Targeted disruption of the neuronal nitric oxide synthase gene. Cell **75:** 1273–1285.
5. SIMMONS, M.L. & S. MURPHY. 1992. Induction of nitric oxide synthase in glial cells. J. Neurochem. **59:** 897–905.
6. WALLACE, M.N. & K. FREDENS. 1992. Activated astrocytes of the mouse hippocampus contain high levels of NADPH-diaphorase. Neuroreport **3:** 953–956.
7. IADECOLA, C.X. XU, F. ZHANG, et al. 1995. Marked induction of calcium-independent nitric oxide synthase activity after focal cerebral ischemia. J. Cereb. Blood Flow Metab. **15:** 52–59.
8. MONCADA, S. & A. HIGGS. 1993. The L-arginine-nitric oxide pathway. N. Engl. J. Med. **329:** 2002–2012.
9. DENNINGER, J.W. & M.A. MARLETTA. 1999. Guanylate cyclase and the ·NO/cGMP signaling pathway. Biochem. Biophys. Acta. **1411:** 334–350.
10. MCANDREW, J., R.P. PATEL, H. JO, et al. 1997. The interplay of nitric oxide and peroxynitrite with signal transduction pathways: implications for disease. Semin. Perinatol. **21:** 351–366.
11. GASTON, B. 1999. Nitric oxide and thiol groups. Biochem. Biophys. Acta. **1411:** 323–333.
12. ISHIROPOULOS, H. 1998. Biological tyrosine nitration: a pathophysiological function of nitric oxide and reactive oxygen species. Arch. Biochem. Biophys. **356:** 1–11.
13. JOURD'HEUIL, D., K.M. MIRANDA, S.M. KIM, et al. 1999. The oxidative and nitrosative chemistry of the nitric oxide/superoxide reaction in the presence of bicarbonate. Arch. Biochem. Biophys. **365:** 92–100.
14. SZABO, C. 1996. DNA strand breakage and activation of poly ADP ribosyltransferase: a cytotoxic pathway triggered by peroxynitrite. Free Radic. Biol. Med. **21:** 855–869.
15. MARLA, S.S., J. LEE & J.T. GROVES. 1997. Peroxynitrite rapidly permeates phospholipid membranes. Proc. Natl. Acad. Sci. U.S.A. **94:** 14243–14248.
16. BECKMAN, J.S., J. CHEN, H. ISCHIROPOULOS & J.P. CROW. 1994. Oxidative chemistry of peroxynitrite. Meth. Enzymol. **223:** 229–240.
17. MCMILLAN-CROW, L.A., J.P. CROW & J.A. THOMPSON. 1998. Peroxynitrite-mediated inactivation of manganese superoxide dismutase involves nitration and oxidation of critical tyrosine residues. Biochemistry **37:** 1613–1622.
18. HADDAD, I.Y., S. ZHU, H. ISCHIROPOULOS & S. MATALON. 1996. Nitration of surfactant protein A results in decreased ability to aggregate lipids. Am. J. Physiol. **270:** L281–288.
19. BRITO, C., M. NAVILIAT, A.C. TISCORNIA, et al. 1999. Peroxynitrite inhibits T lymphocyte activation and proliferation by promoting impairment of tyrosine phosphorylation and peroxynitrite-driven apoptotic death. J. Immunol. **162:** 3356–3366.
20. EISERICH, J.P., A.G. ESTEVEZ, T.V. BAMBERG, et al. 1999. Microtubule dysfunction by posttranslational nitrotyrosination of alpha-tubulin: a nitric oxide-dependent mechanism of cellular injury. Proc. Natl. Acad. Sci. U.S.A. **96:** 6365–6370.
21. CASS, W.A. 1997. Decreases in evoked overflow of dopamine in rat striatum after neurotoxic doses of methamphetamine. J. Pharmacol. Exp. Ther. **280:** 105–113.
22. ALI, S.F., G.D. NEWPORT, R.R. HOLSON, et al. 1994. Low environmental temperature or pharmacologic agents that produce hypothermia decrease methamphetamine neurotoxicity in mice. Brain Res. **658:** 33–38.
23. SIEDEN L.S., M.W. FISCHMAN & C.R. SCHUSTER. 1975. Long-term methamphetamine induced changes in brain catecholamines in tolerant rhesus monkeys. Drug Alcohol Depend. **1:** 215–219.

24. ITZHAK, Y. & S.F. ALI. 1996. The neuronal nitric oxide synthase inhibitor, 7-nitroindazole, protect against methamphetamine-induced neurotoxicity *in vivo*. J. Neurochem. **67:** 1770–1773.
25. HOTCHKIS, A.J. & J.W. GIBB. 1979. The long-term effetcs of multiple doses of methamphetamine on neostriatal tryptophan hydroxylase, tyrosine hydroxylase, choline acetyltransferase and glutamate decarboxylase activities, J. Pharmacol. Exp. Ther. **214:** 257–262.
26. WILSON J.M., K.S. KLAASINSKY, A.I. LEVEY, *et al*. 1996. Striatal dopamine nerve terminal markers in human, chronic methamphetamine users. Nature Med. **2:** 699–703.
27. CADET, J.L. & C. BRANNOCK. 1998. Free radicals and the pathobiology of brain dopamine system. Neurochem. Intl. **32:** 117–131.
28. CADET, J.L., S.F. ALI, R.B. ROTHMAN & C.J. EPSTIEN. 1995. Neurotoxicity, drugs of abuse, and the CuZn-Superoxide Dismutase in transgenic mice. Mol. Neurobiol. **11:** 155–163.
29. GIOVANNI, A., L.P. LIANG, T.G. HASTINGS & M.J. ZIGMOND. 1995. Estimating hydroxyl radical content in rat brain using systemic and intraventricular salicylate: impact of methamphetamine. J. Neurochem. **64:** 1819–1825.
30. GIBB, J.W., M. JOHNSON & G.R. HANSON. 1990. Neurochemical basis of neurotoxicity, Neurotoxicology **11:** 317–321.
31. DE VITO, M.J. & G.C. WAGNER. 1989. Methamphetamine-induced neuronal damage: a possible role for free radicals. Neuropharmacol. **28:** 1145–1150.
32. SUBRAMANIAM, J., B. LANDENHEIM & J.L. CADET. 1998. Methamphetamine-induced changes in antioxidant enzymes and lipid peroxidation in copper/zinc-superoxide transgenic mice. Ann. N.Y. Acad. Sci. **844:** 92–102.
33. HOM, D.G., D. JIANG, E. HONG, *et al*. 1997. Elevated expression of glutathione peroxidase in PC12 cells results in protection against methamphetamine but not MPTP toxicity. Mol. Brain Res. **46:** 154–160.
34. NASH, J.F. & B.K. YAMAMOTO. 1992. Methamphetamine neurotoxicity and glutamate release: comparison to 3,4-methylenedioxymethamphetamine. Brain Res. **581:** 237–243.
35. KARLER, R., I.A. CHAUDHRY, L.D. CALDER & S.A. TURKANIS. 1990. Amphetamine behavioral sensitization and excitatory amino acids. Brain Res. **537:** 76–82.
36. GARTHWAITE, J. 1991. Glutamate, nitric oxide, and cell-cell signalling. Trends Neurosci. **14:** 60–67.
37. IMAM S.Z., J.P. CROW, G.D. NEWPORT, *et al*. 1999. Methamphetamine generates peroxynitrite and produces dopaminergic neurotoxicity in mice: protective effects of peroxynitrite decomposition catalyst. Brain Res. **837:** 15–21.
38. IMAM S.Z. & S.F. ALI. 2000. Selenium, an antioxidant, attenuates methamphetamine-induced dopaminergic-neurotoxicity and peroxynitrite generation. Brain Res. **855:** 186–191.
39. ASAHI, M., J. FUJI, T. TAKAO, *et al*. 1997. The oxidation of selenocystine is involved in the inactivation of glutathione peroxidase by nitric oxide donor. J. Biol. Chem. **272:** 19152–19157.
40. ROTRUCK, J.T., A.L. POPE, H.E. GANTHER, *et al*. 1973. Selenium: biochemical role as a component of glutathione peroxidase. Science **179:** 588–590.
41. WATNABE, C., T. SUZUKI, T. OHBA & Y. DEJIMA. 1990. Transient hypothermia and hyperphagia induced by selenium and tellurim compounds in mice. Toxicol. Lett. **50:** 319–326.
42. SEIS, H., L.O. KLOTZ, V.S. SHAROV, *et al*. 1998. Protection against peroxynitrite by selenoproteins. Z. Naturforsch (J. Biosci.) **53:** 228–232.
43. REITER, R.J., L. RANG, J.J. GARCIA & A. MUNOZ-HOYOS. 1997. Pharmacological actions of melatonin in oxygen radical pathophysiology. Life Sci. **60:** 2255–2272.
44. GILAS, E., S. CUZZOCREA, B. ZINGARELLI, *et al*. 1997. Melatonin is a scavenger of peroxynitrite.Life Sci. **60:** 169–174.
45. ESCAMES, G., J.M. GUERRERO, R.J. REITER, *et al*. 1997. Melatonin and vitamin E limit nitric oxide-induced lipid peroxidation in rat brain homogenates. Neurosci. Lett. **230:** 147–150.

46. REITER, R.J., D. MELCHIORRI, E. SEWERYNEK, *et al.* 1995. A review of the evidence supporting melatonin's role as an antioxidant. J. Pineal Res. **18:** 869–875.
47. ITZHAK, Y., J.L. MARTIN, M. DEAN BLACK & S.F. ALI. 1998. Effect of melatonin on methamphetamine and MPTP-induced dopaminergic neurotoxicity and methamphetamine-induced behavioral sensitization. Neuropharmacol. **37:** 781–791.
48. HOOPER, D.C., O. BAGSARA, J.C. MARINI, *et al.* 1997. Prevention of experimental allergic encephalomyelitis by targeting nitric oxide and peroxynitrite: implications for the treatment of multiple sclerosis. Proc. Natl. Acad. Sci. U.S.A. **94:** 2528–2533.
49. IMAM, S.Z., F. ISLAM, Y. ITZHAK, *et al.* 2000. Prevention of dopaminergic neurotoxicity by targeting nitric oxide and peroxynitrite: implications for the prevention of methamphetamine-induced neurotoxic damage. Ann. N.Y. Acad. Sci. **914:** 157–171.
50. STERN M.K., M.P. JENSEN & K. KRAMER. 1996. Peroxynitrite decomposition catalysts, J. Am. Chem. Soc. **118:** 8735–8736.
51. HUNT, J.A., J. LEE & J.T. GROVES. 1997. Amphiphilic peroxynitrite decomposition catalysts in liposomal assemblies. Chem. & Biol. **4:** 845–858.
52. MARAGOS, W.F., R. JAKEL, D. CHETNUT, *et al.* 2000. Methamphetamine toxicity is attenuated in mice that overexpress human manganese superoxide dismutase. Brain Res. **878:** 218–222.
53. IMAM, S.Z., G.D. NEWPORT, Y. ITZHAK, *et al.* 2001. Peroxynitrite plays a role in methamphetamine-induced dopaminergic neurotoxicity: evidence from mice lacking neuronal nitric oxide synthase gene or overexpressing copper zinc superoxide dismutase. J. Neurochem. **76:** 745–749.
54. ARA, J., S. PRZEDBORSKI, B.N. ALI, *et al.* 1998. Inactivation of tyrosine hydroxylase by nitration following exposure to peroxynitrite and 1-methyl-4-phenyl-1,2,3,6-tetrahydropyridine (MPTP). Proc. Natl. Acad. Sci. U.S.A. **95:** 7659–7663.
55. HANTRAYE, P., E. BROUILLET, R. FERRANTE, *et al.* 1996. Inhibition of neuronal nitric oxide synthase prevents MPTP-induced parkinsonism in baboons. Nat. Med. **2:** 1017–1021.
56. SCHULTZ J.B., R.T. MATTHEWS, M.M. MUQIT, *et al.* 1995. Inhibition of neuronal nitric oxide synthase by 7-nitroindazole protects against MPTP-induced neurotoxicity in mice. J. Neurochem. **64:** 936–939.
57. GOODWILL, K.E., C. SABATIER, C. MARKS, *et al.* 1997. Crystal structure of tyrosine hydroxylase at 2.3Å and its implications for inherited neurodegenerative diseases. Nat. Struct. Biol. **4:** 578.
58. GOODWILL K.E., C. SABATIER & R.C. STEVENS. 1998. Crystal structure of tyrosine hydroxylase with bound cofactor analogue and iron at 2.3Å resolution: self-hydroxylation of Phe300 and the pterin-binding site. Biochemistry **37:** 13437–13445.
59. Y. ITZHAK, C. GANDIA, P.L. HAUNG & S.F. ALI. 1998. Resistance to neuronal nitric oxide synthase deficient mice to methamphetamine-induced dopaminergic neurotoxicity *in vivo*. J. Pharmacol. Exp. Ther. **284:** 1040–1047.

Biomarkers of 3-Nitropropionic Acid (3-NPA)-Induced Mitochondrial Dysfunction as Indicators of Neuroprotection

A.C. SCALLET,[a,b] P.L. NONY,[b] R.L. ROUNTREE,[a] AND Z.K. BINIENDA[a]

[a]*Division of Neurotoxicology, National Center for Toxicological Research, USFDA, 3900 NCTR Drive, Jefferson, Arkansas, U.S.A.*

[b]*Department of Pharmacology and Toxicology, University of Arkansas for Medical Sciences, 4301 W. Markham, Little Rock, Arkansas, U.S.A.*

ABSTRACT: In humans or animals, symptoms of mitochondrial energy dysfunction may be produced by mutations or inborn errors of the necessary enzymes, as well as by enzyme inhibitors or uncouplers of the oxidative phosphorylation process. 3-Nitropropionic acid (3-NPA) is a toxin that is sometimes produced on moldy crops (sugarcane, peanuts, etc.) in amounts sufficient to cause severe neuromuscular disorders when consumed by humans. *In vitro*, 3-NPA irreversibly inactivates SDH, a Complex II respiratory enzyme important for mitochondrial energy production. We have been studying biomarkers of 3-NPA exposure in the expectation that such markers may be useful in the screening process to identify neuroprotective agents against neurotoxicity produced by mitochondrial energy dysfunction. Animals were sacrificed at various times after 3-NPA exposure for histochemical visualization of SDH activity and measurement of immediate postmortem rectal temperature. 3-NPA-treated rats experienced progressive hypothermia that reached a loss of 3°C or more in core body temperature by three hours after dosing. The optical density of the SDH stain in brain was reduced, following a similar time course, most prominently in the cerebellum and least sharply in the thalamus. Some rats were given injections of L-carnitine (an enhancer of fatty acid transport) either alone, or as a pretreatment prior to a dose of 3-NPA. Although L-carnitine deficiency by itself can produce mitochondrial dysfunction, pretreatment with L-carnitine was of limited efficacy at overcoming the effects of 3-NPA on either body temperature or quantitative SDH histochemistry. Body temperature and SDH histochemistry may be useful biomarkers for evaluating the efficacy of neuroprotective agents against lower doses of 3-NPA, against other pharmacological models of mitochondrial dysfunction, or even against genetic mitochondrial diseases.

KEYWORDS: Biomarker; 3-Nitropropionic acid; Mitochondrial dysfunction; Neuroprotection; Succinate dehydrogenase.

Address for correspondence: Andrew C. Scallet, Ph.D., Division of Neurotoxicology, NCTR, 3900 NCTR Drive, Jefferson, AR 72079, U.S.A. Voice: 870-543-7146; fax: 870-543-7745.
AScallet@NCTR.FDA.GOV

INTRODUCTION

Mitochondrial energy-producing enzymes of the tricarboxylic acid (TCA) cycle and the electron transport chain are produced as combinations of protein units, some specified by nuclear DNA and some by mitochondrial DNA. Mutations and/or inborn errors of either the nuclear or mitochondrial portions may have profound effects on energy-dependent tissues including muscle and brain. The result includes such conditions as MERRF (mitochondrial encephalomyopathy with ragged red fibers), MELAS (mitochondrial encephalopathy, lactic acidosis, and stroke), retinitis pigmentosa, and various forms of muscular dystrophy.[1,2] 3-Nitropropionic acid (3-NPA) is a toxin sometimes produced on moldy crops (sugarcane, peanuts, etc.) in amounts sufficient to cause severe neuromuscular disorders when consumed by humans.[3] *In vitro*, 3-NPA irreversibly inactivates succinate dehydrogenase (SDH), a Complex II respiratory enzyme important for mitochondrial energy production.[4] We have been studying biomarkers of 3-NPA exposure, such as body temperature and SDH histochemistry.[5] These endpoints may be useful in a screening approach to identify neuroprotective agents efficacious against cellular energy deficiency caused by mitochondrial poisons or diseases. For example, certain agents (such as L-carnitine, an aid to fatty acid transport) have been proposed to be efficacious in a general way against a range of mitochondrial poisons or diseases.[2,6]

3-NPA is a chemical that contributes to the toxicity of various species of *Astragalus*, sometimes known as "locoweeds" or "milkweeds". After their consumption by livestock[7] lung and CNS lesions may occur, resulting in emphysema and locomotor dysfunction. *Astragalus* toxicity has been responsible for extensive economic losses in the ranching and cattle industries of the Western United States. 3-NPA may also be formed by the fungus *Arthrinium sp.*; for example, an epidemic of acute encephalopathy in children that ate moldy sugarcane containing 3-NPA occurred recently in China.[3] *In vitro*, 3-NPA inhibits cellular respiration by *irreversible* inactivation of succinate dehydrogenase (SDH), a mitochondrial Complex II enzyme responsible for the oxidation of succinate to fumarate in the Krebs cycle and the subsequent transport of electrons in oxidative phosphorylation.[4] Ingestion of 3-NPA *in vivo* may thus also lead to impairment of cellular energy metabolism and the production of cellular hypoxia, which may then ultimately lead to its neurotoxicity.

Inhibition of energy production by ischemia results in the failure of the sodium/potassium exchange mechanism, which requires ATP. Improper ionic homeostasis, the loss of neuronal membrane potential and, eventually, osmotic lysis and neuronal degeneration then occur.[8] Affected regions often include the striatum and hippocampus. Injected in a single dose or chronically, 3-NPA can result in distinctive neuropathological changes particularly evident in the corpus striatum, hippocampus, and thalamus of a subset (up to 75%) of treated rats.[6,9] Exposure to malonic acid, a *reversible* SDH inhibitor *in vitro*, also disrupts mitochondrial energy production by disrupting oxidative phosphorylation. Malonic acid may result in lesions, similar to those produced by 3-NPA, that are also especially prominent in the rat striatum.[10] Chronic administration of 3-NPA to baboons produces cognitive and motor defects that have been compared to those observed in humans with striatal damage caused by Huntington's disease.[11] A single low dose of 3-NPA has also been reported to predispose neurons to degeneration upon subsequent insult with a similarly modest dose of amphetamine, a psychoactive drug known to cause energy depletion and neurodegeneration of striatal terminals.[12]

The inhibition of succinate dehydrogenase observed *in vitro*[4] may be sufficient to explain the location and extent of the neurological damage observed in the *in vivo* studies reviewed above. However, other mediating factors such as excitotoxicity via endogenous glutamate, local actions of nitric oxide release, or oxidative stress, may be required for the complete expression of the neurotoxicity as neuronal death. Thus, studies demonstrating protection from the actions of 3-NPA and malonic acid by the NMDA receptor-blocker MK801 have shown that maintenance of membrane potential can reduce the extent of neuronal death resulting from the energy depletion.[13] The production of nitric oxide by inducible nitric oxide synthase (iNOS) immunopositive neurons has also been implicated in the vulnerability of the striatum to 3-NPA lesions.[14] Additional experiments have been performed to determine the role of oxidative stress as a mechanism of 3-NPA neurotoxicity in the production of selective striatal lesions. Transgenic mice carrying the human copper/zinc superoxide dismutase gene were protected from hydroxyl and peroxynitrite radical formation and had less neurotoxic striatal damage after 3-NPA treatment, compared to control littermates.[15] Further study has shown that 3-NPA causes an increase in lipid peroxidation in rat brain and liver.[16] In addition, acute 3-NPA exposure is associated with an increase in free fatty acids, thought to be a biomarker of oxidative stress.[17] Of course, an increase in fatty acids may also result from their accumulation behind the blockade of the TCA cycle. Thus, a number of different mechanisms may be required to interact with each other to ultimately confer neurotoxic sensitivity to 3-NPA.

A neurohistochemical method[18] was used to evaluate the first step in the actions of 3-NPA in the nervous system: the inhibition of SDH activity following *in vivo* administration to adult rats. In brain, SDH is particularly concentrated in mitochondria-rich axon terminals[19] We hypothesized that, as in the *in vitro* studies, SDH activity would be chronically inhibited in brain until replacement enzyme could be produced.

3-NPA-treated rats are noticeably cold to the touch. Hypothermia might be expected since peripheral production of heat by muscle metabolism is required to maintain normal body temperature. Factors affecting body temperature after 3-NPA exposure might include varying metabolic capacities among brain and peripheral muscle cells stemming from differences in mitochondrial density, SDH concentrations, and regional blood flow. If body temperature followed a similar time course to inhibition of SDH activity, it would be useful as a non-invasive biomarker of 3-NPA exposure.

MATERIALS AND METHODS

Dosing and Temperature Measurement

Twenty singly housed adult male Sprague-Dawley rats were injected subcutaneously (SC) with 30 mg/kg of 3-NPA in 0.1M potassium phosphate buffer (Sigma Chemical Co., St. Louis, MO) and five rats with 1.0 ml/kg of 0.1M potassium phosphate buffer as controls. Five treated animals plus at least one control were sacrificed by decapitation at either 1, 3, 6, or 120 hours after dosing. Immediately after sacrifice, postmortem core temperature was obtained rectally using a Telethermometer (Yellow Springs Instrument Co., Yellow Springs, CO) and brains were quickly dissected out, frozen, and stored at $-150°C$ until sectioning.

A second experiment was conducted with fifteen more singly housed male Sprague-Dawley rats, using five groups of three rats each. One group (controls) received an intraperitoneal (IP) injection of saline followed 30 minutes later by an SC injection of buffer. A second group (high carnitine alone) received an IP injection of 100 mg/kg L-carnitine followed 30 minutes later by an SC injection of buffer. A third group (3-NPA alone) received IP saline followed 30 minutes later by 30 mg/kg 3-NPA (SC, in buffer). The fourth group (low carnitine + 3-NPA) received 50 mg/kg L-carnitine (IP) followed 30 minutes later by 30 mg/kg 3-NPA (SC). Finally, the fifth group (high carnitine + 3-NPA) received 100 mg/kg of L-carnitine (IP) followed 30 minutes later by 30 mg/kg 3-NPA (SC). All animals were allowed to survive for 90 minutes after the second injection. Immediately after sacrifice, postmortem core temperature was obtained rectally using a Telethermometer (Yellow Springs Instrument Co., Yellow Springs, CO) and brains were quickly dissected out, frozen, and stored at $-150°C$ until sectioning.

SECTIONING AND HISTOCHEMISTRY

Frozen sections were cut sagittally at 15 microns with a cryostat, warmed briefly onto gelatin coated glass slides, and refrozen. The sagittal orientation allowed us to stain a single section from each of the twenty-five subjects as a batch under uniform conditions, and still evaluate a wide range of the brain regions of most interest: caudate, cerebellum, hippocampus, cortex and thalamus, all within the same section. Prior to the enzyme assay, tissues were fixed in 100% acetone at $-10°C$. Visualization of SDH location was performed according to Lillie and Fullmer.[20] Briefly, tissues were incubated in a 37°C staining solution containing 0.81 g succinate disodium salt (Sigma) added as the enzyme substrate and one 10 mg tablet of NitroBlue Tetrazolium (NBT, Sigma) in 60 ml of 0.05 M potassium phosphate buffer. Enzyme activity is visualized by the production of an insoluble blue product (a formazan) formed by reduction of NBT at sites of enzyme activity. Tissues were incubated for 45 minutes, removed from the solution, air dried, and coverslipped.

For the second experiment, to maximize the sensitivity to detect any L-carnitine protection against loss of SDH histochemical staining, we incubated an additional set of tissues at 40°C for 60 minutes. The additional sections, collected under conditions designed to maximize their enzymatic activity, were used to quantify SDH.

Densities were measured either in terms of grey scale (0–255 or 8-bit), or relative optical densities (a log-transform of the grey-level data), using an MCID-M1 (Experiment 1) or M5+ (Experiment 2) image analysis system (Imaging Research, Inc., Brock University, Ontario, CAN). The image of each brain section can be digitized through a video camera mounted on a microscope and viewed on a computer monitor. Within each region of interest, a sample of the optical density of the stain was measured. In Experiment 1, six regions of brain were analyzed in each of the five control and 20 treated brain sections: cerebellum (molecular layer), hippocampus (CA1), ventral thalamus, caudate nucleus, occipital cortex, and frontal cortex.

The procedure for optical density measurement was identical for each specimen with respect to tissue magnification, lighting, condensor settings, and size of the area being evaluated (approximately 1 mm^2); all sections were measured in a single session of data collection to ensure no bias via drift in instrument settings. In Experiment 2, only three regions were measured from each brain: the caudate nucleus, the dentate granule cell layer of the hippocampus, and a nearby region of neocortex.

Statistical Analysis

In Experiment 1, there were no apparent differences in body temperature or SDH between rats sacrificed at any of the four different times after they had been dosed with saline. Therefore, the data from these rats was combined to form a single control group ($n = 5$). GraphPAD InStat software was then used to analyze the temperature data employing a one-way ANOVA, with five levels of treatment (control plus the four sacrifice intervals after the 3-NPA treatment). SigmaStat software was used to analyze the optical density measurements for the six regions in each tissue section. A two-way mixed model ANOVA, with one between and one within factor, was performed with the six brain regions as the repeated measures ("within" factor) and the four sacrifice times plus control as the five levels of the "between" groups variable. *Post hoc* individual comparisons were by the Student-Newman-Keuls or Bonferroni approach. The $p < 0.05$ significance level was employed for all statistics.

Experiment 2 was analyzed as a one-way analysis of variance with five treatment levels using Sigmastat software. *Post hoc* comparisons at the 0.05 significance level were made between all pairs of means using Fisher LSD tests, to allow for maximum sensitivity.

RESULTS

Hypothermia

Experiment 1

A marked decrease compared to controls of nearly 3°C in core body temperature was evident at three hours after 3-NPA dosing; this statistically and biologically significant plunge in temperature was sustained at least through six hours after 3-NPA exposure ($F\{4,17\} = 22.2, p < 0.0001$, see FIGURE 1). The hypothermia was accompanied by lethality for both of the two animals that reached temperatures below 36°C. Core body temperature had returned to near normal, in fact slightly elevated, in those 3-NPA-treated animals that were sacrificed at 120 hours.

Experiment 2

The hypothermia caused by 3-NPA was also clearly evident in the second experiment (see FIGURE 4 A, below). There was a strong effect of treatment group ($F(4,10) = 5.92, p < 0.01$), and the *post hoc* tests indicated that each group receiving 3-NPA was significantly ($p < 0.05$) colder than the control group. However, it should perhaps also be noted that, whereas the 3-NPA alone group was significantly colder than the controls at the 0.01 level, the 3-NPA groups that also received carnitine (low or high) were significantly different only at the less stringent 0.05 level.

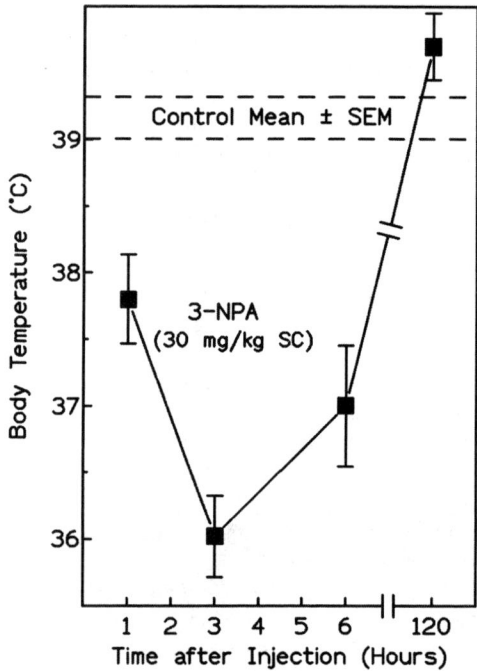

FIGURE 1. The effects of 30 mg/kg 3-NPA on core body temperature at various times of sacrifice after dosing (Experiment 1). Rectal temperatures were obtained immediately postmortem. Data are expressed in degrees Celsius as mean ± SEM based on five separate 3-NPA-treated animals evaluated at each time. The control data was combined from five saline-dosed animals, one or two of which were sacrificed at each time. Therefore, the control data shown is a range demarcated by *dashed lines* encompassing the mean plus or minus one standard error of the mean. Treated rats showed significant decreases in core temperature at all times except 120 hours.

SDH Histochemistry

Experiment 1

The treatment of rats with 3-NPA produced an easily visualized and obvious decrease in SDH staining intensity in brain as shown by a nearly complete absence of blue NBT-derived formazan deposition (see FIGURE 2). This decrease was readily quantified: the imaging and densitometry showed a marked 3-NPA-induced inhibition of SDH, as compared to saline controls, when relative optical density was measured in each of the six regions of interest (see FIGURE 3). The two-way repeated measures ANOVA revealed significant differences among the different times of sacrifice ($F(4,18) = 29.34$, $p < 0.0001$) and among the different regions ($F(5,20) = 10.23$, $p < 0.0001$). The analysis also revealed a significant interaction between time of sacrifice and brain region, signifying that the degree of 3-NPA inhibition of SDH varied between brain regions, depending on the length of exposure to the toxin ($F(20,84) = 3.33$, $p < 0.0001$). The time until maximum inhibition of SDH activity

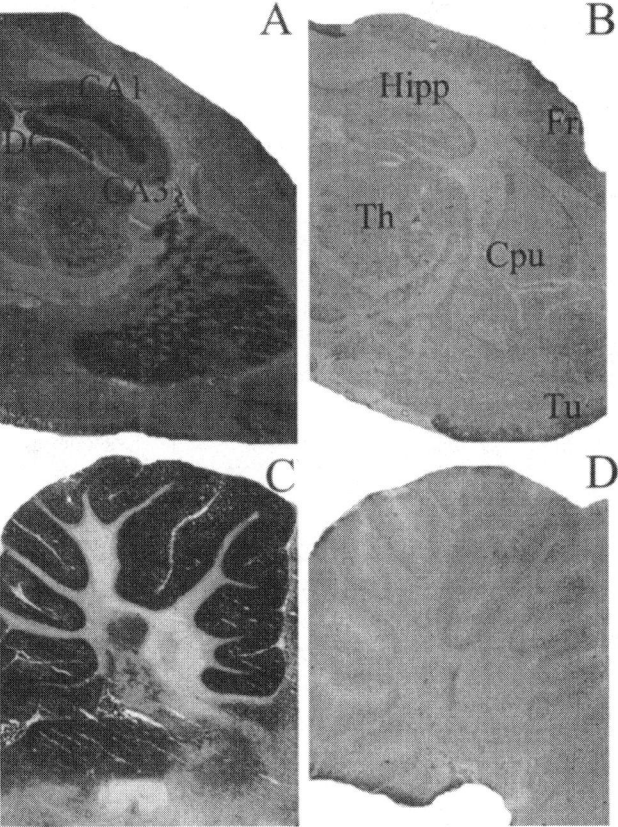

FIGURE 2. NitroBlue Tetrazolium (NBT) stain for SDH activity in sagittal brain tissue sections from control and treated animals (Experiment 1). **(A)** Exemplifies NBT stain intensity in caudate, thalamus, hippocampus, frontal, and occipital cortex of a control animal (2.5× magnification). **(B)** Shows the same regions as **(A)** in a sagittal section of a treated animal from the six-hour group (2.5× magnification). **(C)** Exemplifies the NBT stain in the cerebellum of a control animal (3.3× magnification). **(D)** Shows the NBT stain in the cerebellum of a treated animal from the six-hour group (3.3× magnification). Note the marked decrease in NBT color intensity from the control to the treated sections and the relatively high density of staining in the caudate, dentate granule cell layer, and cerebellum of the control animal. Abbreviations: Fr, frontal cortex; Th, thalamus; Cpu, caudate-putamen; Tu, olfactory tubercle; Hipp, hippocampus; DG, dentate gyrus; CA, cornu ammonis region 1 or 3 of the hippocampus.

for most of the regions was at three hours after dosing (except for cerebellum and thalamus, where SDH was maximally inhibited at one hour). Inhibition of SDH compared to control levels was statistically significant whether measured 1, 3, 6, or even 120 hours after dosing in most regions (Bonferroni *post hoc* comparisons, $p < 0.01$). Exceptions include the occipital cortex, where SDH activity had partially recovered by six hours after dosing (FIG. 3E), even though it remained at least marginally

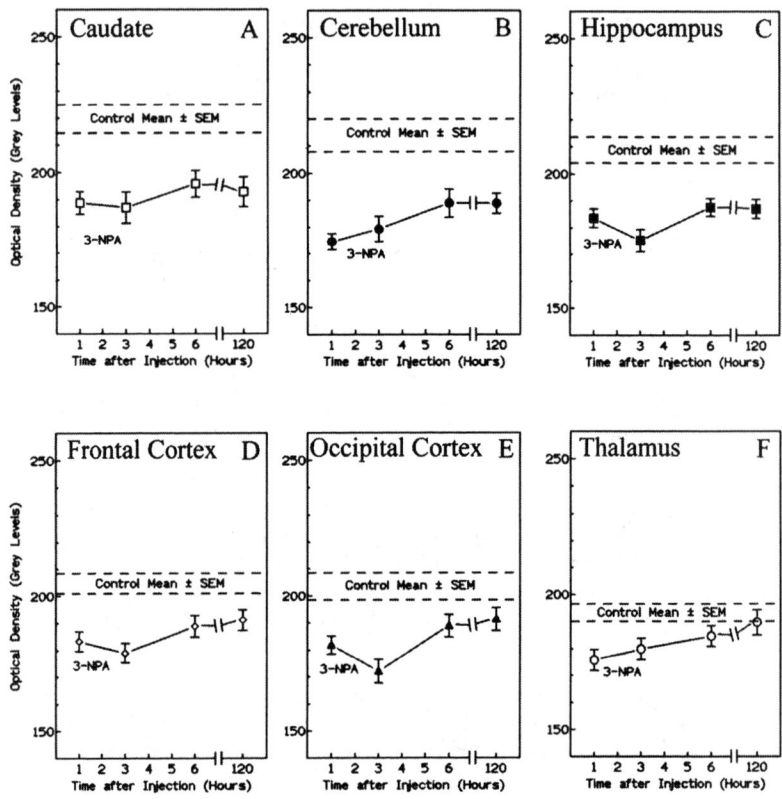

FIGURE 3. Quantitative analysis of NBT stain intensity in control and treated tissue sections (Experiment 1). Regions were analyzed for stain intensity in units of relative optical density (measured in grey levels). Relative optical density values are expressed as the mean ± SEM, based on five separate 3-NPA-treated animals collected at each time point. The control data was combined from five saline-dosed animals, one or two of which were sacrificed at each time. Therefore, the control data shown here is a range demarcated by *dashed lines* encompassing the mean plus or minus one standard error of the mean. Note that NBT stain intensity decreased significantly through 120 hours postdrug exposure in all regions except ventral thalamus (**F**), exemplifying prolonged SDH inhibition by 3-NPA. The relative intensity of SDH activity in the various brain regions of the control rats can be appreciated by comparing the locations of the control ranges. Thus, the caudate had the most NBT staining signifying the greatest SDH activity, whereas the thalamus had the least staining.

suppressed for up to 120 hours ($p < 0.05$). The thalamus was the least sensitive brain region to 3-NPA suppression of SDH, and had returned to levels not significantly lower than the controls ($p > 0.05$) by six hours after dosing. The greatest degree of inhibition of SDH caused by 3-NPA seemed to occur in the cerebellar molecular layer, as shown by the densitometric data (FIG. 3B) and as exemplified by a marked decrease in NBT staining (see FIG. 2, C versus D).

Regional Variation in SDH (Control Rats)

We hypothesized that the neurotoxic effects of 3-NPA treatment might be influenced by the original amount or activity of the SDH present in a given brain region. In order to address the possibility that increased NBT-staining might explain an increased susceptibility (or resistance) to 3-NPA neurotoxicity, we compared the relative optical density data between brain regions in control animals. From inspection of FIGURE 2 and the quantitative data from FIGURES 3, it seems clear that the caudate, cerebellum, and hippocampus (especially the dentate gyrus) are particularly

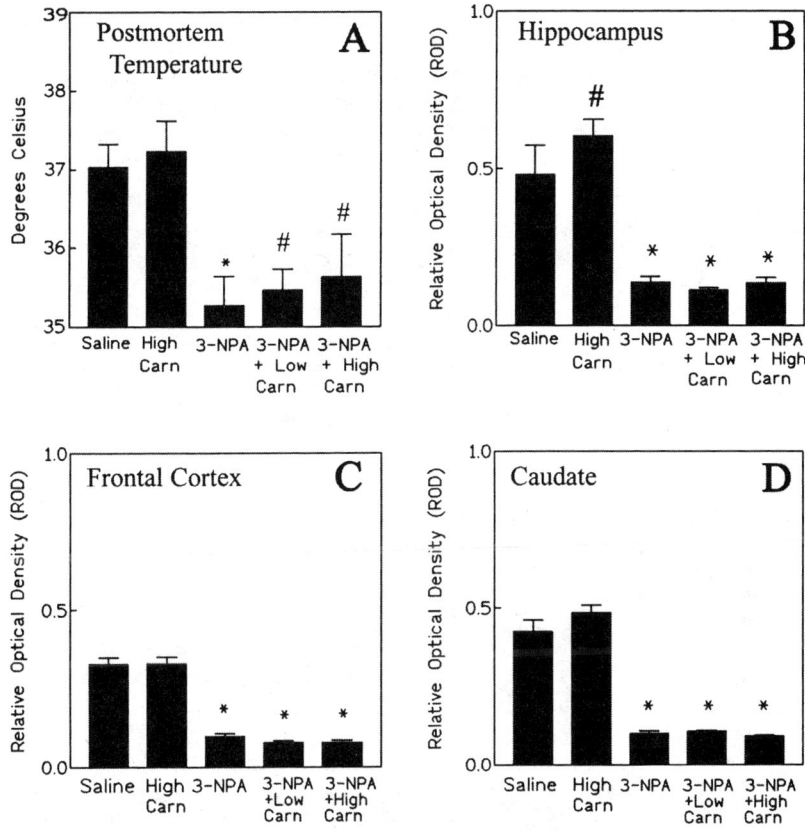

FIGURE 4. Quantitative analysis of the effects of L-carnitine pretreatment on core body temperature and NBT stain intensity (Experiment 2). **(A)** shows that the hypothermia observed from 3-NPA treatment in Experiment 1 was also seen in Experiment 2. There appeared to be a tendency for L-carnitine to increase rectal temperature somewhat, whether or not animals were treated subsequently with 3-NPA. **B–D**, indicate that the strong suppression of NBT staining (a measure of SDH activity) seen in Experiment 1 was also observed in Experiment 2. Moreover, in the hippocampal dentate granule cell layer **(B)**, there was a significant increase of NBT staining from L-carnitine alone, compared to saline controls. #$p < 0.05$ compared to saline controls; *$p < 0.01$ compared to saline controls.

intensely stained. This impression was confirmed by a one-way repeated measures ANOVA of the optical density data, using the six brain regions as repeated measures within the five control animals. The result indicated a statistically reliable difference between the regions ($F(5,20) = 11.1, p < 0.01$). *Post hoc* comparisons between all the means performed pairwise by the Bonferroni approach indicated that caudate, cerebellum, and hippocampus had the highest levels of staining intensity, but were not significantly different from each other (FIG. 3). The thalamus had the lowest levels, but was not reliably less than occipital cortex and frontal cortex.

Experiment 2

As in Experiment 1, 3-NPA again virtually eliminated SDH histochemical staining (not shown), and were readily measured quantitatively in the hippocampal dentate gyrus as an effect of treatment group ($F(4,10) = 42.5, p < 0.001$, see FIGURE 4B). The *post hoc* tests indicated that all the groups receiving 3-NPA had significantly less SDH stain ($p < 0.001$) than controls, and that it made no difference on this measure whether or not they had been pretreated with L-carnitine. It may be of interest to note that in this case, the group that received 100 mg/kg carnitine alone had significantly *more* SDH stain ($p < 0.05$) than controls (FIG. 4B). When SDH staining was measured in neocortical tissue, there were similar effects of 3-NPA ($F(4,9) = 76.7$, p<0.001). Again all groups receiving 3-NPA showed significantly less SDH stain ($p < 0.001$) in their neocortical tissue than did controls, and there were no effects of L-carnitine pretreatment. However, in neocortex unlike in the hippocampal dentate gyrus, there was no observable increase of SDH staining ($p > 0.10$) in the L-carnitine alone group compared to the controls. Finally, the evaluation of striatal (caudate) tissue again reflected the loss of SDH staining produced by 3-NPA ($F(3,9) = 78.1, p < 0.001$). *Post hoc* tests of SDH staining in the corpus striatum revealed the same pattern as observed with the neocortex: all 3-NPA groups less stained ($p < 0.001$) than controls and no effects of L-carnitine.

DISCUSSION

Treatment with 3-NPA causes animals to be noticeably cold to the touch. The present data suggest that inhibition of SDH by 3-NPA is sufficient to block energy production to the extent that hypothermia ensues. For example, 3-NPA produced a progressive hypothermia in treated rats represented by measured decreases in core body temperature over a period of hours. The observed hypothermia was associated with inhibition of SDH that was demonstrated visually as diminished histochemical staining of brain SDH activity *in situ*. Core body temperature returned to normal by five days, suggesting that damaged SDH had been replaced and normal energy production had been restored. The densitometry data, however, show that SDH levels, at least in most brain regions, had still not returned to control levels even by five days. There is some evidence from Experiment 2 to suggest that L-carnitine treatment counteracted somewhat the severity of the hypothermia produced by 3-NPA. Since the anaerobic pathway should remain unaffected by 3-NPA, L-carnitine may increase energy availability via this route, since the aerobic paths through the TCA cycle seem thoroughly blocked. Perhaps the effects of putative neuroprotective agents would be more readily appreciated against a less potent, reversible inhibitor of the TCA cycle such as malonic acid rather than 3-NPA.

The normal regional distribution of SDH may in part explain the location of 3-NPA-induced neurological damage, when it occurs. Caudate-putamen had the most intense labeling, and has also been the area most consistently reported to incur damage.[6,9,12,14,15,21] However, within the hippocampus, which is also widely recognized to be sensitive to 3-NPA-induced damage,[6,21] the CA1 pyramidal neurons are most sensitive to damage. The dentate granule neurons (two synapses away) appear to have the greatest SDH (or at least histochemical staining intensity), but have never been reported to be damaged by 3-NPA. Perhaps the observation from Experiment 2 that this region may have uniquely been able to upregulate its energy generating capacity following L-carnitine exposure can explain its resistance to neuropathological damage. Moreover, the ventral thalamus is very subject to damage,[6] despite its normally low levels of SDH activity. However, thalamus was also the only region to rapidly recover from its depleted SDH levels back to a near-normal amount of enzyme activity. Thus, it is apparent that there may be some other properties of the tissues, besides simply their SDH activity, that can account for the regional selectivity of 3-NPA-induced neuronal cell death.

Experiment 2 indicated that L-carnitine was of limited effectiveness against the actions of 3-NPA; perhaps this could have been expected given the extreme potency of the irreversible blockade by 3-NPA of SDH for at least a week after a single moderate dose. The relatively simple biomarkers of core body temperature and NBT histochemical staining appear to be very useful means of monitoring the capability of the tissues to generate energy. Agents such as L-carnitine and others can be readily tested in the whole animal for their ability to restore energy generation as measured with these biomarkers. Perhaps future studies can further improve these measures by cleanly separate aerobic from anaerobic energy pathways in the brain and periphery. We also intend to address more attention in the future to testing neuroprotective agents against *reversible* antagonists of the TCA cycle, as these should be much more useful pharmacologically.

REFERENCES

1. SCHRODER, J.M. 1993. Neuropathy associated with mitochondrial disorders. Brain Pathol. **3**(2): 177–190.
2. HESTERLEE, S. 1999. Carnitine and coenzyme Q10: miracle cures or money down the drain? Quest **6**(1): (Internet Magazine, Muscular Dystrophy Association. <http://www.mdausa.org/publications/Quest/q61coq102.html>.
3. MING, L. 1995. Moldy sugarcane poisoning—a case report with a brief review. Clin. Toxicol. **33**(4): 363–367.
4. COLES, C.J., D.E. EDMONDSON & T.P. SINGER. 1979. Inactivation of succinate dehydrogenase by 3-nitropropionate. J. Biol. Chem. **254**(12): 5161–5167.
5. NONY, P.A., A.C. SCALLET, R.L. ROUNTREE, et al. 1999. 3-Nitropropionic acid (3-NPA) produces hypothermia and inhibits histochemical labeling of succinate dehydrogenase (SDH) in rat brain. Metab. Brain Dis. **14**(2): 83–94.
6. BINIENDA, Z., D.L. FREDERICK, S.A. FERGUSON, et al. 1995. The effects of perinatal hypoxia on the behavioral, neurochemical, and neurohistological toxicity of the metabolic inhibitor 3-nitropropionic acid. Metab. Brain Dis. **10**(4): 269–282.

7. JAMES, L.F., W.J. HARTLEY, M.C. WILLIAMS & K.R. VAN KAMPEN. 1980. Field and experimental studies in cattle and sheep poisoned by nitro-bearing Astragalus or their toxins. Am. J. Vet. Res. **41**(3): 377–382.
8. IKONOMIDOU, C. & L. TURSKI. 1996. Neurodegenerative disorders: clues from glutamate and energy metabolism. Crit. Rev. Neurobiol. **10**(2): 239–263.
9. HAMILTON, B.F. & D.H. GOULD. 1987. Nature and distribution of brain lesions in rats intoxicated with 3-nitropropionic acid: a type of hypoxic (energy-deficient) brain damage. Acta Neuropathol. **72**: 286–297.
10. GREENE, J.G., R.H. PORTER, R.V. ELLER & J.T. GREENAMYRE. 1993). Inhibition of succinate dehydrogenase by malonic acid produces an "excitotoxic" lesion in rat striatum. J. Neurochem. **61**(3): 1151–1154.
11. PALFI, S., R.J. FERRANTE, E. BROUILLET, et al. 1996. Chronic 3-nitropropionic acid treatment in baboons replicates the cognitive and motor deficits of Huntington's disease. J. Neurosci. **16**(9): 3019–3025.
12. BOWYER, J.F., P. CLAUSING, L. SCHMUED, et al. 1996. Parenterally administered 3-nitropropionic acid and amphetamine can combine to produce damage to terminals and cell bodies in the striatum. Brain Res. **712**: 221–229.
13. ZEEVALK, G.D., E. DERR-YELLIN & W.J. NICKLAS. 1995. Relative vulnerability of dopamine and GABA neurons in mesencephalic culture to inhibition of succinate dehydrogenase by malonate and 3-nitropropionic acid and protection by NMDA receptor blockade. J. Pharmacol. Exp. Ther. **275**(3): 1124–1133.
14. NISHINO, H., I. FUJIMOTO, Y. SHIMANO, et al. 1996. 3-Nitropropionic acid produces striatum selective lesions accompanied by iNOS expression. J. Chem. Neuroanat. **10**(3–4): 209–212.
15. BEAL, M.F., R.J. FERRANTE, R. HENSHAW, et al. 1995. 3-Nitropropionic acid neurotoxicity is attenuated in copper/zinc superoxide dismutase transgenic mice. J. Neurochem. **65**(2): 919–922.
16. FU, Y., F. HE, S. ZHANG & J. ZHANG. 1995. Lipid peroxidation in rats intoxicated with 3-NPA. Toxiconomics **33**: 327–331.
17. BINIENDA, Z. & C.S. KIM. 1997. Increase in levels of total free fatty acids (FFA) in rat brain regions following 3-nitropropionic acid administration. Neurosci. Lett. **230**: 199–201.
18. HAJOS, F. & S. KERPEL-FRONIUS. 1969. Electron histochemical observation of succinic dehydrogenase activity in various parts of neurons. Exp. Brain Res. **8**(1): 66–78.
19. LILLIE, R.D. & H.M. FULLMER. 1976. Enzymes. *In* Histopathologic Technic and Practical Histochemistry, 4th edit. Chapter 10, McGraw-Hill, New York. 471–476.
20. NISHINO, H., Y. SHIMANO, M. KUMAZAKI, et al. 1995. Hypothalamic neurons are resistant to the intoxication with 3-nitropropionic acid that induces lesions in the striatum and hippocampus via damage in the blood-brain barrier. Neurobiol. **3**(3–4): 257–267.

Role of Peroxynitrite in Neurodegeneration and Neuroprotective Role of Antioxidants and Peroxynitrite Decomposition Catalysts

SYED F. ALI,[a] SYED Z. IMAM,[a] GLENN D. NEWPORT,[a] YOSSEF ITZHAK,[b] AND WILLIAM SLIKKER, JR.[a]

[a]*Neurochemistry Laboratory, Division of Neurotoxicology, National Center for Toxicological Research/FDA, Jefferson, Arkansas, U.S.A.*

[b]*Department of Psychiatry and Behavioral Sciences, University of Miami School of Medicine, Miami, Florida, U.S.A.*

KEYWORDS: Peroxynitrite; Neurodegeneration; Antioxidant; Decomposition catalyst.

Oxidative stress, reactive oxygen (ROS) and nitrogen (RNS) species have been known to be involved in a multitude of neurodegenerative disorders such as Parkinson's disease (PD), Alzheimer's disease (AD), and amyotrophic lateral sclerosis (ALS). Both ROS and RNS have a very short half-life, thereby making their identification very difficult as a specific cause of neurodegeneration. Recently, we developed a high performance liquid chromatography/electrochemical detection (HPLC/EC) method to identify 3-nitrotyrosine (3-NT), an *in vivo* biomarker of peroxynitrite production, in brain to evaluate if an agent-driven neurotoxicity is produced by the generation of peroxynitrite. We have shown that a single or multiple injections of methamphetamine (METH) produced a significant increase in the formation of 3-NT in the striatum. Formation of 3-NT correlates with the striatal dopamine depletion caused by METH administration. We also report that pretreatment with antioxidants such as selenium and melatonin can completely protect against the formation of 3-NT and depletion of striatal dopamine. Furthermore, pretreatment with peroxynitrite decomposition catalysts such as 5,10,15,20-tetrakis (*N*-methyl-4'-pyridyl)porphyrinato iron III (FeTMPyP) and 5,10,15,20-tetrakis (2,4,6-trimethyl-3,5-sulfonatophenyl) porphinato iron III (FeTPPS) significantly protected against METH induced 3-NT formation and striatal dopamine depletion. We used two different approaches, pharmacological manipulation and transgenic animal models, in order to further investigate the role of peroxynitrite. We have shown that a selective neuronal nitric oxide synthase (nNOS) inhibitor, 7-nitroindazole (7-NI), significantly protected against the formation of 3-NT as well as striatal dopamine depletion. Similar results were observed with nNOS knockout mouse models. Peroxynitrite also played an important role in apoptotic gene expression. METH administration upregulated the protein expression of p53 and downregulated bcl-2 protein expression in the striatum of wildtype mice but not in the nNOS knockout mice. Together, these data support the hypothesis that the reactive nitrogen species, peroxynitrite, plays a major role in METH-induced dopaminergic neurotoxicity, and selective antioxidants and peroxynitrite decomposition catalysts can protect against METH-induced neurotoxicity. These antioxidants and decomposition catalysts may have therapeutic potential in the treatment of neurodegenerative disorders.

Part IX: Mechanistically Based Neuroprotection

QUESTION FOR DRS. ALI, BINIENDA, AND REITER

From Dr. Maynard

Do we know what the levels of melatonin are in Parkinson's disease patients? Given the remarkable recovery of dopamine in METH-induced toxicity *in vivo* that you have observed, would you support the treatment of Parkinson's disease patients with melatonin?

ANSWER: The reports on the levels of melatonin in Parkinson's disease patients vary widely; some claim the melatonin levels are lower than normal, whereas others assert they are the same as in age-matched controls.

In reference to the use of melatonin as a treatment for Parkinsonism, it is known that in experimental models of this disease in animals melatonin reduces the neural damage and neuronal loss (*Ann. N.Y. Acad. Sci.* **890:** 471–485, 2000). Considering its very low toxicity and high neural efficacy, melatonin would be worthy of a trial to defer the progression of this devastating condition.

(Ali) No, I do not know if there is any report about the level of melatonin in Parkinson's patients. If there is one, Dr. Reiter may know about it. Since our, as well as Dr. Reiter's and others', studies suggest that there is no toxicity, I definitely support the trial treatment of melatonin in Parkinson's patients.

QUESTIONS FOR DR. ALI

From Dr. Lin (Comment)

In my study, 10 mg/kg melatonin was used to prevent iron-induced oxidative injuries in rat. Therefore, the potency of the antioxidative effect of melatonin should be investigated. The therapeutic use of melatonin as antioxidants should be evaluated.

ANSWER: Yes, I agree with you.

From Dr. Manev

Is neuroprotection of antioxidants associated with prevention in methamphetamine-induced increased body temperature?

ANSWER: Yes, we found that antioxidants, such as melatonin and selenium, reduce body temperature as well as methamphetamine-induced hyperthermia.

From Dr. Slikker

Based on your data, oxidative stress plays a major role in METH-induced striatal dopamine decrease. How does this translate into neurotoxicity *in vivo*?

ANSWER: Based on our data, we think that methamphetamine-induced striatal dopamine release begins autooxidation and that you have the formation of free radicals, superoxide and nitric oxide, as shown in some of my slides. Once you have these two radicals, they react together and form a much more powerful and more active free radical, peroxynitrite, that nitrates the tyrosine at the tyrosine hydroxylase active center and forms 3-nitrotyrosine. We consider 3-nitrotyrosine as a biomarker that increases in striatal tissue after meth administration.

From Dr. Reiter

In reference to the hyperthermia mentioned by Dr. Manev, Dr. Ali mentioned melatonin prevented the hyperthermia caused by methamphetamine. I want to add that melatonin also prevents hyperthermia due to zymosan and carragheenan as well.

I presume that the effectiveness of selenium in reducing the toxicity of methamphetamine, which seems to work via $ONOO^-$, is due to the stimulation of glutathione peroxidase, which is reported to be a peroxinitrite reductase as well. Is that the interpretation of your data?

I will add one additional response in reference to melatonin and Parkinsonism. Melatonin has been shown to prevent DA depletion after MPTP and 6-OHDA; thus, in models of Parkinsonism, melatonin has been shown to reduce the neurochemical changes of this condition (Ann. N.Y. Acad. Sci. **890:** 471–485, 2000).

ANSWER: Yes, thank you Dr. Reiter.

QUESTIONS FOR DR. SCALLET

From Dr. Slikker

Knowing the role of mitochondria in brain function and damage, what approaches to mitochondrial manipulation may be used to provide neuroprotection?

ANSWER: An intriguing approach has been proposed employing mitochondrial toxins to produce natural selection. Perhaps the weaker, more mutated subpopulation of aging or genetically deficient mitochondria could be selectively purged. If so, perhaps the remaining healthy subpopulation would be propagated. Then an efficient remaining complement of mitochondria would be neuroprotective when stressed. A similar phenomenon may underly the many observations that pre-treating rats with hypoxia confers subsequent neuroprotection, probably through stimulation of the mitochondria to multiply. A cold, hypoxic winter visit to the top of the nearby Squaw Valley ski area might be wonderfully neuroprotective for those that can afford it! Of course other, non-physiological approaches involving substrate loading with carnitine, coenzyme Q10, and idebenone are being tried. However, it appears that more efficient ways to increase the energy flux through the oxidative phosphorylation chain must be sought.

From Dr. Jonas

I gather that 3-NPA can lower the temperature of an animal by $2\,°C$ in one hour without killing the animal. Dr. Auer has told us that a $2\,°C$ reduction can have a favorable effect on focal cerebral ischemia. Would (a drug with) 3-NPA (-like effect) be useful as an ambulance drug for acute stroke?

ANSWER: That is an intriguing possibility, providing the dose could be carefully limited below a neurotoxic level. However, 3-NPA is an irreversible inhibitor of succinate dehydrogenase (SDH), and its effects last until more SDH is synthesized—up to at least a week or so later in rats. The related compound malonate also inhibits SDH, but reversibly. If it shared the property of lowering body temperature, it might well be a better candidate for this type of application.

QUESTION FOR DR. BINIENDA

From Dr. Slikker

You mentioned that L-carnitine acts to protect against 3-NPA toxicity downstream from SDH. Please speculate where L-carnitine may exert its neuroprotective effect?

ANSWER: The neuroprotective effect of L-carnitine and its acylated form acetyl-L-carnitine against 3-NPA toxicity may be mediated by maintaining an equilibrium between acetylCoA and CoA levels and enhancement of beta oxidation resulting in the increase of ATP levels, but also via the reported effects of carnitine on metabotropic glutamate receptors that would prevent the secondary exicitotoxicity induced by 3-NPA.

QUESTION FOR DRS. SCALLET AND BINIENDA

From Dr. Virmani

The recovery of the SDH enzyme activity following 3-NPA inhibition in different parts of the brain was different. Is it dependent on antioxidant potential in that particular area of the brain? Could the L-carnitine action on the SDH activity following 3-NPA inhibition be in part related to potentiation of the beta oxidation pathway?

ANSWER: (Binienda) We have taken this possible effect into consideration in our study. However, compared to controls, no effect of L-carnitine on the SDH activity in animals injected with L-carnitine alone was observed.

(Scallet) Yes, L-carnitine enhances the ability of long-chain fatty acids to enter the mitochondrial matrix. These fatty acids may then substrate load the beta oxidation (TCA, oxidative phosporylation) pathway. If SDH (a Complex II enzyme) is not completely inhibited or if there are ways to bypass it, then L-carnitine could certainly promote the flux of energy through this pathway.

QUESTIONS FOR DR. YOUDIM

From Dr. Abbracchio

I was intrigued by your data on mitochondrial depolarization and apoptosis. Could you give us more details on the time course of mitochondrial depolarization? Moreover it would be nice to see if, upon addition of SIM-1 to your cells, there are also any changes to capase-δ, which is the mitochondial-associated caspase, and whether these changes are causally related to the activation of caspase-3, which is indeed the "final" effector of caspase-3.

ANSWER: We have noted that 6-hydroxy dopamine, methyl salsolinol and SIN-1 activate caspapse 3 and induce mitochondrial depolarization within minutes. This is presumably the first step in cell death. Indeed rasagiline and its optical isomer not only prevent the mitochondrial depolarization but also caspase-3 activation. This may account for their neuroprotective actions.

From Dr. Sablin

The protective effect of deprenyl in MPTP-induced neurotoxicity is not only due to the inhibition of MAO B, but also due to the inhibition of the DA reuptake system. Is anything known about the interaction of rasagiline with DA-transporters?

ANSWER: We have sufficient data showing that rasagiline is a more potent neuroprotective and MAO inhibitor that deprenyl and does not have significant DA uptake inhibitory activity. Thus neuroprotective action is not dependent on DA uptake. Indeed J.P.M. Finberg examined the effect of rasagiline on DA-transporter and found on chronic treatment it increases it. These data were published.

From Dr. Chuang

We have documented that overexpression and nuclear translocation of glyceraldehyde-3-phosphate dehydrogenase (GAPDH) is involved in neural apoptosis. In your presentation you only briefly touched this issue. Can you comment on the importance of blocking GAPDH nuclear translocation in mediating the neuroprotective effects of rasagiline and its derivatives?

ANSWER: This is a very good question and goes to the heart of what rasagiline does in its neuroprotective activity. Indeed we have very good evidence that inhibition of GAPDH nuclear translocation is one mechanism by which rasagiline (Maruyama, Naoi, and Youdim, submitted for publication) and its derivatives produce their neuroprotective activity. However, they also increase BCL2, SOD and catalase activities. They also prevent the fall in mitochondrial membrane potential.

QUESTIONS FOR DR. STEYN

From Dr. Sablin

MPTP may be metabolized by both MAO A and MAO B, although the catalysis by MAO B is more efficient. In your study you focused on MAO B only. This may be the reason why you do not observe neuroprotective effect by deprenyl.

ANSWER: Classical neuroprotection studies with (R)-deprenyl involved administering (R)-deprenyl before MPTP-treatment to mice in an attempt to prevent the MAO-mediated bioactivation of MPTP. However, in our study, chronic low dose (R)-deprenyl treatment was initiated 72 hours post MPTP administration to C57BL/6 mice. Since the MAO-mediated bioactivation of MPTP is almost complete within two hours in the brain of the C57BL/6 mouse, it is most unlikely that our (R)-deprenyl treatment protocol influenced the MAO-mediated bio-activation of MPTP. By following this reported (R)-deprenyl treatment protocol, we were able to investigate the neuronal restorative properties of (R)-deprenyl in C57BL/6 mice already lesioned by MPTP, which is totally different from classical neuroprotection studies.

From Dr. Chiueh

We have reported the spontaneous recovery of striatal dopamine in MPTP-treated C57Bl6 mice in 1985 at the first MPTP symposium. This spontaneous recovery of dopamine levels in MPTP-treated mice is due to lack of a significant nigral death in these mice. $R(-)$ selegiline can rescue MPP^+ -induced nigral injury if this compound is administered immediately after intracranial infusion of MPP^+ and followed with three to four treatments of small doses (less than 0.4 mg/kg). This reported neurorescue effect of selegiline is different from the present study of the putative neurotrophic effect of selegiline.

ANSWER: Our results on the "recovery" of neostriatal dopamine in MPTP-treated C57BL/6 mice are consistent with similar observations made by others in the past. The important point here is that recovery must be distinguished from "maturation". A major component of recovery in young mice treated with MPTP is likely to be a consequence of maturation. The point concerning protection against MPP^+ administered intracranially by infusion followed immediately by chronic administration of (R)-deprenyl certainly is of interest. The mechanism of such protection (possibly via free radical scavenging), however, is not likely to be relevant to the effects of (R)-deprenyl administered under the conditions of our experiments. The critical result of our study is that, irrespective of mechanism, (R)-deprenyl, when administered chronically in low doses starting three days after treating with a neurotoxic dose of MPTP, does not restore neostriatal dopamine levels. (See : R.E. Heikkila *et al*. Protection against the dopaminergic neurotoxicity of 1-methyl-4-phenyl-1,2,5,6-tetrahydropyridine by monoamine oxidase inhibitors. *Nature* **311**: 467–469, 1984; S.P. Markey *et al*. Intraneuronal generation of a pyridinium metabolite may cause drug-induced parkinsonism. *Nature* **311**: 464–467, 1984; K. Castagnoli *et al*. The neuronal nitric oxide synthase inhibitor 7-nitroindazole also inhibits the monoamine oxidase-B-catalyzed oxidation of 1-methyl-4-phenyl-1,2,3,6-tetrahydropyridine. *Chem. Res. Toxicol.* **10**: 364–368, 1997.)

QUESTION FOR DR. YOUDIM

From Dr. Slikker

You mentioned three or more possible mechanisms for the neuroprotective activity of rasagiline: which one do you think is most important and why?

ANSWER: At present we do not know what is the most important mechanism for rasagiline's neuroprotective activity. I presume it is the combination of these factors that give rasagiline the unusual neuroprotective activity. What we know is that the propargyl group on the molecule is essential and mandatory. Propargyl group reacts with MAO to inhibit the enzyme. But MAO inhibition is not that crucial because the S-isomer of rasagiline, which is not an MAO inhibitor, has similar neuroprotective activity. We are now examining, by microarray gene analysis, what genes they regulate. That may shed light on their mechanism of action.

QUESTIONS FOR DR. YENARI

From Dr. Slikker

The longevity of the gene transfer was described to last for approximately four days. This time course is excellent for stroke treatment but are there other vectors that provide a longer effect and thereby be useful for long-term treatment of neurodegeneration diseases such as Parkinson's or Alzheimer's?

ANSWER: Other vectors such adenovirus or retrovirus may express the transgene for longer periods of time, but no vector seems to express indefinitely. Some reports claim that retroviral vectors can alter *in vivo* phenotype for as long as 36 months. However, retroviruses will not work on postmitotic cells such as neurons. As you point out, this is probably not as critical an issue for stroke treatment as it is for more chronic diseases.

From Dr. Bowyer

Only a limited number of cells in the injection regions are transfected with the desired viral vector. How can these numbers be increased to protect more of the neurons in the injections sites?

ANSWER: Other vectors, such adenovirus, can generate higher titers than herpes simplex. Another approach might be to identify a transgene that acts remotely from the expression site, such that only a few cells need to be transfected to produce a transgene that can affect a greater number of neurons, or better yet, larger areas of brain. This approach has already been examined by Betz *et al.* (*JCBFM,* 1995). This group used an adenoviral vector to transfect ependymal cells that could overexpress an antiinflammatory gene. The transfected ependymal cells then produced sufficiently high levels of the transgene so that when animals were subjected to a stroke, the treated ones actually had smaller infarcts.

COMMENT

From Dr. Chiueh

We reported last week in *FASEB J.* that preconditioning stress induces upregulation of human NOS_1 gene and protein, which, in turn, increases tolerance of brain neurons to oxidative stress (*FASEB J.* **10:** 2000). The induction of NOS_1 also leads to tolerance of brain cells against MPP^+-induced neurotoxicity. We agree with your idea that identification of novel and known genes (i.e., NDNA repair, antioxidative genes) contributes to brain tolerance to oxidative stress and would facilitate the development of gene induction therapy for treating brain disorder caused by oxidative stress.

QUESTIONS FOR DR. CARBONE

From Dr. Slikker

Fever induced by virus—does it exacerbate induced cell death as a result of Borna virus infection?

ANSWER: Although Borna virus disease (BDV) has been shown to cause apoptotic cell death in the developing nervous system, very little is known about fever responses in natural or experimental BDV infection, or the role that fever plays in apoptosis of infected cells. In experimental BDV infections of horses, fever was noted only for a limited time (i.e., around the time of initial presentation of neurological disease).

In virus infections in general, however, there is some evidence that hyperthermia after infection (e.g., with encephalomyocarditis virus in mice) is associated with increased death of infected myocardial cells. However, the data suggested that these cells are dying due to necrotic rather than apoptotic mechanisms. In addition, in experimental Vaccinia virus infection, a virus protein is produced which blocks interleukin-1β-converting enzyme (ICE), with the result of protecting virus-infected cells from TNF and Fas-mediated apoptosis. Notably, the fever due to IL-1β is not blocked by this viral protein.

Clearly, this is an area of great interest that requires additional study.

From Dr. Reiter

Dr. Carbone, you showed nicely the similarity of the pathology of the cerebellum of autistic individuals and the virus-infected animals. You also showed the marked loss of neurons in the dentate gyrus of the virus-infected animals. Is the loss of neurons in the dentate gyrus also common in the brains from autistic individuals?

ANSWER: Controversies concerning changes in the hippocampus in autism exist. In part, the controversies may stem from the limited number of brains from autistic individuals available for study, and from the inability to evaluate changes in the hippocampus over regulated times before, during, and after the development of autistic disease symptoms. Nonetheless, in autistic patients versus controls, abnormalities in microscopic neuropathology, volume and function of the hippocampus have been reported. Neuropathological changes include reduction in neuron size and reduced dendritic branching. Reductions in hippocampus volume, relative to total brain volume, have also been reported. Functional studies have indicated biochemical disturbance in developing autistic brain, including asymmetrical levels of serotonin synthesis in the hippocampus.

In neonatally BDV-infected rats, we see 100% infection of the neurons of the dentate gyrus of the hippocampus, as well as a high degree of virus infection of the CA3 and CA4 regions. We know that this finding is associated with loss of cognitive functions (e.g., spatial learning and memory via the Morris Water Maze or MWM). Over time, infected neurons in the dentate gyrus are lost, and this neuroanatomical finding is associated with increasingly poor performance in the MWM. In addition, we have seen clear abnormalities in serotonin levels in the brains of the infected rats. As an animal model of autism spectrum disorders, BDV infection of the neonatal rat offers us the opportunity to carefully study the neuropathological, behavioral and neurochemical changes in the infected, developing brain over time, in a controlled setting.

Is the degeneration of the pyramidal neurons in virus-infected animals a consequence of the release of glutamate from the degenerating mossy fibers of the dentate neurons?

ANSWER: This is an excellent hypothesis that has yet to be tested. It is known that BDV likes to replicate in areas associated with neuroexcitatory amino acids. *In vitro,* BDV infection reduces the uptake of glutamate by astrocytes, an event that might very well result in neurotoxic damage to the pyramidal neurons. Although this hypothesis has yet to be tested *in vivo,* glutamate toxicity may very well prove to be important in the pathogenesis of BDV-induced neurotoxicity.

From Dr. von Lubitz

Studies have shown that exposure of autistic children to immersive virtual reality environments results in a significant improvement of social interaction skills. Considering the possibility of viral infection as the cause of autism, would you speculate on the nature of the eventual compensatory mechanism that may be involved?

ANSWER: Certainly, the bad news regarding virus infection in the developing brain is that delicate, carefully orchestrated developmental critical periods are very sensitive to damage by virus infection. The good news is that the specificity of the damage by a virus such as BDV might allow the young brain to develop the classic "work arounds" to recover some of the lost neurological functions. How this is accomplished in BDV is not known; however, we do have evidence that the social deficits (e.g., abnormal play) induced in neonatally BDV-infected rats are somewhat amenable to improvement following opportunities to play with uninfected rats. This improvement can be seen rapidly, within 24 hours. Thus, although the mechanisms of this partial recovery are unknown, the presence of this phenomenon suggests that improvement in social skills in virus-infected animals is a feasible goal.

From Dr. Marini

Is the substantia nigra affected in the Borna-treated animals? Are there differences in the pharmacokinetics of the two drugs responsible for difference in responses?

ANSWER: Abnormal dopamine metabolism and receptor levels have been reported in BDV infection, but only in the adult infected rats with frank encephalitis. In neonatally BDV-infected rats without inflammatory responses in the brain, we have measured normal dopamine levels and normal dopamine turnover.

Although BDV infection of the kidney and liver has been reported in rats infected at birth, the infection is low grade and no evidence of functional damage has been seen (and we have looked carefully for evidence of kidney dysfunction in the Lewis rat). Although it is unlikely that BDV infection alters the pharmacokinetics of the two drugs we tested, this question is a good one that needs to be addressed with additional studies.

From Dr. Ali

Have you or anyone else looked at the mechanism of action of this BDV model of autism? Do you think the mechanism may be mediated by excitotoxicity or nitric oxide?

ANSWER: As mentioned above, there are some data that BDV may interfere with the normal handling of excitatory amino acids (e.g., glutamate) and this certainly is a plausible mechanism for at least some of the damage observed. Nitric oxide abnormalities have been reported in rats with encephalitic forms of Borna disease (i.e., rats infected as adults), but have not been seen with rats infected as newborns, in the absence of a significant inflammatory infiltrate.

From Dr. Virmani

How long does the virus persist in the brain and is there potential to treat with antiviral drugs? Have you tried any antioxidant or neurotrophic agents in your models of autism in these animals?

ANSWER: Interestingly, it is rarely recognized that even in most animals/humans there is a normal virus flora of the brain—in humans this includes cytomegalovirus, JC virus, and herpes simplex virus. Nonetheless, BDV cannot be considered "normal" virus flora of the brain, because, in rats, and most animals evaluated, the virus infection persists for the life of the animal with continued replication (i.e., it does not go "latent" like herpes viruses).

Two drugs with apparent antiviral or documented antiviral effect have been used for *in vivo* or *in vitro* treatment of BDV infection. One, amantadine, lacks documented efficacy against virus replication, and is not approved for BDV treatment. The other drug, ribavirin, has shown some anti-BDV efficacy *in vitro*—however, ribavirin has significant toxic side effects and is also not approved for BDV treatment.

We have not yet tried antioxidant or neurotrophic agent treatments in our BDV model. Interestingly, it should be noted that some neurotrophic factors, identified (e.g., nerve growth factor) and as-yet-unidentified (e.g., produced by glial cells), appear to upregulate virus replication; therefore, it is important to consider the effects of neurotrophins on the virus as well as the neural cells.

Emerging Role of Lithium as a Neuroprotective Drug

Therapeutic Implications

DE-MAW CHUANG

Section on Molecular Neurobiology, National Institute of Mental Health, NIH, 10 Center Drive, MSC 1363, Bethesda, Maryland 20892, U.S.A.

KEYWORDS: Lithium; Neuroprotective drug; Therapeutic implications.

Lithium is most commonly used for the treatment of acute mania and the prophylactic therapy for manic depressive illness. Despite long and intensive research, the therapeutic mechanisms of lithium remain poorly understood. It has been recently suggested that there is a decrease in the volume and glial/neuronal number in discrete brain areas of unipolar and bipolar patients. We have attempted to study the neuroprotective effects of lithium in cultured CNS neurons and *in vivo* using animal models of neurological and neuropsychiatric disorders.

We found that lithium robustly protected against glutamate excitotoxicity mediated by NMDA receptors in neurons prepared from rat brains. The protective effects of lithium involved inactivation of NMDA receptors, occurred with an EC_{50} of about 1.3 mM, and required protracted pretreatment. Long-term lithium treatment in these neurons also downregulated proapoptotic p53 and Bax, but upregulated cytoprotective Bcl-2. Moreover, protracted lithium pretreatment blocked glutamate-induced increase of p53 and Bax, as well as cytochrome c release from mitochondria and subsequent caspase activation. Lithium also activated the major cell survival PI 3-kinase/Akt signalling pathway, and suppressed glutamate-induced loss of Akt-1 activity resulting from protein phosphatase activation. Glutamate also rapidly and robustly decreased the levels of phosphorylated CREB, and this effect was largely inhibited by lithium pretreatment.

In a focal ischemia modal of rats using middle cerebral artery occlusion (MCAO), we found that the brain lesion volume and neurological deficits were markedly reduced by pretreatment with therapeutically relevant doses of lithium. Preliminary results show that lithium administered after the onset of MCAO also significantly reduced brain infarct volume. In a Huntington's disease model of rats in which an excitotoxin was injected into the striatum to induce the apoptotic death of medium-sized neurons, we found that lithium pretreatment for one or 16 days markedly reduced striatal lesions, and this protective effect was associated with Bcl-2 overexpression in the striatum and cortex. These results suggest that the neuroprotective action of lithium might contribute to the therapeutic mechanisms of this drug. Moreover, our observations raise the possibility that lithium may have expanded use to treat neurodegenerative diseases, particularly those linked to excitotoxicity.

Antiviral Medications Improve Cerebrovascular Perfusion in HIV+ Non-Drug Users and HIV+ Cocaine Abusers

RONALD I. HERNING, WARREN E. BETTER, KIMBERLY TATE, AND JEAN L. CADET

Molecular Neuropsychiatry Section, National Institute on Drug Abuse, National Institutes of Health, Baltimore, Maryland, U.S.A.

ABSTRACT: Antiviral medications have been useful in delaying the time course of HIV infection. Antiviral medications have also been reported to delay or reduce symptoms associated with AIDS related dementia and to improve cortical perfusion. The mechanism for this improvement is unclear. Thus, this report studies the effects of antiviral medications on cerebral blood flow velocity in HIV+ cocaine abusers, HIV+ control individuals and appropriate control individuals. Thirty-two unmedicated HIV+ individuals (28 cocaine abusers and 4 control individuals), 22 HIV+ individuals using antiviral medications (16 cocaine abusers and 6 HIV+ control individuals), 47 HIV− cocaine abusers, and 27 control HIV− subjects were studied. Blood flow velocities were determined for the anterior and middle cerebral arteries using transcranial Doppler sonography. HIV+ individuals on antiviral medications had lower pulsatility values, suggesting decreased resistance in the cerebral blood vessels, in comparison to HIV+ individuals not taking antiviral medications. HIV+ cocaine abusers and HIV+ control individuals using antiviral medications had pulsatility values similar to HIV− control subjects. Antiviral medications appear to reduce these cerebrovascular perfusion deficits in HIV+ individuals. The antiviral medications appear to have a direct neuroprotective effect in addition to their antiviral effects. The neuroprotective role of antiviral medications requires further investigation.

KEYWORDS: Pulsatility; Blood flow velocity; TCD; Cocaine abuse; Non-drug users; Antiviral medications; HIV-induced dementia; Human immunodeficiency virus; HIV.

INTRODUCTION

Patients with HIV infection exhibit neurological symptoms that develop into dementia.[1–3] Patients with HIV-associated cognitive deficits or dementia have been reported to have cortical and subcortical lesions[4–5] and cerebral perfusion abnormalities on SPECT scans.[6–8] HIV+ patients with no evidence of cognitive deficits also show cerebral perfusion abnormalities.[9] Cognitive deficits[10–13] and cerebral perfusion deficits[14,15] were also observed in cocaine abusers that tested negative for HIV

Address for correspondence: Dr. Ronald I. Herning, Molecular Neuropsychiatry Section, NIDA/IRP, P.O. Box 5180, Baltimore, MD 21224, U.S.A. Voice: 410-550-1551.
rherning@intra.nida.nih.gov

infection. However, HIV+ cocaine-dependent patients who were methadone maintained were more impaired than HIV− cocaine-dependent patients.[16] On the other hand, no differences in cerebral perfusion using SPECT were observed between HIV+ and HIV− drug abusers,[17] nor between cocaine abusers and non-drug-using HIV+ individuals.[18] The limited drug abuse history and the presence of antiviral medication have made comparisons difficult in these studies.

It is possible that antiviral agents may protect against HIV-induced or cocaine-induced. neurological damage, but such evidence in human studies is limited. Atervidine, an antiviral drug, was shown to improve cerebral perfusion as measured by SPECT in two of four patients with AIDS dementia complex.[19] MRI studies have shown that protease inhibitors can decrease white matter and basal ganglia abnormalities in HIV+ patients[3] and AZT can normalize the abnormal EEG observed in HIV+ individuals.[20,21] Given the reported positive clinical efficacy of antiviral agents against HIV infection, we sought to determine if these agents might have direct beneficial effects on the cerebral perfusion of HIV+ cocaine abusers and HIV+ control subjects.

METHODS

Twenty-seven control HIV− subjects (HIV-C), four HIV+ control subjects not on antiviral medications (HIV+C) (CD4: mean 318, SD 211), six HIV+ control subjects on antiviral medications (HIV+CM) (CD4: mean 353, SD 179), 47 HIV− cocaine abusers (HIV−Coc), 16 HIV+ cocaine abusers on antiviral medication (HIV+CocM) (CD4: mean 240, SD 108), and 28 HIV+ cocaine abusers not on antiviral medications (HIV+Coc) (CD4: mean 439 SD 272) were studied. Of the 22 subjects on antiviral medication, 11 were on nucleoside analog(s), one was on protease inhibitor therapy and 10 were also receiving a nucleoside analog(s) and a protease inhibitor. Demographic data and drug use history obtained from the Addiction Severity Index[22] are presented in TABLE 1. Some of the cocaine abusers used other drugs or alcohol.

All individuals underwent extensive medical, psychological, urine toxicology, and laboratory evaluations before being accepted into the study. Exclusion criteria were: (1) major medical or psychiatric illnesses, (2) head injuries with loss of consciousness for greater than five minutes, and (3) evidence of any neurological abnormalities by history or structured examination. HIV+ subjects with current opportunistic infections were excluded. The research protocol was approved by the National Institute on Drug Abuse and Johns Hopkins Bayview Medical Center Institutional Review Boards for Human Research. Informed consent was obtained from all subjects. The 32 HIV+ abusers not on antiviral medications were counseled and referred to treatment for HIV infection.

Cocaine abusers were tested within 72 hours of admission to the research unit. Control subjects were tested as outpatients. Blood flow velocity was determined using a temporal window (zygomatic arch) for four arteries: right and left middle (MCA), and right and left anterior (ACA) cerebral arteries using pulsed transcranial Doppler sonography (Nicolet, Model TC2000). Mean velocity (V_m: cm/s),

TABLE 1. Demographic and drug history measures

	HIV−C n = 27	HIV+C n = 4	HIV+CM n = 6	Coc n = 47	HIV+Coc n = 28	HIV+CocM n = 16
Demographic Measures						
Age years	31.1 ± 8.2[a]	32.5 ± 8.9	36.3 ± 4.4	33.8 ± 5.4	36.5 ± 5.1	39.6 ± 5.2
Education years	13.0 ± 1.7	12.5 ± 1.0	12.7 ± 1.8	11.9 ± 1.5	11.4 ± 1.9	11.8 ± 1.0
Male percent	63.3	75.0	100.0	79.6	80.0	70.6
African American percent	83.3	100.0	66.7	91.8	93.3	94.4
Drug History						
Cocaine days[b]				17.5 ± 5.8	18.9 ± 10.2	10.8 ± 9.8
Cocaine years				8.8 ± 3.6	12.1 ± 6.1	8.9 ± 4.9
Alcohol days	1.1 ± 2.1	0.0	1.7 ± 2.2	9.0 ± 6.2	7.5 ± 7.2	7.9 ± 7.0
Alcohol years	5.0 ± 7.2	3.0 ± 6.0	11.7 ± 11.9	13.9 ± 6.1	15.4 ± 10.0	15.6 ± 10.0
Heroin days				2.1 ± 3.7	7.5 ± 11.3	3.5 ± 7.1
Heroin years				1.6 ± 2.8	6.3 ± 8.4	3.4 ± 5.4
Marijuana days		0.0	0.8 ± 2.0	2.3 ± 4.9	2.1 ± 5.6	3.6 ± 7.7
Marijuana years		0.8 ± 1.5	3.5 ± 7.6	7.9 ± 6.6	8.1 ± 9.0	12.0 ± 8.5
Cigarettes/day	5.1 ± 9.6	1.2 ± 2.5	17.7 ± 12.7	17.3 ± 12.8	13.9 ± 11.3	8.0 ± 7.3
Cigarette years	7.4 ± 11.6	4.8 ± 9.5	17.2 ± 11.3	13.7 ± 7.4	21.6 ± 11.6	16.3 ± 11.2

[a]Mean and standard deviation.
[b]Days indicate the number of days these drug was use during the previous 30 days.

systolic velocity (V_s: cm/s), diastolic velocity (V_d: cm/s), and pulsatility index ($PI = [V_s - V_d]/V_m$) were determined for each artery.

The primary hypothesis (whether the medicated HIV+ subjects differed from the unmediated HIV+ subjects in blood flow parameters) was first tested using medication status (HIV+, HIV+ w/medication) by side (right vs. left) ANOVA. If the later analysis was significant, a group (control vs. cocaine) by type (HIV–, HIV+, HIV+ w/medication) by side (right vs. left) ANOVA was used to test for differences in blood flow parameters among the six groups. Post hoc comparisons among group means were made using the planned comparison method.

RESULTS

The medication status by side analysis indicated that those HIV+ subjects taking antiviral medication had significantly smaller PI values than those subjects not taking medication for the MCA (unmedicated mean 0.956, SD 0.108; medicated mean 0.817, SD 0.097; $F(1,52) = 28.54$, $p < 0.0001$) and ACA (unmedicated mean 0.937, SD 0.122; medicated mean 0.791, SD 0.163; $F(1,52) = 21.28$, $p < 0.0001$). This analysis was not significant for the other blood flow velocity measures. A group by type by side ANOVA was then used to test for differences in PI among the six groups on the MCA and ACA to further pinpoint the nature of the differences. The group by

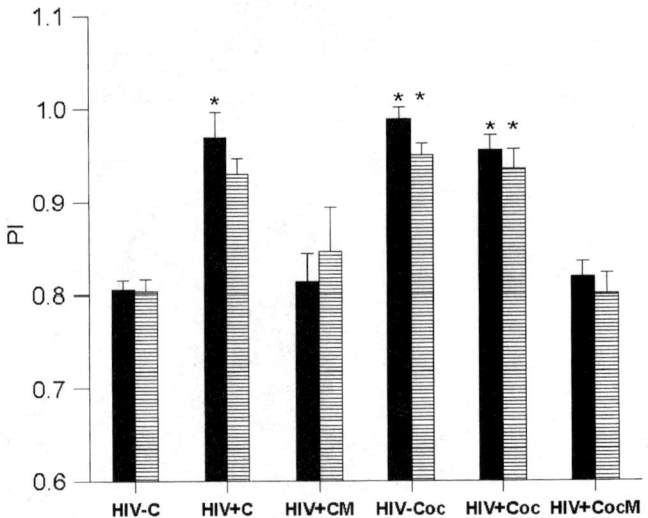

FIGURE 1. The pulsatility (PI) values for the anterior cerebral (*striped bars*) and middle cerebral (*solid bars*) arteries are plotted for the six groups of subjects. The values are averaged over the right and left arteries. The mean for each group and standard error is plotted. **The mean for both anterior cerebral and middle cerebral arteries for the HIV– Coc and HIV+ Coc groups were significantly ($p < 0.05$) higher than the values for the other groups. *The mean for middle cerebral artery for the HIV+C group was significantly ($p < 0.05$) higher than the mean for the HIV–C, HIV+CM, and HIV+CocM groups.

type interaction was significant for PI values from the MCA ($F(2,122) = 5.26$, $p < 0.01$) and approached significance for PI values from the ACA ($F(2,122) = 2.50$, $p < 0.09$). Since the group by type by side interaction was not significant for either the MCA or ACA, the PI values for the left and right side were averaged for presentation in FIGURE 1 and for the planned *post hoc* tests. As can be observed in FIGURE 1, The HIV+C, HIV+Coc, and HIV−Coc groups had larger PI values for the MCA than the HIV−C, HIV+CM, and the HIV+CocM groups. For the ACA, HIV+ groups (cocaine abusers and control) on antiviral medications had smaller PIs as compared to the HIV+ groups not on antiviral medications, but the difference only approached significance for HIV+ control group.

DISCUSSION

The principal finding of the study is that HIV+ cocaine abusers and HIV+ control subjects not on antiviral medications had blood flow velocity deficits. Both HIV+ groups (cocaine abusers and control) not on antiviral medications had significantly higher blood flow pulsatility values than the HIV− control group and their respective HIV+ groups in the middle cerebral arteries. Elevated pulsatility is thought to indicate increased cerebrovascular resistance in other patient populations secondary to vasoconstriction of small cortical vessels.[23–25] Thus, the increases in pulsatility in this study can be explained as increases in cerebrovascular resistance.

Cerebral perfusion deficits in both HIV− cocaine abusers[15,16] and HIV+ non-drug abusers[6–8] have been previously reported. SPECT cerebral perfusion deficits were observed between HIV− cocaine abusers and non-drug using HIV+ individuals, although the status of their treatment with antiviral medication was not indicated.[18] In the present study, perfusion deficits were observed in HIV− cocaine abusers and HIV+ individuals (cocaine abusers and non-drug abusers) who were not taking antiviral medications.

The present study using TCD methodology, however, found that HIV+ cocaine abusers and HIV+ control subjects had elevated pulsatility values whereas our HIV+ cocaine abusers and HIV+ control subjects who were using antiviral medications did not. These results suggest that antiviral medications may improve cerebral hemodynamics in seropositive individuals and possibly in seronegative cocaine abusers. The mechanism for improved cerebral perfusion in HIV+ individuals on medications is not clear, but there is some evidence that these drugs can produce improvement of cognition and cerebral perfusion. Antiviral medications have been reported to delay or reduce symptoms associated with AIDS-related dementia complex.[19,26] Nine of 14 patients with HIV-associated neurocognitive disorders had improved cortical perfusion after taking antiviral medications.[27] Protease inhibitors when added to nucleoside analog therapy improved cognition and MRI white matter signal abnormalities in HIV+ individuals.[3] Atervidine, was shown to improve SPECT perfusion in two of four patients with ADC.[19] These observations are in agreement with our present demonstration that relatively normal perfusion is observed in individuals taking nucleoside analog medications alone or in combination with protease inhibitor therapy.

A common mechanism might account for the poor cerebral perfusion observed in HIV– cocaine abusers, HIV+ cocaine abusers and HIV+ non-drug users. Cocaine-induced vasoconstriction appears to be mediated through the release of endothelin[28] and the inhibition of nitric oxide release.[29] Endothelin and nitric oxide systems may, also, be involved in cerebrovascular vasoconstriction in HIV disease and secondarily in HIV-associated dementia.[30,31] Thus, the perfusion deficits observed in cocaine abusers and in HIV patients may be secondary to similar biochemical phenomena. Nucleoside analogs in addition to their role in the inhibition of DNA synthesis have been found to stimulate nitric oxide production.[32] Nitric oxide has vasodilator effects.[33] Thus, nucleoside-induced nitric oxide production might be responsible for the improved cerebral blood flow in the medicated HIV+ individuals observed in this report.

The study reported here has several limitations. First, the size of the HIV+ control groups is small. A larger sample HIV+ individual is currently being sought to remedy this problem. Second, the dose, duration, and types of antiviral medications varied across the subjects tested in this cross-sectional study. Although a placebo active medication study cannot be ethically undertaken in this area, a study evaluating different HIV medications could be designed to follow patients' cerebral perfusion upon discovery of HIV infection and after a fixed period of medication use. Finally, the similarity of the perfusion deficits in HIV infection and prolonged cocaine use could be a logical conclusion of this research. Clearly, the nature of the perfusion deficits in these two disorders requires further study.

In conclusion, our findings suggest that antiviral medications may reduce the cerebral perfusion deficits in HIV+ individuals and this reduction may prevent or delay AIDS-related dementia. The ability of these agents to promote cerebral perfusion with chronic use also suggest the study of antiviral agents for other neuroprotective applications.

REFERENCES

1. NAVIA, B.A. & R.W. PRICE. 1987. The acquired immunodeficiency syndrome dementia complex as the presenting or sole manifestation of human immunodeficiency virus infection. Arch. Neurol. **44:** 65–69.
2. BENCHERIF, B. & D.A. ROTTENBERG. 1998. Neuroimaging of the AIDS dementia complex. AIDS **12:** 233–244.
3. FILIPPI, C.G., G. SZE, S.J. FARBER, et al. 1998. Regression of HIV encephalopathy and basal ganglia signal intensity abnormality at MR imaging with AIDS after the initiation of protease inhibitor therapy. Radiology **206:** 491–498.
4. BRODERICK D.F., D.F. WIPPOLD, D.B. CLIFFORD, et al. 1993. White matter lesions and cerebral atrophy on MR images in patients with and without AIDS dementia complex. Am. J. Roentgenol. **161:** 177–181.
5. AYLWARD, E.H., P.D. BRETTSCHNEIDER, J.C MCARTHUR, et al. 1995. Magnetic Resonance Imaging measurement of gray matter volume reductions in HIV dementia. Am. J. Psychiatry **152:** 987–994.
6. TRAN DINH, Y.R., H. MAMO, J. CERVONI, et al. 1990. Disturbances in cerebral perfusion of human immune deficiency virus-1 seropositive asymptomatic subjects: a quantitative tomography study of 18 cases. J. Nucl. Med. **31:** 1601–1607.
7. SACKTOR, N., I. PROHOVNIK, R.L. VAN HEERTUM, et al. 1995. Cerebral single-photon emission computed tomography abnormalities in human immunodeficiency virus type 1-infected gay men without cognitive impairment. Arch. Neurol. **52:** 607–611.

8. SACKTOR, N., R.L.VAN HEERTUM, G. DOONEIEF, et al. 1995. A comparison of cerebral SPECT abnormalities in HIV-positive homosexual men with and without cognitive impairment. Arch. Neurol. **52:** 1170–1173.
9. CHANG, L., T. ERNST, M. LEONIDO-YEE & O. SPECK. 2000. Perfusion MRI detects rCBR abnormalities in early stages of HIV-cognitive motor complex. Neurology **54:** 389–396.
10. HERNING, R.I., B.J. GLOVER, B. KOEPPL, et al. 1991. Cognitive deficits in abstaining cocaine abusers. In Residual Effects of Abused Drugs. J. Spencer & J.J. Boren, Eds.: 167–187. National Institute on Drug Abuse Research Monograph 101, U.S. Government Printing Office, Wasshinton, D.C.
11. EASTON, C. & L.O. BAUER. 1997. Neuropsychological differences between alcohol-dependent and cocaine-dependent patients with and without a drinking problem. Psychiatry Res. **71:** 103–107.
12. BOLLA, K.I., J.L. CADET & E.D. LONDON. 1998. The neuropsychiatry of chronic cocaine abuse. J. Neuropsych. **10:** 280–289.
13. BOLLA, K.I., F.R. FUNDERBURK & J.L. CADET. 2000. Differential effects of cocaine and cocaine + alcohol on neurocognitive performance. Neurology **54:** 2285–2293.
14. HOLMAN, B.L., P.A.CARVALHO, J. MENDELSON, et al. 1991. Brain perfusion is abnormal in cocaine-dependent polydrug users: A study using technetium-99m-NMPAO and ASPECT. J. Nucl. Med. **32:** 1206–1210.
15. HERNING, R.I., D.E. KING, W. BETTER & J.L. CADET. 1999. Neurovascular deficits in cocaine abusers. Neuropsychopharm. **21:** 110–118.
16. AVANTS, S.K., A. MARGOLIN, T.J. MCMAHON & T.R. KOSTEN. 1997. Association between self-report of cognitive impairment, HIV status, and cocaine use in a sample of cocaine-dependent methadone-maintained patients. Addict. Behav. **22:** 599–611.
17. BELDARRAIN, M.G., J.C. GARCIA-MONCO, V. LLORENS, et al. 1994. Neuropsychological differences but not regional cerebral blood changes in asymptomatic HIV-1 positive and negative drug addicts. Eur. Neurol. **34:** 193–198.
18. HOLMAN BL, B. GARADA, K.A. JOHNSON, et al. 1992. A comparison of brain perfusion SPECT in cocaine abusers and AIDS dementia complex. J. Nucl. Med. **33:** 1312–1315.
19. SZETO, E.R., J. FREUND, B.J. BREW, et al. 1998. Cerebral perfusion scanning in treating AIDS dementia: a pilot study. J. Nucl. Med. **39:** 298–302.
20. GRASSI, B., M. LOCATELLI, A. LAZZARIN & S. SCARONE. 1996. Temporal lobe electroencephalogram power modifications during olfactory stimulation in HIV-infected. AIDS Res. Hum. Retrovir. **12:** 547–451.
21. BALDEWEG, T. & J.H. GRUZELIER. 1997. Alpha EEG activity and subcortical pathology in HIV infection. Int. J. Psychophysiol. **26:** 431–442.
22. MCLELLAN, A.T., L. LUBORSKY, J. CACCIOLA, et al. 1986. Guide to the Addiction Severity Index: Background, Administration, and Field Testing Results. National Institute on Drug Abuse, Treatment Research Reports. U.S. Government Printing Office, Washington, D.C.
23. MARTIN, P.J., D.H. EVANS & A.R. NAYLOR. 1994. Transcranial color-coded sonography of basal cerebral circulation: reference data from 115 volunteers. Stroke **25:** 390–396.
24. BIEDERT, S., W. HWER & H. FORST. 1995. Multiinfarct dementia vs Alzheimer's disease: sonographic criteria. Angiology **46:** 129–135.
25. CHO, S.J., G.W. KIM & Y.H. SOHN. 1997. Blood flow velocity changes in the middle cerebral artery as an index of chronicity of hypertension. J. Neurol. Sci. **50:** 77–80.
26. WILKINSON, I.D., S. LUNN & K.A. MISZKIEL. 1997. Proton MRS and quantitative MRI assessment of the short term neurological response to antiretroviral therapy in AIDS. J. Neurol. Neurosurg. Psychiat. **63:** 477–482.
27. TOZZI, V., S. GALGANI, P. BALESTRA, et al. 1998. Regression of neurocognitive impairment with protease inhibitor based regimens. Int. Conf. AIDS **12:** 559–560.
28. WIBERT-LAMPEN, U., C. SELIGER, T. ZILKER & R.M. ARENDT. 1998. Cocaine increases the endothelin release of immunoreactive endthelin and its concentrations in human plasma and urine: reversal by coincubation with sigma-receptor antagonists. Circulation **98:** 385–390.

29. Mo, W., A.K. Singh, J.A Arruda & G. Dunea. 1998. Role of nitric oxide in cocaine-induced acute hypotension. Am. J. Hypert. **11:** 708–7124.
30. Giovannoni, G., R.F. Miller, J.S. Heales, *et al.* 1998. Elevated cerebrospinal fluid and serum nitrate and nitrite levels in patients with central nervous system complications of HIV-1 infection: Correlation with blood-brain-barrier dysfunction. J. Neurol. Sci. **1156:** 53–58.
31. Horti, K., P.R. Burd, K. Furuke, *et al.* 1999. HIV-1 infected macrophages induce inducible nitric oxide synthase and nitric oxide (NO) production in astrocytes: astrocytic NO as a possible mediator of neural damage in acquired immunodeficeincy syndrome. Blood **93:** 1843–1850.
32. Zidek, Z., A. Holy & D. Frankova. 1997. Antiretroviral agent (R)-9-(2-phosphonethoxypropyl) adenine stimulates cytokine and nitric oxide production. Eur. J. Pharmacology **331:** 245–252.
33. Toda, N. & T. Okamura. 1996. Neurogenic nitric oxide (NO) in the regulation of cerebroarterial tone. J. Chem. Neuroanat. **10:** 295–265.

Marijuana Abusers Are at Increased Risk for Stroke

Preliminary Evidence from Cerebrovascular Perfusion Data

RONALD I. HERNING, WARREN E. BETTER, KIMBERLY TATE, AND JEAN L. CADET

Molecular Neuropsychiatry Section, National Institute on Drug Abuse, National Institutes of Health, Baltimore, Maryland, U.S.A.

ABSTRACT: We have recorded blood flow velocity in the anterior and middle cerebral arteries by transcranial Doppler sonography in abstinent marijuana abusers ($n = 16$) and control subjects ($n = 19$) to assess the effects of prolonged marijuana use of the cerebrovascular system. The pulsatility index, a measure of cerebrovascular resistance, and systolic velocity were significantly ($p < 0.005$) increased in marijuana abusers compared to the control subjects. These findings suggest that cerebral perfusion observed in 18–30 year old marijuana abusers is comparable to that of normal 60 year-olds. Thus, chronic abuse of marijuana might be a risk factor for stroke.

KEYWORDS: Pulsatility; Blood flow velocity; TCD; Marijuana abuse; Stroke.

INTRODUCTION

Marijuana continues to be abused throughout the world, with increased use among young persons.[1,2] Cognitive impairment,[3] increased emergency department mentions of marijuana,[4] and case reports of marijuana-associated stroke in relatively young individuals[5–7] have been documented. These marijuana-induced cognitive impairments and cases of stroke may be related to reduced cerebral blood flow in chronic marijuana abusers.[8]

We recorded blood flow velocity from cerebral arteries using transcranial Doppler sonography[9] (TCD) from marijuana abusers during a month of monitored abstinence, to determine if the perfusion changes observed in early abstinence were simply part of marijuana withdrawal syndrome or were a residual and perhaps permanent cerebrovascular deficit.

Address for correspondence: Dr. Ronald I. Herning, Molecular Neuropsychiatry Section, NIDA/IRP, P.O. Box 5180, Baltimore, MD 21224, U.S.A. Voice: 410-550-1551.
rherning@intra.nida.nih.gov

METHODS

Subjects

Sixteen 18–30 year-old male marijuana abusers and 19 control males were studied. All marijuana abusers met the DSM-IIIR criteria for marijuana dependence or abuse using the Diagnostic Interview Schedule.[10]

Procedures

Blood flow velocity was recorded for the right and left middle (MCA), and right and left anterior (ACA) cerebral arteries using pulsed transcranial Doppler sonography (Nicolet, Model TC2000). Systolic velocity (V_s: cm/s), diastolic velocity (V_d: cm/s), and pulsatility index ($PI = [V_s - V_d]/V_m$) were determined for each artery. The inpatient marijuana abusers were tested within 72 hours of admission and after 28 to 30 days of monitored abstinence on the closed research ward. The 19 control male subjects were tested once as outpatients.

RESULTS

The marijuana abusers had significantly higher PI than the control subjects on both cerebral arteries. The PI for MCA and ACA did not change during the month of monitored abstinence. FIGURE 1 shows the plots of the mean PI values.

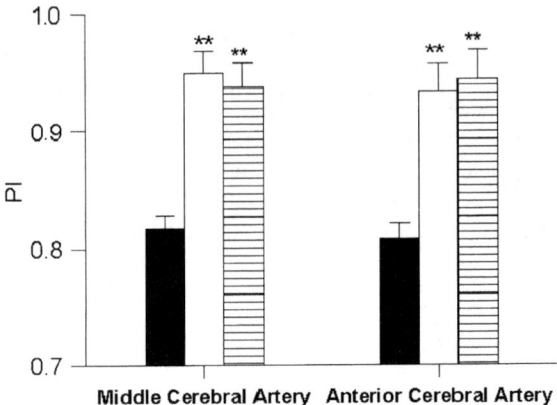

FIGURE 1. The PI mean and standard error are plotted for marijuana and control group. **The mean values for marijuana group on the MCA and ACA at both test times are significantly greater than for the control group ($p < 0.01$). ■, control; □, marijuana, 72 h; ▤, marijuana, 28 days.

DISCUSSION

Our observations using TCD sonography for this sample of young marijuana abusers reflect a prolonged deficit in cerebral hemodynamics. The PI for both the MCA and ACA were increased in marijuana abusers as compared to control subjects. Increased pulsatility in certain patient populations indicate increased cerebrovascular resistance, which was due to vasoconstriction of small and large cortical vessels.[9,11] The elevated PI values observed in 18–30 year old marijuana abusers in the present study were found to be similar to that of 60-year-old individuals.[9] The pulsatility index in the MCA or ACA did not change in the marijuana abusers after a month of monitored abstinence. These deficits in cerebrovascular perfusion were observed beyond the time when acute withdrawal symptoms were previously reported in this population.[12] Thus, these perfusion deficits in marijuana abusers do not appear to be related to a temporary marijuana withdrawal syndrome. Our findings suggest that the vasculature of 18–30 year old marijuana abusers might be comparable to that of 60 year olds and may be of clinical relevance since the elderly are likely to be at risk for stroke.

REFERENCES

1. BAUMAN, A. & P. PHONGSAVAN. 1999. Epidemiology of substance use in adolescence: prevalence, trends and policy implications. Drug Alcohol Depend. **55:** 187–207.
2. HEYMAN, R.B., T.M ANGLIN, S.M. COPPERMAN, *et al.* 1999. American Academy of Pediatrics. Committee on Substance Abuse. Marijuana: a continuing concern for pediatricians. Pediatrics **104:** 982–985.
3. EHRENREICH, H., T. RINN, H.J. KUNERT, *et al.* 1999. Specific attentional dysfunction in adults following early start of cannabis use. Psychopharmacol. **142:** 295–301.
4. COMMUNITY EPIDEMIOLOGY WORK GROUP. 1998. Assessing Drug Abuse Within and Across Communities. National Clearing House for Drug and Alcohol Information, # BKD256, Government Printing Office #017-024-016414-4.
5. ZACHARIAH, S.B. 1991. Stroke after heavy marijuana smoking. Stroke **22:** 406–409.
6. LAWSON, T.M. & A. REES. 1996. Stroke and transient ischemic attacks in association with substance abuse in a young man. Postgrad. Med. J. **72:** 693.
7. WHITE, D., D. MARTIN, T. GELLAR & T. PITTMANN. 2000. Stroke associated with marijuana abuse. Pediatr. Neurosurg. **32:** 92–94.
8. AMEN, D.G. & M. WAUGH. 1998. High resolution brain SPECT imaging of marijuana smoker with AD/AD. J. Psychoact. Drugs **30:** 209–214.
9. MARTIN, P.J., D.H. EVANS & A.R. NAYLOR. 1994. Transcranial color-coded sonography of basal cerebral circulation: reference data from 115 volunteers. Stroke **25:** 390–396.
10. ROBINS, S., J.E. HELZER, L. CUTTLER & E. GOLDING. 1988. National Institute of Mental Health Diagnostic Interview Schedule Version III-R. U.S. Government Printing Office, Washington D.C.
11. GRUBB, B.P., H. HAHN, L. ELLIOTT, *et al.* 1998. Cerebral syncope: loss of consciousness associated with cerebral vasoconstriction in the absence of systemic hypotension. Pacing Clin. Electrophysiol. **21:** 652–658.
12. JONES, R.T., N.L. BENOWITZ & R.I. HERNING. 1981. Clinical relevance of cannabis tolerance and dependence. J. Clin. Pharmacol. **21:** 143S–152S.

Delayed Multidose Treatment with Nicotinamide Extends the Degree and Duration of Neuroprotection by Reducing Infarction and Improving Behavioral Scores up to Two Weeks Following Transient Focal Cerebral Ischemia in Wistar Rats

KENNETH I. MAYNARD, ISSAM A. AYOUB, AND CHIUNG-CHYI SHEN

Neurophysiology Laboratory, Neurosurgical Service, Massachusetts General Hospital and Harvard Medical School, Boston, Massachusetts, U.S.A.

ABSTRACT: A single, delayed dose of nicotinamide (NAm) was shown to be protective against focal cerebral ischemia in rats, but the protection was limited to three to seven days following stroke. The investigation reported here was conducted to examine if the use of multiple doses of NAm, administered after the onset of focal cerebral ischemia, would extend the duration of neuroprotection compared with a single dose treatment regimen. Male Wistar rats were subjected to transient focal cerebral ischemia by occluding the right middle cerebral artery (MCAo) for two hours. Following MCAo, motor and sensory behavioral tests were performed daily and the cerebral infarct volumes were measured at two weeks after sacrifice. Each animal was placed into one of four groups that received either normal saline alone (Group S), one (Group A), two (Group B), or three (Group C) doses of NAm (500 mg/kg). Each animal, therefore, received three treatments over two weeks, with the first dose administered intravenously two hours after the onset of MCAo. Single and multiple doses of NAm reduced the infarction ($p < 0.01$) and improved ($p < 0.05$) the neurologic sensory and motor behavior when compared with the saline-treated animals up to two weeks after stroke. Moreover, animals that received multiple doses of NAm recuperated full motor function not different from normal, preoperative motor behavior. Delayed treatment with NAm given as multiple doses, therefore, further enhances the extent and duration of neuroprotection by significantly reducing cerebral infarct volumes, improving neurologic behavioral scores, and confers a complete motor recovery up to two weeks from the onset of focal cerebral ischemia in Wistar rats.

KEYWORDS: Middle cerebral artery occlusion; Stroke; Niacin; Neuroprotection; Rats.

Address for correspondence: Dr. Kenneth I. Maynard, Neurophysiology Laboratory, Neurosurgical Service, Massachusetts General Hospital, 55 Fruit St., EDR 414, Boston, MA 02114, U.S.A. Voice: 617-724-5329; fax: 617-726-3926.
maynard@helix.mgh.harvard.edu

INTRODUCTION

Stroke is a major public health problem, affecting as many as 700,000 Americans annually.[1] Depending on the size and location of the cerebral infarct, resulting from a stroke a number of neurologic deficits can ensue, including focal motor weakness, sensory loss, visual damage, speech comprehension and/or expression impairment, cognitive, and memory disturbances.

Nicotinamide (NAm), administered two hours after the onset of stroke modeled by permanent or transient focal cerebral ischemia, reduces cerebral infarction and improves the behavioral deficits in various strains of male and female rats.[2–4] A single delayed, intraperitoneal dose of NAm reduced cerebral infarction up to three days, but not at seven days, protected against sensory and motor behavioral deficits, and improved weight gain in rats up to one week postinsult.[3] Furthermore, a single, delayed, intravenous (IV) treatment of NAm (500 mg/kg) improved the degree of neuroprotection observed in comparison with a single, delayed, intraperitoneal (IP) treatment.[4] In the present experiments, therefore, we investigated whether the administration of multiple doses of NAm to animals subjected to transient focal cerebral ischemia would enhance the degree and duration of the neuroprotective effects as compared to a single dose of NAm. We assessed the cerebral infarct volume, the neurologic sensory and motor behavior and the weight of the animals for a followup period of 14 days.

MATERIALS AND METHODS

All experimental procedures were approved by the Subcommittee on Research Animal Care of the Massachusetts General Hospital, whose standards meet that of the Federal and State reviewing organizations.

Animal Preparation and Monitoring

Adult male Wistar rats ($n = 55$) weighing 270–300 grams (Charles River Laboratories, Wilmington, MA, USA) were operated on, of which 45 rats survived and were analyzed for this study. Mortality was, therefore, 18.2% (10/55) and occurred almost equally across the groups of animals. In most cases, death occurred during the first three days postoperation. Animals were allowed free access to food and water before and after surgery. Briefly, rats were anesthetized using halothane anesthesia (1–2% in 50% N_2O/50% O_2) in free breathing animals whose body temperatures were kept stable at $36.5 \pm 0.5°C$ using a feedback regulating heating pad and a rectal probe (Yellow Springs Instruments, OH, USA). The right femoral artery and vein were cannulated for measurement of arterial blood gases, blood glucose, hematocrit, mean arterial blood pressure, and heart rate. These physiological parameters were monitored before and after middle cerebral artery occlusion (MCAo). A venous catheter was kept under the skin for the first treatment only, which was given by intravenous (IV) injection.

Surgery

All rats were subjected to two hours of right MCAo, induced using a well established and modified procedure.[5,6] Under the operating microscope, the right common carotid artery was exposed through a midline incision in the neck. A 4-0 monofilament nylon suture with its tip rounded by heating over a flame and subsequently coated with poly-L-lysine (Sigma Chem. Co., USA) was introduced into the external carotid artery and then advanced into the internal carotid artery for a length of 18–19 mm from the bifurcation. This method placed the tip of the suture at the origin of the anterior cerebral artery, thereby occluding the middle cerebral artery. The suture was left in place for two hours and the animals were allowed to awaken from the anesthesia following closure of the operation sites. The suture was gently removed at two hours after MCAo following another brief period of anesthesia.

Drug Administration

NAm (500 mg/kg) or an equivalent volume (270–300 μL) of saline (vehicle) was administered daily. The first treatment was a bolus intravenous (IV) injection at the time of reperfusion (i.e., two hours after the onset of MCAo). Subsequent injections were given intraperitoneally (IP), everyday, over the 14 days of the study. Groups A, B, and C received saline injections every day, except: Group A ($n = 11$) received one dose of NAm on the day of the surgery (day 0) only; Group B ($n = 11$) received NAm on days 0 and 5, and Group C ($n = 12$) was given NAm on days 0, 5, and 10. Group S ($n = 11$) received saline injections every day.

Infarct Measurement

After anesthesia with ketamine (44 mg/kg, IP) and xylazine (13 mg/kg, IP), the animals were decapitated on day 14 after MCAo. Each brain was rapidly removed, sliced into seven 2-mm coronal sections using a rat matrix (RBM 4000C, ASI Instrument Inc., Warren, MI) and stained according to the standard 2, 3, 5-triphenyltetrazolium chloride (TTC) method.[7] Each slice was drawn using a computerized image analyzer (Bioquant, R&M Biometrics, Inc., Nashville, TN).[8] The calculated infarction areas were then compiled to obtain the infarct volumes per brain (in mm^3). Infarct volumes were expressed as a percentage of the contralateral hemisphere volume to compensate for edema formation in the ipsilateral hemisphere.[9]

Neurobehavioral Assessment

Each animal was tested for neurologic behavioral function immediately before surgery and then once each day throughout the 14 days. Sensory and motor tests were performed in this study. Both tests were modified versions of previously published methods.[6,10] The sensory test evaluated the forward and lateral placements of the affected forelimb and consisted of three grades: 0, no deficit; 1, delayed placing of less than two seconds; 2, delay (greater than 2 sec.) or absence of placement. The lateral and frontal scores were summed for statistical analysis (i.e., scoring 0 to 4). The motor test had five scores: 0, no deficit; 1, forelimb flexion; 2, forelimb flexion and decreased resistance to lateral push; 3, forelimb flexion, decreased resistance to lateral push, and unilateral circling; and 4, forelimb flexion and no spontaneous walking. The body weight of each animal was measured concurrently with the neurobehavioral test.

Statistical Analysis

Physiological and infarction volume data were analyzed by one-way ANOVA, and body weights were analyzed using repeat measures ANOVA. Fisher LSD *post hoc* or Tukey-Kramer tests were used following a significant ANOVA statistical outcome. The neurobehavioral scores were analyzed using the non-parametric Mann-Whitney test.

RESULTS

Physiological Variables

TABLE 1 shows the physiological parameter data obtained from the 45 rats that completed this study. All parameters were kept within normal physiological limits before and after occlusion with no significant differences between the groups.

Infarct Volume

The infarct volumes produced by the two-hour MCAo were measured and expressed as a percentage of the contralateral cerebral hemisphere. These infarcts were significantly reduced ($p < 0.01$) in all the NAm-treated groups (i.e., Groups A, B, and C) compared with the saline-injected group (Group S). There was no statistically significant difference among groups A, B, and C (see FIGURE 1).

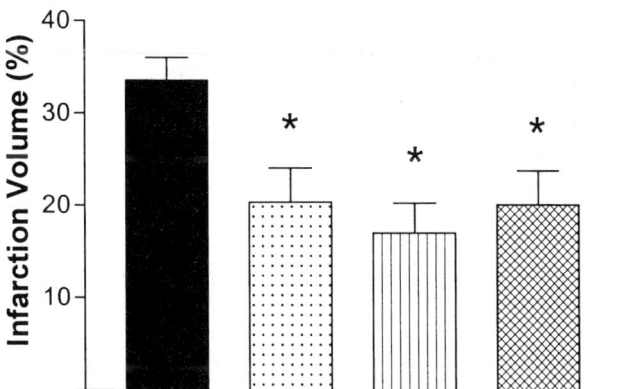

FIGURE 1. Nicotinamide (NAm, 500 mg/kg), injected once (Group A, ▭), twice (Group B, ▭), or three times (Group C, ▭) during 14 days, beginning at two hours after the onset of transient (2-h) middle cerebral artery occlusion significantly reduced the cerebral infarct volume in male Wistar rats. Data are expressed as a percentage of the contralateral (control) hemisphere and represented as mean ± standard error of the mean. Each group consisted of 11 or 12 animals; *$p < 0.05$ compared to the saline-treated animals (Group S, ▬).

TABLE 1. Physiologic parameters

	Group S n = 11	Group A n = 11	Group B n = 11	Group C n = 12
Preischemia				
pH	7.44 ± 0.01	7.48 ± 0.01	7.45 ± 0.01	7.47 ± 0.01
PCO_2 (mmHg)	34.8 ± 1.04	35.2 ± 0.85	34.6 ± 1.05	34.1 ± 0.76
PO_2 (mmHg)	165.0 ± 10.3	171.3 ± 5.1	180.8 ± 7.9	172.7 ± 8.4
Hct (%)	38.3 ± 0.2	38.2 ± 0.2	38.3 ± 0.1	38.5 ± 0.1
Gluc (mg/dL)	111 ± 2.0	114 ± 2.4	114 ± 2.0	116 ± 2.9
MABP (mmHg)	90 ± 1.3	88 ± 2.3	87 ± 2.0	85 ± 2.1
HR (beats/min)	371 ± 6.6	363 ± 8.7	355 ± 5.7	357 ± 6.0
Postischemia				
PH	7.41 ± 0.01	7.44 ± 0.01	7.43 ± 0.01	7.45 ± 0.01
PCO_2 (mmHg)	39.5 ± 1.3	40.5 ± 1.2	40.0 ± 2.0	38.7 ± 1.7
PO_2 (mmHg)	165.2 ± 8.8	164.9 ± 5.9	180.4 ± 5.5	176.5 ± 5.3
Hct (%)	38.1 ± 0.1	38.0 ± 0.1	37.8 ± 0.2	38.0 ± 0.2
Gluc (mg/dL)	110 ± 1.6	110 ± 2.3	110 ± 1.8	115 ± 2.5
MABP (mmHg)	92 ± 1.8	94 ± 1.5	93 ± 2.5	89 ± 2.3
HR (beats/min)	382 ± 7.2	378 ± 7.7	382 ± 5.0	378 ± 4.7

NOTE: Parameters were monitored and maintained within normal ranges across all animal groups. There were no significant differences amongst the groups of animals. Values are mean ± standard error of the mean. Hct, indicates hematocrit; Gluc, blood glucose; MABP, mean arterial blood pressure; and HR, heart rate. All animals were maintained at 36.5 ± 0.5°C rectal temperature. Physiologic parameters were measured prior to the onset of middle cerebral artery occlusion (preischemia) and following reperfusion (postischemia).

TABLE 2. Posttreatment with NAm (500 mg/kg)

	Group S n = 11	Group A n = 11	Group B n = 11	Group C n = 12
Motor behavior	1.7 ± 0.2	0.5 ± 0.2*	0.3 ± 0.1*	0.4 ± 0.1*
Sensory behavior	2.4 ± 0.5	0.5 ± 0.2*	0.2 ± 0.2*	0.2 ± 0.2*

NOTE: Posttreatment improved sensory and motor behavioral scores during 14 days after the onset of transient middle cerebral artery occlusion. Values are mean ± standard error of the mean. There were 11 or 12 animals per group. Animals injected with either two doses (Group B) or three doses (Group C) of NAm recovered their motor deficits to the extent that they were not significantly different from normal, preoperated animals, whereas saline-treated animals (Group S) and those that received one dose of NAm (Group A) were still impaired; *$p < 0.05$ compared to the saline-treated animals (Group S).

Neurobehavioral Outcome and Weight Gain

On day 14 post-MCAo, all three NAm-treated groups showed significantly ($p < 0.05$) improved sensory and motor neurologic scores compared to saline-treated control (Group S) animals. Furthermore, animals injected with more than one dose of NAm (i.e., Groups B and C) recorded scores in motor function that were not statistically different from normal, preoperative neurologic behavior scores observed on day 0. Animals given one treatment with NAm (Group A), however, did show some behavioral motor deficit (see TABLE 2).

Animals in all groups lost weight transiently after surgery and then steadily gained weight from the second or third postoperative day. There was no significant difference in the initial weight loss or subsequent weight gain among any of the animal groups (data not shown).

DISCUSSION

In previous studies, we showed that delayed treatment with NAm protected against focal cerebral ischemia in Wistar rats compared to saline-injected animals using two different models of stroke, permanent and transient MCAo, with animals sacrificed at 24 hours,[2] three days, or one week later.[3] In the transient MCAo model, we showed an improvement in neurologic behavior and weight gain on day seven after MCAo, but a significant reduction in cerebral infarction volume at three days, but not at seven days after stroke. In the present study, NAm treatment, two hours after the onset of MCAo, significantly reduced the infarction volume for up to two weeks after MCAo. Thus, we not only confirm that delayed treatment with NAm is neuroprotective against focal cerebral ischemia reperfusion, but we have extended, by more than four-fold, the period of significant reduction in cerebral infarction.

We previously observed that IV treatment improved the degree of neuroprotection with NAm compared to IP treatment.[4] This observation is confirmed in the present study since the only parameter that could explain the improved degree and duration of neuroprotection in the single-dose-treated NAm group (A) in this study, compared to our previous study where the significant reduction in infarction volume was lost on day seven, is the IV route of administration of NAm, compared to IP in the previous study.[3] The improved reduction in infarction volume is not due to the administration of multiple doses of NAm (500 mg/kg) since the effect was also seen in animals given only a single NAm treatment (Group A). Thus, the IV treatment paradigm not only improved the degree of infarction reduction, but also extended the duration of protection (as assessed by infarct reduction) from three days to two weeks post-MCAo. Since there was no significant difference between single and multiple dose groups in this study, and given that previous data show that a higher single dose of 750 mg/kg or 1000 mg/kg NAm led to the loss of protection in previous studies, it is unlikely that a larger amount of NAm alone could explain the improved neuroprotection observed, as measured by reduction in infarction volume.[2,4]

An improved neuroprotective effect was also demonstrated from the neurologic behavior outcome. Each group treated with NAm exhibited a significantly lesser degree of sensory and motor deficits compared to the saline-injected group on Day

14 after MCAo. These data, therefore, also extend the initial period (seven days)[3] following treatment with NAm during which behavioral scores were improved. The saline-injected group and the group treated with one dose of NAm, both had significantly worse motor scores when comparing the results for Day 0 (no deficit) and Day 14. However, Group B (two doses of NAm) and Group C (three doses of NAm) showed no significant worsening of motor deficit (comparing Day 0 with Day 14). Thus, multiple dose NAm–treated groups showed motor behavior similar to their normal, preoperative scores, whereas the single dose NAm–treated group still exhibited a significant behavioral deficit 14 days after MCAo. Two or three treatments with NAm poststroke are, therefore, better than one because of the significant functional benefit. Furthermore, the present data confirm our previous finding that sensory and motor behavior outcomes are more sensitive measures of stroke endpoints than infarct volume.[3]

The experiments discussed here were not designed to study the mechanism(s) of the neuroprotective effect of NAm in focal cerebral ischemia; however, the data obtained indirectly support certain conclusions. Since the model of ischemia used in this study was a transient MCAo with sacrifice at 14 days following the onset of stroke, it can be argued that NAm might be protecting against reperfusion injury that may accompany models of ischemia reperfusion. Moreover, given the acute[2,4] and chronic[3] (this study) nature of the neuroprotection observed following the delayed treatment with NAm, both ischemia-induced cell death due to necrosis and apoptosis, respectively, may be sensitive to NAm.

More directly, however, in a collaborate study we recently showed that NAm protects against nitric oxide–induced neuronal injury through the blockage and reversal of discrete pathways of programmed cell death *in vitro*.[11] NAm has also been shown to prevent apoptosis induced by tertiary-butylhyroperoxide in the mouse brain *in vivo*.[12] In addition, there are numerous other potential mechanisms by which NAm may be neuroprotective. NAm was reported to prevent the depletion of neuronal ATP and boost the amount of ATP in the brain because it is a precursor of NAD^+.[12–14] Thus, NAm may rectify the initial energy imbalance caused by ischemia. NAm is also a poly(ADP-ribose) polymerase (PARP) inhibitor, similar to other well-known agents, such as 3-aminobenzamide, 3,4-dihydro-5-[4-(1-piperidinyl)butoxyl-1(2H)-isoquinolinone, and GP-6150, all of which also reduce cerebral infarct volume in models of stroke.[15–18] Furthermore, NAm is a reported anticonvulsant,[19] anticoagulant,[20] and angiogenic agent.[21] It has been shown to attenuate lipid peroxidation,[22] stroke-induced increase in lactate in humans,[23] and inhibit inducible nitric oxide synthase mRNA.[24] Huang and Chao[25] reported an increase in regional cerebral blood flow (rCBF) after administration of NAm. However, Brown *et al.*[26] found a decrease in rCBF in normal animals and no change in rat brain tumors with the identical neuroprotective dose of NAm used in the same strain of rats as that tested in our experiments. Consequently, whether NAm also partially protects by increasing rCBF remains to be determined. Thus, there may be numerous ways in which NAm could act to protect against cerebral ischemia. These potential multimechanistic actions of NAm may indeed be the optimal approach, and might at least in part explain the robust neuroprotective effect observed in these studies.

In conclusion, we have shown that NAm, given in multiple doses, two hours and thereafter following transient MCAo, extends the degree and duration of protection against transient focal cerebral ischemia by reducing the brain infarct volume and by improving the neurologic behavioral outcome for up to two weeks after the onset of stroke. The mechanism of the neuroprotection needs to be clarified, but given that NAm is already used clinically in the treatment of pellagra,[27] type I diabetes,[28] and for cellular sensitization in radiotherapy,[29] and that is can be taken in large quantities with mild side effects by humans,[30] it is a very encouraging putative neuroprotective agent against stroke.

ACKNOWLEDGMENTS

K.I.M. is a Minority Scientist Development Awardee of the American Heart Association. This project was also partially funded by a Bridge Award, Harvard Medical School and a Minority Scientist Award, Massachusetts General Hospital. The authors thank Dr. Seth P. Finklestein for the use of equipment and Dr. Christopher S. Ogilvy for his support.

REFERENCES

1. FURIE, K.L., C.S. OGILVY, M. SMRCKA, et al. 1998. Cerebrovascular disease. *In* The Atlas of Clinical Neurology. R.N. Rosenberg, Ed.: 63–64. Current Medicine, Philadelphia.
2. AYOUB, I.A., E.J. LEE, C.S. OGILVY, et al. 1999. Nicotinamide reduces infarction up to two hours after the onset of permanent focal cerebral ischemia in Wistar rats. Neurosci. Lett. **259:** 21–24.
3. MOKUDAI, T., I.A. AYOUB, Y. SAKAKIBARA, et al. 2000. Delayed treatment with nicotinamide (vitamin B_3) improves neurologic outcome and reduces infarct volume after transient focal cerebral ischemia in Wistar rats. Stroke **31:** 1679–1685.
4. SAKAKIBARA, Y., A.P. MITHA, C.S. OGILVY & K.I. MAYNARD. 2000. Post-treatment with nicotinamide (vitamin B_3) reduces the infarct volume following permanent focal cerebral ischemia in female Sprague-Dawley and Wistar rats. Neurosci. Lett. **281:** 111–114.
5. KOIZUMI, J., Y. YOSHIDA, T. NAKAZAWA & G. OONEDA. 1986. Experimental studies of ischemic brain edema. 1. A new experimental model of cerebral embolism in rats in which recirculation can be introduced in the ischemic area. Japan. J. Stroke **8:** 1–8.
6. BELAYEV, L., O.F. ALONSO, R. BUSTO, et al. 1996. Middle cerebral artery occlusion in the rat by intraluminal suture: neurologic and pathological evaluation of an improved model. Stroke **27:** 1616–1623.
7. BEDERSON, J.B., L.H. PITT, S.M. GERMANO, et al. 1986. Evaluation of 2,3,5-Triphenyltetrazolium chloride as a stain for detection and quantification of experimental cerebral infarction in rats. Stroke **17:** 1304–1308.
8. REN, J.M. & S.P. FINKLESTEIN. 1997. Time window of infarct reduction by intravenous basic fibroblast growth factor in focal cerebral ischemia. Eur. J. Pharmacol. **327:** 11–16.
9. SWANSON, R.A., M.T. MORTON, G. TSAO-WU, et al. 1990. A semi-automated method for measuring brain infarct volume. J. Cereb. Blood Flow Metab. **10:** 290–293.
10. BEDERSON, J.B., L.H. PITT, M. TUJI, et al. 1986. Rat middle cerebral artery occlusion: evaluation of the model and development of a neurologic examination. Stroke **17:** 472–476.
11. LIN, S.H., A.M. VINCENT, T. SHAW, et al. 2000. Prevention of nitric oxide-induced neuronal injury through the modulation of independent pathways of programmed cell death. J. Cereb. Blood Flow Metab. **20:** 1380–1391.

12. KLAIDMAN, L.K., S.K. MUKHERJEE, T.P. HUTCHIN & J.D. ADAMS. 1996. Nicotinamide as a precursor for NAD^+ prevents apoptosis in the mouse brain induced by tertiary-butylhydroperoxide. Neurosci. Lett. **206:** 5–8.
13. BEAL, M.F., D.R. HENSHAW, B.G. JENKINS, et al. 1994. Coenzyme Q_{10} and nicotinamide block striatal lesions produced by the mitochondrial toxin malonate. Ann. Neurol. **36:** 882–888.
14. WAN, F.J., H.C. LIN, B.H. KANG, et al. 1999. D-Amphetamine-induced depletion of energy and dopamine in the rat striatum is attenuated by nicotinamide pretreatment. Brain Res. Bull. **50:** 167–171.
15. ELIASSON, M.J.L., K. SAMPEI, A.S. MANDIR, et al. 1997. Poly(ADP-ribose) polymerase gene disruption renders mice resistant to cerebral ischemia. Nature Med. **3:** 1089–1095.
16. ENDRES, M., Z.-Q. WANG, S. NAMURA, et al. 1997. Ischemic brain injury is mediated by the activation of poly(ADP-ribose) polymerase. J. Cereb. Blood Flow Metabol. **17:** 1143–1151.
17. TAKAHASHI, K., J.H. GREENBERG, P. JACKSON, et al. 1997. Neuroprotective effects of inhibiting poly(ADP-ribose) synthetase on focal cerebral ischemia in rats. J. Cereb. Blood Flow Metabol. **17:** 1137–1142.
18. WILLIAMS, L., S. LIANG, S. LAUTAR, et al. 1999. GPI-6150, a potent PARP inhibitor, reduces infarct size following permanent and transient focal cerebral ischemia. Soc. Neurosci. Abst. **25:** 1061.
19. KRYZHANOVSKII, G.N., A.A. SHANDRA, R.F. MAKUL'KIN, et al. 1980. Effect of nicotinamide on epileptic activity in the cerebral cortex. Biulleten Eksperimentalnoi Biologii I Meditsiny **89:** 37–41.
20. CHUMAKOV, V.N. & T.G. STARCHIK. 1991. Effect of nucleotide anti-aggregants (NAD, AMP) and ischemia on the tissue blood coagulation factors. Gematologiya I Transfuziologiya **36:** 9–13.
21. MORRIS, P.B., M.N. ELLIS & J.L. SWAIN. 1989. Angiogenic potency of the nucleotide metabolites: potential role in ischemia-induced vascular growth. J. Mol. Cell. Cardiol. **21:** 351–358.
22. BRASLAVSKII, V.E., V.A. SHCHAVELEV, G.N. KRYZHANOSKII, et al. 1982. Effect of nicotinamide on focal and generalized epileptic activity in the cerebral cortex. Biulleten Eksperimentalnoi Biologii I Meditsiny **94:** 39–42.
23. MAJAMAA, K., H. RUSANEN, A.M. REMES, et al. 1996. Increase of blood NAD^+ and attenuation of lactacidemia during nicotinamide treatment of a patient with the MELAS syndrome. Life Sci. **58:** 691–699.
24. FUJIMURA, M., T. TOMINAGA & T. YOSHIMOTO. 1997. Nicotinamide inhibits inducible nitric oxide synthase mRNA in primary rat glial cells. Neurosci. Lett. **228:** 107–110.
25. HUANG, T.F. & C.C. CHAO. 1960. The effect of niacinamide on cerebral circulation. Proc. Soc. Exp. Biol. Med. **105:** 551–553.
26. BROWN, S.L., J.R. EWING, A. KOLOZSVARY, et al. 1999. Magnetic resonance imaging of perfusion in rat cerebral 9L tumor after nicotinamide administration. Int. J. Rad. Oncol. Biol. Phys. **43:** 627–633.
27. GREEN, R.G. 1970. Subclinical pellagra: its diagnosis and treatment. Schizophrenia **2:** 70–79.
28. GAL, E.A. 1996. Theory and practice of nicotinamide trials in pretype 1 diabetes. J. Ped. Endocrinol. Metabol. **9:** 375–379.
29. MAAZEN, R.W., H.O. THIJSSEN, J.H. KAANDERS, et al. 1995. Conventional radiotherapy combined with carbogen breathing and nicotinamide for malignant gliomas. Radiother. Oncol. **35:** 118–122.
30. STRATFORD, M.R., A. ROJAS, D.W. HALL, et al. 1992. Pharmacokinetics of nicotinamide and its effect on blood pressure, pulse and body temperature in normal human volunteers. Radiother. Oncol. **25:** 37–42.

Antihistamine Agent Dimebon As a Novel Neuroprotector and a Cognition Enhancer

S. BACHURIN, E. BUKATINA, N. LERMONTOVA, S. TKACHENKO,
A. AFANASIEV, V. GRIGORIEV, I. GRIGORIEVA, YU. IVANOV,
S. SABLIN,[a,b] AND N. ZEFIROV

*Institute of Physiologically Active Compounds, 142432,
Chernogolovka, Moscow Region, Russia*

[a]*Selena Pharmaceuticals, Inc., San Francisco, California, U.S.A.*

[b]*Department of Biochemistry and Biophysics,
University of California, San Francisco, U.S.A.*

ABSTRACT: Dimebon, launched earlier in Russia as an antihistamine drug, was evaluated as a representative of a new generation of anti-Alzheimer's drugs that have two beneficial actions: (1) to alleviate symptoms, and (2) to prevent progression of the disease. The drug demonstrated cognition and memory-enhancing properties in the active avoidance test in rats treated with the neurotoxin AF64A, which selectively destroys cholinergic neurons. Dimebon protected neurons in the cerebellum cell culture against the neurotoxic action of β-amyloid fragment (Aβ25–35, EC_{50} = 25 μM). *In vitro*, Dimebon displayed Ca^{2+}-blocking properties (IC_{50} = 57 μM, on isolated rat ileum intestine) and pronounced anticholinesterase activity (IC_{50} = 7.9 μM and 42 μM for butyrylcholine esterase and acetylcholine esterase, respectively). It also exhibited strong anti-NMDA activity in the prevention of NMDA-induced seizures in mice (EC_{50} = 42 ± 6 mg/kg i.p.). A beneficial effect of Dimebon in the therapy of Alzheimer's disease was demonstrated in a pilot clinical trial performed in the Moscow Center of Gerontology. Fourteen patients who participated in the trial were evaluated for their state of personality and for the severity of the disease. The evaluation included orientation (space, place, time, and patient personality), memory for the past and present, life in present, speech, irritability, and so forth. During and after the eight-week therapy with Dimebon, cognitive and self-service functions of patients improved significantly, and psychopathic symptoms, anxiety, depression, tearfulness, and headache were substantially diminished. The results of these studies suggest Dimebon as a new candidate for the therapy of Alzheimer's-like disorders.

KEYWORDS: Alzheimer's disease; Dimebon; Neuroprotector; Cognition enhancer.

INTRODUCTION

Alzheimer's disease is the most common form of dementia affecting millions of people in the world. It is characterized by the progressive degeneration and death of nerve cells in certain brain regions, leading to loss of intellectual abilities, such as

---

Address for correspondence: Sergey O. Sablin, Ph.D., Selena Pharmaceuticals, Inc., 167 Skyview Way, San Francisco, CA 94131, U.S.A. Voice: 415-990-4761; fax: 415-750-6959.
sablin@itsa.ucsf.edu

thinking, memory, learning skills, language, and communication. The disease is also characterized by significant psychiatric and emotional problems, such as agitation, depression, delusions, and delirium. Pathological features include general atrophy of brain with a specific loss of cholinergic neurons resulting in a deficiency of the neurotransmitter acetylcholine. The cause of the disease remains unknown. Moreover, there are no FDA-approved medications that have any effect on the progression of the disease. Conventional therapies that are currently used in clinical practice are aimed at compensating for specific symptoms (cognitive function) associated with the loss of acetylcholine, using drugs to elevate the level of this neurotransmitter, as well as to compensate for associated psychiatric and emotional symptoms. Therefore, pharmacological strategies in searching for new effective therapies are focused on drugs that, not only improve cognitive functions, but also prevent the progression of the disease.

The key event that results in the progressive degeneration of cholinergic neurons in the brain of an Alzheimer's patient is thought to be the formation of β-amyloid peptide (Aβ), which induces the neurotoxic effect of excitatory amino acids, glutamate and aspartate (for a recent review see Ref. 1). The toxic action of glutamate and aspartate is mediated through the activation of N-methyl-D-aspartate (NMDA) receptors. This activation leads to an increase in the intracellular Ca^{2+} concentration that triggers a cascade of pathological reactions causing the death of nerve cells.[2–4] The disturbances of glutamatergic neurotransmission are thought to underlie the pathomechanism and cognitive deficits of Alzheimer's disease. Therefore, antagonists of Ca^{2+} channels and NMDA receptors are currently a focus of interest for their potential use as neuroprotective agents in the treatment of this disease.[1,4–7]

In this report, we introduce Dimebon, which was launched previously in Russia as an antihistamine drug, as a new neuroprotective and cognition-enhancing agent for the potential use in the therapy of Alzheimer's disease. The properties of Dimebon as an antagonist of Ca^{2+} channels and NMDA receptors are evaluated. The results of a pilot open-label clinical study of the Dimebon treatment are also presented.

MATERIALS AND METHODS

Materials

Dimebon was provided by NPO Organika (Novokuznetsk, Russia). β-Amyloid peptide fragment (Aβ25–35) was purchased from Bachem. All other reagents were purchased from Sigma.

General Methods

Effect of Dimebon on Ca^{2+} channels was studied using rat ileum intestine. Ca^{2+}-induced contractions of segments of small intestine were recorded in the absence or in the presence of varied concentration of Dimebon using dynograph R-612, Beckman, USA.

The effects of Dimebon on the catalytic activities of acetylcholine esterase and butyrylcholine esterase were determined using partially purified enzyme preparations.

The activities of both enzymes were measured in the presence of varied concentration of the drug and monitored by the accumulation of thiocholine, a product of the hydrolysis of acetylthiocholine and butyrylthiocholine by acetylcholine esterase and by butyrylcholine esterase, respectively. Thiocholine was detected spectrophotometrically in the reaction with a thiol reagent, 5,5'-dithio(bis)-nitrobenzoic acid, followed by the formation of the product of this reaction with maximum of absorbance at 412 nm.

Study of the anti-NMDA effect of Dimebon was conducted on white pedigreed male mice (20–24 g). A skin flap on the head of each mouse was removed under ether anesthesia, and a hole was drilled through the skull using a fine drill bit. A solution (1.4 μl) containing 0.1 μg of NMDA was injected into the lateral ventricle of the brain using a microsyringe. The solution also contained methylene blue dye for monitoring the accuracy of the injection. The needle was immersed to a depth of 2.5 mm. After the operation, the wound was treated with a 2% novocaine solution. A solution of Dimebon in 0.2 ml of 5% aqueous dimethylsulfoxide (for experimental animals) or physiological solution (for control animals) was injected intraperitoneally 40 minutes prior to the injection of NMDA. Mice were used for pharmacological experiments two to four hours after the operation. Each dose of Dimebon was tested in a group of six to eight animals. The ED_{50} value (the dose of the agent preventing the development of convulsions and the death of 50% of the animals) was determined by a probit analysis method.[8]

Studies on the Neural Cell Culture

Neuroprotective properties of Dimebon against neurotoxicity induced by β-amyloid were studied on a culture of cerebellar granule cells (CGC). The CGC culture is a homogeneous population of nervous cells containing acetylcholine and NMDA receptors.[9] The mature cell culture in the media containing 25 mM KCl was incubated with varied concentrations of either β-amyloid fragment (Aβ25–35), or with Dimebon, or with both. The viability of cells was monitored by computerized count and by lactate dehydrogenase activity.

Memory and Cognitive Skills Test

Studies of memory and cognitive skills were performed in the active avoidance test in rats treated (i.c.v.) with the cholinergic toxin AF64A that selectively destroys cholinergic neurons.[10,11] Male Wistar rats (250–300 g) were anesthetized with ether and placed into a stereotaxic frame. A cannula was inserted through the skull, and the solution of toxin AF64A (3 nmol/3 μl, bilaterally, experimental group) or artificial cerebrospinal fluid (vehicle group) were infused into each of the ventricles at a rate of 2–3 μl/min.

On the second day after the surgery and during the next 10 days rats were injected (i.p.) daily, either with the physiological solution (vehicle group and one of the groups of AF64A treated rats) or with the solution containing 1 mg/kg Dimebon (other two groups of AF64A treated rats). Rats were allowed 12–14 days to recover from the surgery. They were then given 35 acquisition trials to perform an active conditioned avoidance response in two-way shuttle box with light and electrostimulation as conditioned and unconditioned stimulus, respectively (learning test). Within

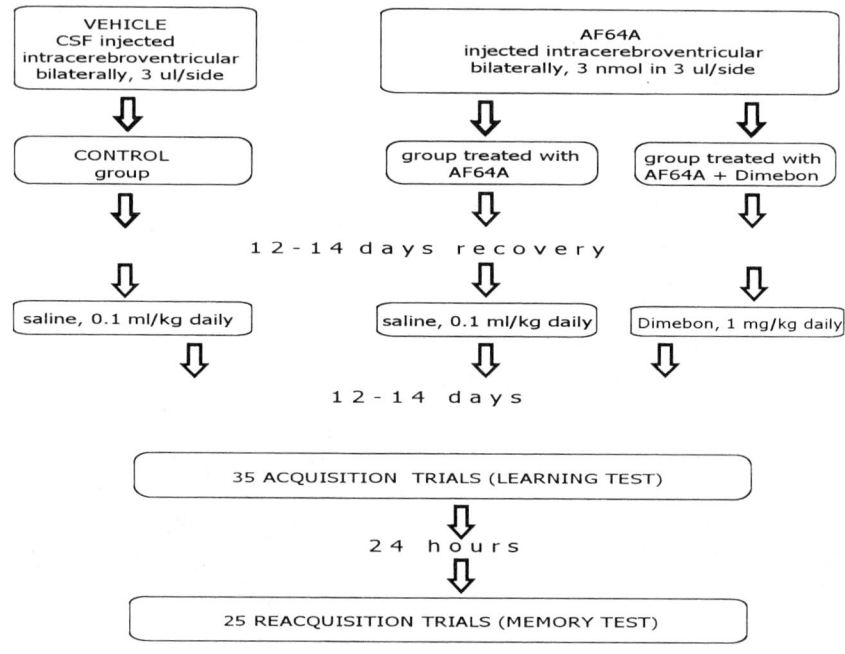

FIGURE 1. Scheme of the experiments on the evaluation of memory and learning skills of rats treated with cholinergic toxin AF64A.

24 hours all rats were given five reacquisition trials to test memory impairment (memory test). The scheme of trials is presented in FIGURE 1.

Clinical Trial

Open label clinical trials were performed in the Moscow Center of Gerontology on 14 patients (1 male, 13 females, age ranging from 64 to 88 years old). All patients who participated in the trial were diagnosed with mild to moderate Alzheimer's disease. The diagnosis was made based on ICD-10 NINCDS ADRDA criteria.[12] The clinical diagnosis was confirmed by the computer tomography of the brain of each patient. Medical records indicated no dramatic change in the condition of patients for two years prior to the trial. Patients were treated for eight weeks with 60 mg (20 mg/dose, three times a day, orally) of Dimebon daily. Before the treatment, at the fourth and at the eighth week of the treatment, and on the eighth week after the treatment was discontinued, all patients were evaluated for the condition of personality and for the severity of the disease. The evaluation was performed according to the Hasegawa scale (similar to the Mini-Mental State examination) and according to the Bukatina scale (introduced in clinical practice, by one of the authors, in the Moscow Center of Gerontology). In brief, the evaluation included the orientation (in space, in place, in time, in the nearest surroundings, and the patient's personality), memory for the past and present, life in present, speech, irritability, articulation, concentration of attention, spirits, delirium, hallucinations, anxiety, headaches, dizziness, tearfulness, spontaneous activity, elementary self-service, and so forth. Statistical analysis was performed based on the Student's t-test and Fisher's ϕ criteria. During the trial patients were under continuous observation by psychiatrists, by the medical staff and by patients' family members.

RESULTS

Interaction of Dimebon with Biochemical Targets

Injection of NMDA into the lateral ventricle of the control mice (0.1 µg per mouse) resulted in hyperactivity of animals, expressed in running and jumping, followed by convulsions, and, ultimately, death. In the experimental group, animals were injected with varying amounts of Dimebon before the NMDA solution was injected. The preinjection of Dimebon prevented both development of convulsions and death of animals. Protection against NMDA toxicity by Dimebon was dose dependent. The value of ED_{50} evaluated in these experiments is presented in TABLE 1.

Calcium channel antagonist activity of Dimebon was determined using Ca^{2+}-dependent contraction of isolated smooth muscles of rat ileum intestine. It was shown that Dimebon suppressed contractions of smooth muscles in Ca^{2+} containing media in a reversible and concentration dependent manner (TABLE 1).

The level of acetylcholine in the brain is regulated through its synthesis, transport, and degradation. Acetylcholine esterase (and to a lesser extent, butyrylcholine esterase) is the key enzyme that provides the degradation of this neurotransmitter. Inhibition of the enzyme results in the elevation of acetylcholine level associated with an improvement in cognitive functions. It was shown that Dimebon is a potent competitive and reversible inhibitor of both acetylcholine esterase and butyrylcholine esterase (TABLE 1).

Effect of Dimebon on Memory Function and Cognitive Skills

Administration of ethylcholine aziridinium ion, AF64A, in laboratory animals induces the degenerative loss of pyramidal neurons in the area CA3 of hyppocampus. This results in chronic cholinergic hypofunction and some neuropathological changes that are similar to changes observed in Alzheimer's disease.[10,11] A number of studies demonstrated that AF64A produced both learning and memory impairments in laboratory animals.[13–15]

TABLE 1. Effect of Dimebon on biochemical targets

Inhibitor		Antagonist	
AChE	BuChE	Ca^{2+} channels	NMDA receptors
IC_{50}, µM	IC_{50}, µM	IC_{50}, µM	ED_{50}, mg/kg, i.p.
42	7.9	57	42

We used rats treated with neurotoxin AF64A to evaluate memory- and cognition-enhancing properties of Dimebon in the active avoidance test. In our experiments, animals treated with AF64A exhibited a decrease in the performance during both the learning test and the memory test (response time to the electrostimulation and light was significantly greater). The treatment of animals in these groups with Dimebon, resulted in a remarkable improvement of learning skills and memory compared to the control group. The improvement was obvious, even in comparison with the experimental group treated with Tacrine under the same conditions (see FIGURE 2).

Protective Effect of Dimebon Against Neurotoxicity Induced by β-amyloid Peptide in Cultured Cerebellar Granule Cells (CGC)

Incubation of mature CGC culture with Aβ25–35 led to a significant decrease in the viability of neural cells. The effect was dose-dependent ($IC_{50} = 25 \pm 8$ μM). Pronounced morphological changes, such as shrinking and fragmentation, of neurons were registered at the third day of incubation with Aβ25–35. Hence, test experiments were carried out at the fourth day of incubation of CGC with 25μM Aβ25–35.

FIGURE 3 shows that Dimebon in concentration of 25 μM increases the viability of CGC and brings the level of the survived cells up to the control.

Clinical Trial

The main objective of the study was to assess the efficacy of Dimebon in the treatment of patients with mild to moderate Alzheimer's disease.

The treatment with Dimebon demonstrated a significant progressive improvement in cognitive functions of all patients participated in the trial. Mean improvements from the baseline were already observed at the fourth week of the trial, as assessed by both Hazegawa's scale and Bukatina scale. The improvements were even more profound after the eighth week of the therapy (see TABLE 2), and were more evident in patients with mild Alzheimer's disease. The improvement during the

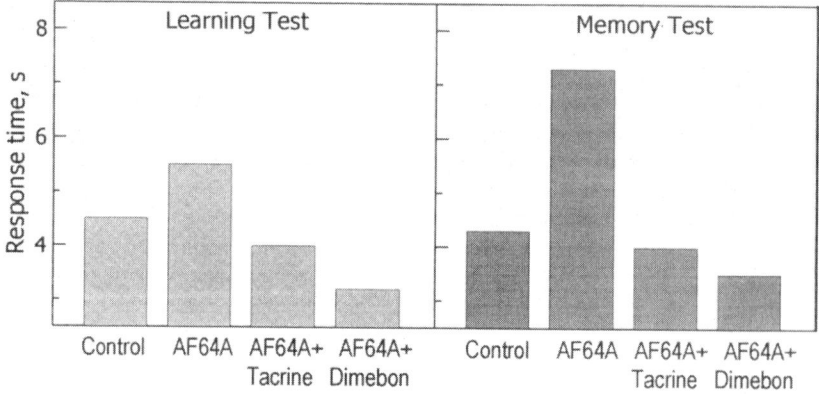

FIGURE 2. Dimebon-dependent improvement in memory and learning skills in rats treated with cholinergic toxin AF64A. Details of the experiment are provided in the text.

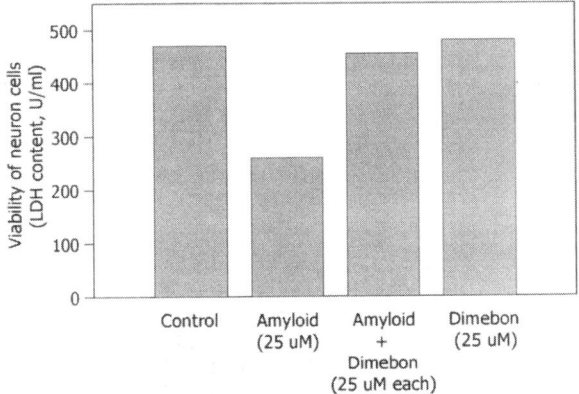

FIGURE 3. Dimebon-dependent increase of the viability of cells incubated with β-amyloid fragment Aβ25–35. Cultured neural cells were incubated for three days with either Aβ25–35, or with Dimebon, or both (25 μM each). The viability of cells incubated with Aβ25–35 is significantly decreased (*second bar*) compared to the control (*first bar*). Dimebon is not toxic to the cells (*fourth bar*) and protects them against the toxic effect of Aβ25–35 (*third bar*).

therapy is more evident on the background of dramatic deterioration in cognitive functions after the Dimebon treatment was discontinued. This was assessed at the evaluation eight weeks after the trial (see FIGURE 4).

Although the overall study was not placebo-controlled, cognitive functions were evaluated against a group of eight patients who received placebo for four weeks. The control group was not given any medication, which could influence cognitive functions of patients. The initial levels of patient dementia in both experimental and control groups were similar. After four weeks of therapy the changes in cognitive functions as evaluated by the Bukatina scale were 0.5 ± 0.14 and 0.12 ± 0.12 ($p < 0.01$) in the experimental and in the control group, respectively.

Before the trial, 12 patients had various depressive symptoms. At the fourth week of therapy with Dimebon, the depressive symptoms abated in five patients. There was not a single case of deterioration or emergency depression. At the eighth week of the treatment, six other patients showed an improvement. One patient showed signs of aggravation. The mean values of depressive manifestations in 12 patients before the trial, and at the fourth and at the eighth week of the therapy were scored as 1.1 ± 0.22, 0.58 ± 0.18 and 0.58 ± 0.14, respectively (TABLE 2). The dynamics of depressive manifestations in patients suffering from evident depression prior to the treatment with Dimebon showed that the abatement of depression was significant at the eighth week of the treatment.

Ten patients complained having headaches during the observation period. Nine patients had headaches prior to the treatment. One patient had the first headache at the eighth week of treatment. Five patients (50% of patients having headaches) showed the abatement or complete cessation of headaches at the fourth week of treatment, and three more (80%) showed the same improvement after eight weeks.

TABLE 2. Results of the Dimebon clinical trial

Function	Scores,[a] mean values			Changes in scores[b]	
	Week 0	Week 4	Week 8	Week 4	Week 8
Cognition[c]	6.29 ± 0.7	5.5 ± 0.79	3.93 ± 0.5	+0.5 ± 0.14	+1.79 ± 0.35
Depression[c]	1.1 ± 0.22	0.58 ± 0.18	0.58 ± 0.14	+0.6 ± 0.11	+0.6 ± 0.11
Orientation					
in locality	1.82 ± 0.19	1.75 ± 0.23	1.57 ± 0.23	+0.11 ± 0.07	+0.18 ± 0.08
in space	0.54 ± 0.16	0.32 ± 0.15	0.25 ± 0.13	+0.11 ± 0.07	+0.11 ± 0.07
Memory					
for past	2.34 ± 0.23	2.29 ± 0.21	1.93 ± 0.24	+0.07 ± 0.07	+0.29 ± 0.1
for present	2.38 ± 0.22	2.21 ± 0.22	1.86 ± 0.23	0 ± 0	+0.29 ± 0.08
Delirium	0.91 ± 0.2	0.57 ± 0.29	0.5 ± 0.19	+0.07 ± 0.09	+0.25 ± 0.17
Irritability	0.57 ± 0.19	0.21 ± 0.15	0.14 ± 0.09	+0.29 ± 0.13	+0.36 ± 0.16
Headaches	1.02 ± 0.27	0.32 ± 0.21	0.64 ± 0.24	+0.54 ± 0.21	+0.21 ± 0.1
Dizziness	0.64 ± 0.19	0.71 ± 0.27	0.68 ± 0.25	−0.14 ± 0.17	−0.18 ± 0.22
Tearfulness	0.36 ± 0.18	0.2 ± 0.18	0.0 ± 0.0	+0.14 ± 0.14	+0.21 ± 0.15

[a]Bukatina scale: score 0 indicates the absence of a symptom; score 4 indicates the highest deterioration.
[b]+ Indicates improvement; − indicates deterioration of function.
[c]Complex parameter that combines results of observations on relevant symptoms.

Before the trial, five female patients suffered from a tearfulness (three patients had slight tearfulness). After four weeks of treatment, tearfulness symptoms were diminished in four female patients, and after eight weeks not a single female patient suffered from tearfulness. Prior to the study, and at the fourth and at the eighth week of the assessment the mean scores of tearfulness were 0.36 ± 0.18, 0.2 ± 0.18, and 0 ± 0.0, respectively (TABLE 2).

Overall, clinically meaningful improvements in cognitive function and reduction of neuropsychiatric symptoms were demonstrated during the trial (TABLE 2). Patients became calmer and more sociable. They started to act and respond more adequately. Psychopathic symptoms (lack of restraint, touchiness, conflict making, evil mindedness, and aggressiveness) that were documented for seven patients before the study were significantly diminished during the first two weeks of the therapy. Two patients showed improvement in spirit, and this effect was attained after eight weeks of the treatment. Four patients became more active and felt that they experienced a sense of cheerfulness and freshness. Eight patients showed a greater interest in their surroundings. Five patients even showed a positive change in their entire appearance, as assessed by the medical staff and by family members.

No pathological changes in the hematological and biochemical parameters were found during the course of the treatment. A reduction in the leukocyte count (although, within a normal limit) was observed at the fourth week of the treatment. However, by the eighth week of the treatment, the leukocyte count returned to normal.

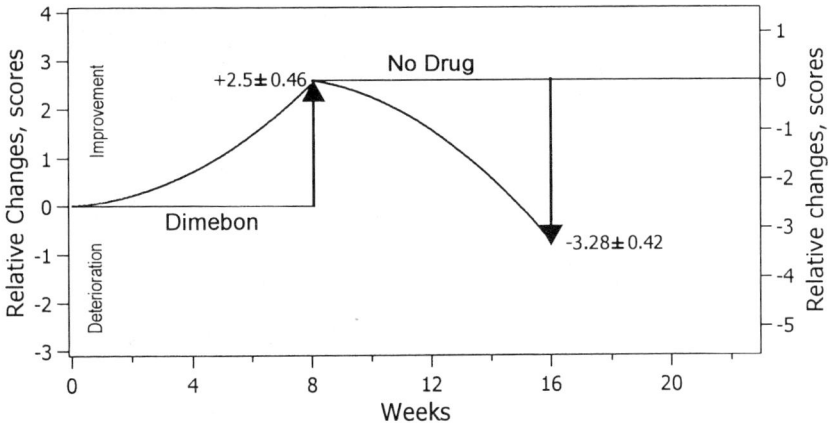

FIGURE 4. The effect of Dimebon treatment on cognitive functions of Alzheimer's patients. Patients were evaluated before the trial, at the fourth week and at the eighth week of Dimebon therapy, as well as eight weeks after the therapy was discontinued.

DISCUSSION

As we mentioned, conventional therapies aimed at improving cognitive functions involve medications that elevate the level of acetylcholine, diminished due to the loss of neural cells. Studies of the mechanism of degeneration suggest that a hyperactivation of NMDA receptors might play a central role in the degeneration of neural cells. Therefore, the development of novel neuroprotective agents is currently focused on antagonists of NMDA receptors and Ca^{2+} channels.

Results from both neurochemical and animal studies presented in this report show that Dimebon combines properties of both a cognition-enhancing agent and a neuroprotective agent. Its neuroprotective properties are explained by its calcium-blocking and anti-NMDA activity. The properties of Dimebon as a cognition-enhancing agent are based on its ability to elevate the level of acetylcholine by inhibition of either acetylcholine esterase or histamine receptors.[16,17] The latter leads to stimulation of the release of the neurotransmitter.

A clinical trial performed on 14 patients with Alzheimer's disease support the results of the neurochemical and animal studies. The Dimebon treatment showed appreciative improvement in cognitive functions and psychiatric and emotional symptoms of all treated patients. A distinct antidepressive effect was also observed in those patients who suffered from depression. This effect most likely results from the ability of Dimebon to inhibit brain monoamine oxidase.[18]

Safety of the therapeutic drugs is especially important when drugs are used with older patients, the main group affected by Alzheimer's disease. Safety studies were performed when Dimebon was first introduced in Russian market in 1983 as an antihistamine agent, showed that the drug has a remarkable safety profile. It has been an over-the-counter drug for almost 20 years, and it has proven to be safe. Toxicological studies showed that administration of the drug for two months in doses that exceeded the therapeutic dose by 100 times, did not produce any changes in physiological parameters or pathological changes in organs of rats, guinea pigs, or dogs.[19,20] Based on the results of these studies, Dimebon was classified as a low toxic substance. A pronounced antiarrhythmic effect of Dimebon and its beneficial effect in certain heart conditions have also been emphasized.[21–23]

In summary, Dimebon has a combination of pharmacological and safety properties that suggest the drug as a candidate for the use in the therapy of Alzheimer's disease. The results of the study reported here show that Dimebon has a great potential as a new medication that may not only improve cognitive functions, but, most importantly, may prevent the progression of the disease. Large-scale clinical trials performed according to the FDA standards are required to confirm its efficacy in the improvement of cognitive functions and in the prevention of the progression of the disease.

REFERENCES

1. HARKANY, T, T. HORTOBAGYI, M. SASVARI, et al. 1999. Neuroprotective approaches in experimental models of beta-amyloid neurotoxicity: relevance to Alzheimer's disease. Proc. Neuropsychopharmacol. Biol. Psychiatry **23:** 963–1008.
2. MATTSON, M.P. 1990. Antigenic changes similar to those seen in neurofibrillary tangles are elicited by glutamate and Ca^{2+} influx in cultured hippocampal neurons. Neuron **2:** 105–117.
3. MILLS, L.R. & S.B. KATER. 1990. Neuron-specific and state-specific differences in calcium homeostasis regulate the generation and degeneration of neuronal architecture. Neuron **2:** 149–163.
4. SCHOESBOE, A., B. BELHAGE & A. FRANDSEN. 1997. Role of Ca^{2+} and other second messengers in excitatory amino acid receptor mediated neurodegeneration: clinical perspectives. Clin. Neurosci. **4:** 194–198.
5. MULLER, W.E., E. MUTSCHLER & P. RIEDER. 1995. Noncompetitive NMDA receptor antagonists with fast open-channel blocking kinetics and strong voltage-dependency as potential therapeutic agents for Alzheimer's dementia. Pharmachopsychiatry **28:** 113–124.
6. KORNHUBER, J. & M. WELLER. 1997. Psychogenicity and N-methyl-D-aspartate receptor antagonism: implications for neuroprotective pharmacotherapy. Biol. Psychiatry **41:** 135–144.
7. PALMER, G.C. & D. WIDZOWSKY. 2000. Low affinity use-dependent NMDA receptor antagonists show promise for clinical development. Amino Acids **19:** 151–155.
8. LITCHFILD, J.T. & F.J. WICOXON. 1949. Pharmacol. Exp. Therap. **96:** 99–114.
9. SCHRAMM, M., S. EIMERL & E. COSTA. 1990. Serum and depolarizing agents cause acute neurotoxicity in cultured cerebellar granule cells: role of the glutamate receptor responsive to N-methyl-D-aspartate. Proc. Natl. Acad. Sci. U.S.A. **87:** 1193–1197.
10. FISHER, A. & I. HANIN. 1986. Potential animal models for senile dementia of Alzheimer's type, with emphasis on AF64A-induced cholinotoxicity. Annu. Rev. Pharmacol. Toxicol. **26:** 161–181.
11. LERMONTOVA, N., N. LUKOYANOV, T. SERKOVA, et al. 1998. Effects of tacrine on deficits in active avoidance performance induced by AF64A in rats. Mol. Chem. Neuropathol. **33:** 51–61.
12. MCKAHN, G., D. DRACHMAN, M. FOLSTEIN, et al. 1984. Clinical diagnosis of Alzheimer's disease: report of the NINCDS-ASRDA Work Group under the auspices of the Department of Health and Human Services Task Force on Alzheimer's Disease. Neurology **34:** 939–944.

13. WALSH, T. & K. OPELLO. 1994. *In* Toxin-Induced Models of Neurological Disorders. 259–279. Plenum Press, New York and London.
14. BAILEY, E.L., D.H. OVERSTREET & A.D. CROCKER. 1986. Effects of intrahippocampal injections of the cholinergic neurotoxin AF64A on open-field activity and avoidance learning in the rat. Behav. Neural. Biol. **45:** 263–274.
15. OGURA, H., Y. YAMANISHI & K. YAMATSU. 1987. Effects of physostigmine on AF64A-induced impairment of learning acquisition in rats. Japan J. Pharmacol. **44:** 498–501.
16. GANKINA, E.M., N.V. PORODENKO, T.I. KONDRATENKO, *et al.* 1993. The effect of antihistaminic preparations on the binding of labelled mepyramine, ketanserin and quinuclidinyl benzilate in the rat brain. Eksp. Klin. Farmakol (Russ.). **56:** 22–24.
17. MATVEEVA, I.A. 1983. Action of Dimebon on histamine receptors. Farmakol. Toksikol. (Russ.). **46:** 27–29.
18. SHADURSKAIA, S.K., A.I. KHOMENKO, V.A. PEREVERZEV & A.I. BALAKLEVSKII. 1986. Neuromediator mechanisms of the effect of the antihistamine agent Dimebon on the brain. Biull. Eksp. Biol. Med. (Russ.) **101:** 700–702.
19. GOLUBEVA, M.I., L.F. SHASHKINA, V.A. PROINOVA, *et al.* 1985. Preclinical study of the safety of the antihistaminic preparation Dimebon. Farmakol. Toksikol. (Russ.) **48:** 114–119.
20. PROINOVA, V.A., I.V. GOLOVANOVA, M.I. GOLUBEVA & L.F. SHASHKINA. 1985. Evaluation of the effect of the antihistaminic preparations Dimebon and diprazin on embryogenesis in rats. Farmakol. Toksikol. (Russ.) **48:** 89–93.
21. GALENKO-IAROSHEVSKII, P.A., I.L. CHEREDNIK, V.V. BARTASHEVICH, *et al.* 1997. Comparative evaluation of the anti-arrhythmia efficacy of fencarol and Dimebon during neurogenic atrial fibrillation. Biull. Eksp. Biol. Med. (Russ.) **124:** 81–85.
22. GALENKO-IAROSHEVSKII, P.A., E.R. MELKUMOVA, V.V. BARTASHEVICH, *et al.* 1996. Comparative evaluation of the effect of Dimebon, obzidan, finoptin, and cordarone on the functional state of the ischemic focus and the size of necrosis zone during experimental myocardial infarction. Biull. Eksp. Biol. Med. (Russ.) **122:** 642–644.
23. GALENKO-IAROSHEVSKII, P.A., Iu.R. SHEIKH-ZADE, O.A. CHEKANOVA, *et al.* 1996. Effect of Dimebon on coronary blood flow and myocardial contractibility. Biull. Eksp. Biol. Med. (Russ.). **121:** 506–508.

Cell Death in Spinal Cord Injury (SCI) Requires *de Novo* Protein Synthesis

Calpain Inhibitor E-64-d Provides Neuroprotection in SCI Lesion and Penumbra

SWAPAN K. RAY, DENISE D. MATZELLE, GLORIA G. WILFORD, EDWARD L. HOGAN, AND NAREN L. BANIK

Department of Neurology, Medical University of South Carolina, Charleston, South Carolina, U.S.A.

ABSTRACT: Degradation of cytoskeletal proteins by calpain, a Ca^{2+}-dependent cysteine protease, may promote neuronal apoptosis in the lesion and surrounding areas following spinal cord injury (SCI). Clinically relevant moderate (40 g-cm force) SCI in rats was induced at T12 by a standardized weight-drop method. Internucleosomal DNA fragmentation or apoptosis in the lesion was inhibited by 24-h treatment of SCI rats with cycloheximide (1 mg/kg), indicating a requirement for *de novo* protein synthesis in this process. To prove an involvement of calpain activity in mediation of apoptosis in SCI, we treated SCI rats with a cell-permeable calpain inhibitor E-64-d (1 mg/kg). Following 24-h treatment, a 5-cm-long spinal cord section centered at the lesion was collected, and divided equally into five segments (1 cm each) to determine calpain activity, as shown by degradation of the 68-kD neurofilament protein (NFP), and apoptosis as indicated by internucleosomal DNA fragmentation. Neurodegeneration propagated from the site of injury to neighboring rostral and caudal regions. Both calpain activity and apoptosis were readily detectable in the lesion, and moderately so in neighboring areas of untreated SCI rats, whereas these were almost undetectable in E-64-d-treated SCI rats, and absent in sham animals. Results indicate that apoptosis in the SCI lesion and penumbra is prominently associated with calpain activity and is inhibited by the calpain inhibitor E-64-d providing neuroprotective benefit.

KEYWORDS: Cell death; Spinal cord injury; Calpain inhibitor; Neuroprotection.

INTRODUCTION

Spinal cord injury (SCI) interrupts motor impulses descending from the brain to other parts of the body causing devastating neurological dysfunctions, such as sensory loss, bowel and bladder problems, and even paralysis.[1] The extent of disruption of communication between the brain and other parts of the body depends upon both the severity and site of injury to the spinal cord. A typical severe injury results in total loss of movement and sensation below the level of injury. An injury to the

Address for correspondence: Naren L. Banik, Department of Neurology, Medical University of South Carolina, Charleston, SC 29425, U.S.A. Voice: 843-792-3946; fax: 843-792-8626.
baniknl@musc.edu

cervical spine may cause paralysis in both arms and legs, leading to quadriplegia, whereas an injury to the thoracic (T) region may affect only the lower body parts including the legs, resulting in paraplegia. Studies of animal models indicate that the pathophysiology of SCI is complex.[2] The primary injury immediately disrupts neural cell membranes, destroys myelin and axons in the longitudinal tracts, and damages microvessels releasing devastating secondary injury factors including electrolytes, metabolites, cytokines, and enzymes from the damaged tissues.[2,3] An active secondary injury process subsequently causes tissue degeneration by a variety of cellular and molecular mechanisms working through complex cascades to spread the damage beyond the site of injury.[4-6] One frustrating aspect of this trauma to the central nervous system (CNS) is the lack of a cure for paralyzed patients experiencing the catastrophic consequences of SCI. It is encouraging, however, that recent studies in cell culture and animal models suggest that neurons with proper treatment may be capable of axonal regeneration and reconnection of injured processes.[7,8] Some investigators have focused on the mechanisms involved in CNS tissue degeneration caused by several secondary injury mediators, including Ca^{2+} accumulation,[9] Ca^{2+}-dependent proteases including calpains,[10,11] inflammatory cytokines,[12] and reactive oxygen species (ROS).[13] This mechanistic approach to understanding the factors mediating spinal cord degeneration after injury is building a foundation for the development of therapeutic strategies. The secondary injury cascade is an active biological process, and thus, it provides a window of opportunity for experimental therapy of SCI using selective antagonists or inhibitors.

Morphologically, there are two distinct forms of cell death, necrosis and apoptosis. Necrosis occurs passively without cellular control, and is likely to be responsible for tissue damage immediately after primary mechanical injury to the spinal cord. There has been an assumption that necrosis is the mode of cell death in the SCI lesion,[3] and this view is consistent with a role of excitotoxicity in the pathogenesis of SCI.[14] Excitotoxicity is characterized by neuronal cell swelling, which has been documented in the study of necrosis of cultured cortical neurons.[15] Necrosis is irreversible, causing damage to the plasma membrane and leakage of cell ingredients into the extracellular space. Although some neurons and glial cells in the lesion die by necrosis immediately after the primary injury, later they die by apoptosis during the secondary injury process in the SCI lesion and surrounding areas. Apoptosis is a complex process, with involvement of several factors initiating an intrinsic suicide program in the cells. Apoptotic cells are subsequently cleared by proteolytic digestion.[16] Recent studies have demonstrated apoptosis following SCI in rats,[17-23] rabbits,[24] monkeys,[25] and humans.[26] The occurrence of apoptosis is an important part in the pathogenesis of the secondary injury process in the spinal cord, because it is an active cell death process mediated by the activation of a number of cysteine proteases, including calpain,[27] and may therefore be preventable with selective inhibitors.

Many investigations have been conducted to elucidate a role for calpain[28,29] in apoptosis, a mode of cell death usually dependent on *de novo* synthesis of macromolecules,[30,31] including proteins and proteases. Inhibition of a biochemical process by cycloheximide[32] indicates that the process is dependent upon *de novo* synthesis of proteins and proteases. Neuronal apoptosis has been shown to depend on the synthesis of new proteins, and be inhibited by cycloheximide.[33,34] Rat CNS cells undergo apoptosis with increased synthesis of calpain in response to diverse chemical stimuli[35,36] and physical injury.[22,23] Activation of calpain following SCI may have detrimental consequences,[37] since many cytoskeletal and membrane proteins in neurons are known to be calpain substrates and are readily degraded by calpain.[38] Intracellular free Ca^{2+} is absolutely required for activation of calpain, and this is increased in SCI.[9] Thus, calpain activated with an increased intracellular free Ca^{2+} level following injury, may participate in apoptosis.[39,40] Neuronal cells experiencing calpain-mediated degradation of cytoskeletal and membrane proteins may become unstable and eventually yield to apoptosis. This investigation with a rat SCI model indicates that (1) cell death in SCI is inhibited by cycloheximide, confirming the mode of cell death as apoptosis, which requires *de novo* protein synthesis; and (2) calpain activity is involved in mediation of apoptosis in the SCI lesion and penumbra because the degradation of 68-kD NFP, a preferred calpain substrate, and occurrence of internucleosomal DNA fragmentation are inhibited by a calpain specific inhibitor, E-64-d. Part of the results of this investigation was presented elsewhere.[41]

MATERIALS AND METHODS

Induction of SCI in Rats

Animals were appropriately maintained and used in this investigation in accordance with the Laboratory Animal Welfare Act and the Guide for the Care and Use of Laboratory Animals of the U.S. Department of Health and Human Services (National Institutes of Health, Bethesda, MD). Interventions were approved by the Institutional Animal Care and Use Committee (IACUC) of the Office of Research Integrity (Medical University of South Carolina, Charleston, SC). Female Sprague-Dawley rats (six weeks old, and 225–250 g weight) were purchased (Charles River Laboratories, Wilmington, MA) for use in the therapeutic model of SCI as described elsewhere.[42] Briefly, the body temperature of the animals was maintained at $37 \pm 1°C$ using a heat lamp regulated by a temperature controller with a thermal probe inserted into the rectum during anesthesia with ketamine (100 mg/kg) and xylazine (5 mg/kg). A single segment laminectomy was made at T12 with the aid of a dissecting microscope. The spine was immobilized with a spinal stereotactic device. SCI was induced by dropping a constant weight of 5 g from a height of 8 cm (equivalent to 40 g-cm force). Our previous studies indicated that SCI induced by 40 g-cm force was optimal for therapeutic trials. Sham animals underwent a laminectomy only. Every effort was made to reduce discomfort experienced by the animals. Urinary bladders were massaged twice a day until the experiment was completed.

Therapeutic Treatment of SCI Rats

Cycloheximide (Sigma Chemical Co., St. Louis, MO) and E-64-d (Calbiochem-Novabiochem, San Diego, CA) were used for experimental therapy of SCI rats. Drugs were dissolved in 100% dimethyl sulfoxide (DMSO) as a vehicle, and diluted with sterile saline (0.85% NaCl) at the time of intravenous (i.v.) administration to SCI rats. A group of four sham animals received nothing. Six SCI animals were used for each drug treatment. A group of SCI rats received vehicle (1.5% DMSO in sterile saline). Another group of SCI rats was treated with the protein synthesis inhibitor cycloheximide (1 mg/kg), and a third group of SCI rats with the calpain inhibitor E-64-d (1 mg/kg). Following 24-h treatment, animals were sacrificed under anesthesia with ketamine (100 mg/kg) and xylazine (5 mg/kg). A 5-cm-long spinal cord section with the lesion at the epicenter was collected and divided equally into five 1-cm segments (distant rostral [S1], near rostral [S2], lesion or injury [S3], near caudal [S4], and distant caudal [S5]). The segments were immediately processed for extraction of genomic DNA and protein.

Detection of Internucleosomal DNA Fragmentation in SCI

We have described elsewhere a non-radioactive agarose gel electrophoresis method for detecting internucleosomal DNA fragmentation in rat SCI tissue.[22,23] Briefly, each spinal cord segment (1-cm long) was homogenized in the DNA homogenization buffer (10 mM Tris, 150 mM NaCl, 50 mM EDTA, pH 8.0) in a 1.5-ml Eppendorf tube using a cordless homogenizer (Kontes Instruments, Vineland, NJ). The homogenate was digested overnight in the digestion buffer (10 mM Tris, 50 mM NaCl, 10 mM EDTA, pH 8.0, 0.5% SDS, 250 ng/µl proteinase K) at 37°C. The contents were extracted twice with equal volume of a mixture (1:1, v/v) of phenol (freshly purchased and pre-equilibrated with equal volume of 500 mM Tris-HCl, pH 8.0) and chloroform, and once with chloroform. To enhance precipitation of internucleosomally fragmented DNA (if any), 1 M $MgCl_2$ was added to the aqueous phase to a final concentration of 10 mM. Total genomic DNA was precipitated by adding two volumes of absolute ethanol and pelleted by centrifugation. The pellet was washed with 70% ethanol, air dried, and dissolved in 20 µl of TE (10 mM Tris-HCl, 1 mM EDTA, pH 8.0) buffer containing RNase A (50 ng/µl). After one-hour incubation at 37°C, the samples were loaded onto 1.6% agarose gels and electrophoresed in 1 × TAE (40 mM Tris-HCl, 40 mM Na-acetate, 1 mM EDTA, pH 8.3) buffer. The gels were stained with ethidium bromide (1 µg/ml) for 30 min, destained in water for a clean background, and photographed on a UV (303 nm) transilluminator using Polaroid film (positive/negative) type 665. The negatives of gel photographs were used to identify internucleosomal DNA fragmentation, the biochemical hallmark of apoptosis.

Western Blot Analysis for Calpain Activity in SCI

Calpain activity in spinal cord segments was measured indirectly by the degradation of 68-kD NFP on Western blots, as described elsewhere.[10,22] Briefly, spinal cord segments (S1, S2, S3, S4, and S5) from sham, vehicle plus SCI, and E-64-d plus SCI groups were homogenized in ice cold (4°C) protein homogenization buffer (50 mM Tris-HCl, 320 mM sucrose, 0.1 mM PMSF, and 1 mM EDTA, pH 7.4). Total proteins were extracted by centrifugation and quantified spectrophotometrically using Coomassie® Plus Protein Assay Reagent (Pierce, Rockford, IL). Heat denatured protein samples (2 µg) in duplicates were resolved on SDS-polyacrylamide linear gradient (4–20%) gels (Bio-Rad, Hercules, CA) by SDS-polyacrylamide gel electrophoresis (SDS-PAGE). After SDS-PAGE, one gel with resolved proteins was stained with Coomassie Blue to monitor for equal amounts of protein loading. Similarly resolved proteins on a sister gel were transferred by electroblotting to Immobilon™-

P membrane (Millipore Corporation, Bedford, MA). Western blots were incubated with (1:800) mouse monoclonal anti-68 kD NFP IgG primary antibody (Sigma Chemical Co., St. Louis, MO) followed by incubation with (1:2,000) alkaline horse radish peroxidase (HRP) conjugated antimouse IgG secondary antibody (ICN Pharmaceuticals, Aurora, OH). Subsequently, specific 68-kD NFP bands were detected by alkaline HRP catalyzed oxidation of luminol in presence of H_2O_2 using the enhanced chemiluminescence (ECL) system (Amersham Life Science, Buckinghamshire, UK), and immediate autoradiography using X-OMAT XAR-2 film (Eastman Kodak Company, Rochester, NY). The extent of 68-kD NFP degradation, as an indication of calpain activity, was estimated by scanning the ECL autoradiograms on a PowerLook Scanner (UMAX, Fremont, CA) using PhotoShop software (Adobe Systems, Seattle, WA) and densitometric analysis of the digitized images using Quantity One software (PDI, Huntington Station, NY).

Statistical Analysis

Differences in 68-kD NFP degradation among the groups were evaluated by one-way analysis of variance (ANOVA) followed by Fisher's protected least significant difference (PLSD) *post hoc* tests at a 95% confidence interval using StatView software (Abacus Concepts, Berkeley, CA). Results were expressed as mean ± SEM of separate experiments ($n = 4$–6). The null hypothesis was rejected at a p value of 0.05 or less. Significant differences (sham vs. vehicle plus SCI, or vehicle plus SCI vs. E-64-d plus SCI) are indicated by $*p < 0.03$; and $**p < 0.01$.

RESULTS

Cycloheximide Inhibited Apoptosis in SCI Lesion

We applied a non-radioactive agarose gel electrophoresis method for detection of internucleosomal DNA fragmentation (apoptosis) in rat SCI tissue. Spinal cord lesion segment (1-cm long) from cycloheximide-treated SCI rats, and corresponding segments from the sham- and vehicle-treated SCI rats, were processed for genomic DNA isolation and electrophoresis on agarose gels (see FIGURE 1). Neither sham animal (lane 1) nor cycloheximide-treated SCI rat (lane 3) showed the 180-bp ladder of internucleosomal DNA fragmentation. Total genomic DNA from vehicle-treated SCI animal (lane 2) showed a clear DNA ladder, indicating the occurrence of neuronal apoptosis in SCI. Since cycloheximide inhibits *de novo* synthesis of proteins in general,[32] the inhibition of internucleosomal DNA fragmentation (apoptosis) in cycloheximide-treated SCI animals suggests that this cell death process is dependent on *de novo* synthesis of proteins and proteases. In addition to non-specific inhibition of *de novo* protein synthesis, cycloheximide also causes prolonged inhibition of DNA synthesis in the rat brain[43] and blocks long-term potentiation.[44] Therefore, the therapeutic value of cycloheximide is limited; and cycloheximide-treated SCI rats may develop impairment of normal physiological functions in the long run. However, prevention of PCD in SCI rats by treatment with an inhibitor of a specific protease such as calpain may prove beneficial and realistic for the development of a clinical therapy.

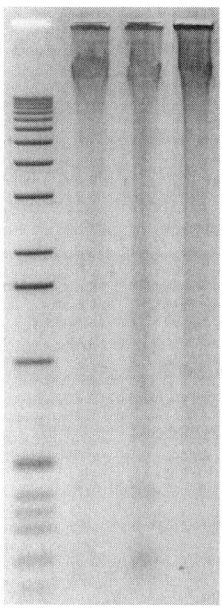

FIGURE 1. Cycloheximide-inhibited apoptosis in SCI rats. A representative agarose gel shows the genomic DNA samples isolated from 1-cm-long lesion segments from different rats. *M*, marker (1-kb ladder); *lane 1*, sham animal; *lane 2*, SCI rat treated with vehicle (1.5% DMSO); and *lane 3*, SCI rat treated with cycloheximide (1 mg/kg). Inhibition of internucleosomal DNA fragmentation (apoptosis) in SCI rats by treatment with cycloheximide indicated the requirement for *de novo* protein synthesis for occurrence of apoptosis.

E-64-d Inhibited 68-kD NFP Degradation in SCI Lesion and Penumbra

The degradation of 68-kD NFP was monitored on Western blots with a mouse monoclonal anti-68-kD NFP antibody that did not crossreact with other intermediate filament proteins. The extent of degradation of 68-kD NFP in the vehicle plus SCI group of rats varied depending on the location of spinal cord segments with respect to the lesion (see FIGURE 2). The highest degradation of 68-kD NFP was 64.6% ($p = 0.0232$) in the lesion segment (S3) from the vehicle plus SCI group of rats compared to the corresponding S3 from the sham group of rats (FIG. 2b). Following E-64-d treatment of SCI rats, degradation of 68-kD NFP was prevented in all these segments. Significant increases in 68-kD NFP were 236.2% ($p = 0.0023$) in S3 and 147.4% ($p = 0.0093$) in S4 from the E-64-d plus SCI group of rats compared to corresponding segments from vehicle plus SCI group of rats (FIG. 2b). The inhibition of degradation of 68-kD NFP, a preferred calpain substrate, indicated the involvement of calpain activity in neurodegeneration in SCI rats.

E-64-d Inhibited Apoptosis in SCI Lesion and Penumbra

Various degrees of internucleosomal DNA fragmentation (apoptosis) were detected in the spinal cord segments (see FIGURE 3). Treatment of sham animals with vehicle alone did not cause apoptosis in the spinal cord (data not shown). Thus, vehicle by itself was unable to cause apoptosis in sham rats or to inhibit apoptosis in SCI animals (FIG. 3). Almost intact genomic DNA was detected in spinal cord segments collected from the sham group, indicating the absence of any recognizable cell

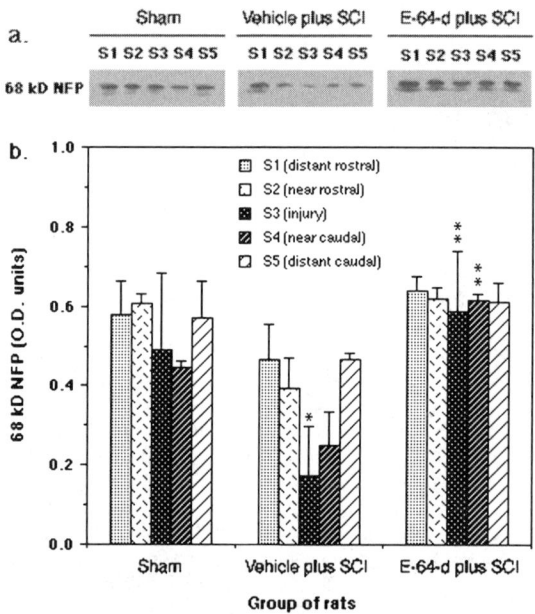

FIGURE 2. Degradation of 68-kD NFP as a measure of calpain activity on Western blots. Spinal cord segments from different groups of rats were used to extract protein samples. **Panel a**, 68-kD NFP bands on the representative Western blots. **Panel b**, Densitometric quantitation of changes in 68-kD NFP levels. (This figure is reproduced, with permission, from Ref. 22.)

death. The spinal cord segments collected from vehicle plus SCI group showed different degrees of DNA degradation depending on their distance from the site of injury. The highest amount of internucleosomal DNA fragmentation occurred in the site of injury (S3). Rostral and caudal segments (S2 and S4) adjacent to the site of injury demonstrated moderate degrees of DNA fragmentation. Distant rostral and caudal segments (S1 and S5) did not show appreciable DNA fragmentation. Treatment of SCI rats with E-64-d prevented DNA fragmentation in all five spinal cord segments tested, indicating the involvement of calpain in neuronal apoptosis in SCI rats.

DISCUSSION

Morphological and Biochemical Changes in SCI

Following primary injury to the spinal cord, several morphological and biochemical changes initiate the devastating secondary injury process leading to neuronal death and dysfunction. The morphological alterations (granular degeneration of axons, vesiculation of myelin, and inclusion of calcium crystals) have been correlated

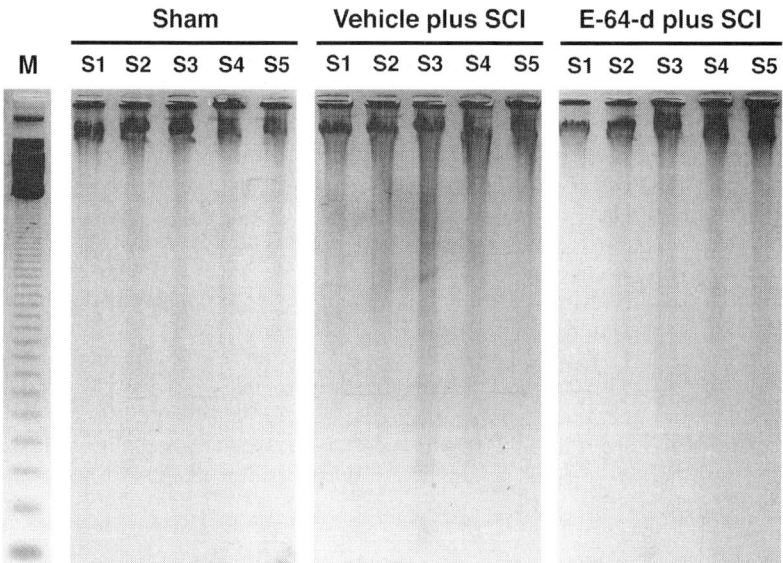

FIGURE 3. Calpain inhibitor E-64-d-attenuated apoptosis in rat SCI lesion and penumbra. Agarose gel electrophoresis of genomic DNA samples isolated from spinal cord segments from different groups of rats. Each agarose gel is representative of at least three separate experiments. Genomic DNA samples were isolated from 1-cm long spinal cord segments: *S1*, distant rostral; *S2*, near rostral; *S3*, lesion or injury; *S4*, near caudal; and *S5*, distant caudal. *M*, marker (123-bp DNA ladder). (This figure is reproduced, with permission, from Ref. 22.)

with biochemical changes (decrease in plasmalogen, activation of plasmalogenase, increase in arachidonic acid and its metabolites, accumulation of quinolinic acid, changes in Na^+ and K^+ levels, and decrease in Na^+-K^+-ATPase activity) in the SCI lesion.[45–49] The most important finding may well be an increase in intracellular total Ca^{2+} levels in the lesion.[9] Subsequently, it was found that the increase in intracellular Ca^{2+} was due to a decrease in the extracellular Ca^{2+} following SCI.[50,51] Progressive degradation of myelin proteins such as myelin basic protein (MBP) and proteolipid protein (PLP) and axonal proteins such as neurofilament proteins (NFPs) and microtubule-associated proteins (MAPs) have been correlated with the structural degeneration of myelin and axons in the spinal cord after trauma.[52] Increased degradation of structural proteins suggested that there was activation of neutral as well as lysosomal proteinases in the SCI lesion.[53–56] Here, we report that: (1) there is a requirement for *de novo* synthesis of proteins and proteinases for cell death in the SCI lesion (FIG. 1); and (2) the Ca^{2+}-activated neutral proteinase calpain is involved in 68-kD NFP degradation (FIG. 2) leading to internucleosomal DNA fragmentation (FIG. 3) in the SCI lesion and penumbra. Thus, damage to the spinal cord does not stay confined to the site of injury or lesion, but rather it extends over time to the rostral and caudal regions.

Various Sources of Calpain in SCI

Since degeneration of myelinated axons and degradation of structural proteins occur before infiltration of inflammatory immune cells, increased calpain activity may be derived early from endogenous neural cells (neurons, oligodendrocytes, astrocytes, and microglia) and later from infiltration of inflammatory immune cells (T cells and macrophages). Increased calpain immunoreactivity has been detected in macrophages, reactive astrocytes, microglia, neurons, and myelin in SCI.[57] Any increase in calpain expression and activity in SCI lesion at an early stage is likely provoked by elevation of intracellular free Ca^{2+} levels. The later increase in calpain activity in SCI probably stems from reactive astrocytes and inflammatory immune cells through production of inflammatory cytokines and activation of the arachidonic acid cascade. Several *in vivo* studies suggest the release of arachidonic acid and its metabolites, and cytokines such as IL-1, IL-6, IL-18, and TNF-α in CNS injuries.[58–60] Furthermore, our *in vitro* studies have revealed an increased synthesis and activity of calpain in glial,[35] neuronal,[36] and lymphoid[61] cells in response to stress. Now, we show that *de novo* synthesis and activation of calpain may occur as a consequence of SCI in rats.

Contribution of Calpain to Apoptosis in SCI

An increase in intracellular free Ca^{2+} levels and free radicals may contribute to neuronal apoptosis in CNS injury, ischemia, stroke, and glutamate neurotoxicity.[62] Intracellular free Ca^{2+} levels may also be increased due to release of Ca^{2+} from intracellular organelles such as mitochondria and microsomes, in response to SCI.[63,64] Increases in intracellular Ca^{2+} levels,[9,65] ROS,[13,66–70] and lipid inflammatory mediators[48,71] can stimulate calpain activity and mediate neuronal apoptosis in SCI. There is a role for calpain in apoptosis of glial[35] as well as neuronal[36,72] cells. Our previous studies,[10,11] showing increased calpain expression and activity, suggested that calpain may be one of the cysteine proteases responsible for neuronal apoptosis in rat SCI. With a sustained increase in intracellular Ca^{2+} levels, calpains are activated by autolytic cleavage of N-terminal peptides and this limited autolysis is the key mechanistic step in the activation of calpains.[73] Structural proteins and other microtubular proteins are partially degraded by calpain.[74,75] Here, we observe the degradation of 68-kD NFP, a preferred calpain substrate, not only in the lesion but also in surrounding areas following SCI in rats (FIG. 2). Loss of 68-kD NFP was high at the site of injury (S3), moderate in adjacent regions (S2 and S4), and low in distant regions (S1 and S5). The extent of degradation of 68-kD NFP (FIG. 2) and internucleosomal DNA fragmentation (FIG. 3) are well correlated, indicating a contribution of calpain activity to apoptosis in SCI. The cell-permeable, calpain-specific inhibitor E-64-d was therapeutic in preventing calpain activity and apoptosis in SCI.

Therapeutic Potential of Calpain Inhibitors in SCI

Many pharmacological agents (corticosteroids, Ca^{2+} entry blockers, clonidine, etc.) have been used for the treatment of acute experimental SCI.[76,77] Specific enzyme inhibitors are useful tools for studying mechanisms and disease processes. Calpain has been implicated in the destruction of the axon–myelin structural unit in CNS trauma, warranting the examination of the effect of calpain inhibitors in SCI.

The endogenous protein inhibitor of calpain, calpastatin (110 kD), is too large to be cell permeable. Furthermore, with an increased calpain:calpastatin ratio, calpain degrades calpastatin as a suicide substrate *in vitro*,[78,79] in cultured cells,[80] and even *in vivo*.[81,82] Therefore, therapeutic use of calpastatin for inhibition of calpain in SCI is not viable. On the other hand, use of calpain-specific inhibitors has been found to be beneficial in a number of pathophysiological conditions. For example, E-64-d (also known as loxastatin) was the first reported cell-permeable calpain inhibitor[83] that reduced muscle degeneration in muscular dystrophy,[84,85] calpeptin and MDL-28170 improved posthypoxic recovery of cells,[86] and AK295 protected neurons in brain ischemia.[87] We previously reported that a combination of calpeptin and methylprednisolone prevented 68-kD NFP degradation and internucleosomal DNA fragmentation in the SCI lesion and penumbra.[88] In the present investigation, we find that E-64-d alone is capable of providing significant neuroprotection by inhibition of calpain activity and apoptosis in the rostral, lesion, and caudal regions following SCI in rats. These findings in animal models of CNS injury strongly suggest that therapeutic use of calpain inhibitors may also provide neuroprotection following SCI in humans.

ACKNOWLEDGMENTS

This research was supported in part by grants from National Institute of Neurological Disorders and Stroke (NS-31622, NS-31767, NS-38146, and NS-41088), National Multiple Sclerosis Society (RG-2130), and American Health Assistance Foundation.

REFERENCES

1. COLLINS, W.F., J. PIEPMEIR & E. OGLE. 1986. The spinal cord injury problem. A review. Central Nerv. Syst. Trauma **3**: 317–331.
2. ANDERSON, D.K. & E.D. HALL. 1993. Pathophysiology of spinal cord trauma. Ann. Emerg. Med. **22**: 987–992.
3. BALENTINE, J.D. 1978. Pathology of experimental spinal cord trauma. Lab. Invest. **39**: 254–266.
4. BANIK, N., E. HOGAN & C.Y. HSU. 1987. The multimolecular cascade of spinal cord injury. Neurochem. Pathol. **7**: 57–77.
5. TATOR, C.H. & M.G. FEHLINGS. 1991. Review of the secondary injury theory of acute spinal cord trauma with emphasis on vascular mechanism. J. Neurosurg. **75**: 15–26.
6. YOUNG, W. 1993. Secondary injury mechanisms in acute spinal cord injury. J. Emerg. Med. **11**: 13–22.
7. LI, Y., P.M. FIELD & G. RAISMAN. 1997. Repair of adult rat corticospinal tract by transplants of olfactory ensheathing cells. Science **277**: 2000–2002.
8. NICHOLLS, J. & N. SAUNDERS. 1996. Regeneration of immature mammalian spinal cord after injury. Trends Neurosci. **19**: 229–234.
9. HAPPEL, R.D., K.P. SMITH, N.L. BANIK, *et al.* 1981. Ca^{2+}-accumulation in experimental spinal cord trauma. Brain Res. **211**: 476–479.
10. BANIK, N.L., D.L. MATZELLE, G. GANTT-WILFORD, *et al.* 1997. Increased calpain content and progressive degradation of neurofilament protein in spinal cord injury. Brain Res. **752**: 301–306.
11. RAY S.K., D.C. SHIELDS, T.C. SAIDO, *et al.* 1999. Calpain activity and translational expression increased in spinal cord injury. Brain Res. **816**: 375–380.

12. BLIGHT, A. 1992. Macrophages and inflammatory damage in spinal cord injury. J. Neurotrauma **9:** S83–S90.
13. BRAUGHLER, J.M. & E.D. HALL. 1992. Involvement of lipid peroxidation in CNS injury. J. Neurotrauma **9:** S1–S6.
14. FADEN, A.I. & R.P. SIMON. 1988. A potential role for excitotoxins in the pathophysiology of spinal cord injury. Ann. Neurol. **23:** 623–626.
15. GWAG, B.J., J.Y. KOH, J.A. DEMARO, et al. 1997. Slowly triggered excitotoxicity occurs by necrosis in cortical cultures. Neuroscience **77:** 393–401.
16. THOMPSON, C.B. 1995. Apoptosis in the pathogenesis and treatment of disease. Science **267:** 1456–1462.
17. LI, G.L., G. BRODIN, M. FAROOQUE, et al. 1996. Apoptosis and expression of Bcl-2 after compression trauma to rat spinal cord. J. Neuropathol. Exp. Neurol. **55:** 280–289.
18. LIU, X.Z., X.M. XU, R. HU, et al. 1997. Neuronal and glial apoptosis after traumatic spinal cord injury. J. Neurosci. **17:** 5395–5406.
19. LOU, J., L.G. LENKE, F.J. LUDWIG & M.F. O'BRIEN. 1998. Apoptosis as a mechanism of neuronal cell death following acute experimental spinal cord injury. Spinal Cord **36:** 683–690.
20. RAY, S., D. DAVIS, D. SHIELDS, et al. 1998. Increased calpain expression, cell death, and neuroprotection in rat spinal cord injury. J. Neurochem. **70:** S63 (Abstr. B).
21. SPRINGER, J.E., R.D. AZBILL & P.E. KNAPP. 1999. Activation of the caspase-3 apoptotic cascade in traumatic spinal cord injury. Nat. Med. **5:** 943–946.
22. RAY, S.K., D.C. MATZELLE, G.G. WILFORD, et al. 2000. E-64-d prevents both calpain upregulation and apoptosis in the lesion and penumbra following spinal cord injury in rats. Brain Res. **867:** 80–89.
23. RAY, S.K., D.D. MATZELLE, G.G. WILFORD, et al. 2000. Increased calpain expression is associated with apoptosis in rat spinal cord injury: Calpain inhibitor provides neuroprotection. Neurochem. Res. **25:** 1191–1198.
24. SAKURAI, M., T. HAYASHI, K. ABE, et al. 1998. Delayed selective motor neuron death and Fas antigen induction after spinal cord ischemia in rabbits. Brain Res. **797:** 23–28.
25. CROWE, M.J., J.C. BRESNAHAN, S.L. SHUMAN, et al. 1997. Apoptosis and delayed degeneration after spinal cord injury in rats and monkeys. Nat. Med. **3:** 73–76.
26. EMERY, E., P. ALDANA, M.B. BUNGE, et al. 1998. Apoptosis after traumatic human spinal cord injury. J. Neurosurg. **89:** 911–920.
27. ZHIVOTOVSKY, B., D.H. BURGESS, D.M. VANAGS & S. ORRENIUS. 1997. Involvement of cellular proteolytic machinery in apoptosis. Biochem. Biophys. Res. Commun. **230:** 481–488.
28. CARAFOLI, E. & M. MOLINARI. 1998. Calpain: A protease in search of a function? Biochem. Biophys. Res. Commun. **247:** 193–203.
29. JOHNSON, D.E. 2000. Noncaspase proteases in apoptosis. Leukemia **14:** 1695–1703.
30. DUKE, R.C., R. CHERVENAK & J.J. COHEN. 1993. Endogenous endonuclease-induced DNA fragmentation: an early event in cell-mediated cytolysis. 1983. Proc. Natl. Acad. Sci. U.S.A. **80:** 6361–6365.
31. WYLLIE, A.H., R.G. MORRIS, A.L. SMITH & D. DUNLOP. 1984. Chromatin cleavage in apoptosis: association with condensed chromatin morphology and dependence on macromolecular synthesis. J. Pathol. **142:** 67–77.
32. OBRIG, T.G., W.J. CULP, W.L. MCKEEHAN & B. HARDESTY. 1971. The mechanism by which cycloheximide and related glutarimide antibiotics inhibit peptide synthesis on reticulocyte ribosomes. J. Biol. Chem. **246:** 174–181.
33. KHARLAMOV, A., J. Y. JOO, T. UZ & H. MANEV. 1997. Cycloheximide reduces the size of lesion caused in rats by a photothrombotic model of brain injury. Neurol. Res. **19:** 92–96.
34. CASTAGNE, V. & P.G.H. CLARKE. 1998. Cooperation between glutathione depletion and protein synthesis inhibition against naturally occurring neuronal death. Neuroscience **86:** 895–902.
35. RAY, S.K., G.G. WILFORD, C.V. CROSBY, et al. 1999. Diverse stimuli induce calpain overexpression and apoptosis in C6 glioma cells. Brain Res. **829:** 18–27.

36. RAY, S.K., M. FIDAN, M.W. NOWAK, et al. 2000. Oxidative stress and Ca^{2+} influx upregulate calpain and induce apoptosis in PC12 cells. Brain Res. **852:** 326–334.
37. BANIK, N.L., D.C. SHIELDS, S.K. RAY & E.L. HOGAN. 1999. The pathophysiological role of calpain in spinal cord injury. In CALPAIN: Pharmacology and Toxicology of Calcium-Dependent Protease. K.K.W. Wang & P.-W. Yuen, Eds.: 211–227. Taylor & Francis, Philadelphia.
38. GUTTMAN, R.P. & G.V.W. JOHNSON. 1999. Calpain-mediated proteolysis of neuronal structural proteins. In CALPAIN: Pharmacology and Toxicology of Calcium-Dependent Protease. K.K.W. Wang & P.-W. Yuen, Eds.: 229–249. Taylor & Francis, Philadelphia.
39. TRUMP, B.F. & I.K. BEREZESKY. 1995. Calcium-mediated cell injury and cell death. FASEB J. **9:** 219–228.
40. NICOTERA, P. & S. ORRENIUS. 1998. The role of calcium in apoptosis. Cell Calcium **23:** 173–180.
41. RAY, S.K., D.C. MATZELLE, G.G. WILFORD, et al. 2000. Calpain inhibitor E-64-d attenuates apoptosis in the SCI lesion and penumbra. Fifth International Conference on Neuroprotective Agents. Lake Tahoe, CA.
42. PEROT, P.L., W.A. LEE, C.Y. HSU & E.L. HOGAN. 1987. Therapeutic model for experimental spinal cord injury in rat central nervous system. Trauma **4:** 149–159.
43. PAVLIK, A. & J. TEISINGER. 1980. Effect of cycloheximide administered to rats in early postnatal life: Prolonged inhibition of DNA synthesis in the developing brain. Brain Res. **192:** 531–541.
44. BAREA-RODRIGUEZ, E.J., D.T. RIVERA, D.B. JAFFE & J.L. MARTINEZ. 2000. Protein synthesis inhibition blocks the induction of mossy fiber long-term potentiation in vivo. J. Neurosci. **20:** 8528–8532.
45. BALENTINE, J.D. 1978. Pathology of spinal cord trauma II: ultrastructure of axons in myelin. Lab. Invest. **39:** 254–266.
46. BRESNAHAN, J.C. 1978. An electron microscopic analysis of axonal alterations following blunt contusion of the spinal cord of the Rhesus monkey (*Macaca mulata*). J. Neurol. Sci. **37:** 92–98.
47. CLENDENON, N.R., N. ALLEN & W.A. GORDON. 1978. Effect of trauma on Na^+-K^+-activated ATPase activities in dog spinal cord. Trans. Am. Soc. Neurochem. **9:** 88.
48. HSU, C.Y., P.V. HALUSHKA, E.L. HOGAN, et al. 1985. Alteration of thromboxane and prostacyclin levels in experimental spinal cord injury. Neurology **35:** 1003–1009.
49. POVOVICH, P.G., J.F. REINHARD, E.M. FLANAGAN & B.T. STOKES. 1994. Elevation of the neurotoxin quinolinic acid following spinal cord trauma. Brain Res. **633:** 348–352.
50. YOUNG, W., V. YEN & A. BLIGHT. 1982. Extracellular ionic activity in experimental spinal cord contusion. Brain Res. **253:** 105–113.
51. STOKES, B.T., P. FOX & G. HOLLINDEN. 1983. Extracellular calcium activity in the injured spinal cord. Exp. Neurol. **80:** 561–572.
52. BANIK, N.L., J.M. POWERS & E.L. HOGAN. 1980. The effect of spinal cord trauma on myelin. J. Neuropath. Exp. Neurol. **19:** 232–244.
53. BANIK, N.L., E.L. HOGAN, J.M. POWERS & L.J. WHETSTINE. 1982. Degradation of neurofilament proteins in spinal cord injury. Neurochem. Res. **7:** 1465–1475.
54. ZIMMERMAN, U.-J.P. & W.W. SCHLAEPHER. 1982. Characterization of a brain calcium-activated neutral protease that degrades neurofilament protein. Biochemistry **21:** 3977–3983.
55. BRAUGHLER, J.M. & E.D. HALL. 1984. Effects of multi-dose methylprednisolone sodium succinate administration of injured cat spinal cord neurofilament degradation and energy metabolism. J. Neurosurg. **61:** 290–295.
56. BANIK, N.L., E.L. HOGAN, J.M. POWERS & K. SMITH. 1986. Proteolytic enzyme in spinal cord injury. J. Neurol. Sci. **73:** 245–256.
57. LI, Z., E.L. HOGAN & N.L. BANIK. 1995. Role of calpain in spinal cord injury: increased mcalpain immunoreactivity in spinal cord after compression injury in the rat. Neurochem Int. **27:** 425–432.
58. GIULAN D. & L.B. LACHMAN. 1985. Interleukin 1 stimulation of astroglial proliferation after brain injury. Science **228:** 497–499.

59. OTT, L., C.J. MCCLAIN, M. GILLESPIE & B. YOUNG. 1994. Cytokines and metabolic dysfunction after severe head injury. J. Neurotrauma **11:** 447–472.
60. PERRY, V.H. 1994. Macrophage and microglia responses in CNS injury. *In* Macrophages and the Nervous System. V.H. Perry, Ed.: 62–86. R. G. Landes Company, Austin.
61. DESHPANDE, R.V., J.M. GOUST, A.K. CHAKRABARTI, *et al.* 1995. Calpain expression in lymphoid cells : increased mRNA and protein levels after cell activation. J. Biol. Chem. **270:** 2497–2505.
62. LIPTON, S.A. & P. NICOTERA. 1998. Calcium, free radicals and excitotoxins in neuronal apoptosis. Cell Calcium **23:** 165–171.
63. FARBER, J.L. 1981. The role of calcium in cell injury. Life Sci. **29:** 1289–1295.
64. HOGAN, E.L., W. MCIVER, A. KRALL & N.L. BANIK. 1985. Subcellular distributions of calcium in spinal cord trauma. Trans. Am. Soc. Neurochem. **16:** 132.
65. DU, S., A. RUBIN, S. KLEPPER, *et al.* 1999. Calcium influx and activation of Calpain I mediate acute reactive gliosis in injured spinal cord. Exp. Neurol. **157:** 96–105.
66. AZBILL, R.D., X. MU, A.J. BRUCE-KELLER, *et al.* 1997. Impaired mitochondrial function, oxidative stress and altered antioxidant enzyme activities following traumatic spinal cord injury. Brain Res. **765:** 283–290.
67. YAMAMOTO, K., T. ISHIKAWA, T. SAKABE, *et al.* 1998. The hydroxyl radical scavenger Nicaraven inhibits glutamate release after spinal injury in rats. NeuroReport **9:** 1655–1659.
68. HALL, E.D. & J.M. BRAUGHLER. 1989. Central nervous system trauma and stroke. II. Physiological and pharmacological evidence for involvement of oxygen radicals and lipid peroxidation. Free Radic. Biol. Med. **6:** 303–313.
69. LIU, D.X., J. LIU & J. WEN. 1999. Elevation of hydrogen peroxide after spinal cord injury detected by using the Fenton reaction. Free Radic. Biol. Med. **27:** 478–482.
70. LEWEN, A., P. MATZ & P.H. CHAN. 2000. Free radical pathways in CNS injury. J. Neurotrauma **17:** 871–890.
71. XU, J.A., C.Y. HSU, T.H. LIU, *et al.* 1990. Leukotriene B4 release and polymorphonuclear cell infiltration in spinal cord injury. J. Neurochem. **55:** 907–912.
72. JORDAN, J., M.F. GALINDO & R.J. MILLER. 1997. Role of calpain and interleukin-1β converting enzyme-like proteases in the ß-amyloid-induced death of rat hippocampal neurons in culture. J. Neurochem. **68:** 1612–1621.
73. BAKI, A., P. TOMPA, A. ALEXA, *et al.* 1996. Autolysis parallels activation of µcalpain. Biochem. J. **318:** 897–901.
74. CROALL, D.E. & G.N. DEMARTINO. 1991. Calcium-activated neutral protease (calpain) system: structure, function, and regulation. Physiol. Rev. **71:** 813–847.
75. SAIDO, T.C., H. SORIMACHI & K. SUZUKI. 1994. Calpain: New perspectives in molecular diversity and physiological-pathological involvement. FASEB J. **8:** 814–822.
76. FADEN, A.I. 1985. Pharmacologic therapy in acute spinal cord injury: experimental strategies and future directions. *In* Central Nervous System Trauma. D.P. Becker & J.T. Povlishock, Eds.: 481–485. National Institutes of Health, Bethesda.
77. HALL, E.D. 1991. The neuroprotective pharmacology of methylprednisolone. J. Neurosurg. **76:** 13–22.
78. MELLGREN, R.L., M.T. MERICLE & R.D. LANE. 1986. Proteolysis of the calcium-dependent protease inhibitor by myocardial calcium-dependent protease. Arch. Bioch. Biophys. **246:** 233–239.
79. NAKAMURA, M., M. INOMATA, S. IMAJOH, *et al.* 1989. Fragmentation of an endogenous inhibitor upon complex formation with high- and low-Ca^{2+}-requiring forms of calcium-activated neutral proteases. Biochemistry **28:** 449–455.
80. NAGAO, S., T.C. SAIDO, Y. AKITA, *et al.* 1994. Calpain-calpastatin interactions in epidermoid carcinoma KB cells. J. Biochem. (Tokyo) **115:** 1178–1184.
81. SAIDO, T.C., S. KAWASHIMA, E. TANI & M. YOKOTA. 1997. Up- and down-regulation of calpain inhibitor polypeptide, calpastatin, in postischemic hippocampus. Neurosci. Lett. **227:** 75–78.
82. BLOMGREN, K., U. HALLIN, A.L. ANDERSSON, *et al.* 1999. Calpastatin is upregulated in response to hypoxia and is a suicide substrate to calpain after neonatal cerebral hypoxia-ischemia. J. Biol. Chem. **274:** 14046–14052.

83. TAMAI, M., K. MATSUMOTO, S. OMURA, et al. 1986. In vitro and in vivo inhibition of cysteine proteinases by EST, a new analog of E-64. J. Pharmacobio-Dyn. **9:** 672–677.
84. KOMATSU, K., K. INAZUKI, K. HOSOYA & S. SATOH. 1986. Beneficial effect of new thiol protease inhibitors, epoxide derivatives on dystrophic mice. Exp. Neurol. **91:** 23–29.
85. TAMAI, M., S. OMURA, M. KIMURA, et al. 1987. Prolongation of life span of dystrophic hamster by cysteine protease inhibitor, loxistatin (EST). J. Pharmacobio-Dyn. **10:** 678–681.
86. ARLINGHAUS, L., S. MEHDI & K. LEE. 1991. Improved post-hypoxic recovery with a membrane permeable calpain inhibitor. Eur. J. Pharmacol. **209:** 123–125.
87. BARTUS, R.T., N.J. HEYWARD, P.J. ELLIOTT, et al. 1994. Calpain inhibitor AK295 protects neurons from focal brain ischemia: effects of post-occlusion intra-arterial administration. Stroke **25:** 2265–2270.
88. RAY, S.K., G.G. WILFORD, D.C. MATZELLE, et al. 1999. Calpeptin and methylprednisolone inhibit apoptosis in rat spinal cord injury. Ann. N.Y. Acad. Sci. **890:** 261–269.

The Anti-Parkinson Drug Rasagiline and Its Cholinesterase Inhibitor Derivatives Exert Neuroprotection Unrelated to MAO Inhibition in Cell Culture and *in Vivo*

MOUSSA B.H. YOUDIM,[a] A. WADIA,[b] W. TATTON,[b] AND MARTA WEINSTOCK[c]

[a]*Technion-Faculty of Medicine, Eve Topf and National Parkinson Foundation Centers, Haifa, Israel*
[b]*Mount Sinai School of Medicine, New York, New York, U.S.A.*
[c]*Hebrew University, Department of Pharmacology, Jerusalem, Israel*

> ABSTRACT: The antiapoptotic and neuroprotective activity of irreversible monoamine oxidase (MAO) B inhibitor, rasagiline [R(+)-N-propargyl-1-aminoindan], its S-isomer (TVP1022) and TV 3219, a novel anti-Alzheimer cholinesterase-MAO inhibitor drug derived from rasagiline were examined in PC12 cells cultures and *in vivo*. We found that these drugs have potent antiapoptotic and neuroprotective activities in response to serum and NGF withdrawal in partially neuronally differentiated PC12 cells and prevent the fall in mitochondrial membrane potential, the first step in cell death. Closed head injury studies in mice have shown that both rasagiline and TVP1022 are neuroprotective. All these compounds possess a propargyl moiety, which normally is responsible for irreversible inactivation of MAO, as is seen with rasagiline. However, neither TVP1022 nor TV3219 are MAO inhibitors, both share the antiapoptotic and neuroprotective actions of rasagiline, indicating that MAO inhibition is not a prerequisite for neuroprotection and that the propargyl moiety exhibits intrinsic neuroprotective pharmacological activity that requires identification.
>
> KEYWORDS: Propargylamine; Monoamine oxidase inhibitor; Bcl-2; Closed head injury; Neuroprotection; Cholinesterase inhibitors.

INTRODUCTION

Rasagiline (N-propargyl-1R-aminoindan), the R-isomer of the racemic AGN1135 (see FIGURE 1)[1–5] is a potent selective irreversible inhibitor of monoamine oxidase (MAO) B in peripheral tissues and the brain of rats, monkeys, and humans. It has structural similarity to selegiline, but unlike the latter is not a sympathomimetic drug, nor is it metabolized to amphetamine, but rather to aminoindan. At selective MAO B inhibitory doses it does not potentiate the sympathomimetic action of tyramine ("cheese reaction"), an adverse reaction found with irreversible MAO A and MAO

Address for correspondence: Professor Moussa B.H. Youdim, Technion-Faculty of Medicine, Eve Topf and NPF Neurodegenerative Disease Centers, Efron Street, Bat Galim, Haifa, Israel.
youdim@tx.technion.ac.il

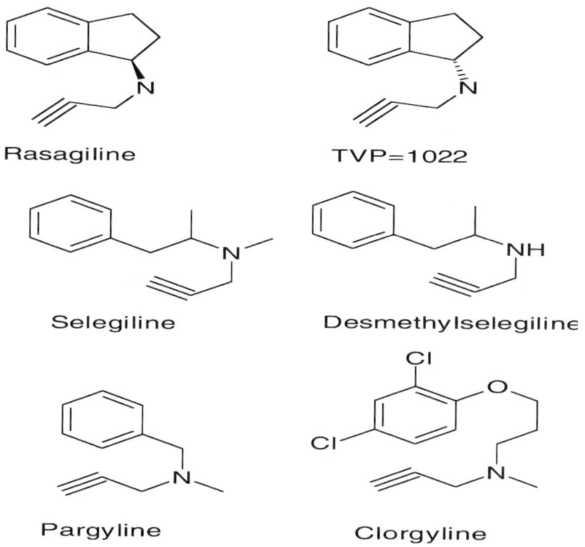

FIGURE 1. The structure of selegline, rasagiline, and other related propargylamine MAO inhibitors.

A-B inhibitors.[2–4] It has been developed as an anti-Parkinson drug. Recent multicenter double blind monotherapy in phase 3 studies by the Parkinson Study Group[6] have shown it to produce significant symptomatic benefit in Parkinsonian subjects at a dose as low as 1 mg. Clinical studies and those in experimental animals show that rasagiline is 5–10 times more potent than selegiline in inhibiting MAO B and as an anti-Parkinson drug. Similar to other MAO B inhibitors, AGN 1135 and rasagiline protect against dopaminergic neurotoxicity by MPTP in mice.[7] Selegiline and other propargylamine derived MAO inhibitors may have some other intrinsic neuroprotective activity, not related to their MAO inhibition,[8–10] since they increase the activities of SOD, catalase and BCL-2. The structural similarity of rasagiline to selegiline (FIG. 1), prompted us to investigate more closely the neuroprotective activity in neuronal cell cultures (PC12 cells) and *in vivo* of rasagiline, its s-isomer (TV1022) and several of its other derivatives, including the racemic (TV3219) form of the novel carbamate containing MAO inhibitor, with cholinesterase inhibition activity, TV3266 (see FIGURE 2), a drug derived from rasagiline and developed for the treatment of Alzheimer's disease (see Weinstock, *et al.*, in this volume, and Ref. 11).

ANTIAPOPTOTIC PROPERTIES OF RASAGLINE AND ITS DERIVATIVES

Like selegiline, both rasagiline and AGN1135 pretreatment protect against dopaminergic neurotoxicity of MPTP in mice whereas the s-isomer (TV1022) of

AGN1135, which is not an MAO-B inhibitor,[4] does not.[12] Selegiline and its major metabolite (−)-desmethyldeprenyl have antiapoptotic and neuroprotective activities in partially neuronally differentiated PC12 cells after serum and NGF.[13–15] However, since both these compounds are inhibitors of MAO B, this activity may be responsible for the antiapoptotic and neuroprotective activities. Rasagiline, TV1022 and TV3219, unlike selegiline, are N-desmethylated propargylamines and the latter two compounds do not inhibit MAO *in vitro* or *in vivo*.[4,11] Therefore, we investigated[16] the antiapoptotic actions of these drugs in comparison to (−)-desmethyldeprenyl in partially neuronally differentiated PC12 cells during serum and NGF withdrawal, according to the method devised by Tatton *et al.*[13,14] to see whether an action other than MAO-B inhibition could contribute to the antiapoptotic effect. Twenty-four

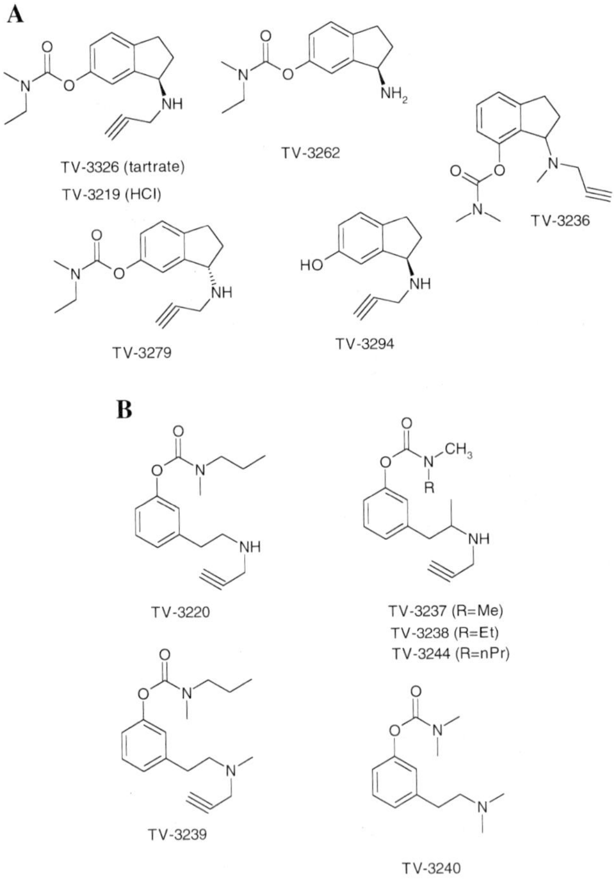

FIGURE 2. The structures of novel carbamate containing cholinesterase-MAO inhibitors derived from **(A)** rasagiline and **(B)** selegline.

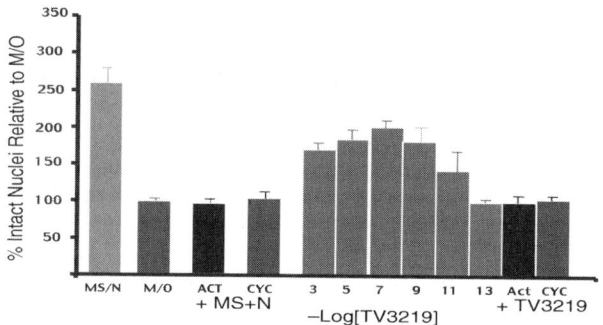

FIGURE 3. The antiapoptotic action of TV3219 was examined in partially neuronally (NGF) differentiated PC12 cells treated with cycloheximide and actinomycin. *M/S+N*, plus serum and NGF: *M/O*, absence of serum and NGF; *ACT*, actinomycin, 3 µg/ml; *CYC*, cycloheximide, 10 µg/ml; TV3219 in absence of serum and NGF; *ACT* or *CYC* plus TV3219 (10^{-7} M).

hours after washing and withdrawal of the serum and NGF, the number of intact nuclei were reduced to about 40% of that found for partially neuronally differentiated PC12 cells that were washed and then replaced in serum and NGF. Rasagiline, TVP1022, and TV3219 showed a dose (10^{-3}–10^{-10} M) dependent inhibition of apoptosis similar to (–)-desmethyldeprenyl (see FIGURE 3), whereas the metabolite of

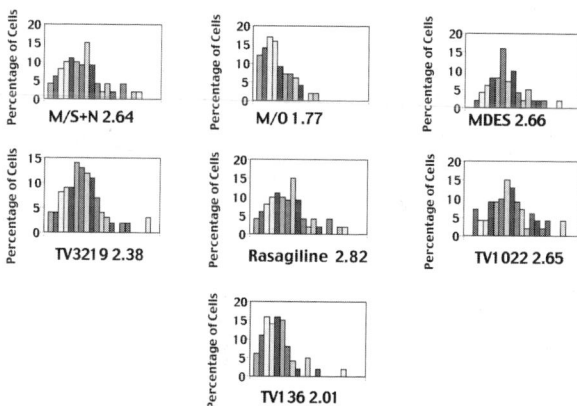

FIGURE 4. Mitochondrial membrane potential ($\Delta\Psi_M$) of partially NGF differentiated PC12 cells 12 hours after serum and NGF withdrawal (*M/O*). Notations are the same as in FIGURE 3. The results are ratio of cytoplasmic/nuclear fluorescence observed with the dye, CMTMR(choloromethyltetra-. methylrhodamine methyl ester). The fluorescence values obtained for the rasagiline, TVP1022, TV3219, and MDES (*N*-desmethyldeprenyl) at a concentration of 10^{-7} M are similar to those with serum and NGF present, but are significantly ($p > 0.05$) different from M/O. TV136, the metabolite of rasagiline, had no significant effect on mitochondrial membrane potential.

rasagiline, TVP136 (aminoindan), showed no such effect. The antiapoptotic activity correlated with prevention of the decrease in mitochondrial potential observed during serum and NGF withdrawal (see FIGURE 4). To image the apoptotic nuclei, we counted the PC12 cells showing chromatin condensation after staining with the molecular probe, YOYO-1. Spontaneous apoptosis, as indicated by the percentage of nuclei with chromatin for cells in presence of serum and NGF, was approximately 0.55%. Withdrawal of serum and NGF increased this value to 4.16 ± 0.34%. Treatment of cells with 0.1 µM rasagiline, TVP1022, or TV3219 significantly reduced the percentage of nuclei with chromatin condensation and the values were 1.41, 1.30, and 1.35%, respectively. The value (3.92 ± 0.54%) obtained for TV136 was not different from that seen with cells in absence of serum and NGF.

THE EFFECT OF TRANSLATIONAL AND TRANSCRIPTIONAL INHIBITORS ON RASAGILINE PREVENTION OF APOPTOSIS

It was previously demonstrated that synthesis of new protein is required for the antiapoptotic action of selegiline in partially neuronally differentiated PC12 cells.[13] This effect can be observed by the use of transcription and translation inhibitors actinomycin (3 mg/ml) and cyclohexiimide (10 mg/ml) without causing the death of PC12 cells. Similar to selegiline and N-desmethyldeprenyl, the antiapoptotic actions of rasagiline and its derivatives, including TV3219, were blocked by cycloheximide and actinomycin (FIG. 4). The results suggest that these compounds act by altering gene transcription. This is supported by our recent studies on gene expression employing a cDNA microarray. The most notable increases in gene expression were those for Bcl-2, SOD, and catalase.[17] Indeed, serum and NGF withdrawal from partially neuronally differentiated PC12 cells results in a decrease in the antiapoptotic enzymes Bcl-2 (see FIGURE 5) and SOD (data not shown), which is prevented by rasagiline and its derivatives (TVP1022 and TV3219), but not by the aminoindan metabolite, TVP136, that lacks the propargyl group. This effect of rasagiline also occurs *in*

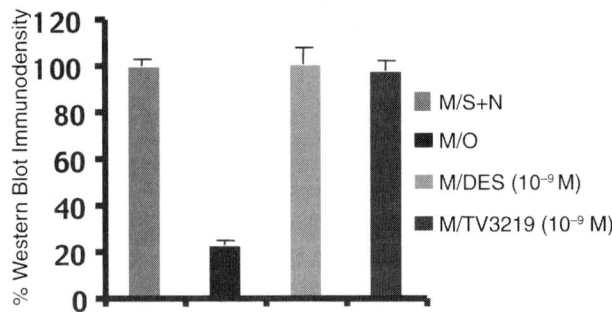

FIGURE 5. The alteration of Bcl-2 protein as measured by Western blot optical density of partially-NGF differentiated PC12 cells six hours after serum and NGF withdrawal (M/O) and compared to M/S+N; M/DES, N-desmethyldeprenyl; and TV3219.

vivo since chronic treatment (four weeks) of rats resulted in a 2–4-fold increase in SOD and catalase in the striatum, cortex, hippocampus, heart, and kidney.[9]

NEUROPROTECTIVE EFFECTS OF RASGILINE AND TVP1022 AGAINST CLOSED HEAD INJURY IN THE MOUSE

The potential neuroprotective effects of rasagiline and its MAO-B inhibitory inactive (S) enantiomer, TVP1022, were examined against the sequelæ of closed head injury in mice induced in the left hemisphere under ether anesthesia.[18] Rasagiline (0.2 and 1 mg/kg) or TVP1022 (1 and 2 mg/kg) injected once, sc, five minutes after injury accelerated the recovery of motor function and spatial memory as measured in a Morris water maze and reduced the cerebral edema by about 40–50%, ($p < 0.01$). Scopolamine (0.2 mg/kg) prevented the neuroprotective effects on motor function and spatial memory, but not on cerebral edema. Rasagiline (0.1 and 1 mg/kg) injected three days after injury and for the next five days accelerated the recovery of spatial memory but not motor function. These studies demonstrate that early administration of rasagiline or TVP1022 can reduce the immediate sequelæ of brain injury. They can also hasten the recovery from injury even if given several days later. The mechanism of action does not appear to result from MAO-B inhibition but could be mediated by the maintenance of cholinergic transmission in particular brain neurons. The mechanism of action of rasagiline and TVP1022 in this model does not appear to result from monoamine oxidase-B inhibition but, as has been suggested by Huang *et al.*,[18] could result from the maintenance of cholinergic transmission in brain neuron since impairments in motor function and memory resulting from brain injury are associated with a deficit in cholinergic activity. However, these drugs have no direct effect on cholinergic receptors, neither do they inhibit cholinesterase. Like selegiline,[19] rasagiline may increase the activity of choline acetyl-transferase. Their neuroprotective activity may also involve other mechanisms related to activation of antioxidant enzymes such as SOD, catalase, glutathione peroxidase and BCl2,[9] since in closed head injury significant evidence for the participation of oxidative stress and inflammatory processes has been provided.[20,21]

DISCUSSION

In the treatment of the progressive neurodegenerative disorders such as Parkinson's and Alzheimer's disease MAO-B (selegiline) and acetylcholinesterase (rivastigmine) inhibitors, respectively, have been employed. However, unlike the cholinesterase inhibitors presently in clinical use, the MAO-B inhibitor selegiline has been reported to have intrinsic neuroprotective activity not necessarily related to its ability to inhibit MAO. It has been suggested that its neuroprotective action may reside in the propargyl moiety of selegiline.[9,18,22] However, until recently it was not possible to test this hypothesis because of the lack of MAO-B selective inhibitors with a similar structure to that of selegiline. We have now synthesized several MAO-B inhibitors having a structural resemblance to selegiline and tested their neuroprotective activities in several models (see FIG. 1). Of the compounds in question, the anti-Parkinson selective MAO-B drug, rasagiline, has received particular attention. Rasagiline exhibits neuroprotective activity in neuronal cell cultures (PC12, S 5) and *in vivo*; it offers an advantage over selegiline in that it is not metabolized to amphetamine, which may be neurotoxic. TABLE 1 summarizes the studies on neuroprotective activity of these propargylamine carried out by various groups. One of the most prominent aspects of our studies has been the demonstration that MAO-B inhibition is not a prerequisite for neuroprotection. This is derived from experiments in which both the R and S-isomers of rasagiline derivatives have been examined. Although rasagiline is a highly potent irreversible inhibitor of MAO-B,[4] its optical S-isomer, TVP1022, and its metabolite aminoindan (TVP136) are at least 100-fold weaker. Similarly, the racemic cholinesterase-MAO inhibitor TV3219 and its R-isomer (TV3326) (see Weinstock, *et al.*, this volume) only inhibits MAO-A and -B *in vivo* after chronic administration, but are inactive *in vitro*, whereas the S-isomer (TV3279) is devoid of MAO inhibitory activity *in vitro* and *in vivo*. However, all the compounds that contain the propargyl moiety, but not the aminoindan metabolite, exhibit neuroprotective and antiapoptotic activities. This is also seen with other molecules containing a propargyl group (Maruyama, *et al.*, submitted). The mechanism by which the propargyl moiety confers neuroprotection is not fully understood.[16] It has been suggested that dimerization and the inhibition of translocation of the anti-apoptotic component GAPDH (glyceraldehyde dehydrogenase) from cytoplasm to the nucleus is a major component. Indeed, in our preliminary studies we have shown that rasagiline inhibits the translocation of GAPDH in dopamine-derived neuroblastoma cells. However, consideration should also be given to the ability of rasagiline and its derivatives to increase brain (striatum, cortex and hippocampus) SOD, Bcl-2, and catalase[9] and to inhibit the permeability of the mitochondrial transition pore.

TABLE 1. Neuroprotective properties of the irreversible monoamine oxidase B inhibitor, rasagiline

1.	Increases SOD, catalase and BCL-2 activities by transcriptional and translational means in PC12 cells and rats (brain, heart and kidney)[9,16]
2.	Prevents peroxynitrite induced activation of caspase 3[8,10,23]
3.	Prevents peroxynitrite induced DNA lathering[8,10,23]
4.	Prevents glutamate and NMDA induced neurotoxicity in hippocampal and cortical cell cultures[24-26]
5.	Prevents peroxynitrite and salsolinol induced fall in mitochondrial membrane potential[8,10,23]
6.	Protects against peroxynitrite induced apoptosis[8,10,23]
7.	Neuroprotective in closed head injury in mice[18]
8.	Protects against cell death induced by ischaemia and by glucose deprivation in PC12 cells[27]
9.	Prevents MPTP (mice) and 6-OHDA (PC12 cells and rats) neurotoxicity[7,23]
10.	Increased survival of dopaminergic neurons[25]
11.	Neuroprotection in models of motor and cognition disorders[28,29]

Overexpression of Bcl-2 and SOD in mice has been shown to make them resistant to neurotoxin-induced neurodegeneration. It is apparent that the mechanism of neuroprotection induced by propargylamines,[8,22,23] involves a complex set of neurochemical events that includes alteration in gene expression.[17]

ACKNOWLEDGMENT

We wish to thank Teva Pharmaceutical Co. (Israel), The Stein Foundation (Philadelphia, U.S.A.), NPF (Miami, U.S.A.) and Golding Parkinson Fund (Technion, Haifa, Israel) for the support they have given to this project.

REFERENCES

1. KALIR, A., A. SABBAGH & M.B.H. YOUDIM. 1981. Selective acetylenic 'suicide' and reversible inhibitors of monoamine oxidase types A and B. Br. J. Pharmacol. **73:** 55–64.
2. FINBERG, J.P., M. TENNE & M.B. YOUDIM. 1981. Tyramine antagonistic properties of AGN 1135, an irreversible inhibitor of monoamine oxidase type B. Br. J. Pharmacol. **73:** 65–74.
3. FINBERG, J.P. & M.B. YOUDIM. 1985. Modification of blood pressure and nictitating membrane response to sympathetic amines by selective monoamine oxidase inhibitors, types A and B, in the cat. Br. J. Pharmacol. **85:** 541–546.
4. YOUDIM, M.B.H., A. GROSS & J.P.M. FINBERG. 2001. Rasagiline [N-propargyl-1R(+)-aminoindant], a selective and potent inhibitor of mitochondrial monoamine oxidase B. Br. J. Pharmacol. **132:** 500–507.
5. FINBERG, J.P., J. WANG, K. BANKIEWICZ, *et al.* 1998. Increased striatal dopamine production from L-DOPA following selective inhibition of monoamine oxidase B by R(+)-N-propargyl-1-aminoindan (rasagiline) in the monkey. J. Neural. Transm. Suppl. **52:** 279–285.
6. PARKINSON STUDY GROUP. 2000. A controlled clinical trial of rasagiline in early Parkinson's disease. Am. Neurol. Assoc. Boston.
7. HEIKKILA, R.E., R.C. DUVOISIN, J.P. FINBERG & M.B. YOUDIM. 1985. Prevention of MPTP-induced neurotoxicity by AGN-1133 and AGN-1135, selective inhibitors of monoamine oxidase-B. Eur. J. Pharmacol. **116:** 313–317.
8. MARUYAMA, W., T. YAMAMOTO, K. KITANI, *et al.* 2000. Mechanism underlying antiapoptotic activity of a (−)deprenyl-related propargylamine, rasagiline. Mech. Age. Dev. **116:** 181–191.
9. CARRILLO, M.C., C. MINAMI, K. KITANI, *et al.* 2000. Enhancing effect of rasagiline on superoxide dismutase and catalase activities in the dopaminergic system in the rat. Life Sci. **67:** 577–585.
10. MARUYAMA, W. & M. NAOI. 1999. Neuroprotection by (−)-deprenyl and related compounds. Mech. Age. Dev. **111:** 189–200.
11. WEINSTOCK, M., T. GOREN & M.B.H. YOUDIM. 2000. Development of a novel neuroprotective Drug (TV3326) for the Treatment of Alzheimer's Disease, with cholinesterase and monoamine oxidase inhibitory activties. Drug Dev. Res. 216–222.
12. SAGI, Y., M. WEINSTOCK & M.B.H. YOUDIM. 2001. Neuroprotective activities of rasagiline, the cholinesterase inhibitor. TV3326, and their optical isomers against MPTP neurotoxicity in mice. 14th International Congress on Parkinson's Disease, Helsinki.
13. TATTON, W.G., W.Y. JU, D.P. HOLLAND, *et al.* 1994. (−)-Deprenyl reduces PC12 cell apoptosis by inducing new protein synthesis. J. Neurochem. **63:** 1572–1575.
14. WADIA, J.S., R.M.E. CHALMERS-REDMAN, W.J.H. JU, *et al.* 1998. Mitochondrial membrane potential and nuclear changes in apoptosis caused by serum and nerve growth factor with drawal: time course and modification by (−)-deprenyl. J. Neurosci. **18:** 932–947.

15. CARLILE, G.W., R.M. CHALMERS-REDMAN, N.A. TATTON, *et al.* 2000. Reduced apoptosis after nerve growth factor and serum withdrawal: conversion of tetrameric glyceraldehyde-3-phosphate dehydrogenase to a dimer. Mol. Pharmacol. **57:** 2–12.
16. YOUDIM, M.B.H., J.S. WADIA & N.A. TATTON. 1999. Neuroprotective properties of the antiparkinson drug rasagiline and its optical S-Isomer. Neurosci. Lett. **54:** S45.
17. DRIGUES, N., T. POLTYREV, M. WEINSTOCK & M.B.H. YOUDIM. 2000. Gene expression and behavioral profile of different types of anti and non-antidepressant drugs. Neurosci. Lett. Suppl. **55:** S15.
18. HUANG, W., Y. CHEN, E. SHOHAMI & M. WEINSTOCK. 1999. Neuroprotective effect of rasagiline, a selective monoamine oxidase-B inhibitor, against closed head injury in the mouse. Eur. J. Pharmacol. **366:** 127–135.
19. KOUTSILIERI, E., C. SCHELLER, S. SOPPER, *et al.* 2001. Selegiline completely restores choline acetyltransferase activity deficits in simian immunodeficiency infection. Eur. J. Pharmacol. **411:** R1–R2.
20. STAHEL, P.F., E. SHOHAMI, F.M. YOUNIS, *et al.* 2000. Experimental closed head injury: analysis of neurological outcome, blood-brain barrier dysfunction, intracranial neutrophil infiltration, and neuronal cell death in mice deficient in genes for pro-inflammatory cytokines. J. Cereb. Blood Flow Metab. **20:** 369–380.
21. FEUERSTEIN, G.Z., X. WANG & F.C. BARONE. 1998. The role of cytokines in the neuropathology of stroke and neurotrauma. Neuroimmunomodulation. **5:** 143–159.
22. YOUDIM, M.B.H., A. WADIA, W. TATTON & M. WEINSTOCK. 2001. The antiparkinson drug rasagiline and its derivatives exert neuroprotection unrelated to MAO inhibition in cell culture and *in vivo*. Neurosci. Lett. Suppl. **55:** In press.
23. MARUYAMA, W., Y. AKAO, M.B.H. YOUDIM & M. NAOI. 2000. Neurotoxin induced apoptosis in dopamine neurons: Protection by propargylamine derivatives, Rasagiline and TV 1022. J. Neural. Transm. Suppl: 147–162.
24. WEINSTOCK, M., C. BEJAR, R.H. WANG, *et al.* 2000. TV3326, a novel neuroprotective drug with cholinesterase and monoamine oxidase inhibitory activities for the treatment of Alzheimer's disease. P. Riederer, D.B. Calne, R. Horowski, *et al.*, Eds.: 139–156. Springer, Wien, New York.
25. FINBERG, J.P., I. LAMENSDORF, M. WEINSTOCK, *et al.* 1999. Pharmacology of rasagiline (*N*-propargyl-1R-aminoindan). Adv Neurol. **80:** 495–499.
26. FINBERG, J.P., I. LAMENSDORF, J.W. COMMISSIONG & M.B.H. YOUDIM. 1996. Pharmacology and neuroprotective properties of rasagiline. J. Neural. Transm. Suppl. **48:** 95–101.
27. ABU RAYA, S., E. BLAUGRUND, V. TREMBOVLER, *et al.* 1999. Rasagiline, a monoamine oxidase-B inhibitor, protects NGF-different PC2 cells against oxygen-glucose deprivation. J. Neurosci. Res. **1:** 456–463.
28. SPEISER, Z., R. LEVY & S. COHEN. 1998. Effects of *N*-propargyl-1-(R)aminoindan (rasagiline) in models of motor and cognition disorders. J. Neural. Transm. Suppl. **52:** 287–300.
29. SPEISER, Z., O. KATZIR, M. REHAVI, *et al.* 1998. Sparing by rasagiline (TVP-1012) of cholinergic functions and behavior in the postnatal anoxia rat. Pharmacol. Biochem. Behav. **60:** 387–393.

Part X: Novel Neuroprotective Agents

QUESTIONS FOR DR. MAYNARD

From Dr. Palmer

In your neurological evaluations after MCAO does motor recovery lag behind sensory recovery and then are both improved to about the same degree after two weeks?

Can you explain why Vit B3 would have such a profound U-shaped dose response curve—it should be non-toxic?

In permanent models of MCAO we have found that it takes higher doses to get protection, any comments?

ANSWER: There did not seem to be any obvious delay in the onset of recovery of the motor compared to the sensory deficits in the period of 1-week following MCAO. We did notice in this and other studies, however, that the sensory deficits were more significantly improved (i.e., the p values were smaller) compared with the motor recovery.

Yes, the U-shaped dose response curve was an issue we addressed in our first publication (Ayoub, *et al.*, 1999), although I do not now believe that the idea we proposed then is the only possible mechanism. Originally we thought that nicotinamide (NAm, vitamin B3), a poly ADPribose polymerase (PARP) inhibitor, could have beneficial or detrimental effects. On the one hand, inhibiting PARP activity would prevent excess usage of ATP used by PARP to repair damaged DNA and hence prevent adding insult to injury due to the energy deficit caused by ischemia itself. On the other hand, inhibiting PARP too much, though it would save energy, would mean lack of repair of the DNA damaged as a consequence of ischemia. Thus, it made sense to us that one needed to find a carefully controlled dosage of NAm that would carefully attain a balance to allow repair of DNA by PARP, but not too much PARP activity lest it uses too much of the ATP also needed for the normal functioning of the neurons themselves.

I have done some literature searching and to my surprise found that many drugs show a U-shaped dose response curve, not only PARP inhibitors. Thus, I think these results may be telling us something more fundamental about neuroprotection or ischemia pathophysiology that we do not yet understand. Clearly U-shaped dose response curves are being obtained with drugs that have other mechanisms of action independent of modulation of PARP activity.

In terms of neuroprotective doses of NAm in the permanent intraluminal suture model of MCAO (pMCAO) versus the transient (2 h) intraluminal suture MCAO, we found that only 500 mg/kg NAm was effective, but not 100, 250, 750, or 1000 mg/kg in both models and in four species of rats, male and female. Thus, in our hands, the protection with NAm is curiously dose-specific and independent of the model of ischemia, species and gender of the rat. The volume of brain damage obtained with the pMCAO model was larger (about 39% infarction) than with the tMCAO model (about 30% infarction). It could, therefore, be expected that pMCAO may require a more robust drug dosage to achieve the same degree of protection in the tMCAO model, but this was not our experience. Clearly, the pMCAO model is more robust than the tMCAO model and other published studies have shown that drugs, which work on tMCAO models, have sometimes failed in pMCAO models.

From Dr. Skaper

Have your looked at biochemical parameters, for example, PARP activation, in your stroke models, and would nicotinamide inhibit PARP activation in the ischemic brain tissue? Have you looked at nicotinamide levels in brain of nicotinamide-treated animals?

ANSWER: Excellent questions. We would like very much to conduct these experiments, which have been discussed, but they have not yet been undertaken. Given the talk of Professor Curry at this conference, however, I was rethinking those experiments. I would probably now want them conducted in animals that have been subjected to MCAO, since the ischemic insult itself may alter the pharmacokinetics and pharmacodynamics of NAm compared with the drug profile in naïve animals that we previously conducted.

From Dr. Paule

Your blood levels of nicotinamide for doses of 250, 500, and 750 mg/kg are all the same—does that not bother you? Can you comment on what you think might account for this?

ANSWER: This was quite surprising to us as well. We were hoping that somehow the 500 mg/kg dose profile would stand out from the others, but really it did not. It was confusing to us that the 750 mg/kg dose had a very similar profile to 250 mg/kg and 500 mg/kg, yet only 500 mg/kg was protective. It was somewhat of an assurance to us regarding the technical aspects of the experiment, however, that 100 mg/kg dose was not similar to any of the higher doses but was similar to that of the control animals. Unfortunately, though I have no idea of what might account for these findings.

We did make an interesting observation, however, that has assisted us in terms of thinking about blood levels of NAm, which we mentioned in a previous publication (Sakakibara *et al.*, 2000). Intravenous administration of NAm improved the degree of neuroprotection compared with intraperitoneal injection in male Wistar rats subjected to pMACo (55% reduction [unpublished data], versus 31% reduction [Ayoub, *et al.*, 1999] in infarction, respectively). Thus, it seems clear that we need to get the drug to the tissue as quickly as possible to achieve maximal benefit. This was to be expected, but we had no idea that it could improve protection by as much as 75%.

QUESTIONS FOR DR. BANIK

From Dr. Slikker

Does tolerance develop to the beneficial methylpredisone or calpain inhibitor effects on spinal cord injury?
Are cells that are saved by calpain inhibition functional?

PART X: QUESTIONS AND ANSWERS 461

ANSWER: To answer your first question, yes, for these short-term experiments to both methylprednisolone and calpain inhibitors. We do not know at this time whether the cells saved by calpain inhibitors are functional. However, we are exploring this (excellent question) in our cell culture studies.

From Dr. Maynard

Do the histologic changes seen in your weight-drop model in animals mimic that seen in patients with spinal cord injury?

ANSWER: Yes, very similar morphological changes are found in patients with spinal cord injury.

From Dr. Skaper

Have you looked at the ability of the calpain inhibitor E-64-d to improve motor function in spinal cord injury?

ANSWER: Yes, but we could not see any improvement of motor function in these animals with acute injury, which was not surprising since they were only treated for 48 hours. We are now looking into the effect of calpain inhibitors in the chronic injury model over a period of 3–28 days following injury to assess functional recovery.

QUESTIONS FOR DR. CHUANG

From Dr. Slikker

The effect of lithium on Bcl_2 or Bax appeared to be about 40–50%. How much change (e.g., 50% increase in Bcl_2) is necessary to have a protective effect against cell death via apoptosis?

ANSWER: The maximal effects of long-term lithium treatment on Bcl_2 and Bax protein levels are about 200–300% and 50% of the untreated control, respectively (Chen and Chuang, *J. Biol. Chem.* **274**: 6039–6042, 1999). When these data are expressed as a Bcl-2/Bax ratio, the values are increased by 3–5-fold after lithium treatment. In our experiments using cerebellar granule cells, an increase in the Bcl_2/Bax ratio by more than two-fold is sufficient to produce highly significant protection against glutamate excitotoxicity.

From Dr. Manev

Does chronic lithium damage the number of NMDA receptors?

ANSWER: Chronic lithium treatment causes a dose-dependent inhibition of NMDA receptor–mediated calcium influx (Nonaka, *et al.*, *Proc. Natl. Acad. Sci. U.S.A.* **95**: 2642–2647, 1998). Our preliminary results show that this loss of receptor function is not associated with significant changes in protein levels of NMDA receptor subunits such as NR1, NR2A, and NR2C. We are currently investigating whether mRNA levels of these receptor subunits are changed by long-term lithium treatment.

From Dr. Bachurin

Do you have any data about lithium cation's effects on calcium permability transition pores (PTP)?

ANSWER: We did not study the effects of lithium on calcium permeability transition pores (PTPs) such as that present in mitochondria. However, since mitochondrial PTPs are known to be regulated by Bax and Bcl-2, which are targets of lithium, one may speculate that PTPs are also regulated by lithium. Consistent with this possibility is our finding that lithium inhibits glutamate-induced cytochrome c release from mitochondrial membranes (Chen and Chuang, *J. Biol. Chem.* **274:** 6039–6042, 1999).

From Dr. Skaper

Lithium is a low millimolar inhibitor of GSK3. Have you looked at GSK3 activation in your *in vitro* and *in vivo* models of excitotoxic injury? In particular, inhibition of the PI 3-kinase/PKB pathway leads to GSK3 activation, and GSK3 activation has been implicated in apoptotic neuronal death.

ANSWER: It is true that GSK-3 activity has been shown to be directly inhibited by lithium at millimolar concentrations (Kleine and Melton, *Proc. Natl. Acad. Sci. U.S.A.* **93:** 8455–8459, 1998) and that GSK-3 is a prominent kinase involved in promoting apoptosis. We have not yet measured GSK-3 activity in our *in vitro* and *in vivo* studies on excitotoxicity. However, our earlier studies did show that lithium is able to activate the PI 3-kinase/Akt signalling pathway, resulting in increased phosphorylation of GSK-3, a condition known to cause inhibition of GSK-3 activity (Chalecka-Franaszek and Chuang, *Proc. Natl. Acad. Sci. U.S.A.* **96:** 8745–8750, 1999). Thus, our research provides an alternative, indirect mechanism whereby lithium can inhibit GSK-3 activity.

QUESTION FOR DRS. CHUANG AND BANIK

From Dr. Bowyer

Are there difference in Bcl_2 levels in various types of neurons *in vivo,* and do these differences predict susceptibility of neurodegeneration?

ANSWER: (Chuang) The levels of Bcl-2 protein in various neuronal cell types seem to differ and this variation may control the vulnerability to apoptotic insults. For example, our collaborator Dr. Zheng-Hong Qin in the NINDS, NIH has found that Bcl-2 protein levels in large-size interneurons in the rat striatum are considerably higher than those found in medium-size neurons. This could be the reason that these large neurons are spared when quinolinic acid or NMDA is injected into the striatum to induce neurodegeneration.

(Banik) Yes, there are differences in the levels of proapoptotic Bcl_2 in different types of neurons. The upregulation of proapoptotic proteins such as Bcl_2 do predict susceptibility to neurodegeneration.

PART X: QUESTIONS AND ANSWERS

QUESTIONS FOR DR. HERNING

From Dr. Bowyer

Are the increases in pulsatile pressures in cocaine abusers due to atheroslerosis or vascular regulation?

ANSWER: Evidence, at this point, seems to indicate cocaine-induced changes in vascular regulation. Dopaminergic neurons innervate and produced vasoconstriction in small blood vessels in the cortex (L.S. Krimer, *et al.*, Dopaminergic regulation of cerebral cortical microcirculation. *Nature Neurosci.* **1**: 286–289, 1998). Cocaine blocks dopamine reuptake. In addition, repeated cocaine administration in rabbits constricted pial arterioles by blocking their bradykinin-induced dilation (J.R. Copeland *et al.*, Repeated cocaine administration reduces bradykinin-induced dilation of pial arterioles. *Am. J. Physiol.* **271**: H1576–1583, 1996).

From Dr. Slikker

Do other agents that affect NO_2 have beneficial effects (e.g., reduction of pulsatility) in HIV+, cocaine-abusing subjects?

ANSWER: Only antiviral agents have been studied in HIV-positive individuals. In this and the other studies mentioned, antiviral agents appeared to improve cerebral perfusion. In HIV-negative cocaine abusers, buprenorphine improved cerebral perfusion; it is unclear whether NO was involved.

GENERAL COMMENT TO DR. MAYNARD

From Dr. Virmani

In the past ten years it has become evident in that drugs with multiple actions seem to be more effective in neuroprotection, especially compounds that seem to act at the metabolic level in addition to their other actions. So I agree with you that the so-called "dirty drugs" or drugs with "multifactorial effects" may be more useful in these neurodegenerative conditions, as you have shown for compounds like nicotinamide, and I suggest compounds such as the carnitine and selenium.

ANSWER: Thank you for your comment. I agree and in this conference, Dr. Russel Reiter had added to that list with melatonin.

QUESTION FOR DR. DE MAW CHUANG

From Dr. Virmani

Do you think that effects of Li could be in part mediated by the depolarizing actions produced by the inhibition of the Na/KATPase?

ANSWER: This is an intriguing question, as depolarization indeed has neuroprotective effects in many neuronal or neurally related cell types. However, *in vivo* rodent studies showed that brain Na$^+$/K$^+$ ATPase is predominantly inhibited by acute, but not long-term, lithium treatment (for a review, see Wood and Goodwin, Psychol. Med. **17:** 579–600, 1987). This does not seem to be consistent with the neuroprotective effect of lithium, which requires long-term treatment.

Neuroprotection

Past Successes and Future Challenges

WILLIAM SLIKKER, JR., SARAN JONAS, ROLAND N. AUER, GENE C. PALMER, TOSHIO NARAHASHI, MOUSSA B.H. YOUDIM, KENNETH I. MAYNARD, KATHRYN M. CARBONE, AND BRUCE TREMBLY

The year 2000 International Conference on Neuroprotective Agents was punctuated with many firsts. The use of imaging and three-dimensional reconstruction to assess the CNS insult and/or spare damage as a result of neuroprotective therapy was demonstrated. The systematic use of gene expression arrays (genomics) to select sensitive genes as targets for neuroprotection was described. Gene therapy as an approach to neuroprotection was also elaborated. In addition, the potential of individual agents (e.g., lithium, L-carnitine, nicotinamide, and calpain inhibitors) to produce neuroprotection was challenged by the concept that several physiological manipulations (e.g., hypothermia, insulin hypoglycemia, and hyperoxemia) also have tremendous neuroprotective potential. A new use strategy for neuroprotective agents was also unveiled; one that moved beyond the cocktail approach and stipulated the time from insult–dependent administration of carefully selected neuroprotective agents.

Intense and sometimes heated discussions are critical and healthy in a rapidly emerging research and drug development arena, such as neuroprotective therapy. Some of these exciting areas are summarized below, from both a historical and forward looking perspective. These closing comments, from some of our esteemed colleagues, capture the essence of several of the most profound issues discussed at the year 2000 International Conference on Neuroprotective Agents.

A BRIEF AND VERY PERSONAL REVIEW OF THE PREVENTION AND TREATMENT OF STROKE, WITH RECOMMENDATIONS FOR THE DIRECTION OF FUTURE INVESTIGATION

Saran Jonas, M.D.

Department of Neurology, New York University School of Medicine,
New York, U.S.A.

I outline for you the history of the development of maneuvers effective in the prevention or treatment of stroke, as seen from my point of view as a stroke neurologist and investigator. I begin the story a half-century ago, when there were no effective maneuvers. I have had the privilege to witness the entire development since that time: I entered medical school in 1952, and first took part in stroke research in 1962.

The prevention of stroke has been a gratifying domain of success: primary stroke occurrence has been reduced substantially, and so has the chance of stroke recurrence after an initial focal event.

The single most powerful maneuver in this regard has been the pharmacologic control of hypertension, which by preventing or reversing hypertensive cerebrovascular disease has had a potent effect in reducing both primary and recurrent hemorrhagic and ischemic strokes. Antihypertensive therapy has also improved the prognosis for unruptured cerebral arterial aneurysms.

The effective treatment of hypertension began in the late 1960s. The handful of drugs capable of lowering blood pressure that were available during my medical student, resident, and early practice years had intolerable side effects and were of little practical clinical value. A typical example: my beloved father-in-law, a physician who had hypertension and diabetes, stopped taking the somewhat effective agent reserpine, given after he had suffered a right occipital stroke, because it had produced profound depression. He told me he would rather die than continue in this manner. He stopped the reserpine, recovered fully from this, the only depression he had ever experienced, and died soon after, of a brain-stem stroke, in June 1960 when he was 63 years old.

Another typical example: an aphasic, right-hemiparetic hypertensive Chinese-speaking patient of mine, whose blood pressure had actually come down well on reserpine, killed himself while under my treatment in 1963. Between his aphasia and the language barrier, I had not recognized that he was depressed despite my keen awareness of the problem, as explained above.

I have told you about bad treatments. It is sobering to recognize that in April 1945 there was no antihypertensive treatment at all at the time that the most powerful man in the world, President of the United States Franklin D. Roosevelt, died at age 63 of a cerebral hemorrhage while gaunt from intractable congestive heart failure consequent to hypertensive heart disease.

How different is the situation today, when any hypertensive person can receive highly effective therapy with minimal nuisance and side effect, at minimal financial cost! (Concerning ACE inhibitors: more later.)

After antihypertensive therapy, probably the next most effective maneuvers, particularly for primary stroke prevention, have been those directed at the prevention of atherosclerotic vascular disease. These have included maneuvers to reverse or prevent hyperlipidemia (diet; recently the statin drugs, about which I comment below), and recognition of the danger of smoking.

Other effective maneuvers have been the primary prevention of embolic stroke by anticoagulation to prevent intracardiac thrombi in association with myocardial infarction, non-valvular atrial fibrillation, and rheumatic heart disease. The latter disease has in fact been wiped out by the antibiotic treatment, primary and prophylactic, of streptococcal infection in children. My first actual involvement in clinical research was as a medical resident in the rheumatic fever clinic at Bellevue Hospital, New York City, where I took part in trials utilizing sulfonamides and penicillin preparations (we also wiped out Sydenham chorea at the same time).

(With respect to antimicrobial treatment, at the time of my birth in 1931 no antibiotic existed. The bacteriostatic sulfonamides became available in the late 1930s. Penicillin, the first systemically useful bactericidal agent, was used initially in 1943).

Another mechanism of embolic stroke for which successful treatment has been developed has been embolization from carotid plaque. Carotid endarterectomy for this was first performed in 1954, while I was a medical student (I was not aware either of the disease or of the treatment at the time). As a neurology resident I took part in 1962 in the first controlled trials of endarterterectomy (I also took part, in the early 1990s, in the final definitive NASCET trials).

A useful contribution to the control of secondary ischemic stroke has been the development of antiplatelet therapy. It is poignant to recognize that such treatment could have been given 100 years earlier if the understanding of pathogenesis and the effect of then newly introduced aspirin had existed.

It should be pointed out that the attempt to use aspirin in primary prevention of stroke in healthy people not selected for risk factors has proved to be a failure.

Another failure has been the use of anticoagulation for prevention of strokes of non-cardiac origin. Recent studies showed that heparin was no better than aspirin for preventing a new stroke during the weeks immediately following an index stroke. Controlled trials in the 1950s and 1960s had failed to show any advantage of warfarin over placebo in long-term prophylaxis. These trials were conducted at a time when neither treatment of hypertension nor brain imaging to rule out occult hemorrhage was available (CT scanning came into clinical use in 1974; we were awestruck. MRI, even more astonishing to us, came in some years later; I still marvel at both as I look back at their invasive, uncomfortable, and dangerous predecessors: pneumoencephalography, myelography, and angiography; and the benign but poor resolution isotope brain scan of the 1950s).

There may yet be another chapter to the story of anticoagulation for stroke prevention. A definitive modern clinical trial, the WARSS—warfarin versus aspirin in recurrent stroke study—trial was completed recently and should be reported within the next year.

Let us now consider stroke treatment: that is, maneuvers other than rehabilitation that are intended to improve the outcome when a stroke was not prevented. The thrombolytic agent tPA, and the defibrillating agent ancrod have both been shown effective in the acute treatment—administration during the first three hours after stroke onset—of selected cases of ischemic stroke.

By contrast, acute stroke treatment with unfractionated heparin, low molecular weight heparin, and pharmacologic neuroprotectants have all failed to ameliorate stroke outcome in clinical trials (mostly with 24–48-hour admission windows).

Retrospective studies do not demonstrate that strokes that occur in people on aspirin are less severe than those that occur in non-takers of aspirin.

The treatment of ruptured cerebral arterial aneurysms has benefited from advances in technique and technology, and in monitoring and postoperative care.

By contrast, no specific treatment changes the outcome of non-aneurysmal hemorrhagic stroke, except for the neurosurgical evacuation of cerebellar hematoma.

In looking over the above, we note that gratifying advances have been made in stroke prevention, but that there has been very limited success in stroke treatment. Can recognition of this disparity guide us to do better with the stroke treatment problem? I think so. I believe that we will do better if we recognize the following: All the successful maneuvers described above are aimed at the arterial system—heart through intracranial circulation—and at the contents of the circulatory system: the blood. It is maneuvers that protect, preserve, or restore the arterial circulation that have made good things happen. The underlying science here is largely that of cardiology and hematology. By contrast, to date none of the successful maneuvers have derived from "deep neuroscience": for example, understanding of the details of the ischemic or apoptotic cascades, neuroimmunology, and up- or downregulation of genes.

Thus, it is not neuroscience that has underlain the described advances, but the basic sciences, techniques, and approaches of internal medicine (especially cardiology and hematology), and epidemiology, neuroradiology, and neuro- and vascular surgery. Although we have been earnestly discussing lipid peroxidation and glutamate receptors (I mean no disrespect to our dedicated and talented neuroscience colleagues) the internists have once again dramatically upstaged those of us in stroke neurology, this time in the domain of primary prevention. They have done this by showing us from their cardiovascular trials that ACE inhibitors and statin drugs reduce stroke occurrence by 20% each (and apparently independently), not only in hypertensives and hyperlipemics, but also in normotensives, and at lipid levels heretofore considered normal. Furthermore, involved in the ACE effect is the ability to suppress the advance of microvascular disease in diabetics, and to inhibit the emergence of diabetes in people considered non-diabetic. The impact of these developments on the burden of stroke in the population is far greater than the impact of tPA or of all the deep stroke neuroscience of the last 20 years.

In conclusion I say to my fellow stroke neurologists: you are rightly proud of the work you have put into clinical trials in the past five decades. Recognize, however, that you have used neurologic techniques to select patients for study and to define the end points, but the biomedical knowledge underlying the successful therapies has come from areas other than neuroscience. Just because we may do the bench work, does not make our work neuroscience: a neurologist who studies platelet aggregation in the laboratory is not doing neuroscience, but rather hematology.

Since it is clear that the successes have not come from maneuvers based on deep neuroscience, but from hemovascular maneuvers, we stroke neurologists should rethink where we want to put near-term effort and resources. I advise that we should look to run clinical trials that use developments in internal (and other fields of) medicine that manipulate the circulatory system or its contents. Without massive evidence projecting success, we should prevent the premature launching of effort- and resource (importantly including money)-wasting clinical trials based on the present state of "deep" neuroscience alone. (In the development of the idea of avoiding the devotion of resources to clinical trials unlikely to be of value, I am indebted to Russell Andrews, M.D., Professor of Neurosurgery, Texas Tech University, El Paso Texas, who publicly explicated the concept at the Fifth International Conference on Neuroprotective Agents.)

For my respected neuroscience colleagues: you have produced a wonderful body of data giving insight into how the brain breaks down. Through no fault of your own, these results by themselves have as yet had little or no direct therapeutic value. If you wish to do work that could lead to *near-term* clinical value,[a] take the following principle to guide your selection of problems for investigation: *Will the solution to this problem lead to the enhancement of maneuvers that manipulate the circulation?*

---

[a] It should be noted that recently Dr. Michael Chopp has shown in rats that intravenous infusion of bone marrow stem cells as late as seven days after two hours of middle cerebral artery occlusion improved function at 35 days (Jieli Chen, et al., *Stroke*, **32:** 1005–1011, 2001). This is an exciting area for far-term investigation.

SYNOPSIS OF THE STATE OF THE ART

Roland N. Auer, M.D., Ph.D., FRCPC

Department of Pathology, University of Calgary,
Health Sciences Center, Calgary, Alberta, Canada

The failure of pharmacologic monotherapy in clinical stroke trials of neuroprotection is addressed by Auer in this volume. Problems include transferring data to humans from animal models that are very sensitive and show treatment effects easily due to (1) the inbred nature of experimental animals and (2) operator control over important physiologic parameters that reduce variability. Increased awareness of the importance of experimental control of physiologic parameters begets even more positive animal data that cannot be translated to humans, the most outbred species on the earth. Auer argues that it is those very physiologic parameters that should be manipulated intentionally, not merely controlled, in stroke trials of neuroprotection. If any or all of hypothermia, insulin hypoglycemia, hyperoxemia, iatrogenic hypertension, or hypermagnesemia are found effective, drugs can be added later, if they, too, are found effective.

FACTORS AFFECTING STROKE THERAPY: HOW LONG MUST A DRUG BE GIVEN BEFORE IT IS EFFECTIVE?

Gene C. Palmer, Ph.D.

AstraZeneca R&D Boston, 3 Biotech,
Worcester, Massachusetts, U.S.A.

As was discussed repeatedly during the course of the symposium, several factors must be considered when determining both the preclinical and clinical effectiveness of a drug for treatment of stroke, namely: (1) characteristics of the lesion: (a) type (e.g., hemorrhagic or occlusive—either permanent or transient), (b) size, (c) location, (d) experimental model, and (e) animal strain; (2) timing of the initial dose; (3) combination therapy (administered as a cocktail or sequentially); (4) metabolism and pharmacokinetics (especially maintenance of adequate plasma/brain levels); (5) relevant functional tests for assessment of efficacy (e.g., magnetic resonance imaging [MRI] or evaluation of skilled motor behavioral and/or cognition); (6) possible adverse drug reactions; and (7) duration of treatment.

The latter point, duration of treatment, the subject of this discussion, is often neglected. The scientific literature abounds with examples of experiments of an acute nature demonstrating efficacy of various classes of therapeutic agents. The most striking example typically consists of a rat model of focal ischemia (either a permanent or transient occlusion of the middle cerebral artery) for a specified period of time followed by administration of a single dose of test compound. Drug efficacy is measured within 24–48 h of the lesion using histopathology techniques. In many instances the reporting laboratory will suggest, "This compound should be considered for future testing in humans." Unfortunately, recent investigations indicate that such acute therapy may only delay ultimate neuronal death in time (Cregan et al.,

1997; Colbourne *et al.*, 1997, 1999). A limited number of investigations have, however, addressed this problem. Examples of such treatments administered at varying times following experimental stroke coupled with long-term functional/neuropathological assessments include: (1) the GABA agonist, chlormethiazole (Marshall *et al.*, 1999); (2) the low affinity use-dependent NMDA antagonists, AR-R15896AR or remacemide (Bialobok *et al.*, 1999; Marshall *et al.*, in press; Ordy *et al.*, 1992); and (3) hypothermia (Colbourne *et al.*, 1997).

I would like to address the problem of duration of therapy in further detail using examples of our work with AR-R15896AR. Glutamate-induced excitotoxicity occurs immediately after the initiation of stroke. Therefore, glutamate antagonists, in order to be effective, need to be administered within a short time (1–2 h) of the event. Furthermore, in the rat positive therapeutic outcome is sensitive to several factors, namely selection of the experimental model of ischemia, the strain of animals and duration of ischemia (Palmer *et al.*, 1999). Efficacy following acute administration of AR-R15896AR has been demonstrated in a variety of experimental models and species (Cregan *et al.*, 1997; Bialobok *et al.*, 1999; Palmer *et al.*, 1999; Sutherland, *et al.*, 2000). However, when AR-15896AR was administered in an acute fashion beginning at 30 min following 120 min of transient MCAO in a spontaneous hypertensive rat model, long-term outcome of drug efficacy was equivocal. In this experiment, AR-R15896AR afforded limited neuroprotection as assessed by: (1) T2-weighted MRI (at two and seven days post-MCAO); (2) improvement in the deficit of ipsilateral, but not contralateral skilled use of the forepaw at 45 days post-MCAO; but not by (3) standard histopathological examination of the brain (Cregan *et al.*, 1997).

Expanding on the information obtained from the previous experiment, we next determined if maintenance of effective plasma levels of AR-R15896AR for one week following a two-hour transient MCAO in Wistar rats (monofilament model) would produce neuroprotection to eight weeks. Plasma levels of about 2,300 ng/ml of AR-R15896AR were maintained via an i.v. loading dose given shortly after initiation of MCAO followed by insertion of an Alzet minipump calibrated to deliver the appropriate concentration. Four separate groups of rats were examined; neuroprotection was assessed histologically from sections taken at 1, 2, 4, or 8 weeks post-MCAO. The procedure did indeed demonstrate that maintenance of appropriate plasma levels of AR-R15896AR for one week was sufficient to provide neuroprotection out to two months (Bialobok *et al.*, 1999). Recently, Marshall and coworkers (in press) using a marmoset model of permanent MCAO demonstrated neuropathological protection by AR-R15896AR to 10 weeks. In this case, dosing began shortly after the stroke and plasma levels were maintained with an Alzet minipump at 1500 ng/ml for 48 h.

The take-home message provided by the examples given in this short presentation is that duration of treatment with appropriate (e.g., functional) assessments of efficacy should be among the major determinants (e.g., dose–response relationships indicating acute efficacy, therapeutic window, pharmacokinetics, and safety) for nomination of an experimental drug as a candidate for clinical trials in stroke.

REFERENCES

BIALOBOK, P., E.F. CREGAN, S.G. SYDSERFF, *et al.* 1999. Efficacy of AR-R15896AR in the rat monofilament model of transient middle cerebral artery occlusion. J. Stroke Cerebrovasc. Dis. **8:** 388–397.

COLBOURNE, F., H. LI & A.M. BUCHAN. 1999. Continuing postischemic neuronal death in CA1. Influence of ischemia duration and cytoprotective doses of NBQX and SNZ-111 in rats. Stroke **30:** 662–668.

COLBOURNE, F., G. SUTHERLAND & D. CORBETT. 1997. Postischemic hypothermia-A critical appraisal with implications for clinical treatment. Mol. Neurobiol. **14:** 171–201.

CREGAN, E.F., J. PEELING, D. CORBETT, *et al.* 1997. [(S)-Alpha-phenyl-2-pyridineethanamine dihydrochloride], a low affinity uncompetitive *N*-methyl-D-aspartic acid antagonist is effective in rodent models of global and focal ischemia. J. Pharmacol. Exp. Ther. **283:** 1412–1424.

MARSHALL, J.W.B., A.J. CROSS & R.M. RIDLEY. 1999. Functional benefit from clormethiazole treatment after focal cerebral ischemia in a nonhuman primate species. Exp. Neurol. **156:** 121–129.

MARSHALL, J.W.B., E.J. JONES, K.L. DUFFIN, *et al.* 2001. AR-R15896AR, a low affinity, use-dependent, NMDA antagonist, is neuroprotective in a primate model of stroke. J. Stroke Cerebrovasc. Dis. In press.

ORDY, J.M., B. VOLPE, R. MURRAY, *et al.* 1992. Pharmacological effects of remacemide and MK-801 on memory and hippocampal CA1 damage in the rat four-vessel occlusion (4-VO) model of global ischemia. *In* The Role of Neurotransmitters in Brain Injury. M. Globus & W.D. Dietrich, Eds.: 83–92. Plenum, New York.

PALMER, G.C., E.F. CREGAN, P. BIALOBOK, *et al.* 1999. The low-affinity, use-dependent NMDA receptor antagonist AR-R15896AR: An update of progress in stroke. Ann. N.Y. Acad. Sci. **890:** 406–420.

SUTHERLAND, G.R., J.T. PERRON, P. KOZLOWSKI & D.J. MCCARTHY. 2000. AR-R15896AR reduces cerebral infarct volume following focal ischemia in cat. Neurosurgery **46:** 710–720.

RESPECTING THE DIVERSITY AND COMPLEXITY OF NEUROTOXICITY

Toshio Narahashi, Ph.D.

John Evans Professor of Pharmacology,
Alfred Newton Richards Professor
Northwestern University Medical School,
Department of Molecular Pharmacology and Biological Chemistry,
Chicago, Illinois, U.S.A.

Neuroprotective research implies dealing with a broad range of complex mechanisms that, individually or combined, can initiate and/or promote neuronal death. True advances in this field require dealing effectively with the diversity and complexity of this problem. That is,

- Respecting the diversity of the neuropathologies leading to neuronal death. For example, neurotoxic mechanisms associated with acute insults (e.g., trauma, ischemia) are unlikely to be similar to those associated with neurodegenerative disorders. Within this context, the validity of "unifying" mechanisms of neuronal death such as excitotoxicity and apoptosis appears questionable and potentially misleading.

- Respecting the complexity of the neuropathology under study. As much as possible, models should mimic specific pathologies, both qualitatively and quantitatively. Exposure of cultured neurons to glutamate is not a model of ischemia, and the fact that MPTP produces rapidly some of the features of Parkinson's disease does not imply identical neuropathogenesis.

It is essential to match the wide range of very powerful tools now available (e.g., molecular genetics, brain monitoring techniques) with new models and preparations that are relevant and specific to individual neurological disorders.

Effective neuroprotection is inherently linked to early diagnosis and treatment. For acute insults to the brain, this means working towards drugs that are safe enough for self-administration or prevention. With regard to progressive neurodegenerative disorders, critical objectives include: genetic determinants, detrimental environmental conditions, and early biological markers for specific diseases.

IS NEUROPROTECTION POSSIBLE IN NEURODEGENERATIVE DISEASES?

Moussa B. H. Youdim, Ph.D.

Director: Eve Topf and National Parkinson Foundation,
Centers of Excellence for Neurodegenerative Diseases Research, and
Department of Pharmacology, Technion-Faculty of Medicine,
Bat Galim, Haifa, Israel

The mechanism of neurodegeneration (neuronal cell death) is a complex process in which many of its steps have not been identified and these may vary in each individual neurodegenerative diseases (Parkinson's disease, Alzheimer's disease, Huntington's chorea, ischemia, and stroke). However, it is now apparent that in all diseases where cell death occurs, common mechanisms are involved that have subtle differences accounting for the specificity of the diseases. Nevertheless, current studies on cell death and cytoprotection or neuroprotection have shown that most investigators are doing the same type of experiments and observing very similar data and mechanisms and in fact are employing similar so-called neuroprotectants in their *in vitro* and *in vivo* studies. The consequences of these have been that a significant number of drugs have been developed that in animal studies exhibit neuroprotection, but all have failed in the clinic. Nowhere is this more obvious than in the treatment of ischemia and neurotrauma. The progressive neurodegenerative diseases such as Parkinson's and Alzheimer's diseases have their own inherent problems since the progression of neurodegeneration is relatively slow and we do not currently have rapid and sensitive procedures to measure their progression. The few neuroprotective studies so far completed have not met with any success. The failure to induce neuroprotection in the clinic versus in the laboratory with currently available drugs would suggest that either the animal models we are employing are not truly representative of the disease state, or that a single drug would not be sufficiently active to do so and thus a cocktail of drugs is to be employed. This is not far-fetched since in AIDS, cancer, cardiovascular complications, and neuropsychiatric disorders multidrug therapy is employed. The latter may have an explanation in what is becoming obvious, that during neurodegeneration, a cascade of neurochemical events (domino effect) occurs in which neither the initial neurotoxic event nor all the neurochemical steps have been fully established. Briefly these include oxidative stress, activation of inflammatory transcription factors such as NFκB, increased cytotoxic cytokines including TNF-α, nitric oxide and glutaminegric excitotoxicity, compensatory increase of neurotrophic factors, abnormal iron accumulation, and prostaglandin metabolism. Since each of these mechanisms has its own series of biochemical cascades, it would not be expected that a single drug would be effective. Therapeutic employment of a cocktail of drugs would have its inherent problem of drug interaction in the CNS as well as of complicated regulatory approval. One approach to overcome these problems would be to develop drugs with multiple pharmacological actions (dirty drug). Indeed we have made a start in this direction by developing a series of drugs for treatment of Alzheimer's disease (see the paper by Weinstock *et al.* in this volume) that have several neuroprotective activities, cholinesterase and monoamine oxidase inhibitory actions as well as antidepressant activity.

The advent of cDNA array– or gene microchip–based methods may offer powerful tools for investigating neurodegeneration in which differential gene expression patterns of thousands of genes in normal and diseased tissue can be studied. This procedure would give an overview of the neurochemical events as indicated by gene expression. Employing this technique we (Youdim *et al.*, 2000; Mandel *et al.*, 2000; Grunblatt *et al.*, 2000) have recently examined nigrostriatal gene expression in the acute and chronic MPTP (*N*-methyl-4-phenyl-1,2,3,6-tetrahydropyridine) (10 days) mouse model of Parkinson's disease and neuroprotection afforded by a number of drugs (apomorphine, green and black tea extract, blueberry extract, rasagiline, TV3326, D-penicillamine, desferrioxamine, and non-steroid antiinflammatory drugs), previously shown to be neuroprotective in neuronal cell culture and *in vivo*. We have so far shown that chronic MPTP induces gene expression (up and down) changes in some 51 genes, many of which were not previously identified or suspected to participate in MPTP-induced neurodegeneration, enlarging the cascade of neurochemical events thus far identified by biochemical means. Pretreatment of mice with the neuroprotective drugs, apomorphine, or green tea extract that induce neuroprotection in neuronal cell culture and *in vivo,* prevented the expression of many, but not all, of the genes. It remains to investigate what is the contribution of normalized gene expression in apomorphine-induced neuroprotection to the process of either neurodegeneration or neuroprotection.

One obvious aspect of neurodegeneration that has escaped study by most investigators is the early events in the process of neuronal cell death. We consider this to be of primary importance in setting the clock for initiation of neurodegeneration. This can be examined with cDNA micro array and, indeed, our gene expression studies with acute MPTP (3, 6, 12, and 24 hours) have shown a profoundly different gene expression as compared to the chronic MPTP studies. Briefly, important genes are highly expressed or suppressed during the early time course and they disappear within 24 hours when little dopaminergic neurodegeneration occurs, suggesting that the early cascade of events is the more important. Thus, consideration should be given to drug development to prevent the expression of these early events.

Gene array technology has already enlarged our horizon in neurodegeneration. It can contribute to investigations as to how closely the various cell models or animal models employed for neurodegenerative disease resemble the clinical syndrome. For example, we are now studying nigrostriatal gene expression in the 6-hydroxydopamine model of Parkinson's disease and comparing it to MPTP and nigrostriatal samples obtained at autopsy from control and Parkinsonian brains. By mapping the similarities and differences between them we hope to establish how much homology there is between the clinical syndrome and the animal model. We suspect that this procedure may also point to as yet unidentified genes that play a role in the cascade of events leading to neurodegeneration as well as help to develop new therapeutic strategies with new types of drugs.

REFERENCES

GRUNBLATT, E., S. MANDEL & M.B.H. YOUDIM. 2001. Gene expression analysis in MPTP mice model of Parkinson's disease using cDNA microarray. J. Neurochem. In press.

GRUNBLATT, E., S. MANDEL & M.B.H. YOUDIM. 2001. Early gene changes in MPTP treated mice employing cDNA microarray. Exp. Neurology. In press.

MANDEL, S., E. GRUNBLATT & M.B.H. YOUDIM. 2000. CDNA microarray to study gene expression of dopaminergic neurodegeneration and neuroprotection in MPTP and 6-hydroxydopamine models. Implications for idiopathic Parkinson's disease. J. Neural Transm. Suppl. **60:** 117–125.

YOUDIM, M.B.H., E. GRUNBLATT, Y. ROYAL-LEVITES, *et al.* 2001. Molecular events in neurodegeneration and neuroprotection in MPTP model of Parkinson's disease employing CDNA microarray. *In* Brain Diseases: Therapeutic Strategies and Repair. A. Miller & O. Abrahamsky, Eds. Marcel Dekker, Boca Raton. In press.

NEUROPROTECTION RESEARCH FOR STROKE: A SYNOPSIS OF POINTS TO CONSIDER

Kenneth I. Maynard, M.Sc., Ph.D.

Neurosurgical Service, Massachusetts General Hospital
and Harvard Medical School, Boston, Massachusetts, U.S.A.

Combination Treatment versus Combination Therapy

The deleterious cascades that ensue following ischemia are many. It, therefore, stands to reason that an approach to neuroprotection that targets a single mechanism of action (e.g., either calcium, glutamate, free radicals, sodium, or caspase activation) is unlikely to be of maximal benefit. Consequently, some investigators have looked to *combination treatments*—that is combining agents—to try to achieve a beneficial synergistic or additive effect. This, however, is separate from (and may be part of) *combination therapy*, which may be seen as a series of treatments made at different stages following the onset of stroke. Combination therapy may include, initially, a neuroprotective agent that preferably acts at multiple sites in the ischemic cascades (e.g., magnesium, lubeluzole, topiramate, and nicotinamide) to protect the brain from subsequent injury. This may be followed by thrombolysis (e.g., tissue plasminogen activator, urokinase, and streptokinase), if, for example, it is determined that the stroke is of an ischemic and not a hemorrhagic nature, and the patient has presented within an acceptable time frame. Following thrombolysis, treatment may be needed to protect against reperfusion injury (e.g., an antiintercellular adhesion molecule, antiselectin, and CD11/CD18 antibodies), and finally we might administer treatment to improve rehabilitation and improve plasticity in the brain (e.g., growth factors, nootropic agents, gene therapy, or stem cells).

Pharmacokinetics

At the conference, work presented by our laboratory and that of Professor Stephen Curry (AstraZeneca) highlights two important points. First, the optimal neuroprotective dose is not necessarily the maximal tolerable dose. Second, the pharmacokinetics of a drug may be completely different when tested in a naïve animal versus an animal subjected to an insult—for example, stroke. As a consequence, it is crucial to perform preclinical dose–response studies, since many drugs tested exhibit inverse or U-shaped dose–response curves. Furthermore, pharmacokinetic studies need to be performed in animals that have been subjected to the insult under study—for example, middle cerebral artery occlusion procedure for the study of stroke—in order to determine the correct plasma or serum levels of the putative neuroprotective drug. This index could assist clinicians in reasonably determining the optimal dose for treatment in humans.

Testing a Variety of Models, Animal Species, Strains, and Both Sexes

Some agents that exhibited neuroprotective effects when tested in a model of transient middle cerebral artery occlusion (MCAO) failed to be protective in models of permanent MCAO, in which there were more severe lesion volumes, or vice versa (e.g., anti-ICAM, dextromethorphan, and FK506). Since patients are likely to suffer from both types of stroke—transient and permanent occlusions—any neuroprotective agent that is expected to be efficacious in the clinic should be shown to be protective in both permanent and reperfusion models of stroke. Moreover, neuroprotective agents should also be tested in models of hemorrhagic stroke, since, although approximately 80% of strokes are ischemic in nature, the 20% of strokes that are hemorrhagic are often more harmful to patients. Given that "time is brain", any nearby person should be able to administer an ideal neuroprotective agent to a stroke victim. In order for this to become a reality the ideal neuroprotective agent must be shown to be efficacious in models of hemorrhagic, as well as ischemic stroke.

Vascular anatomy, neurochemistry, and immunology are not always consistent from species to species, or from one strain of rats to another. In addition, given that estrogen has been shown to be neuroprotective, it has also now become imperative that neuroprotective agents are tested not only in different animals and strains, but also in both sexes. Furthermore, stroke patients after have comorbid factors (e.g., diabetes, hypertension, and obesity) that may alter the effect of an otherwise efficacious neuroprotective drug, and which should be tested preclinically in an appropriate animal strain (e.g., diabetic, hypertensive, or obese rats).

Imaging as a Surrogate Marker in Stroke

Recent developments in magnetic resonance imaging (MRI) technology have led to the ability of neuroradiologists to view the development of impending infarction in humans using diffusion-weighted (DW)-MRI within one hour of the onset of stroke. This methodology has been applied to *in vivo* animal models of stroke, and it is now well accepted through validation by comparison with standard histologic markers, such as 2,3,5-triphenyltetrazolium chloride, hematoxylin, and eosin, that DW-MRI can be used as a surrogate marker to study the development of infarction during ischemia. This technology is a powerful tool for use in preclinical research because it is non-invasive and allows for continuous measurements of the endpoint of cerebral infarction over time and space without sacrificing numerous animals at various times. Since it is likely to become the modality of choice for imaging stroke patients in hospitals, the more information we can provide preclinically with our putative neuroprotective agents using MRI as an endpoint, the better will be the application of our putative neuroprotective agents and strategies to the clinic.

NEUROPROTECTION OF THE DEVELOPING BRAIN

Kathryn M. Carbone, M.D.

Chief, Laboratory of Pediatric and Respiratory Viral Diseases,
DVP/OVRR/CBER/FDA
Associate Professor, Departments of Medicine and Psychiatry,
Johns Hopkins University School of Medicine, Baltimore, Maryland, U.S.A.

Protecting the nervous system from the effects of aging—for example, from cerebrovascular accidents and idiopathic neurological degenerative disorders—is a traditional goal of neuroprotection researchers. However, since the mature nervous system is more resistant to insults that can cause severe neurodevelopmental damage early in life, the aging process can act as a neuroprotectant to reduce the susceptibility of the nervous system to environmentally associated damage and disease. Environmental exposures that are relatively innocuous in people with mature nervous systems (e.g., rubella virus infection, low-level X irradiation, or many drugs) can be devastating to the fetal brain. Moreover, since substantial brain development continues after birth, the infant nervous system remains an important target for neuroprotection.

Neuroprotection of the developing brain can have an enormous impact on the quality of life of an individual and on public health. In the absence of appropriate neuroprotection, severe injury of a fetal or infant brain will result in lifelong disability—that is, 80 or so person–years at high level of care for the severely brain damaged child. Caring for the disabled child into adulthood can be extremely expensive; for example, in California, it is estimated to be around two million dollars for lifelong care of a severely autistic child.

In addition to the well known types of early brain damage that occur in children (e.g., cerebral palsy, or autism), it is tempting to speculate that some chronic, degenerative diseases of older adults may result from long-term stresses to areas of continuing neurogenesis in adult brain, such as dementia associated with hippocampal sclerosis. In addition, early insults to the young nervous system from which the individual appears to recover, may reappear during later life; for example, individuals exposed to poliovirus during childhood who attain full or partial recovery of motor functions, but then redevelop signs of paralysis in later life.

Neuroprotection of the immature brain is feasible, since short-term neuroprotection during critical periods of brain development can have life long effects in preserving normal neurological function. However, identification of the critical periods for neuroprotection against various insults to various brain regions and their functions remains an area in need of continued intensive study. Furthermore, understanding why neurodevelopmental processes are so susceptible to perturbation is also an important area for ongoing research.

There are other host factors that affect susceptibility to developmental brain damage. For example, host genetic background can alter vulnerability to specific neurological insults, and may explain at least some of the variability in outcomes following nervous system exposure to environmental teratogens. Thus, we may be able to better protect the immature nervous system if we can identify those groups at risk for the environmentally induced poor outcomes, by either preventing exposure or aggressively treating high risk individuals.

In summary, neuroprotection of the immature nervous system is critical and may have tremendous impact on individual and public health consequences. Much more work is needed in this area in order to better understand the pathogenesis of neurodevelopmental damage and how to prevent and treat it.

FINAL COMMENT

W. Slikker

As we enter the new millennium, it appears more evident than ever before that understanding the molecular and pathophysiological basis of neurotoxicity is the key to neuroprotection. The complexity of the cascade of events leading to cell death following ischemic insult mandates a greater understanding of the underlying biology in order to prevent time course–dependent neurodegeneration. Even with the advent of powerful new tools such as genomics, imaging and knockout animal models, the desired end result of neuroprotection is still beyond our current capability.

The complexity and diversity of the lesion, from the initial insult to the time-related manifestation of the enduring outcome, have thus far prevented an easy solution. As we painstakingly elucidate the details of the neurodegenerative process, a sequential cascade of events reflecting the diversity of cell types and innumerable functions of the nervous system, it seems logical that the inevitable achievement of neuroprotection will be equally complex.

Part XI: Summary and Conclusions

GENERAL COMMENT

From Dr. Weinstock

Neurodegeneration occurs both after a stroke and during slowly progressive disorders such as Alzheimer's or Parkinson's diseases. However, the vastly different rates at which neurodegeneration occurs in the former and latter conditions could indicate the necessity for different therapeutic measures. Therapy for stroke is aimed at giving the drug(s) early enough to prevent further rapidly developing cell death. In Alzheimer's and Parkinson's diseases degeneration develops over many years and compensatory mechanisms (as yet poorly understood) could be recruited continuously. This may necessitate a different therapeutic approach designed to increase the survival of remaining vulnerable neurons over a longer period of time.

QUESTION FOR PANEL (DRS. PALMER, AUER, AND GROTTA)

From Dr. Maynard

There are similarities between stroke, head trauma and neurodegenerative diseases in terms of deleterious cascades, yet these are clearly different diseases. What would the panel recommend as criteria for neuroprotection research in each of these disease areas?

ANSWER: (Palmer) For stroke and head trauma immediate acute treatment is necessary; how long to administer this treatment and whether to administer a cocktail of neuroprotective agents in a sequential manner are other questions. For chronic insidious neurodegenerative disorders the goal of treatment will be to begin when first diagnosed and then a drug will have to be shown to be safe enough to be given over a lifetime—more like treatment of epilepsy. Compliance will be another issue.

QUESTIONS FOR DR. MAYNARD

From Dr. Palmer

When treating intracranial hemorrhage we have tried three different classes of drugs including a neuronal NOS inhibitor—none worked. I think we need to rethink our approach to this mechanistically instead of randomly testing compounds in hopes of getting a hit.

ANSWER: I agree. This is a crucial area of stroke research that is not addressed as much as ischemic stroke. Twenty percent of stroke patients have hemorrhagic strokes and it is by no means certain that drugs that are neuroprotective in models of ischemic stroke, will work in cases of hemorrhagic stroke. Yet, little work is done in terms of testing neuroprotective agents in both types of stroke models. We are presently conducting studies to examine whether NAm will help in a model of intracranial hemorrhage and hope to move on to subarachnoid hemorrhage, but unfortunately it is too early to say whether there is any effect.

From Dr. Slikker

Reduction in infarct volume, increases in Bcl_2, or almost any other parameter (increased motor score, etc.) appear to be about a 50% change. Why not an improvement exceeding 50%?

ANSWER: I think the degree to which we can show protection is a combination of the time-to-treatment (the earlier one treats the insult, the larger is the area that can be saved), plus the extent to which the drug(s) is acting on the many deleterious cascades.

I have been suggesting for some years now that a single targeted approach will give us a clean drug with a known mechanism of action and perhaps little side effects, but it is unlikely to result in optimal neuroprotection because there are too many deleterious cascades ongoing, each with a different mechanism capable of causing cell death following stroke. In order to achieve a greater than 50% degree of neuroprotection, therefore, we need to give the drugs as early as possible and the treatment needs to be multimodal in terms of the mechanism of action. Incidentally, we used a pharmacologic cocktail of six different agents for a series of proof-of-concept studies and showed a 100% functional neuroprotection in an *in vitro* rabbit retina preparation (Ames, *et al.*, 1995) when the drugs were given at the onset of oxygen/glucose deprivation used to mimic ischemia. Moreover, the same combination of agents, again given at the moment of the onset of the insult, caused an approximately 80% reduction in infarction volume *in vivo* in a model of focal cerebral ischemia in the rabbit (Maynard, *et al.*, 1998).

A practical question—how do you get the neuroprotective agents to the target if a clot is in the vascular system? How do you get the drug across the placenta for *in utero,* or to the early postnatal newborn (PK could be quite different as compared to adult)?

ANSWER: I am not sure I can address the problem in neonates, but certainly in typical stroke patients one is dependent on the collateral circulation, of which there is plenty in the cortical mantle, particularly in the anterior circulation of the Circle of Willis of primates and humans. The subcortical brain structures, such as the basal ganglia, have less collateral perfusion, although in humans, there are so-called perforator vessels that also improve the collateral circulation, more so than in rodents.

GENERAL COMMENTS TO PANEL

From Dr. Herning

There have been several mentions in this meeting suggesting that stroke is a disease of old age. With increased use of abused drugs in young individuals, there has been an increase of stroke among young individuals (late teens to twenties and thirties).

From Dr. Banik

I am glad to hear that you brought up the subject to investigate white matter in stroke or CNS injury because different cells may have different susceptibilities. For example, oligodendrocytes may die due to any insult to the brain. We do not know much about the mechanisms by which white matter is degenerated. It is therefore, very important to explore the mechanism of white matter degeneration.

From Dr. Auer

On lumping and splitting: during periods of intense scientific activity there is a tendency to splitting rather than lumping. We must, in our quest for neuroprotection, respect the difference between the more than 1,000 neurologic diseases. If common mechanisms are then found, we can then lump, after initially splitting, in the study of neuroprotection.

Index of Contributors

Abbracchio, M.P., 54–63, 63–74
Acuña-Castroviejo, D., 200–215
Afanasiev, A., 425–436
Aiyagari, V., 257–267
Ali, S.F., 359–365, 366–380, 393–394
Allen, J.W., 23–27
Allen, R.R., 237
Andrews, R.J., 101–113, 114–125
Anzivino, M.J., 358
Aschner, M., 23–27
Auer, R.N., 271–282, 465–477
Ayoub, I.A., 416–424

Bachurin, S., 190–199, 219–236, 425–436
Banik, N.L., 436–449
Baskin, I., 219–236
Bejar, C., 148–161
Berenstein, A., 126–136
Better, W.E., 405–412, 413–415
Binienda, Z.K., 359–365, 381–392
Bordoni, F., 74–84
Brambilla, R., 54–63
Bukatina, E., 425–436
Burkhardt, S., 200–215

Cadet, J.L., 366–380, 405–412, 413–415
Camurri, A., 63–74
Carbone, K.M., 318–319, 465–477
Caso, V., 162–179
Castagnoli, Jr., N., 330–339
Castagnoli, K., 330–339
Cattabeni, F., 63–74
Ceruti, S., 63–74
Chen, C.-F., 33–44
Choi, J., 238–253
Chuang, D.-M., 404
Corbett, D., 283–296
Corsi, C., 74–84
Curry, S.H., 297–308

Decker, D., 358
Del Bigio, M.R., 283–296
Deletis, V., 126–136, 137–144
Dumas, T.C., 340–357

El-Yazal, J., 366–380

Facci, L., 11–22
Falzano, L., 63–74
Figueroa, M., 257–267
Fiorentini, C., 63–74

Gaetani, F., 162–179
Giammarioli, A.M., 63–74
Grigoriev, V., 219–236, 425–436
Grigorieva, I., 425–436
Grotta, J., 309–310

Hammond, T., 237
Herning, R.I., 405–412, 413–415
Ho, L.-T., 33–44
Hogan, E.L., 436–449
Hudzik, T.J., 283–296

Imam, S.Z., 366–380, 393–394
Itzhak, Y., 366–380, 393–394
Ivanov, Yu., 425–436

Jacobson, K.A., 63–74
Jonas, S., 257–267, 268–270, 465–477
Jones, M.L., 318–319

Kirschbaum-Slager, N., 148–161

Labutta, R., 238–253
Lai, G.-H., 33–44
Latini, S., 74–84
Lazarovici, P., 148–161
Lee, K.S., 358

Lermontova, N., 219–236, 425–436
Lin, A.M.-Y., 33–44
Lin, R.C.S., 85–96
Lukoyanov, N., 219–236

Malorni, W., 63–74
Manev, H., 45–51
Manev, R., 45–51
Marini, A.M., 238–253
Martini, C., 63–74
Maruyama, W., 320–329
Matzelle, D.D., 436–449
Maynard, K.I., 416–424, 465–477
Melani, A., 74–84
Moran, T.H., 318–319
Mukhina, T., 219–236
Mutkus, L.A., 23–27

Naoi, M., 320–329
Narahashi, T., 179–186, 465–477
Newport, G.D., 366–380, 393–394
Niimi, Y., 126–136
Nony, P.L., 381–392

Obrenovitch, T.P., 1–10
Oxenkrug, G., 190–199

Palmer, G.C., 283–296, 465–477
Palyulin, V., 219–236
Paule, M.G., 237
Pearson, E., 237
Pedata, F., 74–84
Peeling, J., 283–296
Petrova, L., 219–236
Pletnikov, M.V., 318–319
Popke, E.J., 237
Proshin, A., 219–236

Quartermain, D., 268–270

Ray, S.K., 436–449
Reiter, R.J., 200–215
Requintina, P., 190–199

Rossi, S., 162–179
Rountree, R.L., 359–365, 381–392
Rubin, S.A., 318–319
Russo, F., 162–179

Sablin, S., 425–436
Sadovova, N.V., 359–365
Sala, F., 126–136, 137–144
Sapolsky, R.M., 340–357
Scallet, A.C., 359–365, 381–392
Shen, C.-C., 416–424
Shoham, S., 148–161
Simpson, K.L., 85–96
Skaper, S.D., 11–22
Slikker, Jr., W., xi, 366–380,
    393–394, 465–477
Spadoni, A., 162–179
Steinberg, G.K., 340–357
Steyn, S.J., 330–339
Strijbos, P.J.L.M., 11–22
Su, Y., 33–44

Tan, D.-X., 200–215
Tate, K., 405–412, 413–415
Tatton, W., 450–458
Tkachenko, S., 219–236, 425–436
Trembly, B., xi, 465–477
Trincavelli, L., 63–74

Ustinov, A., 219–236
Uz, T., 45–51

Vieira, D., 257–267
Virmani, M.A., 162–179
Von Lubitz, D.K.J.E., 85–96

Wadia, A., 450–458
Weinstock, M., 148–161, 450–458
Wilford, G.G., 436–449

Yeh, J.Z., 179–186
Yenari, M.A., 340–357

Youdim, M.B.H., 148–161, 320–329, 450–458, 465–477
Yunoki, M., 358

Zefirov, N., 219–236, 425–436
Zhang, Z., 45–51
Zhao, X., 179–186